Professor C. H. Bamford studied at Cambridge, being awarded first class honours in the Natural Sciences Tripos, Part II (Chemistry) in 1934, and a Ph.D. in 1938, by which time he was a Fellow of Trinity College, and Director of Studies in Chemistry, Emmanuel College. He worked for the Special Operations Executive from 1941 to 1945, when he joined the fundamental research laboratory of Courtaulds Ltd. at Maidenhead, becoming its head in 1947. The Sc.D. (Cambridge) was conferred upon him in 1954, being followed ten years later by election to the Fellowship of the Royal Society.

Since 1962, he has been Campbell-Brown Professor of Industrial Chemistry at the University of Liverpool, and was Dean of the Faculty of Science from 1965 until 1968.

Professor Bamford is co-author of two books: *Synthetic Polypeptides* (1956), and *The Kinetics of Vinyl Polymerization by Radical Mechanisms* (1958).

Dr. C. F. H. Tipper was awarded first class honours in chemistry at the University of Bristol in 1945, and a Ph.D. two years later. After carrying out postdoctoral research for two years at the Universities of Bristol and Liverpool, he took up a post as a lecturer in physical chemistry at the University of Edinburgh. In 1953 he moved to a similar post at Liverpool University, and was appointed senior lecturer in 1961. The degree of D.Sc. was conferred upon him by the University of Edinburgh in 1959.

Dr. Tipper is co-author of two books: *The Chemistry of Combustion Reactions* (1962), and a *Laboratory Manual of Experiments in Physical Chemistry* (1967). He is also editor of the series *Oxidation and Combustion Reviews*.

COMPREHENSIVE

Section 1. THE PRACTICE AND THEORY OF KINETICS

Volume 1 The Practice of Kinetics

Volume 2 The Theory of Kinetics

Volume 3 The Formation and Decay of Excited Species

Section 2. DECOMPOSITION AND ISOMERISATION REACTIONS

Volume 4 The Homogeneous Decomposition of Inorganic and Organometallic Compounds

Volume 5 The Homogeneous Decomposition of Organic Compounds

Section 3. INORGANIC REACTIONS

Volume 6 Homogeneous Reactions of Compounds of Non-metals

Volume 7 Homogeneous Reactions of Metals and their Compounds

Section 4. ORGANIC REACTIONS (6 volumes)

Section 5. POLYMERIZATION REACTIONS (2 volumes)

Section 6. OXIDATION AND COMBUSTION REACTIONS (2 volumes)

Section 7. SELECTED ELEMENTARY REACTIONS (2 volumes)

Additional Sections

HETEROGENEOUS REACTIONS

SOLID STATE REACTIONS

KINETICS AND TECHNOLOGICAL PROCESSES

CHEMICAL KINETICS

EDITED BY

C. H. BAMFORD

M.A., Ph.D., Sc.D. (Cantab.), F.R.I.C., F.R.S.

Campbell-Brown Professor of Industrial Chemistry,

University of Liverpool

AND

C. F. H. TIPPER

Ph.D. (Bristol), D.Sc. (Edinburgh)

Senior Lecturer in Physical Chemistry,

University of Liverpool

VOLUME 1

THE PRACTICE OF KINETICS

ELSEVIER PUBLISHING COMPANY

AMSTERDAM - LONDON - NEW YORK

1969

ELSEVIER PUBLISHING COMPANY
335 JAN VAN GALENSTRAAT
P.O. BOX 211, AMSTERDAM, THE NETHERLANDS

ELSEVIER PUBLISHING CO. LTD.
BARKING, ESSEX, ENGLAND

AMERICAN ELSEVIER PUBLISHING COMPANY, INC.
52 VANDERBILT AVENUE
NEW YORK, NEW YORK 10017

LIBRARY OF CONGRESS CARD NUMBER 68-29646.
STANDARD BOOK NUMBER 444-40673-5

WITH 161 ILLUSTRATIONS AND 32 TABLES.

PRINTED IN THE NETHERLANDS

COMPREHENSIVE CHEMICAL KINETICS

Contributors to Volume 1

L. BATT	Department of Chemistry, University of Aberdeen, Aberdeen, Scotland
D. N. HAGUE	Department of Chemistry, University of Kent, Canterbury, England
D. MARGERISON	The Donnan Laboratories, University of Liverpool, Liverpool, England
D. SHOOTER	Heavy Organic Chemicals Division, Imperial Chemical Industries Ltd., Billingham, England (*Now* Arthur D. Little, Inc., Acorn Park, Cambridge, Mass., U.S.A.)
R. P. WAYNE	Physical Chemistry Laboratory, University oi Oxford, Oxford, England

Preface

The rates of chemical processes and their variation with conditions have been studied for many years, usually for the purpose of determining reaction mechanisms. Thus, the subject of chemical kinetics is a very extensive and important part of chemistry as a whole, and has acquired an enormous literature. Despite the number of books and reviews, in many cases it is by no means easy to find the required information on specific reactions or types of reaction or on more general topics in the field. It is the purpose of this series to provide a background reference work, which will enable such information to be obtained either directly, or from the original papers or reviews quoted.

The aim is to cover, in a reasonably critical way, the practice and theory of kinetics and the kinetics of inorganic and organic reactions in gaseous and condensed phases and at interfaces (excluding biochemical and electrochemical kinetics, however, unless very relevant) in more or less detail. The series will be divided into sections covering a relatively wide field; a section will consist of one or more volumes, each containing a number of articles written by experts in the various topics. Mechanisms will be thoroughly discussed and relevant non-kinetic data will be mentioned in this context. The methods of approach to the various topics will, of necessity, vary somewhat depending on the subject and the author(s) concerned.

It is obviously impossible to classify chemical reactions in a completely logical manner, and the editors have in general based their classification on types of chemical element, compound or reaction rather than on mechanisms, since views on the latter are subject to change. Some duplication is inevitable, but it is felt that this can be a help rather than a hindrance.

Since all kinetic work commences with the accumulation of experimental data, this first volume deals with the methods used for determining the rates of "slow", "fast" and heterogeneous reactions, together with those for the detection and quantitative determination of labile intermediates. A chapter is also devoted to the processing of the primary data — where appropriate, with the aid of statistical methods.

Finally, the Editors wish to express their sincere appreciation of the advice so readily given by the members of the Advisory Board.

Liverpool
November, 1968

C. H. Bamford

C. F. H. Tipper

Contents

Chapter 1

Experimental Methods for the Study of Slow Reactions

L. BATT

1. Introduction

The labelling of reactions as "fast" or "slow" is obviously arbitrary, though convenient, and for the purposes of this chapter it will be supposed that the latter have half-lives of greater than several seconds (usually of the order of minutes). The main emphasis will be placed on gas-phase systems, since the experimental difficulties are generally more acute, and in fact the methods of following the course of reactions in the liquid phase are usually (with appropriate modifications) the same.

The primary aim of a chemical kinetic study is the determination of the mechanism by which a system changes in composition. This involves tests for heterogeneity as well as the determination of the order of a reaction in terms of the concentration of reactant or reactants, and an analysis of the products. Estimation of the rate coefficient for a particular elementary step as a function of pressure and temperature provides, in the first instance, Arrhenius parameters, which should be rationalised in terms of the thermodynamics of the reverse process, and may finally lead to particular bond energies.

The essential apparatus for pressure measurement and analysis, and other important aspects such as furnaces and temperature control, are reviewed for thermal, photochemical and radiochemical systems. The latter two also involve sources of radiation, filters and actinometry or dosimetry. There are three main analytical techniques: chemical, gas chromatographic and spectroscopic. Apart from the almost obsolete method of analysis by derivative formation, the first technique is also concerned with the use of "traps" to indicate the presence of free radicals and provide an effective measure of their concentration. Isotopes may be used for labelling and producing an isotope effect. Easily the most important analytical technique which has a wide application is gas chromatography (both GLC and GSC). Intrinsic problems are those concerned with types of carrier gases, detectors, columns and temperature programming, whereas sampling methods have a direct role in gas-phase kinetic studies. Identification of reactants and products have to be confirmed usually by spectroscopic methods, mainly IR and mass spectroscopy. The latter two are also used for direct analysis as may UV, visible and ESR spectroscopy. NMR spectroscopy is confined to the study of solution reactions

References pp. 104–111

apart from the identification of derivatives. A combination of chemical and spectroscopic techniques occurs for example in the gas-phase titration for hydrogen atoms by the addition of NO and spectroscopic estimation of the visible emission from the hot nitroxyl. Apart from the above techniques, which are of general application, there are other methods such as calorimetry, gas density measurement, polarography, the damped oscillation of a quartz fibre, ultrasonics, conductivity and refractive index measurements, which have limited application. However, when the general techniques prove unsatisfactory, these techniques will have a particular significance.

2. Determination of a mechanism

The essential experimental task for the gas kineticist is a determination of the mechanism of a homogeneous chemical reaction. This must explain the order of the reaction and the observed products. The first step is a test for heterogeneity, achieved by drastically altering the surface area to volume ratio (S/V) and the composition of the surface of the reaction vessel (RV) in which the reaction is studied. For partly heterogeneous reactions, milder variations of (S/V) are obtained by using spherical, cylindrical or octopus[1]-shaped RV's. In this case, a value for the homogeneous rate (W') is obtained from a graphical plot of the overall rate (W) against S/V, extrapolating to zero S/V. The problem is treated precisely by Hudson and Heicklen[2]. The order of the reaction (n) is determined from an observation of the variables in the equation

$$\log W_i = \log k + n \log c_i \tag{A}$$

where W_i is the initial rate of the reaction at an initial concentration c_i and k is the rate coefficient. A similar equation may be used for a single run where the rate is found as a function of time. However, the two orders are not always the same for a given reaction. For the decomposition of acetaldehyde (AcH), Letort[3] found that the order with respect to initial concentration was $\frac{3}{2}$ whereas that with respect to time was 2. For reactions where there is more than one reactant, the isolation method may be used, where one or more reactants are kept in great excess, while the remaining is subjected to the variables in equation (A). However, in many cases it does not yield reliable results[4].

Provided that there is a change in the number of moles upon reaction, an obvious measure of the extent of a reaction is given by the change in pressure. The latter has to be related to the stoichiometry of the reaction by quantitative analysis of the products and reactant or reactants and by material balance. Abnormal pressure effects sometimes occur due to adiabatic reactions, unimolecular reactions which are in their pressure-dependent regions (particularly in flow systems) and hot molecule or hot radical reactions. For free radical chain reactions the mode of initiation and termination has to be determined, sometimes a difficult task. The former may possibly be assessed from the use of a suitable radical trap. Surface initiated and terminated reactions may be studied using sensitive calorimetric techniques. Some idea of the complexity of initiation and termination may be realised by considering the thermal decomposition of AcH, which has been recently reviewed by Laidler[5]. The rate of initiation is dependent on the square of the AcH concentration but independent of the concentration of added inert gases. Termination, involving the combination of methyl radicals, is pressure-dependent whereas the kinetics of the decomposition of ethane under the same conditions of temperature and pressure would predict the combination rate to be pressure-independent. These conclusions are based on the possibly unsound premise that H_2 and C_2H_6 formation are direct measures of initiation and termination respectively. Information about the elementary steps that comprise a mechanism is produced by the study of a sensitised decomposition such as

$$\mathrm{Me\cdot + AcH \rightarrow CH_4 + Ac\cdot} \tag{1}$$

Very often a particular reaction is studied in order to obtain its Arrhenius parameters. An example of this is given by reaction (1), although probably more interesting is the fate of the acetyl radical. The activation energy (E_f) is related to the standard heat of reaction $\Delta H°$ by the relationship,

$$\Delta H° = E_f - E_r + \Delta nRT \tag{B}$$

where E_r is the activation energy for the reverse process, Δn is the change in the number of moles upon reaction, and R and T have their usual meaning. $\Delta H°$ may be separately determined from equation (C):

$$\Delta H° = \Sigma H_f^\circ(\text{products}) - \Sigma H_f^\circ(\text{reactants}) \tag{C}$$

If E_r is zero, E_f may be related to a particular bond dissociation energy via equations (B) and (D). For the reaction

$$\mathrm{X\cdot + CH_4 \rightarrow XH + Me\cdot} \tag{2}$$

where X· is an atom or a radical, the relevant bond dissociation energies are related by equation (D):

$$\Delta H° = D(\text{Me–H}) - D(\text{X–H}) \tag{D}$$

The pre-exponential factor (A_f) is related to that for the reverse process (A_r) by the expression,

$$\ln A_f/A_r = \Delta S^\circ_{(p)}/R \tag{E}$$

$\Delta S^\circ_{(p)}$ is the standard entropy change for the reaction in pressure units. If a change in the number of moles occurs upon reaction as in reaction (3), and A_r is in concentration units, $\Delta S^\circ_{(p)}$ will have to be converted into concentration units. The

$$t\text{-BuO}\cdot \rightarrow \text{Me}\cdot + \text{Me}_2\text{CO} \tag{3}$$

relation between $K_{(p)}$ and $K_{(c)}$, the equilibrium constants for reaction (3) in pressure and concentration units respectively, leads to the result,

$$\Delta S^\circ_{(c)} = \Delta S^\circ_{(p)} - \Delta nR \ln RT \tag{F}$$

where $\Delta S^\circ_{(c)}$ is the standard entropy change in concentration units. In this case[6],

$$\ln A_f/A_r = (\Delta S^\circ_{(c)}/R) - \Delta n = (\Delta S^\circ_{(p)}/R) - \Delta n(1+\ln RT) \tag{G}$$

ΔS° may be determined from the expression,

$$\Delta S^\circ = \Sigma S^\circ(\text{Products}) - \Sigma S^\circ(\text{Reactants}) \tag{H}$$

Hence the rate of the reverse reaction may be calculated from equations (B) and (G). Perhaps more important from the point of view of this chapter, this also provides a check on the determined values for E_f and A_f and hence the proposed mechanism. Laidler and Polanyi have reviewed empirical methods for predicting E_f via various relationships between it and the heat of reaction[7].

ΔH° and ΔS° may be calculated from values of H°_f and S° listed in various books[8-13], or calculated from Atom or Group Additivity Rules[8,14].

If the particular reaction studied is the unimolecular decomposition of a free radical, such as (3), then the use of a "trap" will enable the effective concentration of the radical to be measured. A radical trap will indicate the presence or absence of a free radical reaction and may sometimes provide evidence for a partly or entirely molecular reaction. Rate data for free radical reactions are derived assuming the occurrence of a steady state concentration of radicals. The time required to produce a steady state concentration of methyl radicals in the pyrolysis of AcH is shown for various temperatures in Fig. 1. Realistic values for rate coefficients may be obtained only if the time of product formation is long compared to the time to achieve the steady state concentrations of the radicals concerned. Thus deductions from the results from the bromination of isobutane[15], neopentane[16], and toluene[17] have been criticised on the grounds that a steady state concentra-

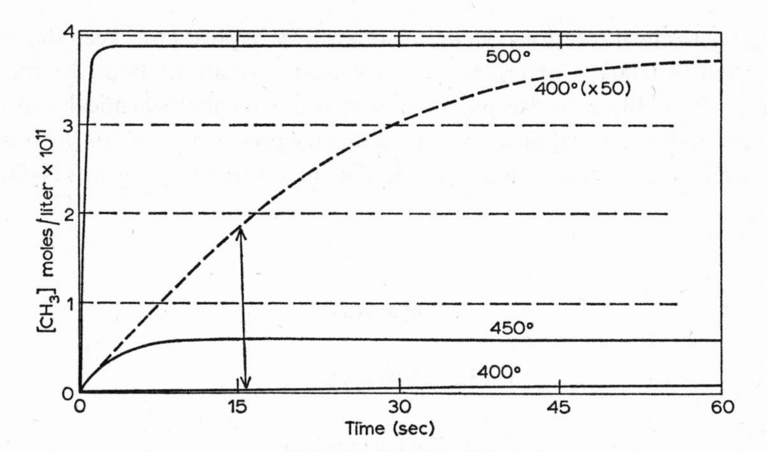

Fig. 1. Approach to the steady state in the pyrolysis of acetaldehyde; time-dependence of ($\dot{M}e$) for three temperatures; the dotted curve represents a 50-fold expansion of the 400° data; calculations of W. J. Probst. From ref. 9.

tion of the bromine atoms was not achieved under the chosen experimental conditions[18].

Two additional complications may be present in photochemical reactions, the presence of hot molecules and hot radicals referred to earlier in thermal systems, and the possible physical and chemical primary processes that may occur. Compounds such as biacetyl, O_2, NO and olefins are particularly efficient quenchers of electronically excited species. In accordance with the Wigner spin conservation rule that the total spin in a quenching process is conserved, triplet state acetone (^3A) is quenched very efficiently to the ground state (^1A) by olefins (O)[19].

$$^3A + {}^1O \rightarrow {}^1A + {}^3O \tag{4}$$

Larson and O'Neal have used HBr as a diagnostic test for the presence of triplet state acetone[20]. Double trapping results in the production of isopropyl alcohol, *viz.*

$$^3A + HBr \rightarrow Me_2\dot{C}OH + Br\cdot \tag{5}$$

$$Me_2\dot{C}OH + \begin{matrix} HBr \\ \text{or} \\ RH \end{matrix} \rightarrow Me_2CHOH + \begin{matrix} Br\cdot \\ \text{or} \\ R\cdot \end{matrix} \tag{6}$$

Information about primary chemical processes and the ensuing secondary processes may be obtained by using suitable radical traps thus isolating molecular primary processes. Direct information about primary processes may be obtained by flash and matrix photolysis. Radical traps may also be used to good advantage in radiochemical reactions. Here it is often very important to keep the extent of conversion to 0.002 % or less to obtain valid information about the primary processes[21]. In addition to the processes that occur in photochemical and thermal systems, radiochemical primary and secondary processes result in the production of ions. Electrons may be removed by efficient electron traps such as N_2O, SF_6 and CCl_4.

3. Apparatus

3.1 GENERAL

3.1.1 The vacuum line

General techniques for high vacuum systems have been covered in detail in various reviews[22], in particular conventional high vacuum taps and their lubri-

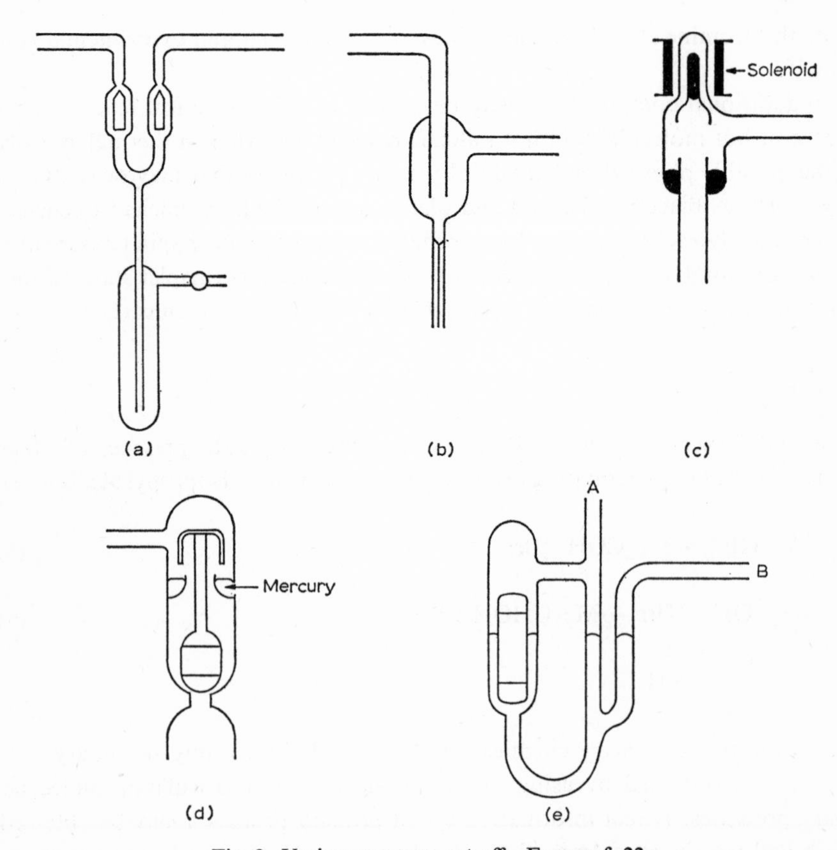

Fig. 2. Various mercury cut-offs. From ref. 22c.

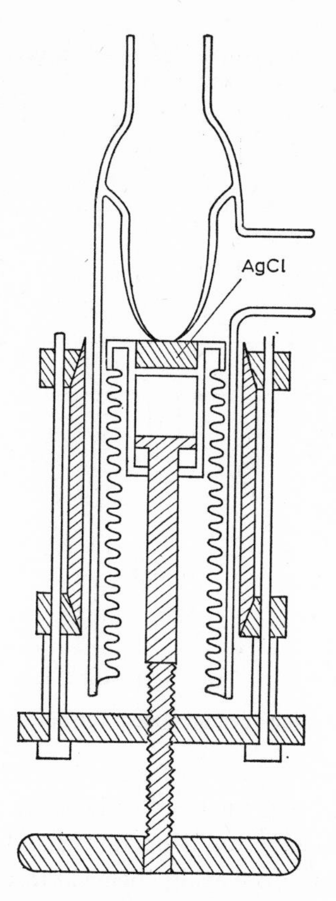

Fig. 3. Ramsperger's greaseless valve. From ref. 22c.

cants, used for the isolation of gases. A problem associated with the use of lubricants is their ready absorption of vapours and usually the inability to "bake" the taps. The Echols type of tap[23] avoids these problems by using graphite as the lubricant and sealing with mercury. Apart from this, alternatives to the conventional taps are various greaseless valves[22]. The simplest of these are mercury cut-offs shown in Fig. 2. There may be some difficulty if there is a large pressure differential between either side of the U-tube. Polythene or teflon keys may be used with glass tubing to form a tap[24]. However, they are subject to degassing and may have their own vapour absorption problems. This also applies to taps using elastomer diaphragms. Rapid operation of the latter type, either manually or electro-magnetically, has been achieved by Verdin[25]. A modification of the Bodenstein all-glass tap[26] is due to Ramsperger[27] (Fig. 3). The tap is closed by seating a plug of AgCl into a Pyrex tube, of which it wets the surface. Seating is achieved by a screw attached to silver bellows coated with AgCl. This is an excellent tap provided no chemical reaction occurs between its components and the gases involved. A similar tap has

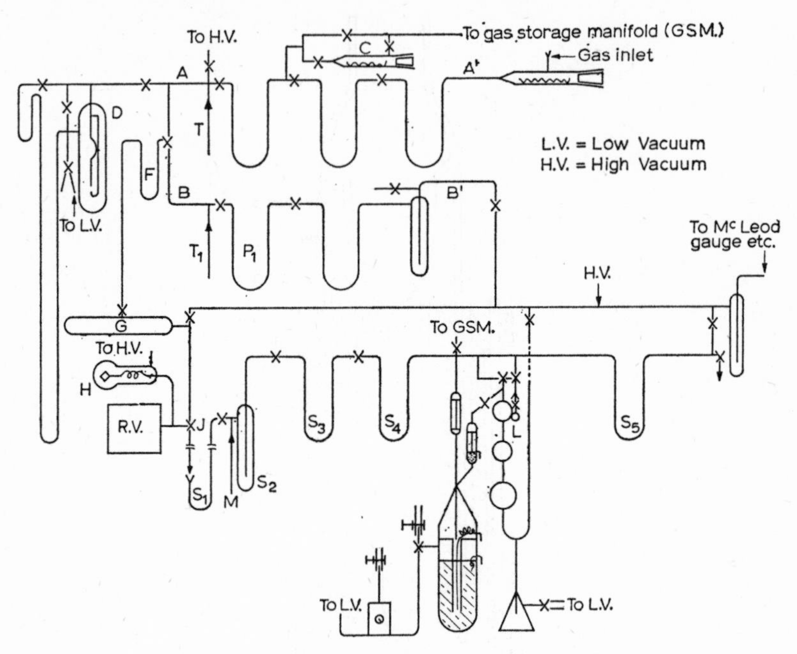

Fig. 4. High-vacuum system for kinetic studies.

been constructed by Judge and Luckey[28]. All-metal taps are available commercially which are bakeable, have rapid operation and provide good sealing; one draw-back is their price, and once again chemical reaction may occur.

Details of a typical vacuum line for a gas phase kinetic study are given by Maccoll[23]. A useful set-up is to have a manifold of large-diameter glass tubing leading from the pumping system, to which the different sections, storage and introduction, mixing and reaction vessels and analytical section, are connected (Fig. 4). This manifold system has the advantage that each section may be evacuated independently of the others.

3.1.2 Temperature control

There are three general methods for maintaining a RV at a particular temperature: thermostats, furnaces and vapour baths[29].

At temperatures close to $20 \pm 20°$, water is a suitable liquid for a thermostat. At higher temperatures it is more convenient to use a fluid such as silicone oil, or above 200°, molten metals. A typical furnace is shown in Fig. 5. A silica former is wound with nichrome wire such that the pitch decreases from one end to the centre and then increases again to the other end. Since heat losses are greatest at the ends this provides a rough correction. An inconel tube also evens out the tempera-

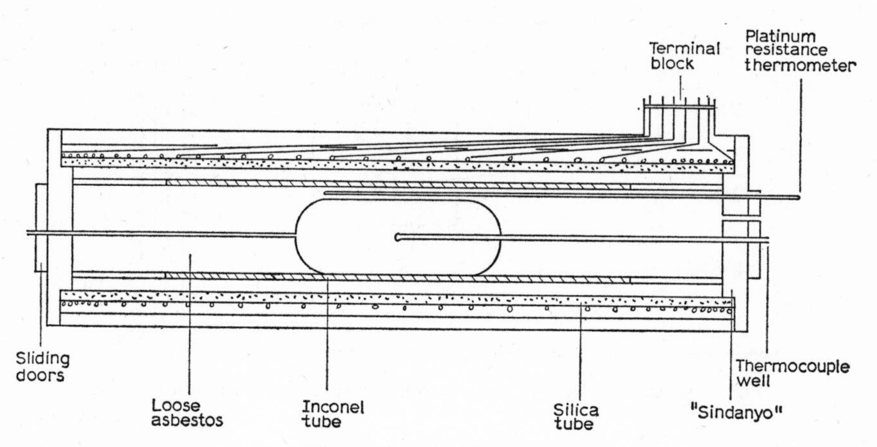

Fig. 5. A typical furnace.

ture along the length of the furnace. Tappings are taken at regular intervals along the winding. The wire is covered with asbestos paper and cement and placed in an asbestos tube or box and packed into 3 in. of kieselguhr or other insulating material. Shunts are connected at the taps in the required places and adjusted until the temperature profile at say 200° is ± 0.5° along the length of the RV. The total resistance of the furnace should be such as to allow a maximum output of $\frac{1}{2}$–1 Kw depending on the temperature range to be covered. For photochemical or radiochemical studies the design has to be modified such that the furnace rests on an optical bench and an uninterrupted path is provided for the source of radiation. This is achieved by splitting the furnace in half longitudinally or laterally. The latter is described in detail in the excellent treatise on photochemistry by Calvert and Pitts[9]. The former presents some difficulty because winding has to be carried out over a semicircular cross-section but can be achieved by drilling holes or having pegs at regular intervals at the extremities of the asbestos tube halves. The inconel tube is replaced by two aluminium half-tubes with horizontal sections such that one half sits on the other when they are slotted and clamped together. Entry tubes to the RV may be led in at the join of the two halves. This design is particularly useful for flash photolysis studies.

A circuit for heating and controlling the temperature of a thermostat or a furnace is shown in Fig. 6. Just the right amount of current from a constant voltage transformer and a variac transformer is fed into the furnace connected in series with a small variable resistance $\sim \frac{1}{5}$–$\frac{1}{10}$ of the value of the furnace resistance, connected across a relay. For thermostats, the relay may be operated by a toluene/mercury or a mercury thermo-regulator, depending on the temperature. Platinum resistance thermometers or thermocouples, forming part of a Wheatstone bridge network, are used to operate the relay for electric furnaces. This type of regulator is available commercially. A strict control of current is required. In this way it is

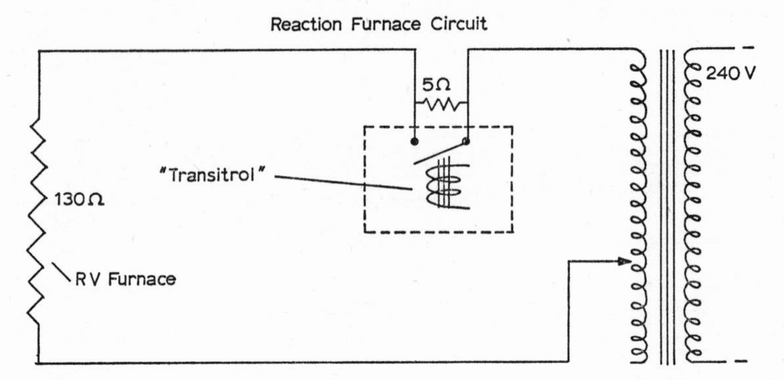

Fig. 6. Circuit for heating and controlling the temperature of a furnace.

possible to control at 200° C to ± 0.2° and at 600° C to ± 2°. Variations of this type of control are described in ref. 22c.

Temperature measurement is almost invariably made using thermocouples[22c]. The latter must be constructed from fine wire[30] and have a fast response[31] such that even a very small temperature change may be measured precisely[32].

3.2 THERMAL SYSTEMS

3.2.1 Static method

(a) Reaction vessels

Reaction vessels are usually spherically or cylindrically shaped, vary in size from 200–1000 cm^3, and have a thermocouple well in the centre. The tip of this well is thin-walled and sometimes a drop of fluid, such as silicone oil, is used to provide good thermal contact. The material most often used is Pyrex glass (maximum temperature 600° C) and fused silica (maximum temperature 1200° C). At room temperature the two materials are practically impermeable to all gases except helium. At higher temperatures hydrogen diffuses through the glasses and particularly through silica[33]. At still higher temperatures oxygen and nitrogen may diffuse through silica[34]. The mechanism appears to involve adsorption at the glass surface followed by passage through the glass. Silica reaction vessels should not be heated to high temperatures when in contact with metals. Iron or nickel, for instance, will diffuse into the quartz and give rise to heterogeneous reactions. Many gases are strongly adsorbed on glass surfaces and hence all glass apparatus, particularly the RV, should be well degassed at elevated temperatures. Both silica and Pyrex glass are resistant to most gases at room temperature except HF. However, HCl, HBr and probably HI are adsorbed on the surface and at high temperatures react with the glass[35]. Adsorption occurs via dissolution in a thin layer of adsorbed water which is held at the surface even at 400°. Atomic hydrogen[36] and chlorine[37] react with glass and silica and this may also occur with atomic iodine and silica[38].

The surfaces of RV's are often coated in some way either to minimise heterogeneous processes (or, in the case of oxidation studies[39], to affect chain termination, *e.g.* by peroxy radicals) or to minimise adsorption and diffusion of gases such as hydrogen[33,40] and water[40]. These surfaces may be films of KCl† or H_3BO_3 produced by rinsing with solutions containing them[41], carbonaceous films produced by the pyrolysis of alkyl halides[42] or nitrites[43], or polymeric coatings produced by the polymerisation of olefins in the RV[44] or washing with solutions of polymers[45]. Very often "aging" of the RV is achieved by several preliminary experiments. This may sometimes be associated with the production of carbonaceous films.

As mentioned previously (p. 2), the homogeneous extent of a partly heterogeneous reaction may be determined by mild variations of S/V using spherical, cylindrical or octopus-shaped vessels[1]. The homogeneous rate (W'_h) is obtained from a graphical plot of the overall rate (W) against S/V and extrapolating to zero S/V (see Hudson and Heicklen[2]). More drastic variation of S/V, by packing the RV with thin-walled glass tubing with fire-polished ends, may isolate the heterogeneous reaction[43]. However, the effect of variation of S/V may be complex. A free radical reaction may be both surface initiated and terminated. An increase of S/V may increase both these processes simultaneously giving no net effect[45]. Alternatively, in the H_2+O_2 system[46], with surface initiation and termination and gas-phase branching, a moderate increase in S/V decreases the rate due to increased surface termination. This effect is observed when packing the RV with tubing. However, if the packing is powdered glass an increase in the rate is observed due to increased heterogeneous reaction. Hence the packing method is not always reliable. A more satisfactory method for investigating the influence of surface is that of differential calorimetry[47]. Two fine-gauge thermocouples are mounted respectively at the centre and the wall of the RV. The total initial temperature rise (ΔT_i), made up of homogeneous and heterogeneous contributions, is proportional to the heat evolved or absorbed and hence to the initial rate (W). Measurement of ΔT and W allows a value for the heterogeneous contribution to be calculated††.

In very exothermic[48] or endothermic[49] reactions it is possible for thermal gradients to be produced in the system, depending on the geometry of the RV. The temperature distribution with respect to time, $\partial T/\partial t$, is given by

$$\partial T/\partial t = \nabla(\lambda)\, T/\rho c_v + \lambda \nabla^2 T/\rho c_v + QH/\rho c_v \qquad \text{(I)}$$

† This film should be quite thick, otherwise the original surface might be exposed due to volatilisation of the salt (see ref. 39, p. 28).

†† This is discussed more fully in the analysis section.

where λ is the coefficient of thermal conductivity, ρ is the density, c_v is the heat capacity of the gas at constant volume, Q is the specific rate of the reaction and H is the molar heat of reaction. If concentration and density gradients are small, this equation reduces to

$$\partial T/\partial t = \lambda \nabla^2 T/\rho c_v + QH/\rho c_v \quad \text{(J)}$$

Here it is assumed that λ and c_v do not change appreciably with time or position in the vessel and that convection is negligible. For a spherical vessel of radius r, the maximum temperature difference ΔT_{max} is given by[50]

$$\Delta T_{max} = QHr^2/6\lambda \quad \text{(K)}$$

Averaged over the vessel, this becomes

$$\Delta T_{av} = 0.4\Delta T_{max} = QHr^2/15\lambda \quad \text{(L)}$$

If r is 5 cm and we consider the decomposition of di-*tert.*-butyl peroxide at 160° C[1] (H = 40 kcal.mole^{-1}, at 5×10^{-3} M (100 mm Hg), $Q = 2.5 \times 10^{-9}$ mole. cm^{-3}. sec^{-1} and $\lambda = 10^{-4}$ cal. cm.$^{-1}$. sec^{-1}), we find a significant temperature difference given by

$$\Delta T_{av} = r^2/15 = 1.67° \quad \text{(M)}$$

The time taken for the thermal gradient to be set up is given by

$$t = r^2 pC_v/\pi^2 RT\lambda \quad \text{(N)}$$

where p is the pressure at a temperature T, C_v is now the molar heat capacity at constant volume, and R is the gas constant in cal.mole^{-1}.deg^{-1}. With C_v = 6 cal. mole^{-1}.deg^{-1}, a thermal gradient is established in one second. However, if r is reduced to 0.5 cm, ΔT_{av} turns out to be two powers of ten lower, which is negligible. This effect was studied by Batt and Benson[1], when it was shown that a significant difference occurs between the two sets of Arrhenius parameters obtained using spherical and octopus-shaped RV's respectively.

(*b*) *Pressure measurement*

Direct pressure measurements may be made with conventional manometers using mercury, silicone oils, hydrocarbon oils, butyl phthalate and other fluids. Apart from the danger of their affecting the reaction studied directly, by chemical reaction, vapour adsorption or modifying the surface of the RV, they produce a varying "dead space". It is important to keep the "dead space" between 1 and 4 %

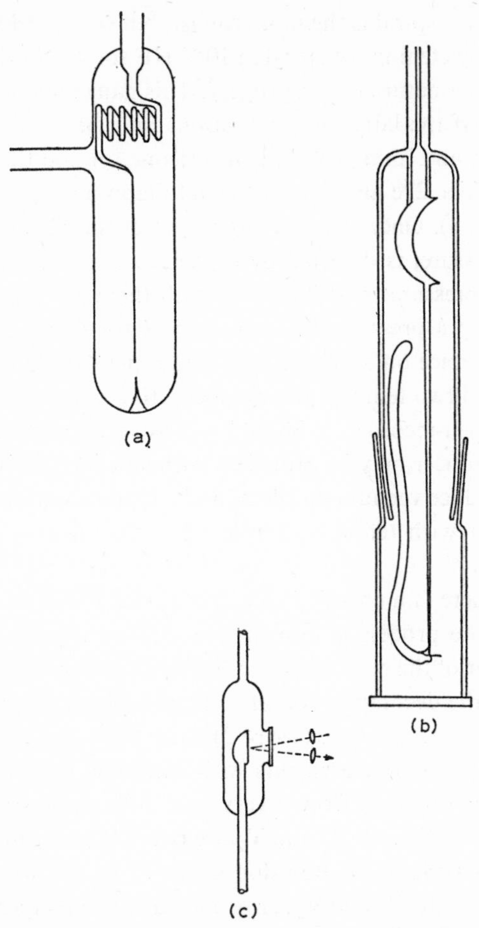

Fig. 7. Mechanical gauges. (a) Spiral gauge; (b) Bourdon spoon gauge; (c) Click gauge.

of the RV volume. More convenient methods are to use glass mechanical gauges of which there are three types: diaphragm[51], Bourdon[52], and spiral[53]. Diagrams of each of these gauges are shown in Fig. 7†. All may be used as null instruments either by means of the movement of a pointer with respect to a fixed reference point, observed with a low-power microscope (Bourdon and spiral) or by the sound of a click at the balance point caused by the change of configuration of the glass diaphragm. Bourdon and spiral gauges may also be used for direct pressure measurement, usually by means of an optical lever. Other methods using Bourdon gauges include capacitance measurement[54] and photoelectric measurements[55].

† For a detailed discussion see ref. 22c, p. 81.

Of the two gauges the spiral is the more robust, but is less sensitive. The sensitivity may be improved by etching the spiral in 10% HF solution, which does not drastically affect the robust nature of the gauge[56]. This gauge may be mounted horizontally or vertically. If the latter, any vibrations may be eliminated by attaching a float to the spindle and immersing this in silicone oil. All three of the gauges are discussed fully in refs. 22c and 56b. All-metal diaphragm gauges have also been constructed[57] (Fig. 8). Once again, a movement of the diaphragm, corresponding to a change in pressure, is detected by capacitance measurement. An alternative to this gauge is a pressure transducer of which there are several commercial versions. In one set-up a pressure difference across the two sides of the transducer produces an off-balance on a Wheatstone bridge network consisting of two strain gauge windings and two temperature-compensating windings. The voltage output may be fed to a pen-recorder[58]. Should corrosive gases be used, the resistance spirals of the transducer may be protected with silicone oil[58] or diaphragms. The latter usually introduce vacuum problems and a better solution is to seal the spirals in glass or to coat with teflon or a silicone in the manner described by Egger and Benson[44].

Provided that there is a change in the number of moles upon reaction and the stoichiometry of the process is known, pressure measurements may be used to determine the order of the reaction according to equation (A). Thus Letort[3] found that the order for the decomposition of AcH was $\frac{3}{2}$ with respect to initial concentration and 2 with respect to time (see p. 2). Such direct conclusions cannot usually be drawn from pressure measurements with oxidation reactions. However, direct information may be obtained from a very neat differential system devised by Dugleux and Frehling[59] (Fig. 9). V_1 and V_2 are two RV's, of different size connected to the inside and outside of the Bourdon gauge J. R_1 allows simultaneous introduction of mixtures into V_1 and V_2. Any fluctuation in temperature of the furnace is thus compensated for. Rapid reactions and the direct effect of promoters and inhibitors on an oxidation may be studied. This apparatus may well be useful with other systems.

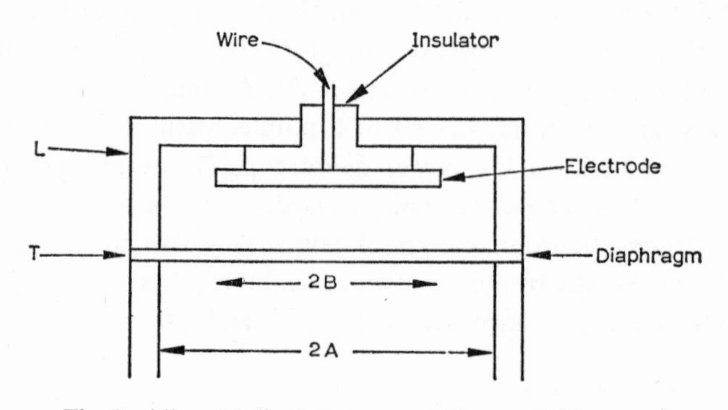

Fig. 8. All-metal diaphragm gauge. From ref. 22c.

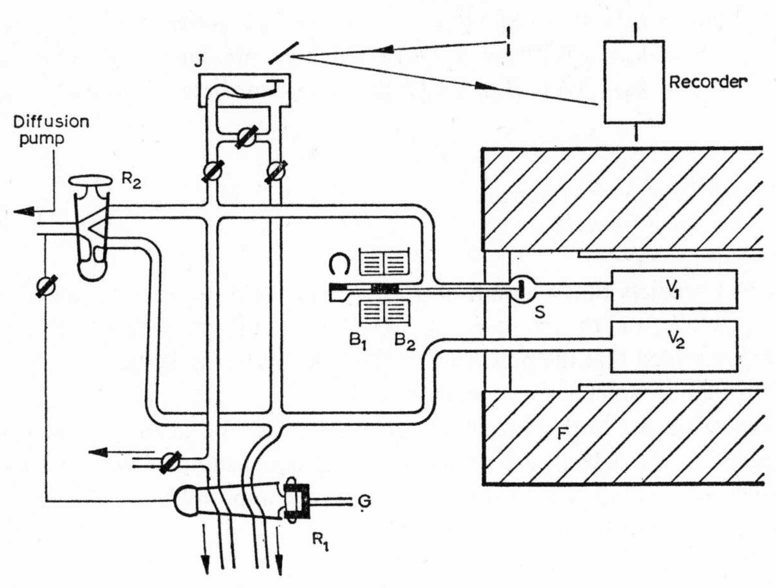

Fig. 9. Apparatus for differential pressure measurement. From ref. 59.

Pressure effects are observed on the dissociation of diatomic molecules and small polyatomic species such as $\dot{C}HO$ and HNO since the decomposition occurs in a bimolecular process. In reaction (7)

$$Br_2+M \rightarrow 2\dot{B}r+M \tag{7}$$

M is Br_2 or any other gas that is present. By the principle of microscopic reversibility[60], the reverse processes are also pressure-dependent.† A related pressure effect occurs in unimolecular decompositions which are in their pressure-dependent regions (including unimolecular initiation processes in free radical reactions). According to the simple Lindemann theory[61], the mechanism for the unimolecular decomposition of a species A is given by the following scheme (for more detailed theories see ref. 47b, p.283)

$$A+A \rightarrow A^*+A \tag{8}$$
$$A^*+A \rightarrow A+A \tag{9}$$
$$A^* \rightarrow \text{Products} \tag{10}$$

A* represents a molecule activated by collision. The rate of decomposition of A is given by

† For qualifying remarks see ref. 4, pp. 110 and 329.

$$-\mathrm{d}[\mathrm{A}]/\mathrm{d}t = k_8[\mathrm{A}]^2 = k_0[\mathrm{A}]^2 \quad \text{at low pressures} \quad (\mathrm{O})$$
$$= k_8 k_{10}[\mathrm{A}]/k_9 = k_\infty[\mathrm{A}] \quad \text{at high pressures} \quad (\mathrm{P})$$
$$= k_8 k_{10}[\mathrm{A}]^2/(k_{10}+k_9[\mathrm{A}]) \quad \text{at intermediate pressures} \quad (\mathrm{Q})$$

and

$$1/k_{\mathrm{exp}} = 1/k_8[\mathrm{A}]+k_9/k_8 k_{10}$$
$$= 1/k_0[\mathrm{A}]+1/k_\infty \quad (\mathrm{R})$$

Equation (R) predicts that a graph of $1/k_{\mathrm{exp}}$ versus $1/[\mathrm{A}]$ should be linear and that the slope is $1/k_0$ and the intercept $1/k_\infty$. Useful data may be obtained from this type of plot provided that the pressures at which the particular system is studied is not close to that at which the rate coefficient is k_∞.

In an opposite sense to these reactions, pressure effects occur in systems where hot molecules or hot radicals are produced[62]. Here quenching of the hot species may be followed in competition with spontaneous decomposition or isomerisation of the hot species (A*).

$$\mathrm{A}^* \rightarrow \text{products} \quad (11)$$
$$\mathrm{A}^*+\mathrm{M} \rightarrow \mathrm{A}+\mathrm{M} \quad (12)$$
$$R_{\mathrm{A}}/R_{\mathrm{products}} = k_{12}[\mathrm{M}]/k_{11} \quad (\mathrm{S})$$

Provided k_{12} can be estimated, a graph of the LHS of the equation versus [M] yields a value for k_{11} which may be compared with the rate coefficient for spontaneous decomposition or isomerisation derived from the classical HRRKM theory[63] or the quantum mechanical theory[64].

All three of these types of process are intimately concerned with the use of energy transfer agents†. Experimental work on energy transfer in gases has been reviewed by Cottrell and McCoubrey[65], and with particular reference to unimolecular reactions by Trotman-Dickenson[66] (see also Volume 3, last chap.). The most common energy transfer agents that are used are N_2, CO, CO_2, SF_6 and the inert gases. However, should they react chemically in the particular system studied, their effect as an energy transfer agent will be obscured and any deductions may be completely erroneous. Thus in the di-*tert.*-butyl peroxide/SF_6 system, it is found that above 140°, significant reaction of the hexafluoride occurs with methyl radicals[67]. The relative efficiencies of various energy transfer agents for several reactions are shown in Table 1‡. A general conclusion about the efficiency of energy transfer is that it increases with increasing complexity of the molecule. (For translational–translational energy transfer, maximum efficiency occurs when collisions occur between species of equal mass.) The chances are high that the most efficient is the

† For a recent theoretical treatment see ref. 47b, p. 346.
‡ See also ref. 4, p. 171.

parent molecule or the reactant concerned, although other molecules sometimes achieve this efficiency.

Pressure effects also occur in adiabatic reactions, in particular for very exothermic reactions. The rate coefficient for the unimolecular decomposition of di-*tert.*-butyl peroxide (dtBP) increases with the pressure in a 500 ml spherical reaction vessel[68]. Batt and Benson have shown that this increase can be directly associated with the production of a thermal gradient[1]. A similar increase in the rate coefficient occurs in the case of di-*tert.*-amyl peroxide[68] but here it may be difficult to separate the exothermic effect from that due to sensitised decomposition. The thermal gradient may be reduced or removed by the addition of diluents or by the use of an octopus-shaped RV (see p. 12). The diluent may be an inert transfer agent or may be taking part in the reaction. Inert transfer agents play a dual role in oxidation systems, in particular the H_2+O_2 system[69]. At low pressures the global rate coefficient is decreased by the addition of foreign gases due to enhanced gas phase termination; at higher pressures, the global rate increases due to gas phase heating.

TABLE 1

RELATIVE EFFICIENCIES OF ENERGY TRANSFER IN UNIMOLECULAR REACTIONS

Reactant species	F_2O [a] 250° C	N_2O [b] 653° C	Me_2N_2 [c] 310° C	Cyclopropane [d] 492° C	Cyclobutane [e] 448° C	*i*-PrO [f] radical
Parent species	1.0	1.0	1.0	1.0	1.0	1.0
He	0.40	0.66	0.07	0.05	0.07	—
Ne	—	0.47	—	—	0.12	—
Ar	0.82	0.20	—	0.07	0.21	0.08
H_2	—	—	—	0.12	0.10	0.2
D_2	—	—	0.46	—	—	—
N_2	1.01	0.24	0.21	0.07	0.21	0.12
O_2	1.13	0.23	—	—	—	—
F_2	1.13	—	—	—	—	—
CO	—	—	0.03	0.08	—	0.14
CO_2	—	1.32	0.25	—	—	0.38
H_2O	—	1.50	0.46	0.7	0.44	—
CH_4	—	—	0.20	0.24	0.38	0.31
SiF_4	0.88	—	—	—	—	—
$C_6H_5CH_3$	—	—	—	1.10	1.12	—

[a] W. KABLITZ AND H. J. SCHUMACHER, *Z. Physik. Chem.*, *B*, 20 (1933) 406.
[b] M. VOLMER AND H. FROEHLICH, *Z. Physik. Chem.*, *B*, 19 (1932) 85; AND M. BOGDAN, *B*, 21 (1933) 257; AND H. BRISKE, *B*, 25 (1934) 81.
[c] D. V. SICKMAN AND O. K. RICE, *J. Chem. Phys.*, 4 (1936) 608.
[d] H. O. PRITCHARD, R. G. SOWDEN AND A. F. TROTMAN-DICKENSON, *Proc. Roy. Soc.* (*London*), *A*, 217 (1953) 563.
[e] H. O. PRITCHARD, R. G. SOWDEN AND A. F. TROTMAN-DICKENSON, IBID, A, 218 (1953) 416.
[f] D. L. COX, R. A. LIVERMORE AND L. PHILLIPS, *J. Chem. Soc.*, *B*, (1966) 245.

(*c*) *Explosion limits*

For a given temperature, at high enough pressures, the exothermicity of the reaction leads to explosion. The critical concentration for the explosion limit in a spherical RV is given by[70]

$$f(C_c) = ShRT_0^2/VHk_0Ee \tag{T}$$

where $f(C_c)$ is a function of the critical concentration equal to C_c^n for an *n*th order reaction, S/V is the surface to volume ratio, h is the heat transfer coefficient, R is the gas constant (per mole), H is the exothermicity of the reaction, k_0 is the rate coefficient for the reaction at the temperature T_0 and E is the activation energy. For laminar convection h is given by

$$h = 2e\lambda/r \tag{U}$$

where λ is the coefficient of thermal conductivity of the gas mixture and r is the radius of the spherical RV. Substituting for h, S and V, we have

$$f(C_c) = 6\lambda RT_0^2/r^2k_0EH \tag{V}$$

This differs from the equation of Rice[71] and Frank-Kamenetskii[72] by a factor of 1.81. Their equation considers pure heat conduction only, and is as follows

$$f(C_c) = 3.32\lambda RT_0^2/r^2k_0EH \tag{W}$$

These equations predict the explosion limit for dtBP as 590 or 330 torr at 150° C in a spherical RV with a radius of 5 cm[73].

The phenomenon of explosion limits is best illustrated by reference to the H_2+O_2 system[69,74]. The rate of reaction of a stoichiometric mixture of H_2 and O_2 is shown as a function of pressure in Fig. 10. At a pressure of $\sim$ 2 torr, a slow steady reaction occurs. At a certain critical pressure, a few torr, explosion occurs, this being the first or lower limit P_1 or P_L. Explosion is favoured by an increase in the diameter of the RV and by added inert gases, since the surface plays an important role in chain termination processes and these gases hinder diffusion to the walls. The nature of the surface will also be important (see p. 11). Starting at 200 torr and reducing the initial pressure, normal reaction proceeds until at $\sim$ 100 torr explosion again occurs. This is the second or upper limit P_2 or P_U. At very much higher pressures another limit is reached which is known simply as the third limit P_3. Here explosion is probably both a chain-branching and thermal process, and occurs at several hundred torr. The limits are shown as a function of temperature and pressure in Fig. 11.

There are three main techniques for measuring explosion limits in static systems.

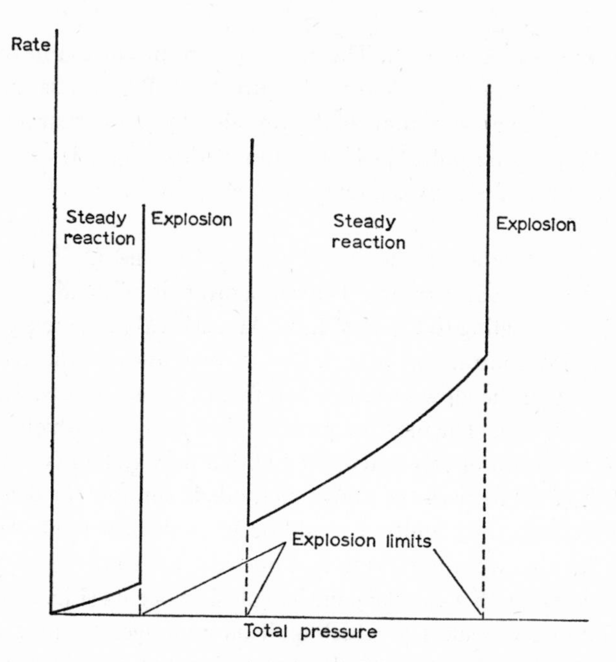

Fig. 10. Rate of reaction for a stoichiometric mixture of H_2 and O_2 at 550° as a function of the total pressure. From ref. 7.

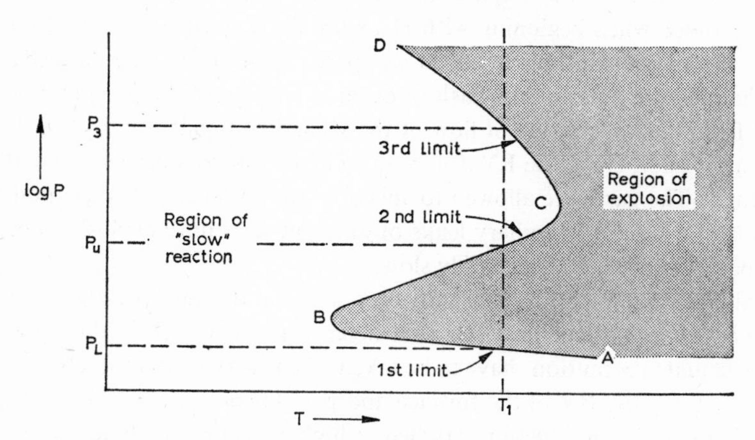

Fig. 11. Critical explosion limits for a typical branching chain explosion showing explosion peninsula. ABCD represents explosion limits. The region ABC is called the explosion peninsula. From ref. 8.

The first two are concerned with the first limit and the third with the second and third limits†.

---

† Techniques are reviewed in ref. 22c and general references are given in refs. 39 and 40.

Pipette and compression methods. The reactants are pre-mixed in the burette B (Fig. 12) and admitted to the RV from the pipette P. Pressure is measured as a function of time, the explosion limit being revealed by the appearance of a flame, a click or a sharp pressure pulse[75]. A variation of this (Fig. 13) is to pre-mix the reactants in the RV and slowly compress the gases by means of the mercury in the burette B[76].

Heating method. Once again the reactants, say CO and O_2[75a], are pre-mixed in the RV but at a low temperature. The temperature is gradually raised when the pressure increases according to Charles' Law. At a critical pressure a blue glow appears. At this point contraction occurs due to reaction. As the temperature is raised, the intensity of the glow increases and finally a flash occurs. This is accompanied by a pressure kick. The pressure limit is taken as that at which the blue glow first appears. A variation of this is also used with a flow system (see p. 24).

Capillary method. Neither of the above methods is suitable for determining the second and third limits since ignition would occur at the first limit. An alternative (of which there are two variations) is to add one reactant to the reaction vessel followed by the next through a capillary until explosion occurs (Fig. 14)[76]. The maximum and minimum pressures at which ignition no longer occurs give values for the first and second limits respectively. Increasing the pressure, the minimum pressure for ignition gives the third limit (for the H_2/O_2 system). Here H_2 is added first, otherwise instantaneous explosion occurs; *i.e.* explosion below the second limit. It is avoided when beginning with H_2, since for a given stoichiometric mixture twice the pressure of gas is added compared with O_2[77]. Alternatively, the time for the pressure pulse or the flash to occur is noted and the pressure limit is calculated from the known rate of flow of the second component into the RV[76b]. A variation is to add O_2 to the RV, followed by hydrogen, to a pressure above the second limit[78]. The gases are allowed to mix for one minute, and then pumped through one of a series of capillary leaks of different sizes until explosion occurs. Reaction before explosion is negligibly slow.

Conventional manometers have usually been used but the ideal pressure-measuring device is the transducer provided that heterogeneous reactions due to it are avoided. Adequate precautions have to be taken when carrying out these explosion studies by encasing the RV in the furnace and possibly putting a shield round the grid‡. This also applies to reactions studied at high pressures. Techniques for this type of kinetic measurement are amply covered in ref. 22c, pp. 347–358.

(d) *Mixing*

If more than one reactant is being used in a system, it is essential that the gases are well mixed prior to admittance to the RV. This may be achieved by convection, producing a thermal gradient[79], by stirring using a soft iron core and rotating

‡ See ref. 22c, p. 417.

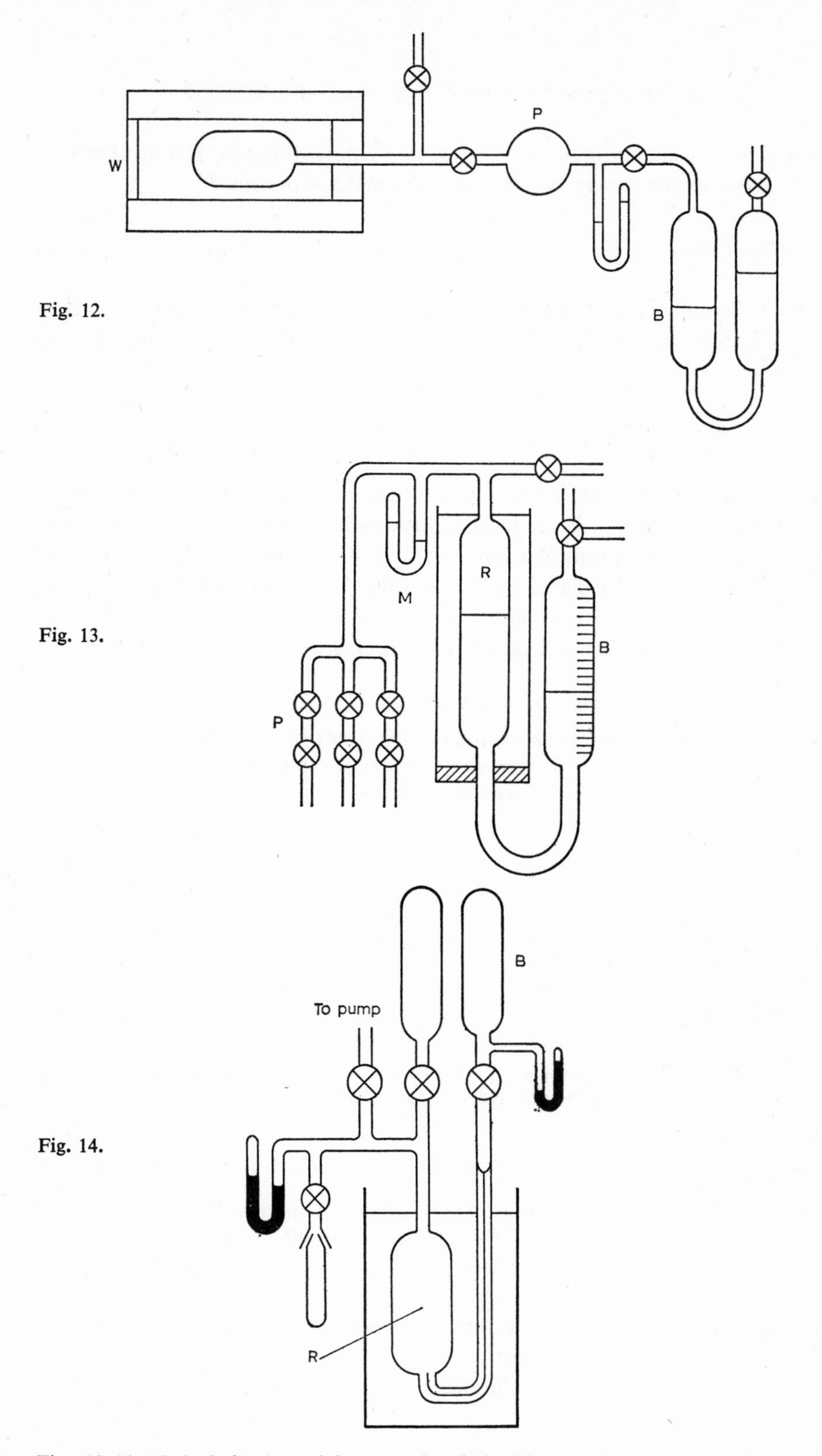

Figs. 12–14. Methods for determining explosion limits. Fig. 12. Pipette method. Fig. 13. Compression method. Fig. 14. Capillary method. From ref. 22c.

magnets or by using an induction motor, or by pure diffusion. For a cylindrical vessel, diffusion times may be calculated from the expression[80]

$$t = 3l^2/4D_{1,2} \tag{X}$$

where t is the time in seconds, l is the length of the mixing vessel in cm and $D_{1,2}$ is the diffusion coefficient for species (1) and (2) in cm^2.sec^{-1}. $D_{1,2}$ may be calculated from Chapman's equation[81]

$$D_{1,2} = 3\bar{c}_{1,2}/32n_t\sigma_{1,2}^2 \tag{Y}$$

where $\bar{c}_{1,2}$ is the relative mean velocity of the two species, n_t is the total number of molecules per cc and $\sigma_{1,2}$ is the mean collision diameter of the two species. The importance of allowing sufficient time for mixing is emphasised by the tabulated values in Table 2. t is calculated on the basis that l is 25 cm. Since $D_{1,2}$ varies inversely as the total pressure, t will be increased by a factor of 10 at 500 torr. This gives the range of t for normal experimental pressures.

TABLE 2

MIXING TIMES FOR VARIOUS GASES TOGETHER WITH THE RELEVANT PARAMETERS IN EQUATION (X), AT A TOTAL PRESSURE OF 54 torr

Gases	$\sigma_{1,2}$ (A)	$10^{-4}\bar{c}_{1,2}$ (cm. sec^{-1})	$D_{1,2}$ at 70° (cm^2. sec^{-1})	t (sec)
CO_2+HBr	3.15 [a,b]	5.04	3.14	149
CO_2+Br_2	4.2 [a,b]	4.58	1.61	291
dtBP+HBr	7.0 [b]	3.61	0.44	1067
dtBP+HCl	7.0 [b]	4.97	0.63	745
dtBP+SF_6	8.9 [b,c]	3.14	0.31	1510

$\bar{c}_{1,2}$ is calculated from the expression $\bar{c}_{1,2} = (8kT/\pi\mu)^{\frac{1}{2}}$, where k is the Boltzmann constant, T is the absolute temperature and μ is the reduced mass of the two species.

dtBP = di-*tert*.-butyl peroxide.

[a] *Handbook of Chemistry and Physics*, Chemical Rubber Publishing Co., 45th edn., 1964–1965, F88.

[b] Estimate of $\sigma_{1,2}$ based on measurement of the appropriate Catalin model.

[c] G. H. CADY, *Advances in Inorg. and Radiochem.*, 2 (1960) 105.

(*e*) *Dilatometry*

The analogue of measurement of pressure change at constant volume for a gas phase reaction is the measurement of volume change at constant pressure for a reaction in the liquid phase by dilatometry[81a]. This is used extensively in polymerisation studies.

A simple dilatometer merely consists of a glass bulb fused to a length of precision bore capillary tubing. Good thermostatting is necessary, and care must be taken with a process having a high heat of reaction that heat exchange is efficient; a dilatometer of annular design may be used. When calculating the degree of reaction from the volume change, it is usually adequate to assume that the molar volumes of reactants and products are additive, but the possibility that they may vary appreciably with concentration should be borne in mind.

(f) Liquid systems involving dissolved gas

Some reactions in solution involve dissolved gas, *e.g.* oxygen, as one of the reactants, or a gas as a product. In such cases, the process may be conveniently followed by measuring at constant pressure the volume either of reactant gas absorbed, or of product gas evolved, by means of a gas burette connected to the thermostatted reaction vessel, usually with a device for shaking or stirring the solution[81b]. Care must be taken to ensure that the rate of absorption or evolution of gas is governed by the rate of reaction and not by the rate of, say, dissolution. This can be done, for instance, by using conditions under which the rate of absorption is independent of the rate of shaking.

3.2.2 Flow method

(a) Some flow systems

"*Straight through*" *system.* In the simplest case (Fig. 15) the reactant A, main-

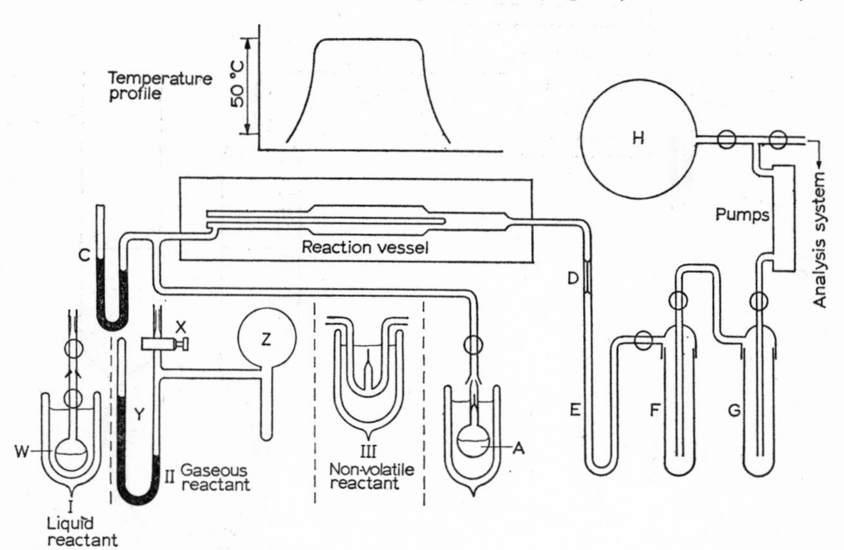

Fig. 15. "Straight-through" flow system. From ref. 66.

tained at a given temperature to provide a constant vapour pressure (liquid), or a constant flow obtained by means of a needle valve (gas), passes through the RV and condensable species are trapped at E, F and G. The flow rate is controlled by capillaries at A and D and measured by manometers such as C at the entry and exit of the RV. The latter measurement also allows the average pressure in the RV to be measured. Non-condensable products are passed into the storage vessel H by diffusion or Töpler pumps. If a carrier gas is used, such as in the toluene carrier technique[82], A now becomes the carrier gas. The reactant is fed into the carrier gas stream from systems I (liquid reactant) or II (gaseous reactant). In the case of the liquids, the bulbs W and A are weighed before and after the experiment to determine the amount used in a run. The flow rate of the reactant is determined by a capillary at the exit point and the temperature of the bath surrounding W. System II is used for a gaseous reactant stored in the bulb Z of known volume such that the amount of reactant used may be determined from the change in pressure of the manometer Y. System III is used for a reactant of low volatility such that it is picked up from the U-tube, held at different temperatures, by the carrier gas. A flow system has been used to determine the explosion limit at different flow rates and pressures[83]. The mixture in the gas burette G (Fig. 16) is passed into the RV at a rate and pressure determined by the taps V_1 and V_2 and measured by the manometer M. The levels of the water in the U of the gas burette are kept the same by automatic adjustment of C. At the required flow rate and pressure the temperature of the furnace is raised until explosion occurs.

Circulating system. In some cases a circulating system may be used[84]. The car-

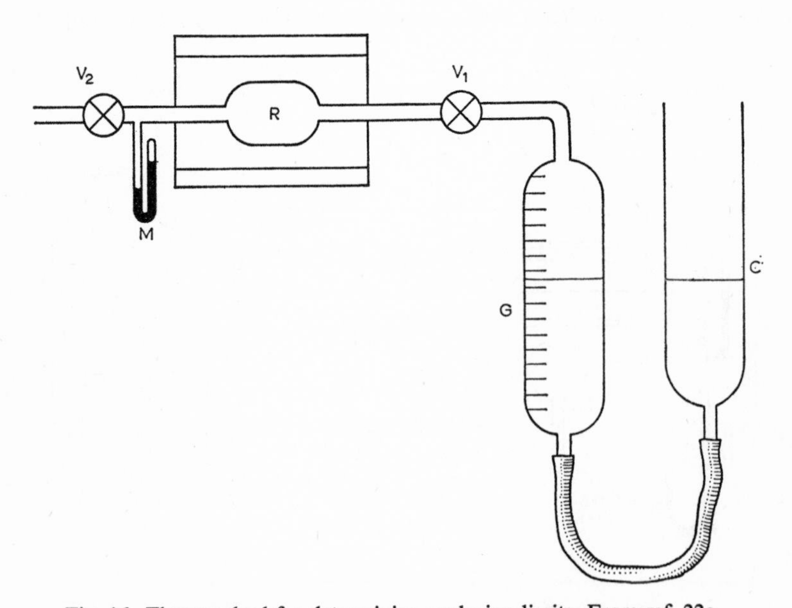

Fig. 16. Flow method for determining explosion limits. From ref. 22c.

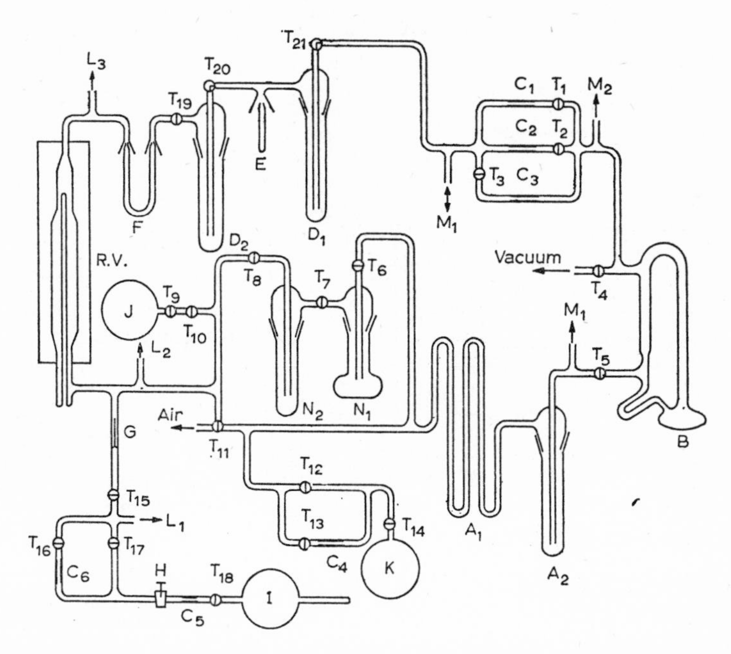

Fig. 17. Circulation flow system. The technique was first used by E. T. Butler and M. Polanyi, *Trans. Faraday Soc.*, 39 (1943) 19.

rier gas is circulated through the RV (Fig. 17) by a mercury diffusion pump[85]. The reactant is picked up from the Warhurst double trap system, N_1N_2[86]. N_1 is kept at a temperature 10–15° above that of N_2 such that a supersaturated mixture with respect to this trap passes into N_2. This ensures that the carrier gas stream entrains a constant vapour pressure of reactant, provided N_2 is kept at a constant given temperature. Having passed through the RV, the reactant and products are, similarly to the above technique, trapped in F, D_2 and D_1. The flow rate may be varied by using the capillaries C_1, C_2 or C_3 and measured using the manometers M_1 and M_2. The latter also give a value for the pressure in the RV. Other gases may be added from the dosing system T_9, T_{10} and kept in J, or through capillary and needle valve systems such as C_4 or C_5, C_6 and H (needle valve). Measurements of flow and the pressure of these gases in the RV are made by manometers L_1, L_2 and L_3.

Very low pressure pyrolysis (VLPP). It is possible to use pressures in flow systems that are very much lower than those in static systems. In VLPP systems the pressures are of the order of 10^{-4}–10^{-5} torr. Previous work has been reviewed and the technique highlighted by Benson and Spokes[87]. Here energy transfer is predominantly via gas–wall collisions. The method provides a new kinetic tool for a detailed study of unimolecular reactions, energy transfer, bimolecular gaseous reactions and heterogeneous reactions[88]. The reactant flows from a 5 l reservoir

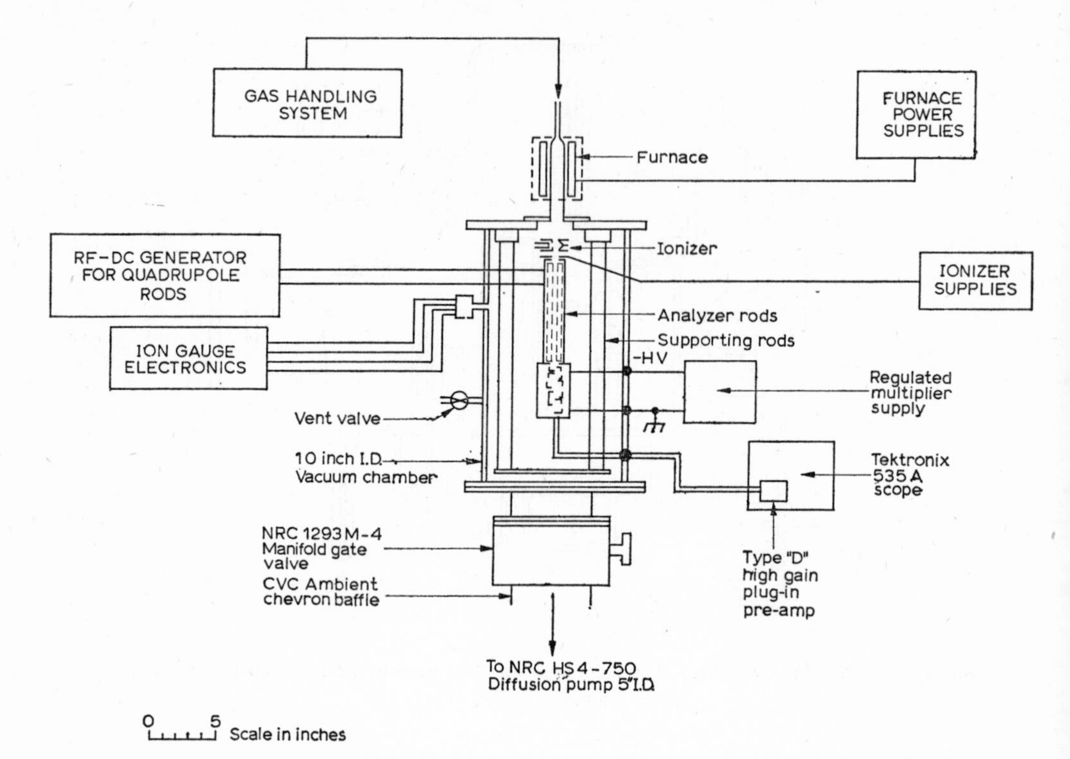

Fig. 18. Schematic drawing of the VLPP apparatus. From ref. 87.

at a few torr through a variable leak valve and a short length of capillary tubing and from there to the RV. Gases which effuse from the reactor pass to the ionisation chamber of a quadrupole mass spectrometer, which enables the extent of reaction to be determined. A schematic drawing of the apparatus is shown in Fig. 18.

(*b*) *Reaction vessels*

Typical RV's are shown in Fig. 19 (thermal systems) and Fig. 20 (photochemical systems). They may be subjected to the usual treatments of coating, packing and conditioning (*cf.* static RV's). Inherent problems attaching to flow systems are uncertainty of RV volume and residence time in the RV, preheating, pressure drop across the RV, Poiseuille shearing of flow across the diameter of the RV, and partial mixing of reactants, products and carrier gas, when "plug" flow is assumed. Gilbert[89] has shown that partial mixing occurs in the decomposition of hydrazine using the toluene carrier gas technique[90]. These difficulties are largely overcome if a stirred-turbulent or capacity-flow RV is used. This system was devised by Bodenstein and Wohlgast[91]. Since then, it has been used extensively for reactions in the liquid phase, but not in the gas phase until recently[92-97]. In particular, the technique has been reviewed and the theory developed by Denbigh[98]. Stir-

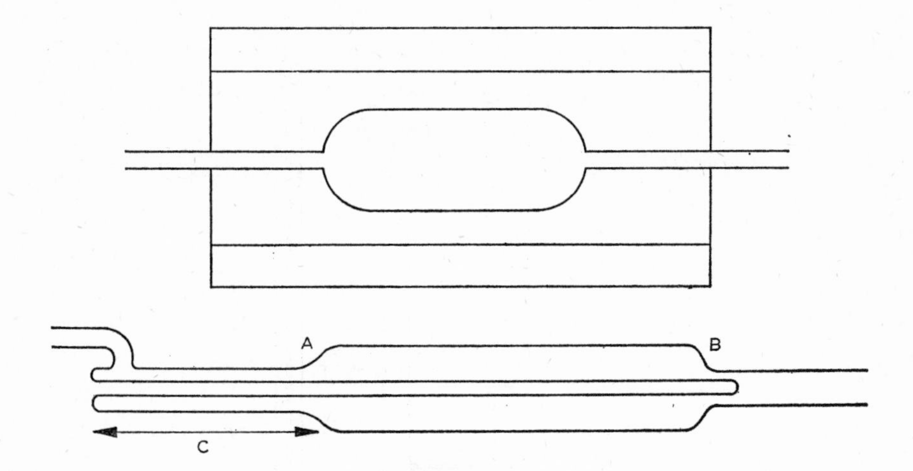

Fig. 19. Typical RV's—thermal systems. From ref. 22c.

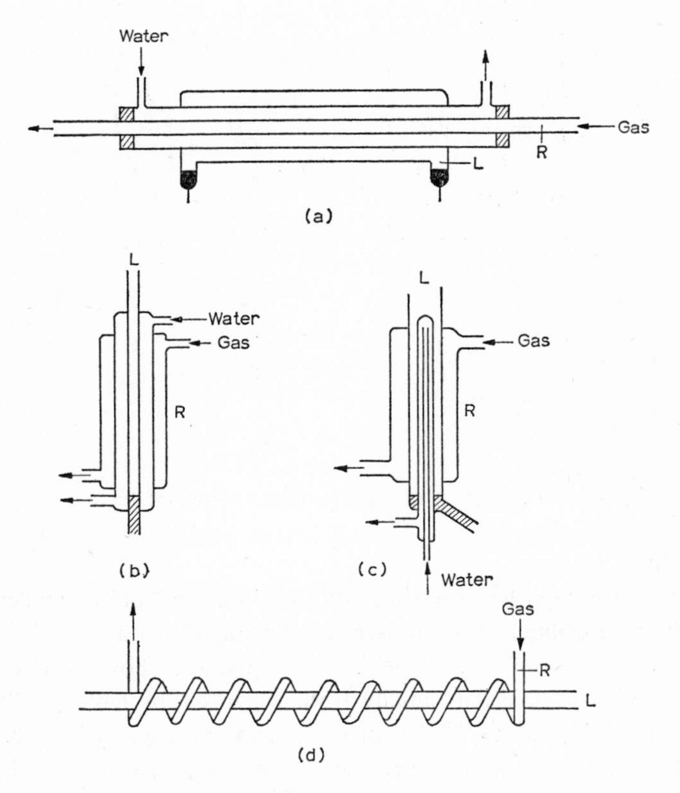

Fig. 20. Typical RV's—photochemical systems. R = vessel, L = light. From ref. 22c.

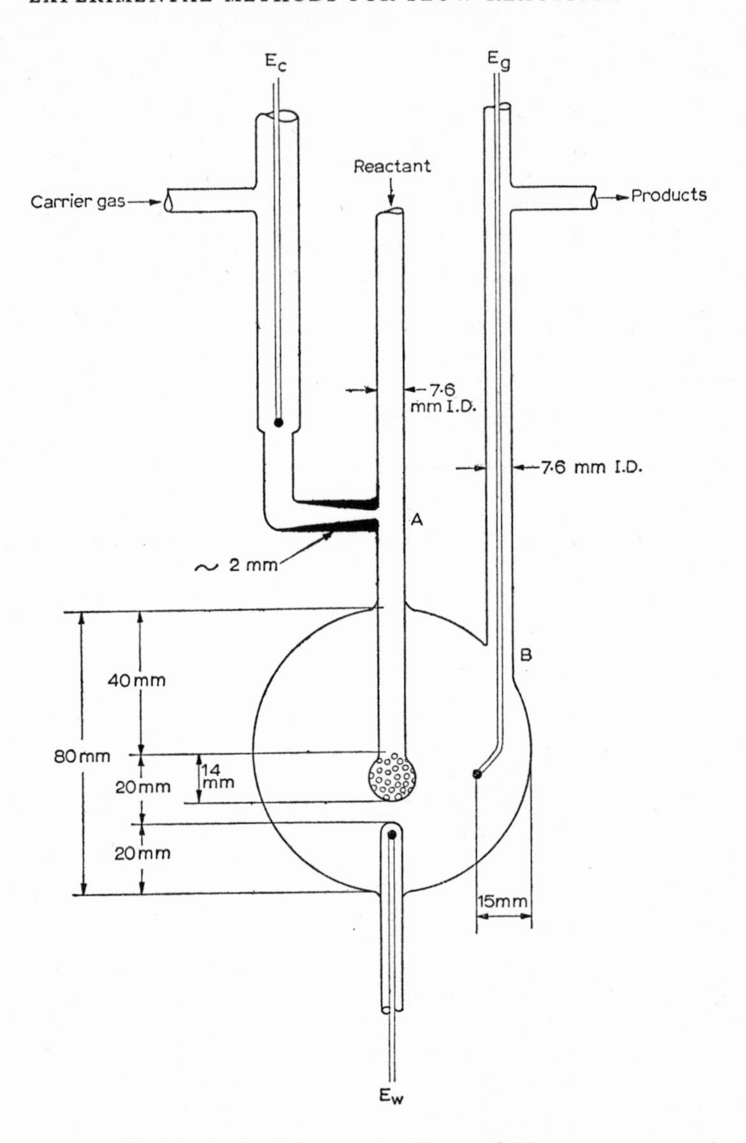

Fig. 21. A stirred flow reactor. From ref. 95.

ring has been accomplished mechanically[96], by diffusion[94] or by turbulent flow[93,95]. Turbulent flow is obtained by Houser and Bernstein[93] with a "cyclone" RV similar in design to a cyclone separator[99]. In the apparatus shown in Fig. 21, reactant and carrier gas are injected radially from a pyrex tube having a "pepper-pot" head, into the spherical RV[100]. The products flow out tangentially at B. The temperature of the gases can be measured by the thermocouples E_w and E_g. Complete mixing has been checked by GLC analysis[94,97,101].

(c) Control and measurement of flow, and measurement of pressure

As seen in Figs. 15–17, flow rates may be controlled by capillaries of different diameter or by needle valves. By measuring the pressure at both ends of the capillary, and assuming streamline flow, the rate of flow may be calculated from Poiseuille's equation

$$dV/dt = (p_1^2 - p_2^2)R^4/16l\eta p_0 \qquad \text{(Z)}$$

where p_1 and p_2 are the pressures at entry and exit of the capillary, p_0 is the pressure at which the rate is measured, l the length of the capillary (radius R) and η the coefficient of viscosity of the gas or the gaseous mixture. Alternatively, and more reliably, the rate of flow is determined by measuring the volume of gas from an aspirator delivered by the capillary at a known pressure. This method may also be used for known, reproducible positions of the needle valve. The pressure in the RV will be a mean of that at entry and exit points. Conventional U-manometers, using mercury or oil for high or low pressures are suitable pressure-measuring devices. These may often be shielded from reactants by strategically placed cold traps. As an alternative for low pressures, the manometers can be replaced by McLeod gauges of low compression ratios or oil–mercury magnifying manometers[43,102] (Fig. 22). The magnification factor depends on the angle θ and the diameter of the inclined and vertical tubes. For diameters of $\sim$7 and $\sim$25 mm and $\theta \sim 15°$, the magnification factor with respect to a mercury manometer is 14.

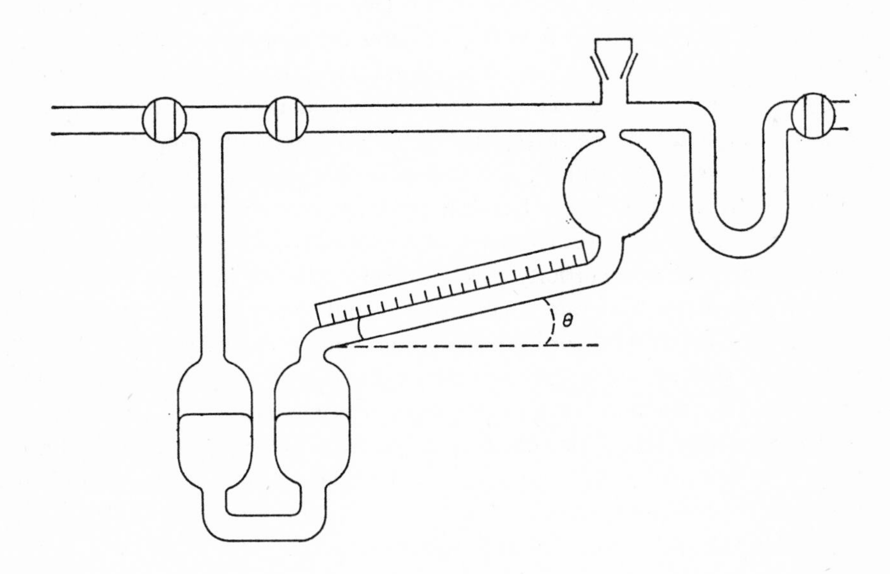

Fig. 22. Oil–mercury magnifying manometer. From ref. 22c.

3.2.3 *Comparison of flow and static systems*

In a static system, a finite time is required for the reactant or reactants to fill the RV and acquire the RV temperature. This renders initial pressure determination (by graphical extrapolation) and initial rate measurements somewhat suspect. (Heating may be assisted by having a heated dosing or mixing vessel.) In practice, filling and heating take 5–15 sec, depending on the conditions, and have to be very much faster than the rate of the reaction studied. This limits the study of the decomposition of dtBP, for example, to a maximum temperature of 160–170° C. Leads to and from the RV and pressure-measuring devices contribute to the production of a "dead space". Ideally, this dead space should amount to no more than 4 % of the RV volume. Corrections for dead space have to be made with pressure measurements and subsequent kinetic expressions[103]. However, for low extents of reaction no (or very simple) corrections are necessary. For adiabatic reactions in large diameter RV's, errors in rate coefficient values may be significant[1], due to the production of thermal gradients. Very often, for a partly heterogeneous reaction, it is difficult to isolate or to determine the rate of the homogeneous process[2].

Flow systems, with low contact times, have the advantage that higher temperatures may be used, thus minimising or obviating heterogeneous processes. Also, percentage conversion may be kept low, while still retaining precision in determining the extent of reaction, since a run can last until sufficient product has accumulated, thus minimising any further reactions of the initial products. The inherent errors associated with rate coefficients determined using conventional flow systems[104] have led to relatively little use of this technique in spite of its advantages. However, the use of a stirred flow RV obviates most of these difficulties[95] and together with a static system allows a reaction to be studied over a wide pressure and large temperature range. Two difficulties remain. In spite of heat exchange the heat capacity effect of the gases entering the RV lowers the temperature of the RV centre relative to the walls[95], and instantaneous cooling still does not occur upon leaving the reaction zone. Thermal gradients will also still be present due to the endothermicity or exothermicity of a reaction. Low pressures should be used to minimise the heat capacity effect. However, the use of low pressures may lead to pressure-dependent rate coefficients[105] and therefore make bond dissociation energies, determined for example by the toluene carrier gas technique[106], suspect. Also, side reactions may not be insignificant[107]. Conversely, the flow system is ideal for the study of pressure-dependent decompositions or isomerisations of free radicals, whose rate coefficients are close to their high pressure limits at normal pressures with static systems. At the higher temperatures used in the flow system, the transition pressure from second to first order kinetics for a unimolecular reaction will also occur at a higher pressure†.

† The transition pressure varies as the $(s-\frac{1}{2})$th power of the absolute temperature, where s is the number of oscillators contributing to the decomposition (see ref. 108).

3.2.4 Analytical section (flow and static systems)

This section usually contains traps, sample bulbs, a gas burette and an automatic Töpler pump or a Sprengel pump. The latter is rarely used nowadays. Two types of automatic Töpler pumps are shown in Fig. 23. One of them combines a McLeod gauge and Töpler pump for the accurate measurement of pressures of small quantities of gases[109]. Three tungsten contacts in conjunction with the mercury in the pump connected to the circuit (Fig. 24a) operate a solenoid valve which is connected from the Töpler pump to either air or vacuum, thus allowing gases to be pumped over to the gas burette. An alternative circuit shown in Fig. 24b incorporates a photoelectric device[110]. This avoids the problem of fitting and possible sparking of tungsten electrodes, particularly when explosive gases are used.

The products and undecomposed reactants may be separated from one another by low temperature distillation using various slush baths[111], produced from various liquids and liquid nitrogen (see Table 3). It is important to maintain these

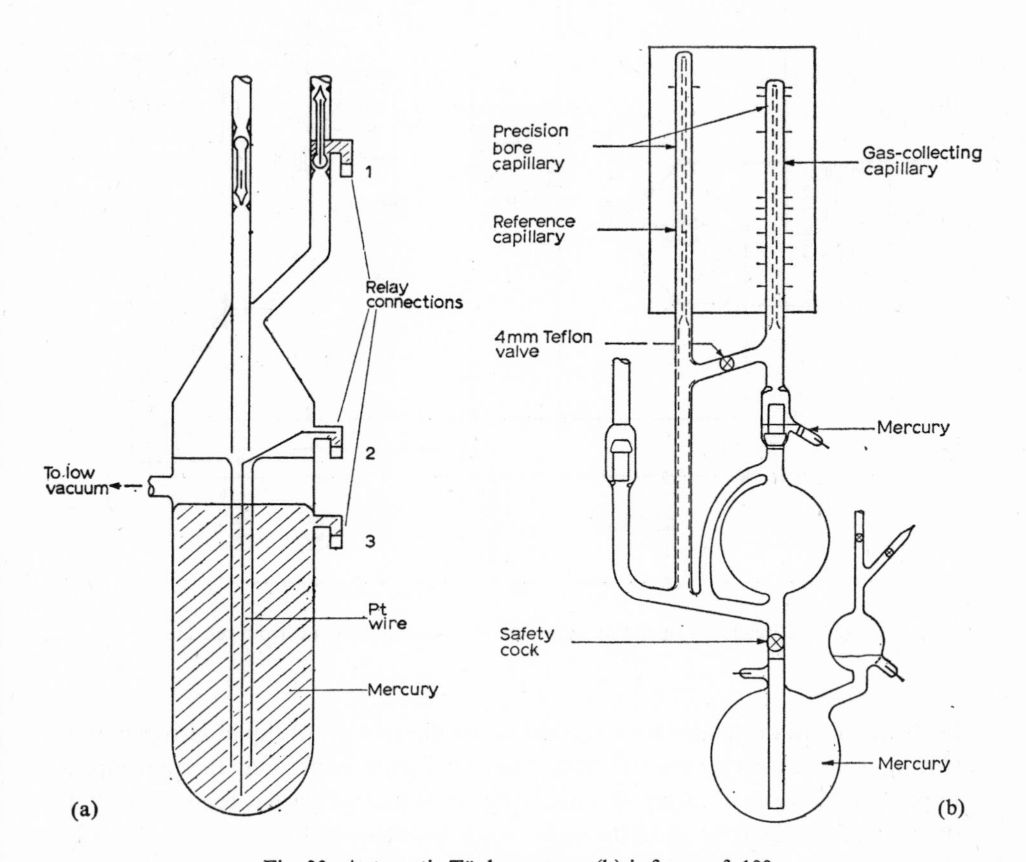

Fig. 23. Automatic Töpler pumps. (b) is from ref. 109.

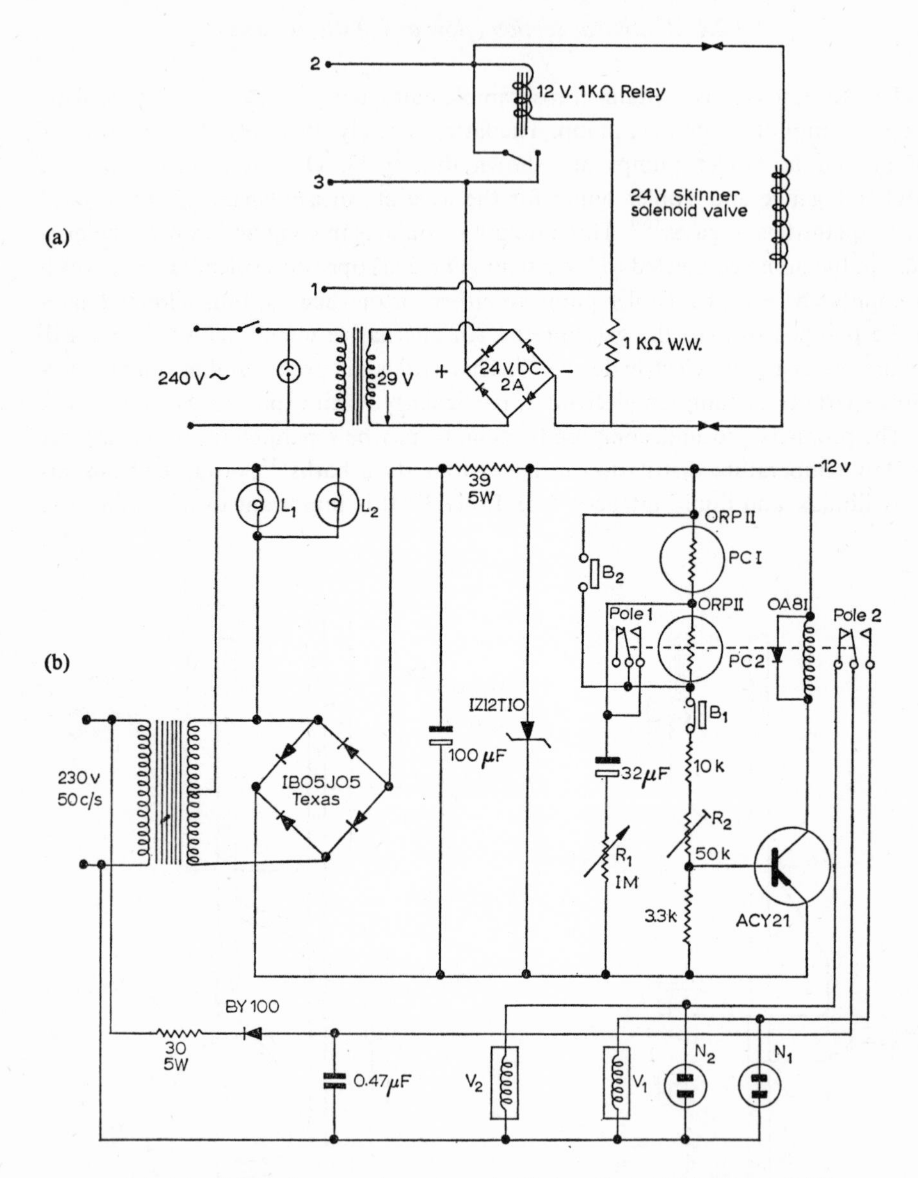

Fig. 24. Circuits for operating automatic Töpler pumps. (b) is from ref. 110.

baths at a constant temperature by adding small amounts of liquid nitrogen and stirring, while observing the temperature with a copper/constantan thermocouple, to prevent "carry over". Care also has to be taken that no gas has been trapped in the remaining liquid or solid by trap-to-trap distillation. Even so, it is difficult to obtain clear cut separations with C_4 and C_5 hydrocarbons for example[112]. Le-

TABLE 3

LIQUIDS FOR SLUSH BATHS[22c]

	Temperature °C		Temperature °C
Aniline	−6.6	Ethyl bromide	−119.0
Methyl salicylate	−9	*n*-Propyl chloride	−122.8
t-Amyl alcohol	−12.0	*n*-Butyl chloride	−123.1
Benzaldehyde	−14.0	Methyl cyclohexane	−126.3
Diethyl carbonate	−15	*n*-Propyl alcohol	−127.0
Benzyl alcohol	−15.3	Allyl alcohol	−129.0
Butyl benzoate	−20	Isobutyl chloride	−131.2
Carbon tetrachloride	−22.9	*n*-Pentane	−131.5
Bromobenzene	−30.6	Allyl chloride	−136.4
Ethylene dichloride	−35.6	Ethyl chloride	−138.7
Anisole	−37.3	Chloroform (19.7 %)+	
Diethyl ketone	−42.0	ethyl bromide (44.9 %)+	
Tetrachloroethane	−43.8	*trans*-1:2-dichloro-	
Chlorobenzene	−45.2	ethylene (13.8 %)+	
n-Hexanol	−48.0	trichloroethylene (21.6 %)	−139
Ethyl malonate	−50.0	Chloroform (14.5 %)+	
Diacetone alcohol	−55	methylene chloride (25.3 %)+	
Chloroform	−63.5	ethyl bromide (33.4 %)+	
Carbon tetrachloride		trans-1:2-dichloro-	
(49.4 %)+		ethylene (10.4 %)+	
chloroform (50.6 %)	−81	trichloroethylene (16.4 %)	−145
Ethyl acetate	−83.6	Chloroform (18.1 %)+	
Toluene	−95.0	ethyl chloride (8.0 %)+	
Methylene chloride	−96.7	ethyl bromide (41.3 %)+	
Methanol	−97.8	trans-1:2-dichloro-	
Carbon disulphide	−111.6	ethylene (12.7 %)+	
n-Butyl bromide	−112.4	trichloroethylene (19.9 %)	−150
Isopropyl chloride	−117.0	Isopentane	−160.5
Isoamyl alcohol	−117.2		

Roy has constructed a variable temperature still (Fig. 25)[113]. A, B and C are 6 mm, 12 mm and 25 mm OD respectively where C (~35 cm long) terminates in a B34 joint. B has four copper/constantan thermocouples at 0, 2, 8, and 14 cm from the bottom. The surface is wound tightly with strips of lead or silver foil up to D. This layer is wound with heating coil (10–15 Ω) with a 10 mm pitch, the lead-in wire being insulated with a glass tube. The thermocouple and heater wires are brought out through W to a cap which is sealed to the tubing with araldite. By having a coolant around C and altering the pressure of air between B and C, the temperature of the trap may be varied. Here the trap has a vertical temperature gradient. By suitable winding, the temperature may be fixed over the length of the trap[†]. Usually two of these traps have to be used in conjunction with one another.

† See also ref. 22c, pp. 156 and 242.

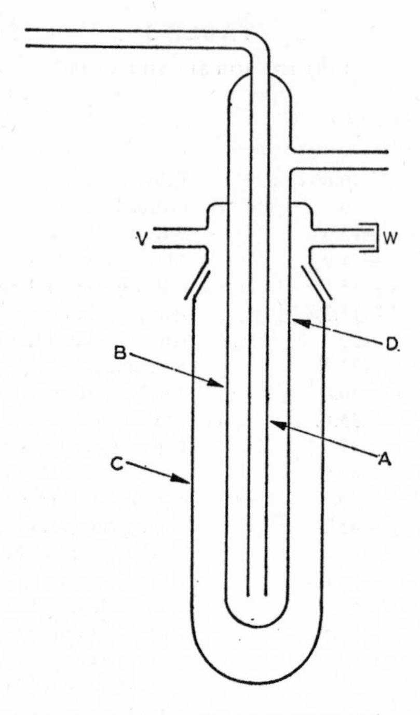

Fig. 25. LeRoy still. From ref. 113.

The different fractions obtained in this way may be analysed later by various methods. Liquids, sealed in sample tubes, and gases may be analysed by GLC or GSC; gases such as halogens, hydrogen halides or CO_2 may be absorbed on various solids and subsequently analysed quantitatively by potentiometric titration techniques. Alternatively, analysis may be carried out *in situ* or small samples withdrawn for direct GLC or GSC analysis. These techniques will be discussed in detail later.

3.3 PHOTOCHEMICAL SYSTEMS (*see also Volume 3, Chapter 1*)

3.3.1 Introduction

Most of the experimental techniques for thermal systems apply here. One important exception is that for experiments using Hg 2537A radiation, the vacuum system must be completely free from mercury, *i.e.* no mercury diffusion pumps or McLeod gauges should be used, unless mercury-sensitised reactions are being studied. Even the use of iodine, gold or similar amalgamating metals does not completely free a particular section of the vacuum system from mercury. There is an excellent treatise on photochemical reactions by Calvert and Pitts[9].

Apart from the two fundamental laws of photochemistry, a very important relationship for the experimentalist is the combined Beer–Lambert law. This describes the extent of absorption of monochromatic light by a homogeneous system, and is given by equation (A′)

$$\log I_0/I = \varepsilon c l \quad \text{(A')}$$

Here I_0 is the intensity of incident monochromatic radiation, I is the intensity of radiation at a distance l cm, and ε is the decadic molar extinction coefficient of an absorbing species (concentration, c mole. l^{-1}). This law is strictly valid only if molecular interactions are unimportant at all concentrations. Deviations occur for a variety of reasons; this means that the validity of the law should be checked under the particular experimental conditions. An initial determination of the absorption spectrum of the compound under investigation is obligatory. This produces immediate qualitative information, particularly about the usefulness of the source of radiation. Banded, diffuse or continuous spectra give direct information about the complexity and variety of primary processes that may occur. Further information will be gained from the effect of radical traps such as O_2 or NO, and of various energy transfer agents.

3.3.2 The optical set-up

A typical optical system is shown in Fig. 26. A lens of short focal length (7–10 cm) projects a nearly parallel beam of radiation from the source A through a filter F, to remove unwanted radiation. The stop S_1 prevents unfiltered radiation from reaching the RV. It is sometimes useful to converge the beam slightly with a second lens (focal length ~40 cm) such that the beam reaches its smallest diameter in the centre of the RV. The latter may be divided into two compartments, one of which contains a compound used for actinometry, or alternatively the beam is focused by the lens L_3 onto a photocell or thermopile P. The intensity of the beam is suitably reduced by the density filter F_2. To provide maximum possible intensity

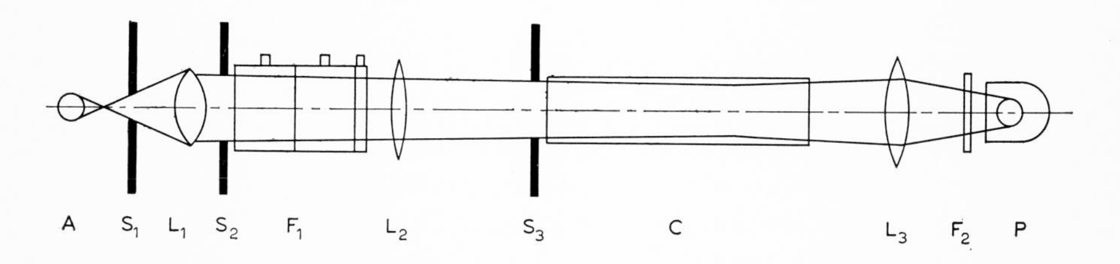

Fig. 26. Optical set-up for photochemical experiments. From ref. 9.

References pp. 104–111

of radiation the source should be as close to the RV as is practicably possible and all windows and lenses should be perpendicular to the path of the radiation. Alignment may be checked roughly using a tungsten lamp or more precisely using a filter soaked with a fluorescent material such as anthracene[114]. Full details for the construction of RV's and the sealing of windows to the cells are given in refs. 9 and 22c.

3.3.3 Sources of radiation

Care should be taken to prevent undue (or indeed any) absorption of radiation by the operator, especially by the eyes and particularly with laser sources. The range of wavelengths of radiation for photochemical reactions extends from 1200 A to 7000 A. When studying photochemical reactions it is particularly useful to be able to convert wavelength (λ) in A to energy (E) in kcal. Einstein^{-1}. This is given by the expression

$$E = 2.8591 \times 10^5/\lambda \text{ kcal. Einstein}^{-1} \tag{B'}$$

The wavelength range is divided into several sections which are shown in Table 4. As far as this discussion of sources is concerned, the range is divided into two; the vacuum UV region is treated separately.

TABLE 4

THE PHOTOCHEMICAL RANGE

A	*Region*	*Energy (kcal. Einstein^{-1})*
1200		238
	Vacuum UV	
2000		143
	Far UV	
3000		95
	Near UV	
4000		71
	Visible	
7000		41

(*a*) *Visible to far ultra-violet*

Mercury lamps

The most common sources of radiation for kinetic photochemical studies are mercury lamps of which there are three types. The low-pressure lamp is used mainly for mercury-sensitised studies. For general photochemical studies the most useful lamp is the medium-pressure lamp having several "lines" of reasonable intensity in the UV and visible regions. The high-pressure lamp tends to have less radiation in the UV than the medium-pressure lamp but is a very intense source of radiation. In order to see how these lamps operate it is instructive to look at the electronic transitions for mercury. The ground state configuration is $(6s^2)^1S_0$. The selection rules for the electronic transitions of atoms (Russell–Saunders coupling) are

$$\Delta S = 0, \quad \Delta L = 0, \pm 1, \quad \Delta J = 0, \pm 1 \quad \text{but} \quad S = 0 \nleftrightarrow J = 0$$
$$\Delta n = 0, 1, 2, \ldots$$

S is the total spin quantum number $= s_1+s_2+s_3 \ldots$, L is the total angular momentum quantum number $= l_1+l_2+l_3 \ldots$, J is the vector sum $S+L$, and n is the primary quantum number. The rule $\Delta S = 0$ is obeyed fairly strictly for the lighter elements but breaks down for heavier elements. One may look at this as some mixing of the various states, *i.e.* the singlet and triplet states have partly triplet and singlet characters, respectively. However, the selection rule is partly reflected in the lifetime for the states due to the relationship

$$\tau = 10^{-4}/\varepsilon$$

where ε is the extinction coefficient. Hence ε is a good deal less for singlet–triplet

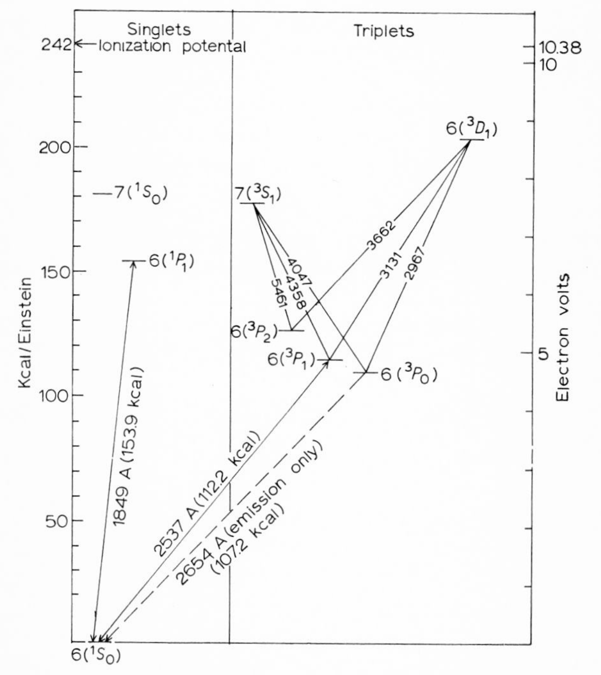

Fig. 27. Lower excited states of the mercury atom. From ref. 9.

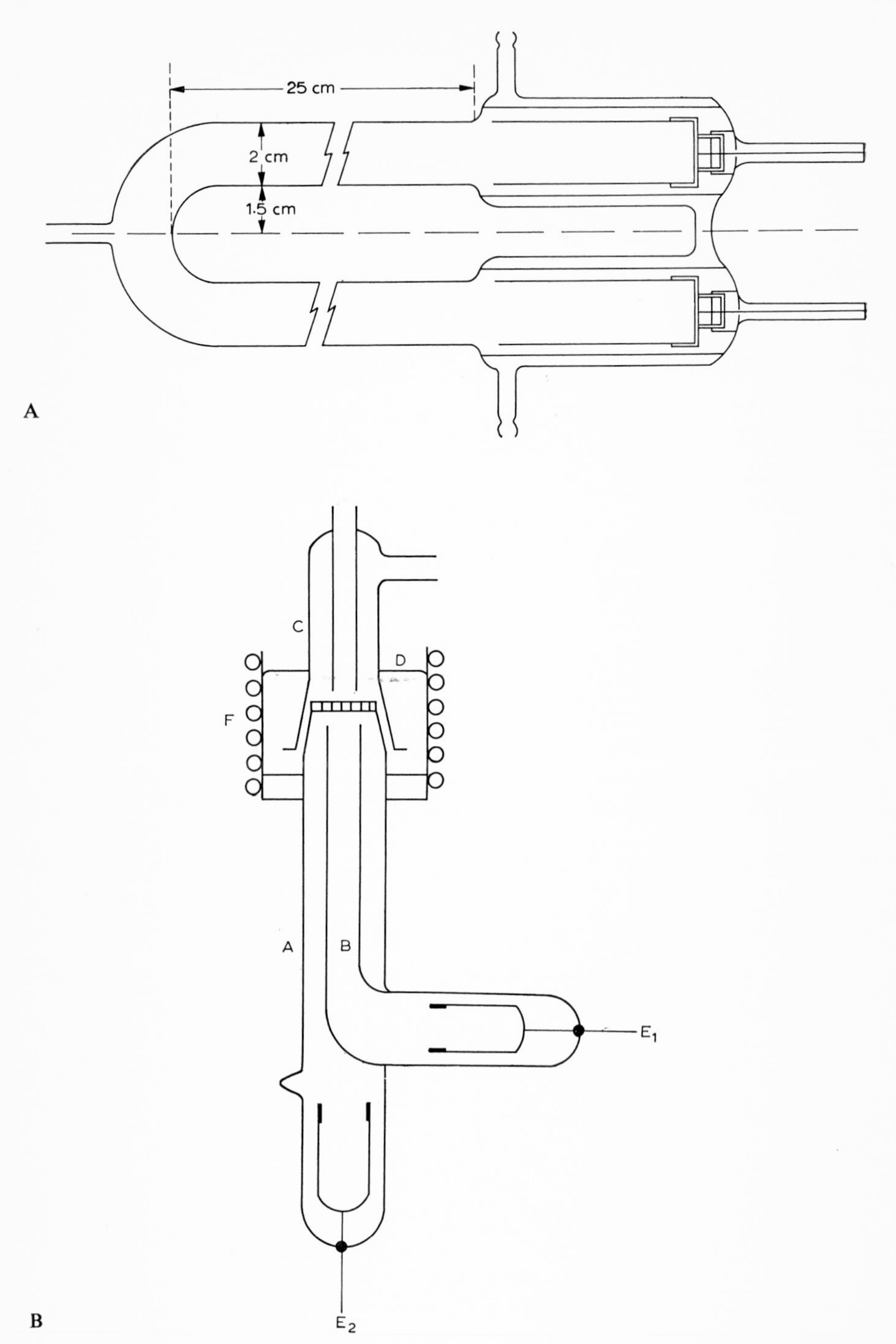

Fig. 28. Low-pressure mercury lamps; (a) from ref. 115, (b) from ref. 116, (c) from ref. 117.

than for singlet–singlet transitions. The lower excited states for mercury are shown in Fig. 27. Immediately above the ground state there are three triplet and one singlet P states. Above these are the $(6s7s)^1S_0$, $(6s7s)^3S_1$ and $(6s7d)^3D_1$ states. The 1849A and 2537A lines are resonance radiations since they appear in absorption from, and emission to, the ground state. Double excitation occurs from the P states giving rise to lines such as 3130A, 3662A and 4358A.

Low-pressure lamps. These lamps operate at or close to room temperature. This means that the vapour pressure of Hg is $\sim 10^{-3}$ torr. In addition to mercury about 6 torr of inert gas, usually neon, is added. This makes for easy firing but there is some evidence that a higher intensity is produced due to reaction (13).

$$\text{Ne}^* + \text{Hg}(^1S_0) \rightarrow \text{Ne} + \text{Hg}(^3P_1) \qquad (13)$$

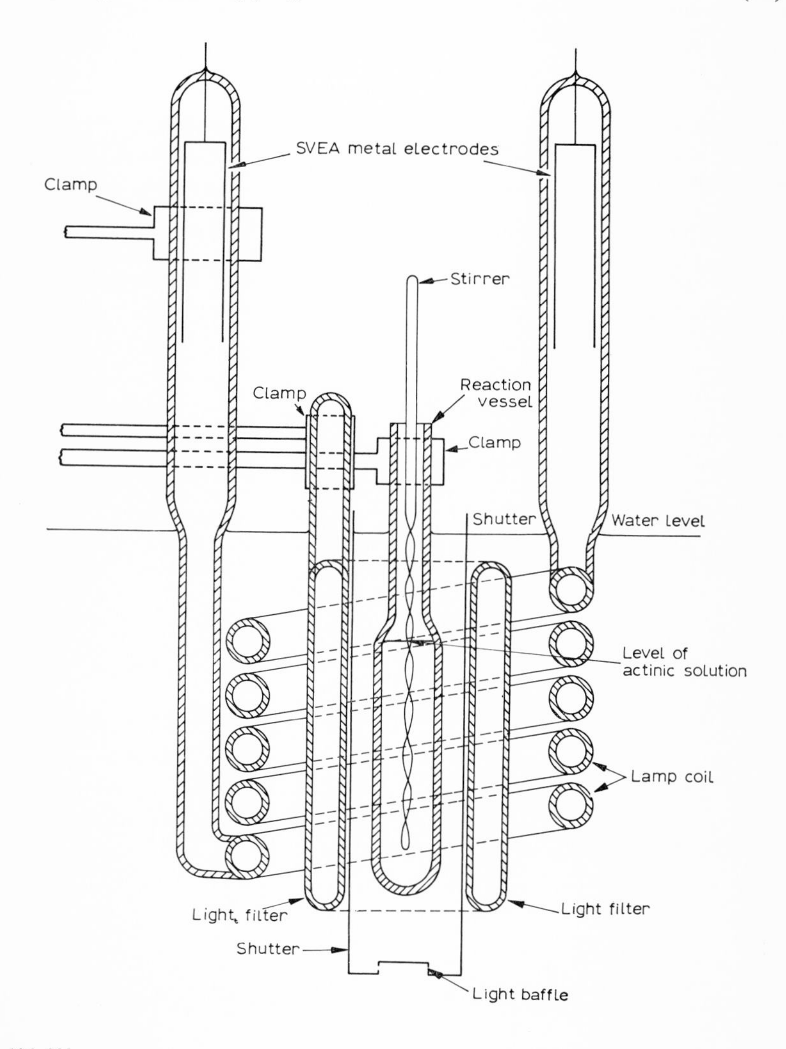

C

TABLE 5

COMPARISON OF ULTRAVIOLET OUTPUT AND OPERATING CONDITIONS FOR SEVERAL COMMERCIAL HIGH-, MEDIUM-, AND LOW-PRESSURE MERCURY ARCS[9]

Lamp designation	Manu-facturer	Arc operating conditions			Useful arc length (inches)	Total UV output (watts) <3800A	UV Energy per inch of arc length	Efficiency of UV generation (%)	Approx. useful lifetime of arc (h)
		Volts	Amperes	Input energy (watts)					
(i) Low-pressure mercury lamps									
G15T8 (heated cathode)	a, b	55	0.30	16.5	14	3.0	0.21	18	2500
G30T8 (heated cathode)	a, b	100	0.34	34.0	32	7.5	0.23	22	2500
2852Q (cold cathode)	c	450	0.030	13.5	24	3.3	0.14	24	12000
WL-782-20 (cold cathode)	d	325	0.055	17.9	20	2.0	0.10	11	4500
WL-782L-30 (cold cathode)	d	410	0.050	20.5	30	5.2	0.17	25	12000
ST46A22 (cold cathode)	b	200	0.120	24.0	11	3.4	0.31	14	12000
ST30A32 (cold cathode)	b	300	0.120	36.0	25	7.0	0 28	19	12000
(ii) Medium-pressure mercury lamps									
SH(616A)	c	100	1.2	120	1.7(U)	6.3	3.7	5	1000
UA-2	a	92	3.1	285	3	30.6	10.2	20	1000
UA-3	a	135	3.1	419	6	38.8	6.5	9	1000
A (673A)	c	145	4.5	653	4.5	95	21.1	15	1000
LL(189A)	c	285	4.7	1340	12	339	28.3	25	1000
UA-11	a	450	3.1	1395	17.7	240	13.6	17	1000
UA-15	a	2000	1.58	3160	48	810	16.9	26	1000
PIS(57A)	c	1260	4.5	5670	46.5	1066	22.9	19	1000
(iii) High-pressure mercury lamps									
AH6 (quartz jacket)	a	840	1.4	1176	1	195	195	17	75
(iv) Compact mercury or mercury–xenon point source lamps									
537B9	c	65	18	1000	0.26	87	335	9	200
PEK107	e	20	5	100	0.12				100
PEK200	e	58	3.5	200	0.10				200

[a] General Electric Company, Lamp Division, Cleveland, Ohio, U.S.A.
[b] Sylvania Electric Products, Inc., Salem, Massachusetts, U.S.A.
[c] Hanovia Lamp Division, Engelhard Industries, Newark, N.J., U.S.A.
[d] Westinghouse Electric Corporation, Lamp Division, Bloomfield, N.J., U.S.A.
[e] PEK, Inc., Palo Alto, California, U.S.A.
[f] The lifetimes shown are approximate lower limits; when the medium-pressure mercury arcs are operated in proper ventilated housings, and with fairly continuous operation, this rating may be increased by several thousand hours.

At these low pressures pure 1849 and 2537A resonance lines are obtained. Three designs are shown in Fig. 28[115–117]. They are constructed from conventional or suprasil quartz and have iron electrodes. The latter are welded to nickel followed by tungsten, which is sealed to the silica with lead or with molybdenum ribbon. Water cooling keeps the pressure low and hence prevents line reversal. If the temperature of the circulating water is thermostatted, the main light-emitting tube may be included in the furnace for the RV, when the output intensity is practically independent of temperature up to 600° C. Suprasil or sapphire windows allow maximum transmission of the 1849A line[116]. The lamps may be operated in three ways: with a heated cathode at a low voltage; with a cold cathode at much higher starting voltages; with an electrodeless discharge excited by microwave frequencies. The heated cathode has the longest life whereas the electrodeless type provides the narrowest "lines". Table 5 gives the characteristics of several commercial low-

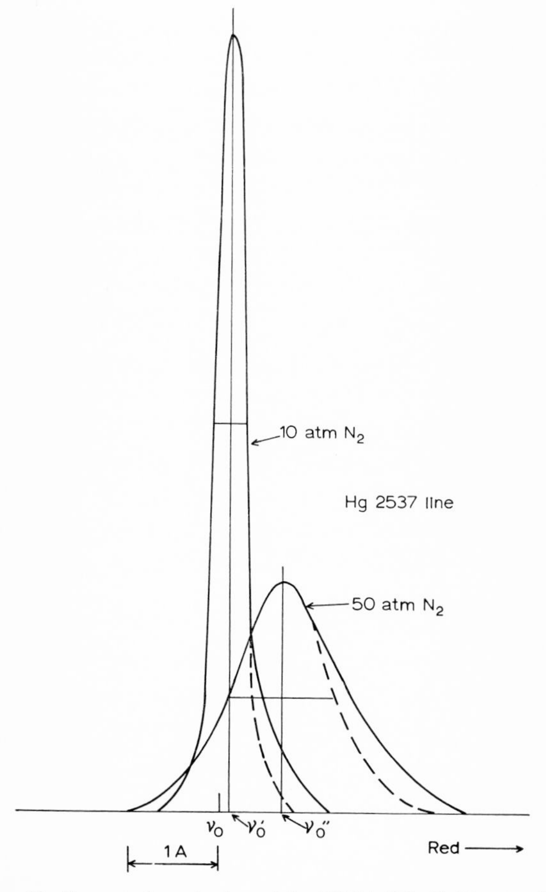

Fig. 29. Pressure broadening of the 2537A "line". From ref. 9.

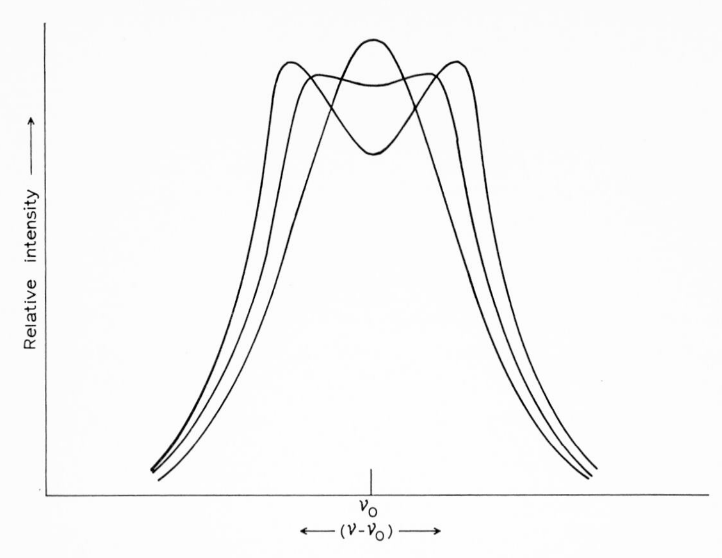

Fig. 30. Change in spectral profile of an emission "line" as a function of increasing self-absorption. From ref. 121.

pressure lamps. Home-made varieties are described in ref. 22c. Since these lamps are of low intrinsic brilliance, they are often coiled about the RV or two or more placed parallel to the RV (Fig. 20).

Medium-pressure lamps. At higher pressures of mercury than that for the low-pressure lamp, "line" broadening[118–122] becomes very important and "line" reversal and double excitation occur. The effect of pressure broadening is shown in Fig. 29. The effect of self-absorption causing "line" reversal is shown in Fig. 30. This radiation "line" is now useless for mercury sensitisation studies, but is suitable for direct photolysis and, in particular, may be used with a vacuum system containing mercury without causing sensitisation. Due to the high population of $^3P_1$ atoms, double excitation is possible resulting in the production of a number of "lines" (see Table 6). Some of the $^3P_1$ atoms are deactivated to the metastable $^3P_0$ state, from where they may be collisionally reactivated or emit 2654A radiation.

A typical lamp is shown in Fig. 31. The high intrinsic brilliance of these lamps and the large number of "lines" make them particularly suitable for kinetic studies.

High-pressure lamps. The most intense sources of UV radiation are the high-pressure lamps. Pressure and temperature broadening are increased under the operating conditions (800° C and 100 atm). At very high pressures, the emission is almost continuous in nature (Fig. 32). The output is 10 times that of the medium-pressure lamp and 1000 times that of the low-pressure lamp per unit length ($\lambda \sim$3800A). The high-pressure lamp is also rich in visible light. The emission is projected from a small capillary quartz envelope (Fig. 33). Since these lamps operate at very high temperatures, water or forced air cooling is required to prevent the melting of the quartz envelope. This indicates a short life for the lamp. For water cooling, distilled water should be used to avoid absorption at shorter wavelengths.

Circuits and operating characteristics for mercury lamps

Since the voltage drop across all lamps decreases as the current increases, a sufficiently large resistance (DC) or reactance (inductance and capacitance, AC) must be placed in series with the lamp to produce the stable condition of an increase in voltage with increase in current. All circuits should include a voltmeter and an ammeter so that readings may be recorded while the lamp is in use, to detect any

TABLE 6

ENERGY DISTRIBUTION IN LOW- AND MEDIUM-PRESSURE MERCURY ARCS[9]

Wavelength, A	*Relative energy*	
	Low-pressure mercury arc[a]	*Medium-pressure mercury arc*[b]
13,673	...	15.3
11,287	...	12.6
10,140	...	40.6
5770–5790	10.14	76.5
5461	0.88	93.0
4358	1.00	77.5
4045–4078	0.39	42.2
3650–3663	0.54	100.0
3341	0.03	9.3
3126–3132	0.60	49.9
3022–3028	0.06	23.9
2967	0.20	16.6
2894	0.04	6.0
2804	0.02	9.3
2753	0.03	2.7
2700	...	4.0
2652–2655	0.05	15.3
2571	...	6.0
2537	100.00	16.6[c]
2482	0.01	8.6
2400	...	7.3
2380	...	8.6
2360	...	6.0
2320	...	8.0
2224	...	14.0

[a] Hanovia Lamp Division, Engelhard Industries, Newark, N. J., SC-2537 lamp.
[b] Hanovia's Type A, 673 A, 550 W lamp.
[c] Reversed radiation.

deterioration in the lamp and to follow the intensity, respectively. Melville[115] found that, with different pressures of inert gas in the low-pressure lamp, maximum intensity of output was obtained at 180mA (Fig. 34). Heidt and Boyles report that to maintain an intensity to $\pm 1\%$, the current through the lamp must be controlled to $\pm 0.6\%$ at 120–130 mA and $\pm 0.8\%$ at 15mA, and the temperature controlled to $\pm 0.1°$ at 130mA and $\pm 0.05°$ at 15mA[117]. However, the lamp may be heated to 600° C without serious change in light intensity[115]. Gomer and Kistiakowsky[123] discuss the control of intensity for medium-pressure lamps. All lamps are subject to some "ageing" as far as intensity is concerned, and the latter may be seriously affected if the quartz or other material is not kept clean by wiping with dilute acetic acid or absolute alcohol.

The lamps may be operated on AC or DC, but for polymerisation studies and radical lifetime measurements DC must be used. For a low-pressure lamp the intensity falls to nearly zero 120 times per second for a 60 cycles AC source[124]; the high-pressure lamp also tends to be unstable with AC[125]. Circuits (Fig. 35) are shown for medium-pressure[126] and high-pressure[127] lamps; operating procedures are given quite clearly in these last two references. The low-pressure lamp reaches stable operating conditions quite quickly, but the other two types require some time to do so (10–15 min). The high-pressure lamp also has a very short lifetime.

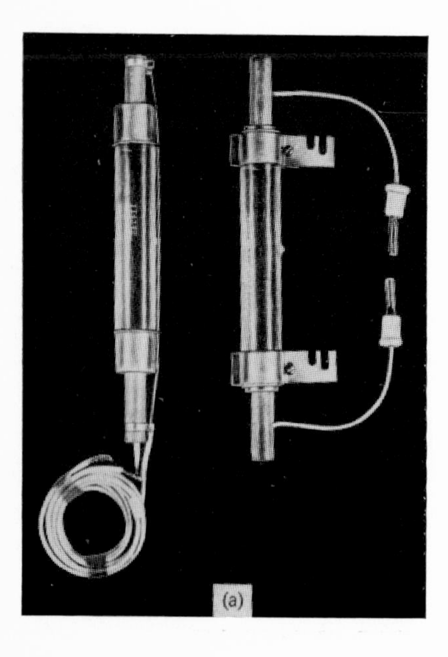

Fig. 31. Typical medium-pressure lamp. Hanovia Chemical and Manufacturing Co., 500 W. See ref. 9, p. 698.

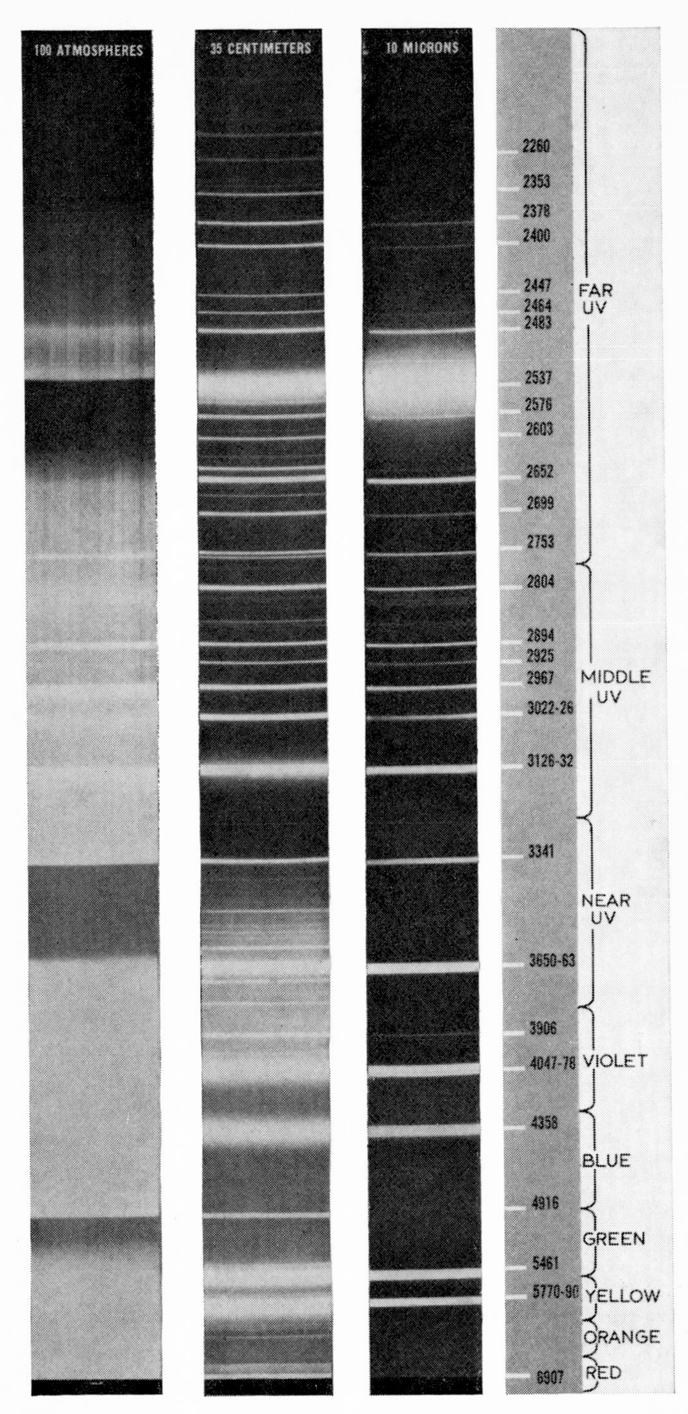

Fig. 32. The emission spectra of the low-pressure (10 μ), medium-pressure (350 torr) and high-pressure (100 atm) mercury arcs. Courtesy Westinghouse Electric Corp.

References pp. 104–111

Fig. 33. High-pressure lamp (it has a water-jacketed mounting). General Electric Co. AH6, 1000 W.

Point sources. An advantageous alternative to the high-pressure lamp for quantitative work is the high-pressure point-source lamp[128,129], having a longer lifetime and a better stability (Fig. 36). These involve a high-powered discharge in xenon, mercury or a mercury–xenon mixture at 20 atm. or more. The bulb-like design obviates the need for water cooling. Either a third electrode, to which a high voltage pulse is applied, or a special power supply for the two electrode model, is used to start the lamp. Figure 37 shows that they are excellent high-intensity continua sources (xenon lamps) and line sources (mercury-containing lamps). Similarly to the other lamps they may be operated with AC or DC. Since the lamps represent a radiation hazard, a special lamp enclosure is used to protect the operator against the intense UV radiation. Vertical operation is obligatory, with the cathode at the top. At 100 watts, ventilation should be controlled so that a chromel/alumel thermocouple (d = 4 thou) at the quartz bulb indicates 800±25° C. Since sputtering occurs with consequent electrode and quartz envelope damage, the warming-up time should be as short as possible. More detailed instructions are given by Calvert and Pitts[9] and by Anderson[130].

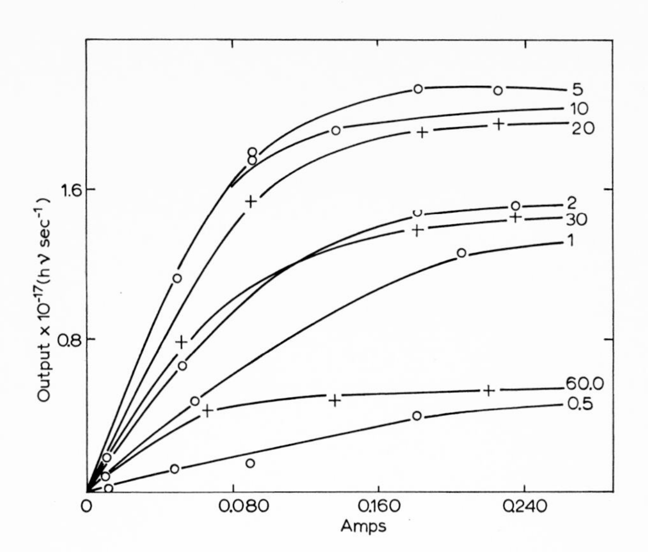

Fig. 34. Effect of current and mean pressure (torr) on output for a low-pressure lamp. From ref. 115.

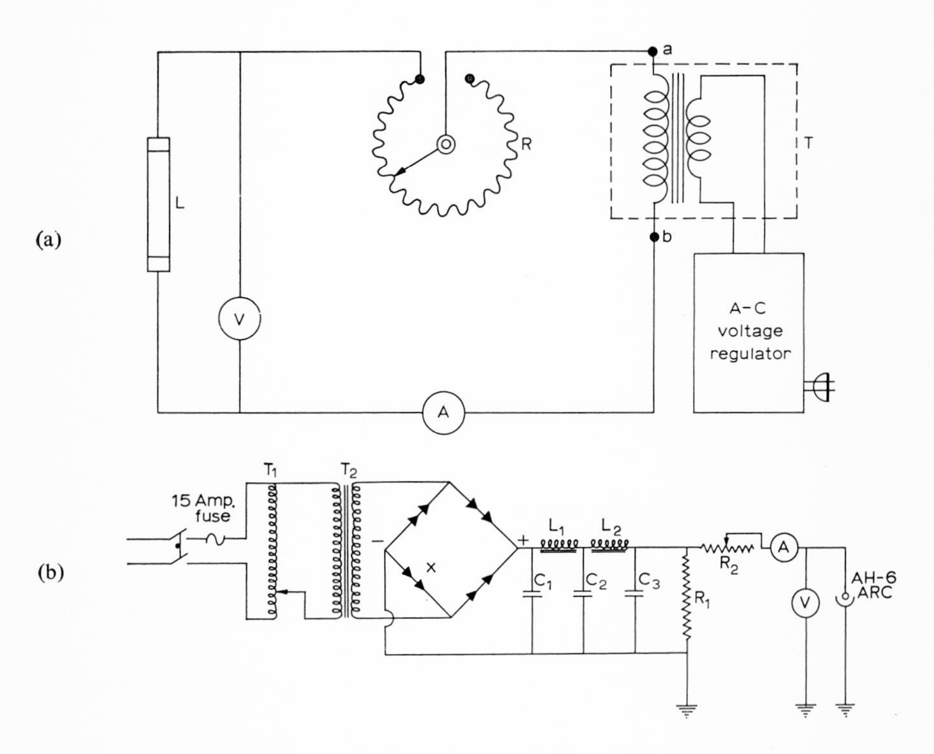

Fig. 35. Circuits for medium- and high-pressure lamps. Key to (b): T_1–10 amp Variac; T_2–59G37 General Electric transformer; L_1, L_2–7h, 1.5 amp dc; C_1, C_2, C_3–8 μf, 1500 DCWV; X–selenium rectifiers (Federal Telephone and Radio Corp. 131H24AX1); R_1–50000 **Ω**, 150 W; R_2–500 **Ω**, 100 W rheostat. (a) From ref. 9; (b) from ref. 127.

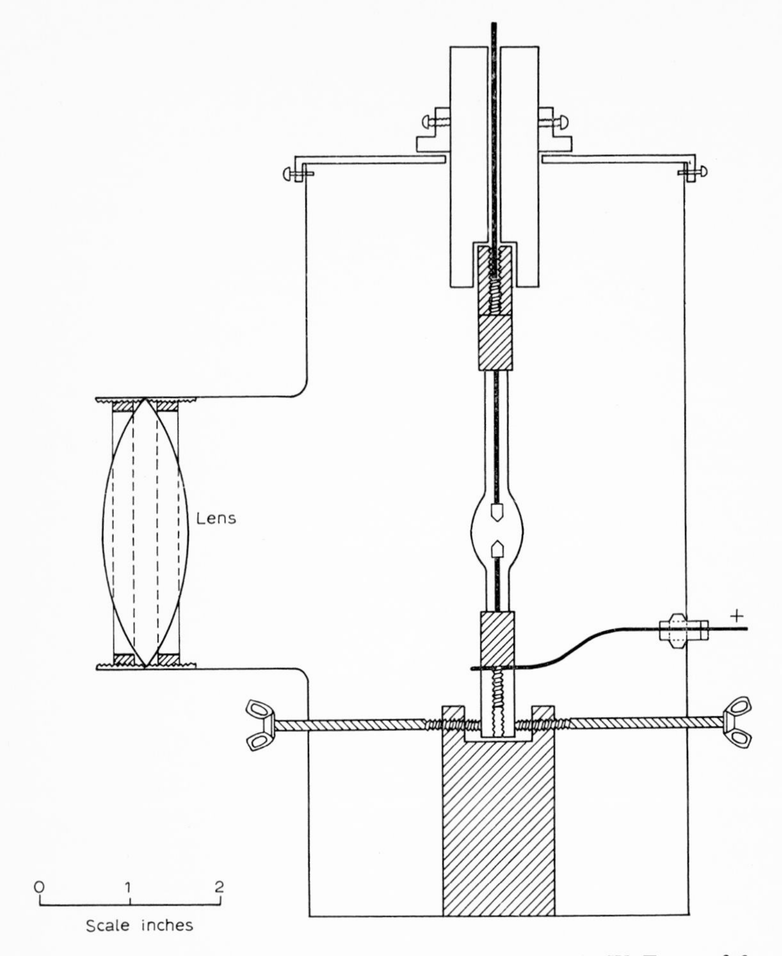

Fig. 36. Point source, high-pressure mercury arc, 200 W. From ref. 9.

Other lamps

Resonance lamps of the alkali metals, zinc and cadmium have been constructed. Laidler[131] shows how Hg, Zn and Cd lamps may be put to very good use to place limits on bond energies and to determine the mechanisms of some photo-sensitised reactions. However, none of these lamps has been used extensively and have limited (or no) photochemical use.

Intermittent radiation

It might be useful at this point to consider a very important photochemical technique, that involving intermittent radiation. This may be used to determine the average life-times of active intermediates in reactions, *e.g.* radical polymerisation, and rate coefficients for elementary processes, *e.g.* radical abstraction reactions and particularly dimerisations such as 2Me·, 2*i*-Pr·, 2*t*-Bu· etc. The technique was first applied to gas phase reactions by Briers and Chapman[132] and has been very clearly reviewed by Burnett and Melville[133a]. A "square" light wave (instantaneous switching from light to dark) is generated by inserting a segmented disc, which may be rotated at varying speeds, in front of the light source. The disc may be solid plastic with segments painted black (visible and near UV) or of metal with segments cut out (far UV), to achieve the desired ratio of dark to light time (r) which usually has the value of three. In order to obtain a stable light flash period, the rate of rotation of the sector has to be precisely controlled. Circuits for this purpose are given by Burnett and Melville[133a] and Kwart *et al.*[134]. The flash duration is measured stroboscopically, with a photocell + electronic scaling unit assembly, or for slow speeds, directly.

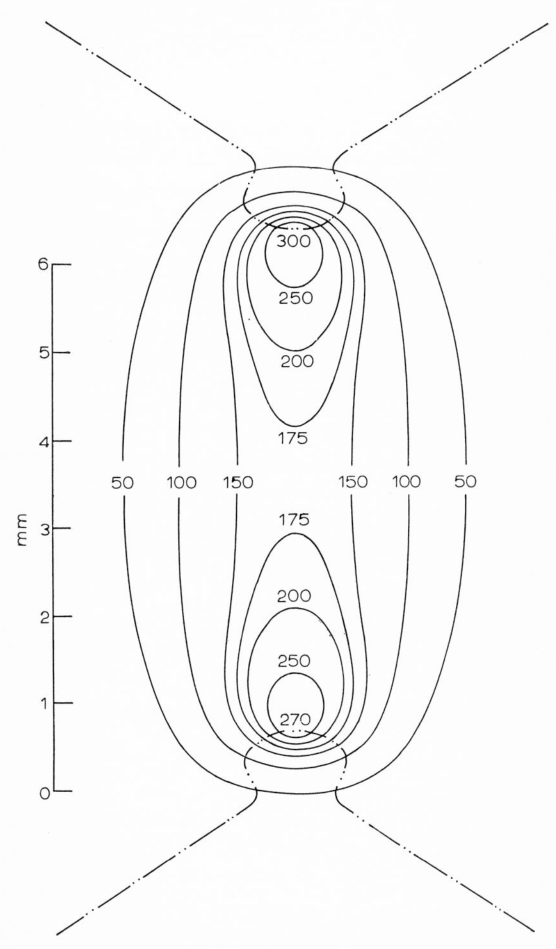

Fig. 37. The brightness distribution in candles per square mm for a 1000 W Xe–Hg point-source lamp with AC operation. From ref. 9.

The rate of formation of some intermediate in steady illumination (R_{st}) is given by equation D′

$$R_{st} = A(I_a)^n \tag{D'}$$

where A is a proportionality constant, I_a is the light absorbed and $0 < n < 1$. Since r is the ratio of dark to light periods, the fraction of the time that the light is on is $1/(r+1)$. If at low sector speeds the time between light periods is long compared with the life-time of the intermediates, the rate of formation of these (R_{slow}) is the time average of the rate under steady illumination (R_{st}) and the zero rate for dark periods, *i.e.*

$$R_{slow} = A(I_a)^n/(r+1) \tag{E'}$$

At fast sector speeds, when the flash period is short compared to the life-time of the intermediates, the effective light intensity is $I_a/(r+1)$ and R_{fast} is given by equation (F'):

$$R_{fast} = A\{I_a/(r+1)\}^n \tag{F'}$$

Hence

$$R_{slow}/R_{st} = 1/(r+1) = B \tag{G'}$$

and

$$R_{fast}/R_{st} = 1/(r+1)^n = C \tag{H'}$$

When $n = 1/2$ and $r = 3$, $B = 1/4$ and $C = 1/2$. The transition between these two values occurs when the period of time between flashes is of the order of magnitude

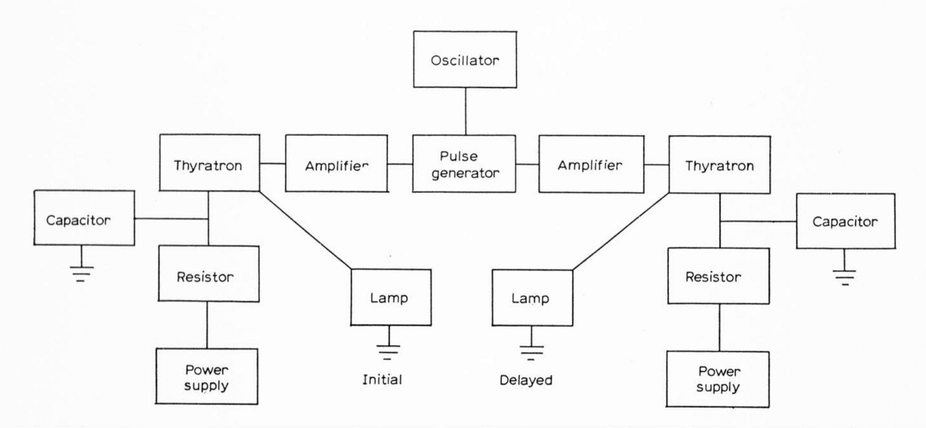

Fig. 38. Block diagram of the lamp circuit for pulsed illumination studies. From ref. 133b.

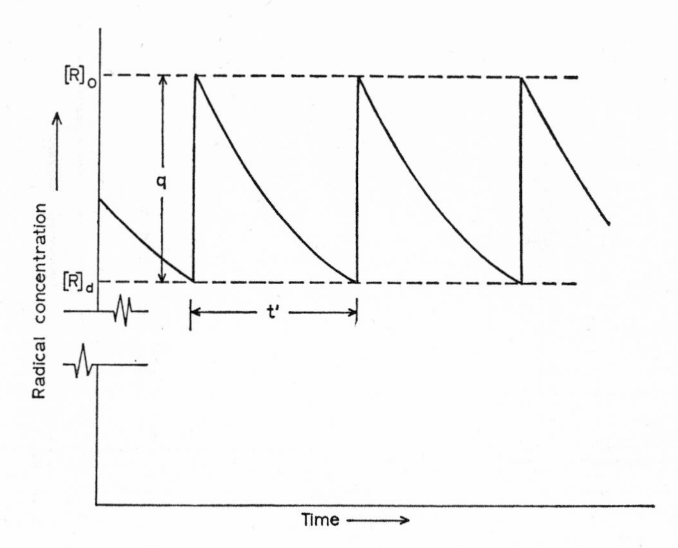

Fig. 39. Variation of radical concentration under quasi-stationary state conditions in the single-pulse-train method. From ref. 133b.

of the life-time of the intermediates. The theory may be developed from here to obtain a value for these life-times†.

A very similar technique is that of pulsed illumination[133b], which corresponds very closely to flash photolysis except that the intensities of the flashes are 10^{-8} times those in the latter technique. The method has been used to determine the rate coefficients for reactions (14) and (15).

$$Me\cdot + Me\cdot \rightarrow C_2H_6 \quad (14)$$

$$Me\cdot + AcMe \rightarrow CH_4 + AcCH_2\cdot \quad (15)$$

A block diagram of the lamp circuit is shown in Fig. 38. Each pulse produced 10^{-5}–10^{-6}% decomposition. They were fired at the rate of 30–40 pulses per second, sufficient to produce a quasi-stationary state. The radical concentration following each pulse falls to the same value $[R]_d$ prior to the pulse (Fig. 39). A consideration of the rate of formation of C_2H_6 and CH_4 leads to an expression from which k_{14} and k_{15} may be determined by varying the length of the dark period t'. Paired pulses with a varying t' should lead to direct evidence for the participation of hot radical reactions.

(b) *Vacuum ultra-violet*

The 1849A resonance line from the low-pressure mercury lamp has already been referred to (p. 41). Suprasil or sapphire windows allow the transmission of this line[116] or alternatively the lamp may be placed inside the RV[135]. The intensity of the

† For a detailed discussion see ref. 133a, p. 1113.

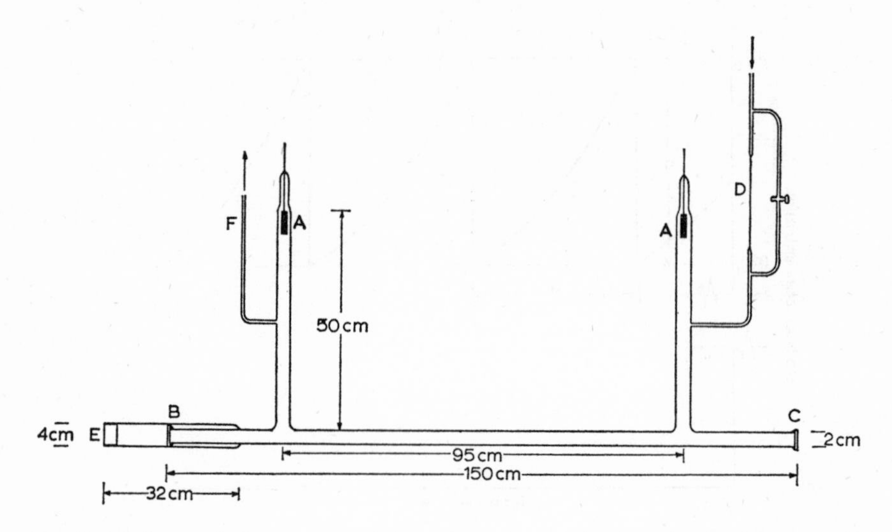

Fig. 40. Hydrogen discharge tube with iron electrodes and fluorite windows. From ref. 136.

line often approaches that of the 2537A line, *viz.* 10^{16} quanta.sec^{-1}. Vaughan and Noyes[136] used a hydrogen discharge tube with iron electrodes and fluorite windows (Fig. 40), which emits from 1900–1700A and has an intensity of 6×10^{16} quanta.sec^{-1} at 1750A. A windowless source from a low-pressure discharge in hydrogen has been found to be very rich in the 800–2000A region[137]. Sauer and Dorfman[138] used a xenon resonance lamp, originally developed by Harteck and Oppenheimer[139] (Fig. 41). The lamp system was filled with a mixture of xenon (0.8%) and neon to a pressure of 0.8-1 torr. Gas circulation was achieved by convection through a ballast volume of 2 litres, to minimise impurity effects. The lamp is run at 120 mA, and its intensity is increased by a factor of 6 if a large horse-shoe magnet is placed such that the maximum field (1700 gauss) is just behind the window[140]. In this case the lamp intensity is 10^{14} quanta.sec^{-1}. Xenon provides 1470A and 1296A lines, whereas a similar lamp using krypton provides 1235A and 1165A lines[141]. Argon or nitrogen provide radiation from 500 to 1650A[142]. More details of similar lamps are given in refs. 9 and 22c.

Very successful vacuum UV sources involve electrodeless microwave excited discharges in low pressures of inert gases (Xe, Kr and Ne): for spectroscopic sources see refs. 143,144; and for photochemical sources refs. 145, 146. The design of this lamp is very simple and it is easy to keep clean and change the gas or window (Fig. 42). A chemical "getter"(Ba–Al–Ni alloy), prepared by flash heating under high vacuum, or a refrigerant is used to remove water, which is an undesirable impurity since it produces so many lines in the region 1500–2000A. Fig. 43 shows the emission from Kr- and Xe-filled lamps with and without a "getter", refrigerant

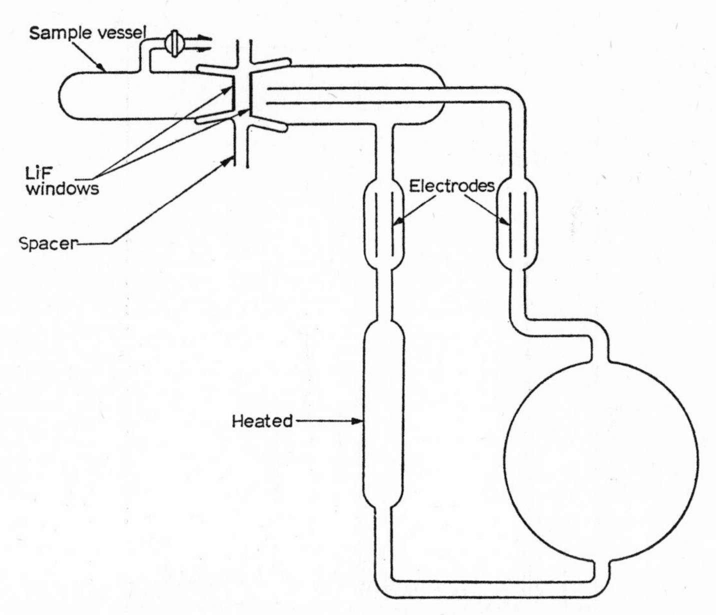

Fig. 41. Xenon resonance lamp in a photolysis system. From ref. 139.

or baking to remove water. The lamp is operated by a 2450Mc.sec^{-1} discharge, 125 watt microwave power, triggered by a Tesla discharge. Maximum intensity (5×14 quanta.sec^{-1}) is obtained with 0.7 torr (Xe) and 1.0 torr (Kr) respectively. A mixture of rare gas and helium gives more intense light than does the pure gas. The average life of the lamp is 10 hours. For a review of vacuum UV photochemistry see McNesby and Okabe[147].

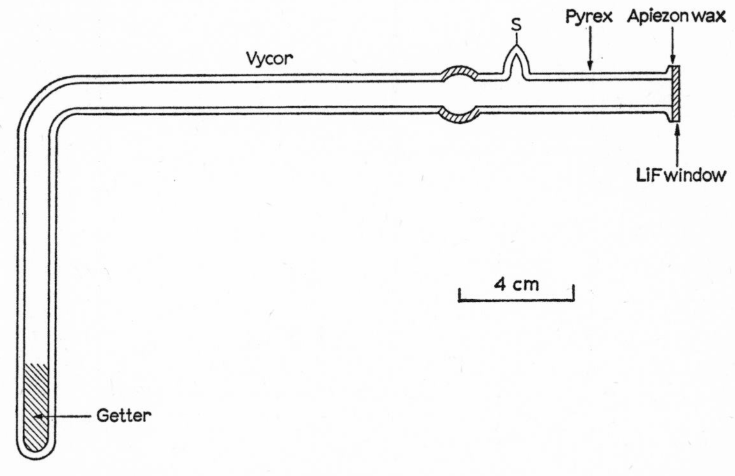

Fig. 42. A simple electrodeless rare-gas resonance lamp for use in a microwave field. From ref. 147.

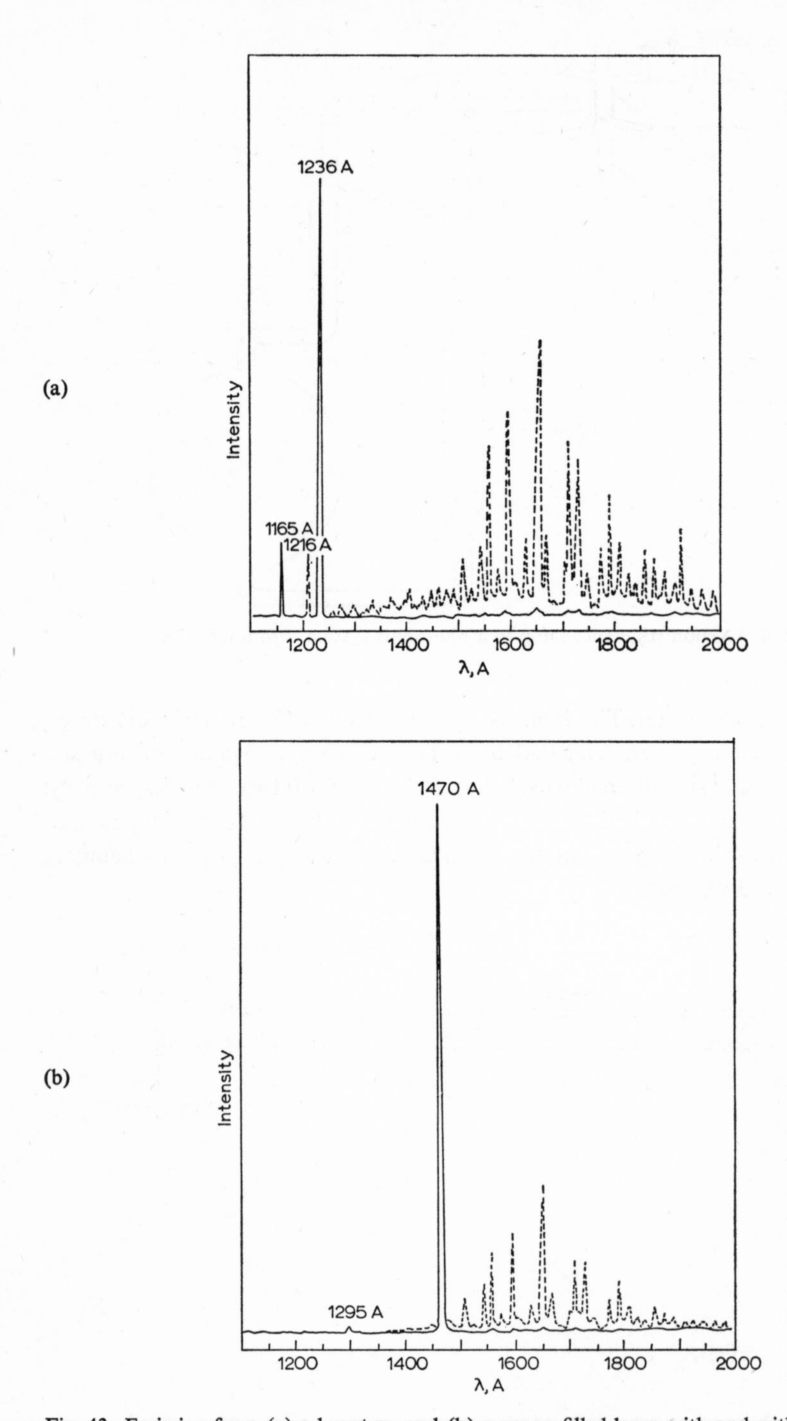

Fig. 43. Emission from (a) a krypton- and (b) a xenon-filled lamp with and without a "getter", refrigerant or baking to remove water. From ref. 147.

3.3.4 *The production of "monochromatic" radiation*

Most quantitative photochemical studies are carried out using "monochromatic" radiation, since the wavelength very often affects the quantum yield (ϕ) of a reaction and the relative extents of the possible primary processes. Also, the use of monochromatic light allows the accurate determination of ϕ, the intensity of the radiation, and the fraction absorbed by the reacting medium. The use of different monochromatic sources will indicate whether there are any hot radical effects, since the excess energy available will vary.

(a) Lasers

An ideal monochromatic source is the laser or uvaser. There is little or no published work on continuous photolysis using these sources although we must be on the threshold of this happening. Porter and Steinfeld[148] have flash-photolysed phthalocyanine in the vapour phase using a Q-switched ruby laser. This emits at 6943A, and 2–3 J are dissipated in 20 μsec. Photolysis occurs because of a two-photon process.

The unique properties of lasers are their coherence, monochromaticity and high intensity. However, as yet, the intensities of commercial lasers emitting radiation below 6000 A are very low. Their photochemical potential is reviewed by Turro[149]. An excellent account of the theoretical background has been given by Smith and Sorokin[150]. Also, Levine[151] has started a useful series on lasers. Both of these reviews, however, are concerned solely with the spectroscopic potential of lasers. In another connection, stimulated emission from vibrationally excited HCl produced in reaction (16)

$$H\cdot + Cl_2 \rightarrow HCl^* + H\cdot \qquad (16)$$

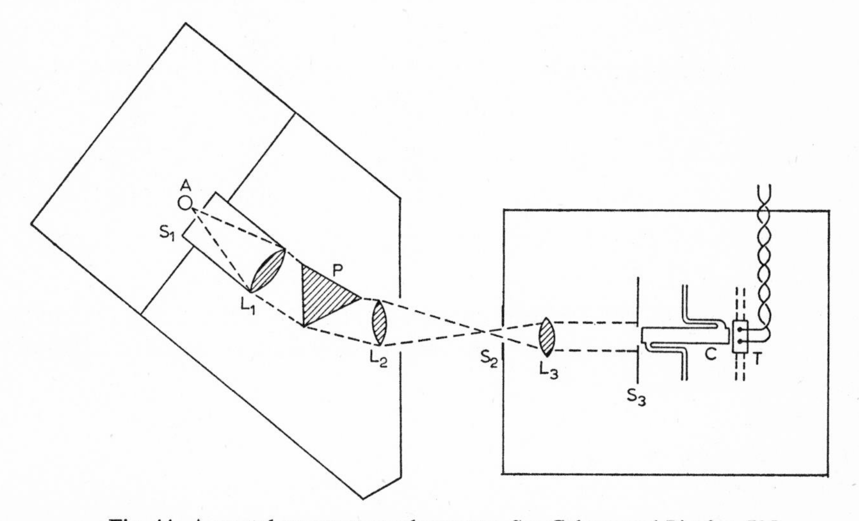

Fig. 44. A crystal quartz monochromator. See Calvert and Pitts[9] p. 725.

References pp. 104–111

TABLE 7

APPROXIMATE WAVELENGTH LIMITS FOR TRANSMISSION OF VARIOUS OPTICAL MATERIALS AND WATER NEAR ROOM TEMPERATURE

Material	*Thickness (mm)*	*Approximate λ (A) for % transmission indicated*		
		50 %	*30 %*	*10 %*
Window glass (standard)	1	3160	3120	3070
	3	3300	3230	3140
	10	3520	3420	3300
Optical (white crown) glass	1.8	3270	3200	3090
Pyrex (Corning 774)	1	3060	2970	2800
	2	3170	3090	2970
	4	3300	3190	3100
Corex D [Pyrex, Corning 9-53 (9700)]	1	2780	2670	2500
	2	2880	2800	2670
	4	3040	2920	2810
Corex A	2.9	2480	2430	2400
Vycor 790	2			>2540
Vycor 791	1	2150	2130	2120
	2	2230	2170	2130
	4	2360	2250	2170
Quartz, crystal	5	1850		
	10	1930	1920	1860
Quartz, clear fused (General Electric Co.)	10	1940	1810	1720
Suprasil (Englehard Industries, Inc.)	10	1700	1680	1660
Sapphire synthetic (Linde Air Products)	0.5			1425
Fluorite (CaF_2), natural	5	1350		
	10	1570	1450	1380
CaF_2, synthetic (Linde Air Products)	3			1220
Lithium fluoride (synthetic)	5	1070		
	10	1420	1270	1150
Plexiglas (polymethylmethacrylate)	2.5	3220	3100	2970
	5.0	3380	3250	3110
	10.0	3500	3420	3260
Water (distilled)	20	1880	1860	1850
	40	1920	1880	1860
	80	2020	1940	1880

apart from being the first reported chemical laser, provides a means of determining the energy distribution among the degrees or freedom associated with this[152] and similar reactions[153].

(b) *Monochromators*

The best way of isolating monochromatic radiation from a source emitting several lines is by means of a prism or diffraction grating monochromator. This

TABLE 8

CHEMICAL FILTERS

<table>
<tr><th>Hg lines</th><th></th><th></th><th></th><th></th></tr>
<tr><td>2480
2537</td><td rowspan="6">145 g $NiSO_4 \cdot 6\text{–}7\ H_2O$+41.5 g $CoSO_4 \cdot 7H_2O$ in 1 l H_2O; 10 cm</td><td rowspan="2">Gaseous Cl_2 (1 atm); 3 cm</td><td colspan="2">0.108 g I_2†+ 0.155 g KI† in 1 l H_2O; 1 cm</td></tr>
<tr><td>2652
2700</td><td colspan="2">CCl_4† 2 mm; or $HgCl_2$‡ (45 g.l^{-1}) in H_2O; 1 cm</td></tr>
<tr><td>2750
2804</td><td colspan="3"></td></tr>
<tr><td>2896
2925
2970
3030</td><td></td><td></td><td rowspan="3">Oxalic acid† sol. in water (20 g. l^{-1}); 1 cm. Or $CuSO_4 \cdot 5H_2O$ (15 g. l^{-1}); 1 cm</td></tr>
<tr><td>3126–31</td><td></td><td rowspan="2">KH phthalate† (5 g. l^{-1}); 1 cm</td></tr>
<tr><td>3340</td><td>Uric acid (sat. sol. in water); 1 cm</td></tr>
<tr><td>3650–63</td><td rowspan="3">4.4 g $CuSO_4 \cdot 5H_2O$+ 150 ml NH_4OH (0.88) in 1 l H_2O; 10 cm</td><td colspan="3">2–3 mm Chance's black "ultra-violet" glass</td></tr>
<tr><td>4046</td><td colspan="3">I_2 in CCl_4 (7.5 g.l^{-1}); 1 cm. Quinine hydrochloride in water (10 g.l^{-1}); 2 cm</td></tr>
<tr><td>4358</td><td colspan="3">75 g $NaNO_2$ in 100 ml water;‡ 1 cm</td></tr>
<tr><td>5461</td><td rowspan="2">13 g $CuSO_4 \cdot 5H_2O$+0.449 g $K_2Cr_2O_7$+ 50 ml conc. H_2SO_4 in 1 l H_2O; 10 cm</td><td colspan="3">Corning glass 512; 5 mm</td></tr>
<tr><td>5770
5790</td><td colspan="3">Corning glass 344; 3.4 mm</td></tr>
</table>

† Renew frequently. ‡ Renew occasionally.

Taken from E. J. BOWEN, *The Chemical Aspects of Light*, Oxford University Press, 1946.

presents "monochromatic" radiation of very high intensity. One design is shown in Fig. 44[154]. Other designs are referred to by Calvert and Pitts[9].

(c) *Chemical and glass filters*

Monochromatic radiation can be obtained quite simply by using glass or chemical filters. Indeed, very often a higher intensity of residual radiation than from monochromators is obtained. The cut-off regions for various glasses provide one method of filtering out unwanted radiation. These regions are given in Table 7. Conversely, these decide what material may be used for the RV and lenses for a particular experiment. Commercial glass filters are also available.

Various chemical filters and their useful regions and stability are shown in Table 8. Fuller details are given in refs. 9 and 22c and also by Nicholas and Pollak[155]. Some chemical filters can be unsatisfactory since their properties are time-dependent simply due to decomposition. An alternative to these are interference filters. Two partially reflecting surfaces produced by vacuum deposition of metallic films on glass plates are separated by a thin transparent spacer with a thickness of the order of the wavelength of the radiation, such that multiple reflection occurs between the two mirrors, which are arranged so that only the required radiation is transmitted. The unwanted radiation is reflected away. These have very stable filtering properties. Filters are commercially available, for the UV, visible and IR wavelengths.

(*d*) *Variable monochromatic filter for the visible spectrum*

The Christiansen filter gives radiation of different wavelengths in the visible region depending on the temperature[156]. It is based on the principle that a finely divided solid suspended in a liquid will have a maximum transmission where the refractive indices of the liquid and solid coincide. The refractive indices vary at a different rate with temperature and hence the maximum transmission position also varies as a function of temperature. This is shown in Fig. 45 for powdered crown glass and methyl benzoate[157]. The filter itself is shown in Fig. 46. G_1 and G_2 are plate-glass plates, L_1 and L_2 are water–air lenses and F holds the mixture of glass and ester. Other ranges may be obtained[9] by using different materials for F.

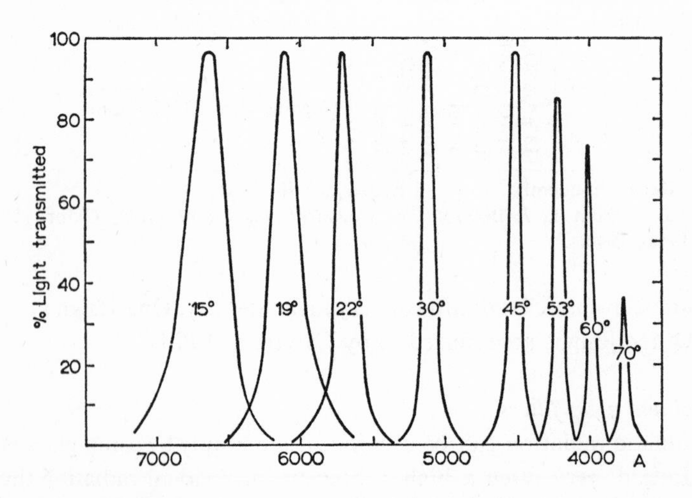

Fig. 45. Maximum transmission wavelength variation with temperature. From ref. 156.

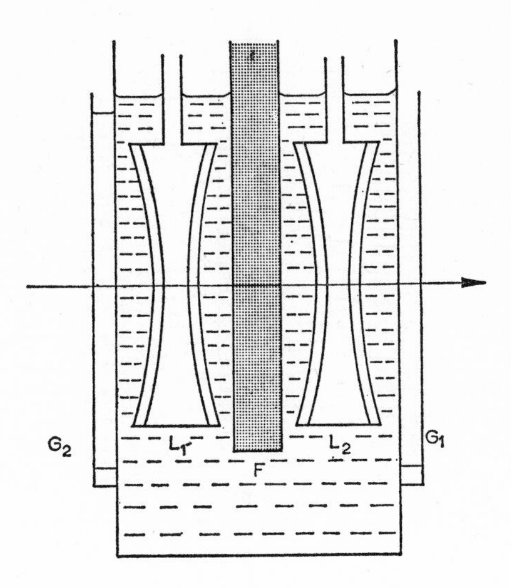

Fig. 46. Christiansen filter. From ref. 157.

3.3.5 *Uniform density filters*

Uniform density filters provide the means of varying the intensity of photochemical radiation. They are made by the vacuum deposition of thin metal films on quartz plates.

3.3.6 *Measurement of the intensity of radiation*

(*a*) *Thermopiles*

An absolute method for measuring intensities is to use a calibrated thermopile, bolometer or radiometer and galvanometer system; the first of these is the most suitable. Calibration may be carried out at the National Physical Laboratories and The National Bureau of Standards, or by using one of their carbon filament lamps. The thermopiles consist of a bank of interconnected fine-wire thermocouples (Fig. 47). One circuit uses a photoelectric amplifier for the measuring galvanometer (Fig. 48)[158]. The image of a tungsten filament F, is projected onto a photocell C by means of a lens L and the galvanometer mirror M. The cell is divided into two parts with a common electrode. When the galvanometer is in its null position the filament image is symmetrically placed with respect to the two halves. If M is deflected, the symmetry is destroyed and C registers the effect. Fuller details are given in refs. 9 and 22c.

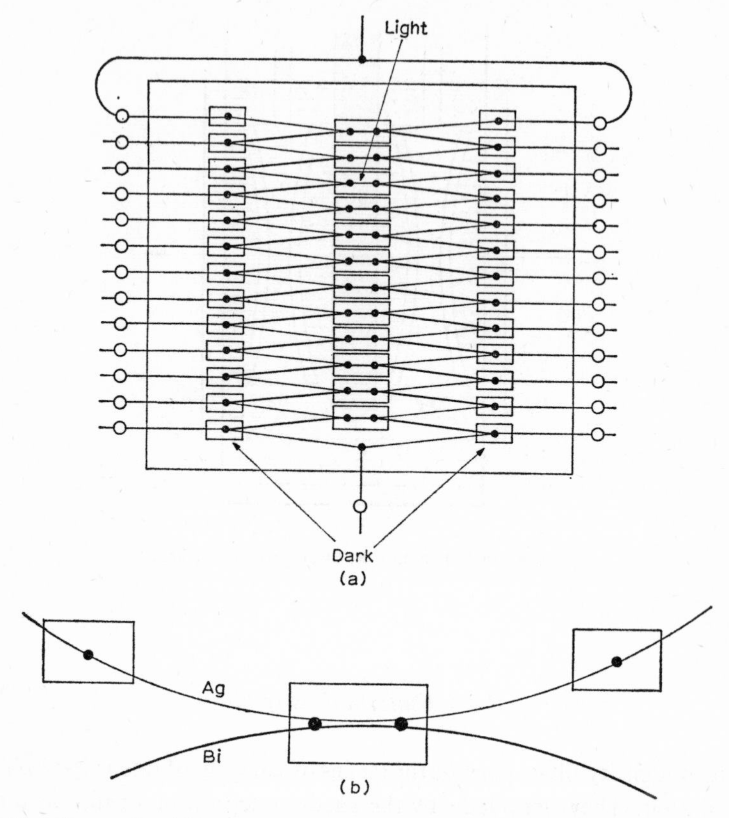

Fig. 47. A line thermopile; (a) arrangement of thermocouple wires, (b) close-up. From ref. 9.

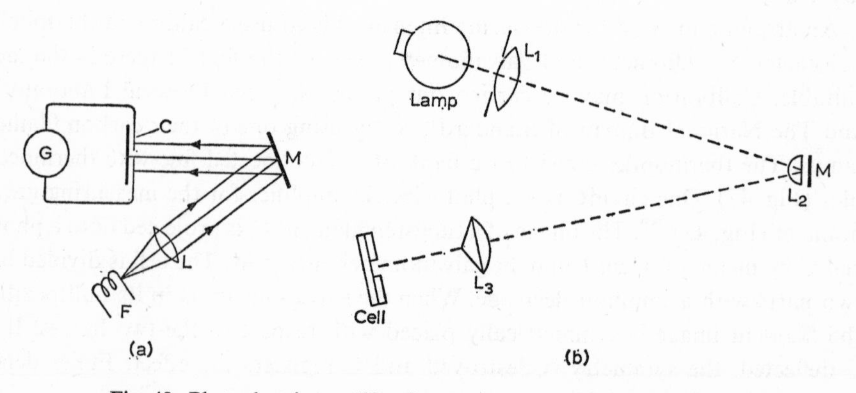

Fig. 48. Photoelectric amplifier for a galvanometer. From ref. 158.

TABLE 9

GASEOUS ACTINOMETERS

Reaction	*Quantum yield*	*Wavelength (A)*	*Comments*
$(CH_3)_2CO \rightarrow CO + 2CH_3\cdot$	$\phi_{CO} = 1$	3200–2500	p > 50 torr; t > 125° C; CO determined by GSC[a]
$(C_2H_5)_2CO \rightarrow CO + 2C_2H_5\cdot$	$\phi_{CO} = 0.93 \rightarrow 1.27$	3200–2500	As above. CO only gaseous product at liquid N_2 temp.[b]
$CH_3CHO \rightarrow CH_4 + CO$ $\rightarrow CH_3 + CHO$	$\phi_{CO} = 0.30 \rightarrow 1.01$	3130–2380	200 torr; 30° C; alternative to above for temp. below 125°. Not very reliable.[c]
$2HI \rightarrow I_2 + H_2$	$\phi_{H_2} = 2.0$	3000–1800	0.1 torr to 3.5 atm (of N_2); ϕ temp.-independent up to 175° C; must be Hg-free system.[d]
$2HBr \rightarrow Br_2 + H_2$	$\phi_{H_2} = 1.0$	2500–1800	For conversions of 1 %, ϕ invariant over a range of conditions. 100 torr; 25° C.[e]
$CO_2 \rightarrow CO + \frac{1}{2}O_2$	$\phi_{CO} = 1.0$	1600–1200	Fast flow system must be used, otherwise back reactions complicate the issue.[f]
$3O_2 \rightarrow 2O_3$	$\phi_{O_3} = 2.0$	1900–1300	As for CO_2.[g]
$N_2O \rightarrow N_2 + \frac{1}{2}O_2$	$\phi_{N_2} = 1.44$	1849–1470	Transparent to 2573A radiation.[h]
$2NOCl \rightarrow 2NO + Cl_2$	$\phi_{NO} = 2.0$	6350–3650	Useful for Hg-free systems.[i]
cis-2-butene $\rightleftarrows$ *trans*-2-butene	$\phi_{t-c} = 0.5$		Hg ($^3P_1$)-sensitised reactions. ϕ independent of pressure above 30 torr.[j]

[a] D. S. Herr and W. A. Noyes, Jr., *J. Am. Chem. Soc.*, 62 (1940) 2052.
[b] K. O. Kutschke, M. H. J. Wijnen and E. W. R. Steacie, *J. Am. Chem. Soc.*, 74 (1952) 714.
[c] F. E. Blacet and R. K. Brinton, *J. Am. Chem. Soc.*, 72 (1950) 4715.
[d] G. K. Rollefson and M. Burton, *Photochemistry*, Prentice-Hall, New York, 1946, p. 190.
[e] G. S. Forbes, J. E. Cline and B. C. Bradshaw, *J. Am. Chem. Soc.*, 60 (1938) 1431.
[f] B. H. Mahan, *J. Chem. Phys.*, 33 (1960) 959.
[g] W. E. Vaughan and W. A. Noyes, Jr., *J. Am. Chem. Soc.*, 52 (1930) 559.
[h] G. A. Castellion and W. A. Noyes, Jr., *J. Am. Chem. Soc.*, 79 (1957) 290.
[i] G. B. Kistiakowsky, *J. Am. Chem. Soc.*, 52 (1930) 102.
[j] R. B. Cundall, in *Progress in Reaction Kinetics*, Vol. 2, G. Porter (Ed.), Pergamon, Oxford, 1964, p. 167.

(b) Chemical actinometers

Chemical actinometers are simpler to use than thermopiles and are often more reliable because of the extreme care needed to calibrate and set up the latter. However, they first have to be calibrated for their useful wavelength range and the corresponding quantum yield ϕ. Several gaseous actinometers are given in Table 9 together with the relevant details of ϕ and wavelength. Any chemical actinometer is most conveniently used when it and the system being studied can be considered as completely absorbing. Apart from one exception, the actinometers cited are considered to be the most reliable over the vacuum–near UV range. There is a distinct lack of actinometers for the visible region. Only one reaction has been cited as such. As a liquid phase actinometer, the long-standing uranyl oxalate actinometer is now being replaced by the 1000 times more sensitive ferrioxalate actinometer[159], although Porter and Volman[160] claim that with GSC (FID) estimation of CO production, the uranyl oxalate system is the most sensitive available.

(c) Photocells

Like chemical actinometers, photocells have to be calibrated against a thermopile–galvanometer system; this has to be done frequently as there tends to be some variation with time. Under these conditions, they can be used to measure the absolute intensity of monochromatic light. The cell best suited for photochemical studies is the photoemissive type, which operates via the photoelectric emission of electrons from an irradiated surface. The metallic cathode is mounted either in a vacuum or in a small pressure of one of the inert gases. The cell may involve a single phototube or a multielement photomultiplier. An amplification of about 10^6 is achieved with the latter.

Glass bulbs are suitable only for the visible part of the spectrum but can be extended to the UV by fitting quartz windows. However, the sensitivity drops with decreasing wavelength. A much better technique is to coat the face of the phototube with a layer of fluorescent material which extends the useful range to 850A[161]. Sodium

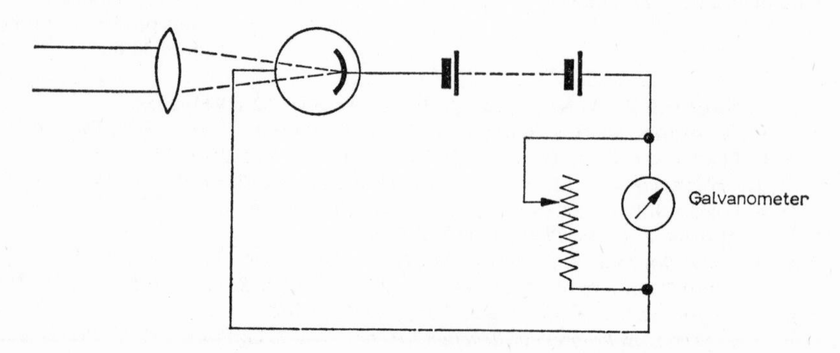

Fig. 49. Basic photocell circuit. From ref. 22c.

salicylate gives the best film, as it combines high sensitivity to the extreme UV with low sensitivity to stray long wavelength light. The basic circuit for a photocell is shown in Fig. 49. This may be used for intensities of 0.01 foot-candle or more when the galvanometer is a sensitive one. (See refs. 9 and 22c for amplification procedures when lower intensities are encountered.)

The procedures employed for the determination of the fraction of light absorbed and of quantum yields with these measuring systems are clearly described by Calvert and Pitts[9].

3.4 RADIOCHEMICAL SYSTEMS (RADIATION CHEMISTRY) (*see also Volume 3, Chapter 2*)

3.4.1 Introduction

As with photolytic systems radiolysis may be carried out by continuous or pulsed techniques. The latter will be dealt with in Chapter 2. The elucidation of the mechanism for a radiochemical system involves determining the extent of ion and hot molecule reactions, leading to the formation of radicals or intramolecular elimination products, and of radical processes. Very often a striking similarity is found between photolytic and radiolytic systems. The value of $k_{17}/k_{14}^{\frac{1}{2}}$ for the reactions (17) and (14)

$$\text{Me}\cdot + \text{MeN}_2\text{Me} \rightarrow \text{CH}_4 + \dot{\text{C}}\text{H}_2\text{N}_2\text{Me} \quad (17)$$
$$\text{Me}\cdot + \text{Me}\cdot \rightarrow \text{C}_2\text{H}_6 \quad (14)$$

in the γ-radiolysis of azomethane[162] are in excellent agreement with values obtained from photolytic experiments. Hence, CH_4 and C_2H_6 must be formed by the reactions of thermalised methyl radicals. The mechanism probably involves the initial formation of an electronically excited molecule of azomethane, either directly or by ionisation followed by neutralisation.

Vacuum UV photolysis experiments also have a direct bearing, since ionisation is possible in certain systems. Becker *et al.*[163] investigated the photolysis of propylene at 1470A and 1236A, *i.e.* above and below the ionisation energy (9.73eV). In spite of the fact that at 1236A 30 % of the absorbed photons result in the formation of $C_3H_6^+$, there is no appreciable difference in the product distribution at these two wavelengths. They conclude that radical production is important in this system together with the following intramolecular elimination processes

$$\text{C}_3\text{H}_6 \rightarrow \text{CH}_4 + \text{C}_2\text{H}_2 \quad (18)$$
$$\rightarrow \text{H}_2 + \text{C}_3\text{H}_4 \quad (19)$$
$$\rightarrow \text{CH}_2 + \text{C}_2\text{H}_4 \quad (20)$$

If an electric field is applied across two electrodes in the mixture being irradiated[164], in the saturation region the passage of electrons through the mixture does not cause increased ionisation, but *does* increase the number of excited molecules formed. Fast ion–molecule processes are unaffected by an electric field. In this way a clear distinction can often be made between the products of ion–molecule reactions and hot molecule processes[165].

Several books on radiation chemistry have appeared recently[166,167,168]; one deals specifically with gases[169]. Two reviews have appeared on the radiation chemistry of hydrocarbons[170,171]. Radiation chemistry has also been the subject of many articles in *Annual Reports of the Chemical Society* and *Annual Reviews of Physical Chemistry*. Two recent contributions are by Collinson[172] and Ausloos[173]. *Radiation Research Reviews*[340] is devoted solely to reviews in this general field, and the first issue contains another review by Ausloos.

3.4.2 Experimental technique

(a) General

Typical thermal and photolytic techniques may be used but a common method is to fill an ampoule, fitted with a break-seal, with the reactant or reactants, dose, quench in liquid nitrogen, and reattach to the vacuum. This is followed by conventional analysis. Facilities for heating the sample at the radiation source can be employed. RV's are constructed from Pyrex or metal[174] and may in some cases be coated with graphite. However, with different surfaces, different products can result[175]. Thin-walled RV's have to be used for α-particle-induced reactions[169]. In the radon-sensitised reaction of H_2 and O_2, Lind has shown that the rate coefficient is inversely proportional to the square of the diameter of a spherical RV.

The ionisation chamber of a mass spectrometer can also be used as a RV for the study of ion-molecule reactions. Recent techniques have increased the maximum pressure from the conventional 10^{-6}–10^{-5} torr to 0.15 torr[176,177]. These types of reaction have been reviewed recently [178,179].

(b) Radiation sources

Radiation sources are treated in the above bibliography. One book[168] deals exclusively with this subject; a section[180] is devoted to health hazards and protection from radiation. Two types of source are available, those from natural or artificial radioisotopes produced in nuclear reactors and those from a particle accelerator. The first includes nuclear recoil processes which are used for studying hot atom displacement reactions in the gas phase.

Particle accelerator sources. Various particle accelerators are listed in a table modified from Spinks and Woods[167]. Pulsed beams are out of context here, but

TABLE 10

PARTICLE ACCELERATORS[167]

Accelerator	Particles accelerated or produced	Energy (MeV)	Comments
X-ray machine	X-rays	0.05–0.3	Pulsed beam unless constant potential power supply used, continuous energy spectrum.
Resonant transformer	X-rays	0.1–3.5	Pulsed beam, continuous energy spectrum.
Cockroft–Walton accelerator	Positive ions	0.1–1.5	Continuous beam, monoenergetic radiation.
Van der Graaff accelerator	X-rays	1.5	Continuous beam, continuous energy spectrum
	Electrons and positive ions	1.5	Continuous beam, monoenergetic radiation.
Betatron	X-rays	10–300	Pulsed beam, continuous energy spectrum.
	Electrons	10–300	Pulsed beam, monoenergetic radiation.
Cyclotron	Positive ions	10–20	Essentially continuous beam, monoenergetic radiation.
Linear electron accelerator	X-rays	3–630	Pulsed beam, continuous energy spectrum.
	Electrons	3–630	Pulsed beam, essentially mono-energetic radiation
Linear ion accelerator	Positive ions	4–400	Pulsed beam, essentially mono-energetic radiation.

are included for the sake of completeness. All the accelerators which give high energy electrons can be used as X-ray sources by stopping the electrons with a heavy metal target such as tungsten. The range of various accelerated particles are given in Table 11.

Radioactive isotope sources. These may be subdivided into the two types natural and those produced in nuclear reactors. The characteristics of reactors are discussed by Bopp and Parkinson[181]. Some commonly used isotopes are shown in Tables 12 and 13.

Other sources are listed in the various references above.

Hot atom chemistry. The subject of hot atom chemistry[182] finds itself here because, together with molecular beam and photochemical methods, nuclear recoil processes have been used extensively as hot atom sources. These reactions lead variously to abstraction, addition and displacement processes both with saturated species such as CH_4[183] and unsaturated species. Table 14 shows various sources of hot atoms.

TABLE 11

RANGE FOR ACCELERATED PARTICLES[167]

Particle	*Energy (MeV)*	*Range in air (cm) at 15° C and 760 torr*	*Range in Al (mm)*	*Range in water (mm)*
Electron	1	405	1.5	4.1
	3	1400	5.5	15
	10	4200	19.5	32
Proton	1	2.3	0.013	0.023
	3	14	0.072	0.14
	10	115	0.64	1.2
	30	820	4.3	8.5
Deuteron	1	1.7	0.0096	
	3	8.8	0.049	0.088
	10	68	0.37	0.72
	30	480	2.65	5.0
Helium nucleus	1	0.57	0.0029	0.0053
	3	1.7	0.0077	0.017
	10	10.5	0.057	0.11
	30	71	0.375	—

TABLE 12

α AND γ EMITTERS

Isotope	*Reaction*	*Energy (MeV)*	*Half-life*
α Emitters			
^{238}U	Decay	4.18	4.55×10^9 year
^{234}U	Decay	4.763	2.3×10^5 year
^{226}Ra	Decay	4.78 (94.2 %) 4.59	1620 year
^{222}Rn	Decay	5.486	3.823 day
γ Emitters			
^{137}Cs-^{137}Ba	Fission	0.662	30 year, 2.6 month
^{140}La	Fission	0.09–3.0	40 hour
^{60}Co	^{59}Co (n, γ)	1.17–1.33	5.3 year
^{182}Ta	^{181}Ta (n, γ)	0.04–1.2	117 day

(*c*) *Units*

The units for an absorbed dose are discussed in refs. 167 and 168. The fundamental unit, in terms of the energy absorbed per gram, is the rad, which is 100 erg. g^{-1} and is equivalent to 6.24 . eV g^{-1} or $6.24 \times 10^{13} \times \rho$ eV. cm^{-3}, where ρ is the density of the material. This unit is required when making dosimetric measurements. For X- or γ-radiation, the exposure dose is used. The unit is one roentgen

TABLE 13

β EMITTERS

Isotope	*Reaction*	*Max. β energy (MeV)*	*Half-life*
^{14}C	$^{14}N(n,p.)$	0.156	5670 day
^{90}Sr-^{90}Y	Fission	0.61 and 2.18	28 year, 2.54 day
^{3}H	$^{6}Li(n,\alpha)$	0.018	12.46 year
^{32}P	$^{32}S(n,p.)$	1.7	14.3 day

TABLE 14

NUCLEAR REACTIONS FOR THE PRODUCTION OF HOT ATOMS[182]

Isotope and half-life	*Reaction*	*Recoil energy (MeV)*
^{3}H(12.3 year)	$^{6}Li(n^{\circ}, \alpha)^{3}H$	2.7
	$^{3}He(n^{\circ}, p)^{3}H$	0.19
^{11}C(20.5 month)	$^{12}C(n, 2n)^{11}C$	~ 1
	$^{12}C(\gamma, n)^{11}C$	~ 1
	$^{12}C(n)^{11}C$	> 10
^{14}C(5760 year)	$^{14}N(n^{\circ}, p)^{14}C$	0.045
^{13}N(10.1 month)	As for ^{11}C	0.1–1
^{15}O(2.1 month)	As for ^{11}C	0.1–1
^{18}F(1.7 hour)	As for ^{11}C	0.1–1
^{30}P(2.6 month)	As for ^{11}C	0.1–1
^{32}P(14.3 day)	$^{35}Cl(n, \alpha)^{32}P$	0.1–1
	$^{32}S(n, p)^{32}P$	
^{35}S(87 day)	$^{35}Cl(n^{\circ}, p)^{35}S$	0.017
^{34}Cl(33.2 month)	As for ^{11}C	0.1–1
^{80}Br	$^{79}Br(n,\gamma.)^{80}Br$	0.00357 max.

n° indicates reaction will proceed with thermal neutrons.

(r). This is defined as the quantity of radiation that produces one e.s.u. of charge in one cc of dry air at NTP. This may be referred to charged particles if this quantity liberates ions carrying 1 e.s.u. in 1 cc of dry air at NTP[184]. Other reference units are given in refs. 166 and 167.

(*d*) *Yields*

Yields were originally measured in terms of the number of molecules M decomposed per ion pair formed by absorption of radiation. This ion pair yield is expressed as M/N†, where N, the number of ion pairs formed, is usually indeterminate in a condensed system. In this case, the yield is referred to the total energy

† First referred to by Lind in 1910, ref. 169, p. 6.

absorbed, and is given the symbol $G(X)$†, where X is some measurable product. $G(X)$ is defined as the number of molecules produced per 100 eV of energy absorbed. The G value and M/N are related by the expression

$$G = 100\ M/WN \tag{I'}$$

W is the average energy required to form one ion pair. Since W is usually of the order of 35 eV,

$$G \simeq 3\ M/N \tag{J'}$$

G values have been predominantly used during the last two decades. Recently, however, authors have returned to M/N since it is the more fundamental unit[186]. Also, G depends upon the measurement of the total energy absorbed which is difficult or impossible (with γ-radiation) to determine in a gaseous system. When the ionisation energy of the compound investigated is lower than the energy of the photon absorbed, the product yields can be expressed unambiguously in terms of the number of molecules formed per ion pair (M/N).

(e) *Dosimetry*

The most direct method of dosimetry is by measurement of the temperature rise in a system. Calorimetry is difficult experimentally and ΔH for the reaction must be known to make the required correction. Experimental techniques are described by Spinks and Woods[167]. The method, although not suitable for routine use, may be used for calibrating chemical dosimeters, for which Spinks and Woods[167] list the essential requirements. By far the best liquid phase one is the ferrous ammonium sulphate or Fricke dosimeter[187]. A 10^{-3} M solution of ferrous ammonium sulphate is used together with 0.1 or 0.8 N H_2SO_4. Extreme care has to be taken with the purity of this solution. $G(Fe^{3+})$ is measured spectrophotometrically at 2750A, where the extinction coefficient is temperature-independent. The values for α-particles and for various types of radiation are listed by Ramaradhya and Freeman[188] and by Swallow[166], respectively. Gaseous acetylene polymerises to cuprene and benzene at a rate dependent only on the rate of absorption of energy by the reactant[189]. Measurement of the rate of disappearance of C_2H_2 allows the energy absorbed by a system to be calculated[190]. Janssen *et al.*[191] use carbon disulphide measuring the drop in pressure manometrically. This was found to be independent of dose, dose rate and CS_2 pressure with γ-rays, electrons and protons. Johnson[192] has reviewed the use of chemical dosimeters and recommends N_2O.

A more acceptable method of dosimetry with gaseous systems is by means of an ionisation chamber. The saturation ionisation current can be converted to ab-

† For a discussion of the term G and its origin see ref. 166, p. 36 and Burton[185].

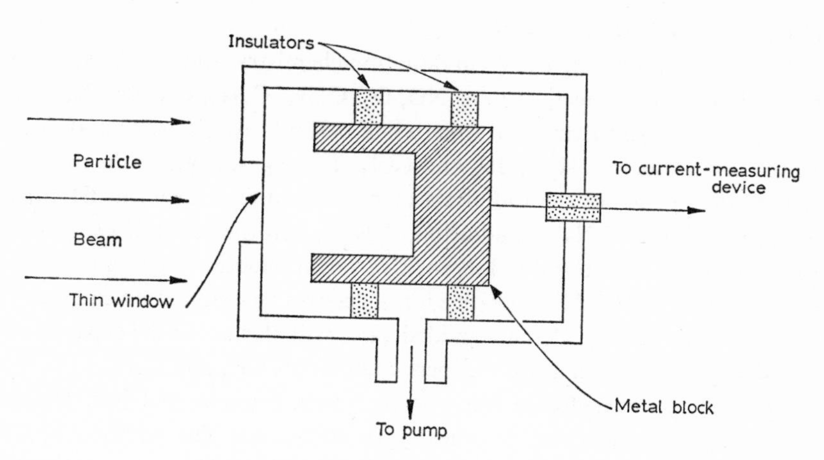

Fig. 50. Faraday cage. From ref. 167.

sorbed energies once the value of W is known[193,194]. Inert gases are suitable reference substances[193,195]. Ionisation chambers are generally used as secondary instruments and calibrated against a Faraday cup or cage. This consists of a metal block thick enough to stop a beam of charged particles (Fig. 50). The block acquires the charge of every absorbed particle, and the current flowing from it is a direct measure of the number of particles entering the chamber. If the energy per particle is known, from a well characterised source or from magnetic deflection measurements, the intensity of the radiation I is determined from the expression in equation (K′)

$$I(\text{erg. cm}^{-2}.\ \text{sec}^{-1}) = N(\text{particles. cm}^{-2}.\ \text{sec}^{-1}) \times E(\text{erg. particle}^{-1}) \quad (\text{K}')$$

Ionisation chambers have been described by Boag[196a], Laughlin[196b] (electron beams), and Birge *et al.*[197] (heavy charged-particle beams). Provided the Bragg–Gray principle is valid, the absorbed dose is given by equation (L′)

$$D_{\text{medium}} = 0.877 \times Q \times S_{\text{medium}}(\text{rad}) \quad (\text{L}')$$

where Q is the charge (e.s.u.) produced in 1 cc of dry air at NTP, S_{medium} is the stopping power and the value 0.877 is derived assuming $W_{\text{air}} = 34$ eV per ion pair. This value is valid for X- and γ-rays and electrons but for heavier charged particles is modified to $0.877 \times W_{\text{air}}/34$, where W_{air} is obtained for the particular system. Meisels[194] has determined W values for the partial absorption of 1 MeV electrons by organic and inorganic compounds. The stopping powers were used to calculate relative energy absorption in various gases. Dosimetry has been the subject of a symposium[198].

(f) Addition of foreign gases

The addition of foreign gases in radiation chemistry studies achieves several purposes. Radical scavengers such as NO, O_2, C_2H_4, C_3H_6, Cl_2, Br_2, I_2, HBr and HI effectively remove thermalised free radicals. Similarly, electrons may be removed by scavengers such as N_2O, SF_6 and CCl_4. These gases modify the charge neutralisation of an ion. In the radiolysis of propane, the decrease in $G(H_2)$ together with the value of $G(N_2)$ on addition of N_2O, and the effect of propene, leads to the conclusion that positive ion/positive ion interaction does not contribute to the formation of H_2. Also, positive ion + electron recombination can give H atoms but not molecular H_2[199]. These added gases will also act in the capacity of energy transfer agents, as well as do the inert compounds CF_4 and C_2F_6.

Apart from conventional trapping reactions, propene and NO, especially the latter, may also take part in other types of process. The addition of 0.25 % of deuteropropene to cyclohexane causes all parent ions produced in the radiolysis to undergo H_2^- transfer[200]

$$C_6H_{12}^+ + CD_3CDCD_2 \rightarrow CD_3CDCD_2H_2 + C_6H_{10}^+ \quad (21)$$

The charge transfer process of $C_6H_{12}^+$ with NO (ionisation potential 9.25 eV)

$$C_6H_{12}^+ + NO \rightarrow C_6H_{12} + NO^+ \quad (22)$$

competes very effectively with this reaction but the same process with O_2 (ionisation potential 12.1 eV) does not. Similar charge transfer processes occur with the inert gases. Xe and Kr inhibit the formation of HD in the radiation-induced exchange between H_2 and D_2 due to the reactions[201]

$$H_3^+ + D_2 \rightarrow H_2 + HD_2^+ \quad (23)$$

$$H_3^+ + Kr \rightarrow H_2 + KrH^+ \quad (24)$$

Inert gases also cause sensitised radiolysis in gaseous system. For methane, the xenon sensitisation is via energy transfer, whereas with Kr and Ar an ionic mechanism is operative[202]. The considerable number of ions produced in the α-particle radiolysis of ethylene is greatly reduced by irradiating Xe–C_2H_4 (50:1) mixtures. In the xenon-sensitised experiments, $C_2H_4^+$ is the major ion formed by charge transfer[203]. Hence a much simpler system is produced.

It is obvious from the above experimental results, that the addition of foreign gases to radiolytic systems may lead to detailed information about mechanisms for these processes. However, as with thermal systems, the results can be misleading. In the radiolysis of propane to low conversions (0.002 %), the yields of unsaturated products are very much higher in the absence than in the presence of added radical scavengers, indicating reaction between these products and the scavengers or their production by the scavengers in processes other than radical reactions[21]. This is particularly important when NO is used.

4. Analysis

4.1 INTRODUCTION

This section is divided into four main sub-sections. The first three deal with equally important methods of analysis. The last is devoted to other methods that are not of general application. However, this does not mean to say that such methods have necessarily a minor role to play in the kinetic study of reactions. An obvious example is the use of the calorimetric method for determining the extent of a heterogeneous component of a gas-phase process.

4.2 CHEMICAL METHODS

4.2.1 *General*

Chemical methods involving derivative formation have now been largely superseded by physical methods of analysis, particularly gas chromatography. However, under some circumstances it may be impossible to use the latter techniques. Alternative chemical methods of product analysis are described by Melville and Gowenlock[204], and are particularly necessary in the field of combustion[205]. For instance, formaldehyde may be estimated by the chromotropic acid method[206], and glyoxalic acid and glyoxal by colorimetric methods[207,208].

Hydrogen is difficult to analyse by gas chromatography; it may be done by GSC using nitrogen as the carrier gas, but with low sensitivity. However, H_2 can be selectively absorbed on activated silica gel at liquid oxygen temperatures from a "liquid nitrogen" fraction containing CO and CH_4, followed by later desorption at elevated temperatures and a PVT determination. Carbon dioxide is very conveniently adsorbed and estimated by passing through soda-lime.

It is sometimes necessary to remove halogens and halogen acids from a gaseous reaction mixture prior to quenching in a liquid nitrogen trap; acetone, for example, polymerises in the presence of hydrogen halides[209]. Various solids absorb these gases very efficiently[210]. Sometimes, selective absorption may be achieved, such that halogen may be separated from halogen halide. Quantitative estimation may be subsequently made by potentiometric titration techniques.

Any substance that has paramagnetic properties will catalyse the *ortho–para* hydrogen conversion. This therefore represents a method for the detection of free radicals and atoms. Virtually pure *para*-H_2 is metastable up to 500° C. The con-

References pp. 104–111

version has been used to detect radicals in the photolysis of MeI[211], HI[212], C_2H_6[213], and NH_3[214]. The *cis-trans* isomerisations of maleic acid, allocinnamic acid and *cis*-1,2-dichloroethylene are also catalysed by free radicals or atoms[215].

4.2.2 Radical traps

An important early method (1929) for the detection of gaseous free radicals was the metallic mirror technique of Paneth[216]. Since then, many compounds have been used to determine the presence and extent of free radical reactions in the gas phase decomposition of organic compounds. The subject of inhibition in gas-phase reactions has been reviewed by Ashmore[217] and by Gowenlock[218].

The most extensively used radical trap is nitric oxide, first used by Staveley and Hinshelwood[219]. There is no doubt that NO is a very efficient radical trap. However, ensuing reactions make the overall process very complex. The further reactions that a nitroso compound (RNO) may undergo, after the combination of NO with a radical $\dot{R}$, have been discussed in detail by Gowenlock[218]. Assuming $\dot{R}$ is Me·, at or close to room temperature, combination is followed by dimerisation of MeNO, *viz.* [220]

$$\dot{M}e + NO \rightarrow MeNO \rightarrow \tfrac{1}{2}(MeNO)_2 \tag{25}$$

At temperatures above 100° C, dimerisation becomes less important.

With excess methyl[221] or NO[222] the following reactions occur

$$2Me\cdot + MeNO \rightarrow Me\text{—}\underset{}{\overset{\displaystyle Me \atop |}{N}}\text{—}OMe \tag{26}$$

$$2NO + MeNO \rightarrow Me\text{—}\overset{\displaystyle O \atop {\| \atop {N \atop |}}}{N}ONO \xrightarrow{\text{surface}} MeN_2NO_3 \tag{27}$$

$$MeN_2NO_3 \rightarrow MeNO_3 + N_2 \rightarrow MeO\cdot + NO_2 + N_2 \tag{28}$$

Above 200° C, MeNO isomerises to CH_2NOH, homogeneously, heterogeneously, and possibly by a chain mechanism[222]. In a free radical system, sensitised decomposition may occur

$$Me\cdot + CH_2{=}NOH \rightarrow \cdot CHNOH + CH_4 \rightarrow HCN + CH_4 + \cdot OH \tag{29}$$

replacing methyl by the more reactive ·OH, the exothermicity of the process increasing the overall rate.

Direct decomposition of CH_2NOH also occurs at these temperatures, in some cases explosively[223,224]. In addition to inhibition, at high concentrations of NO catalysis also occurs, and with acetaldehyde involves the following initiation process[225]

$$NO + AcH \rightarrow Ac\cdot + HNO \tag{30}$$

If R· is Et· or a higher alkyl radical, we also have the possibility of disproportionation with NO[226]

$$\dot{E}t + NO \rightarrow HNO + C_2H_4 \tag{31}$$

Thus, although the reaction R·+NO is very efficient, NO is not in fact a good radical trap, because of the complicating reactions due to the instability of the products, except possibly at the lower temperatures of 100–200° C. Thus, Phillips *et al.*[227] have used NO in the study of the pressure-dependent decomposition of the isopropoxy radical. There are some similarities with the reactions of O_2, such as $R\cdot + O_2 \rightarrow RO_2\cdot$ and $O_2 + RH \rightarrow HO_2\cdot + R\cdot$. Here, inhibition takes the special form of oxidation.

Of all the olefins† used as free radical traps, propene is the favourite. It was first used at about the same time as NO[228]. Propene tends to be 1/10 to 1/100 as efficient as NO depending on the conditions. The trapping occurs in two ways. Hydrogen abstraction from the olefin takes place, producing the (hopefully) unreactive allyl radical

$$R\cdot + MeCH{=}CH_2 \rightarrow RH + \dot{C}H_2CH{=}CH_2 \tag{32}$$

However, there is some evidence that the allyl radical itself is capable of abstracting hydrogen atoms particularly at high temperatures[229]. Secondly, addition across the double bond occurs. Providing the resulting radical is stable, we have a useful trapping process. However, several complications may arise. Should hydrogen atoms add, displacement is almost certain to occur and may well occur with alkyl radicals, *viz.*

$$H\cdot + MeCH{=}CH_2 \rightleftarrows n\text{-}\dot{P}r \text{ or } i\text{-}\dot{P}r \rightarrow Me\cdot + C_2H_4 \tag{33}$$

$$Et\cdot + MeCH{=}CH_2 \rightleftarrows Me\underset{}{\overset{\overset{Et}{|}}{C}}HCH_2\cdot \rightarrow EtCH{=}CH_2 + Me\cdot \tag{34}$$

In both these systems and particularly in those where species like the bromine

† For a list see ref. 218, pp. 174 and 198.

atom are present, the trapping efficiency will depend on the equilibrium constants, $K_{a,b}$ and $K_{c,d}$, in reactions such as (35) and (36)

$$Br\cdot + MeCH{=}CH_2 \underset{b}{\overset{a}{\rightleftarrows}} Me\dot{C}HCH_2Br \quad (35)$$

$$Br\cdot + MeCH{=}CH_2 \underset{d}{\overset{c}{\rightleftarrows}} HBr + \dot{C}H_2CH{=}CH_2 \quad (36)$$

In flow systems, phenol[95], toluene, benzene, aniline and cyclopentene[230] have been used both as radical traps and as carrier gases. The major reaction in each case seems to be hydrogen abstraction but some radical addition almost certainly occurs. These compounds tend to be less efficient than NO and the aliphatic straight chain olefins.

Another class of compounds that have been used are the halogens and halogen acids, Cl_2[231], I_2[232], HCl[233], HBr[234] and HI[209]. These react with radicals to give molecular products that are relatively stable (with X_2), or very stable (with HX), where X stands for halogen. Clearly, the trapping reactions with methyl radicals will be (37) and (38)

$$\dot{M}e + X_2 \rightarrow MeX + \dot{X} \quad (37)$$
$$\dot{M}e + HX \rightarrow CH_4 + \dot{X} \quad (38)$$

TABLE 15

ARRHENIUS PARAMETERS AND ENTHALPY CHANGES FOR THE REACTIONS (37) AND (38)

X_2 or HX	*log A* (*A, l. mole^{-1} sec^{-1}*)	*E* (*kcal. mole^{-1}*)	*ΔH** (*kcal. mole^{-1}*)
F_2[a]	—	~0	−74
Cl_2[b]	9.54	2.5	−23
Br_2[c]	8.75	0.9	−21.5
I_2[d]	9.70	0[f]	−17.5
HF[e]	9.2	34	33
HCl[e]	8.14	4.9	1.1
HBr[c]	8.95	2.9	−14.4
HI[d]	9.47	1[f]	−30.5

* L. BATT AND F. R. CRUICKSHANK, *J. Phys. Chem.*, 70 (1966) 723.
[a] P. D. MERCER AND H. O. PRITCHARD, *J. Phys. Chem.*, 63 (1959) 1468.
[b] Estimated from k_{37}/k_{38}. A. F. TROTMAN-DICKENSON, *Gas Kinetics*, Butterworths, London, 1955.
[c] L. BATT AND F. R. CRUICKSHANK, *J. Phys. Chem.*, 71 (1967) 1836.
[d] Taken or estimated from H. E. O'NEAL AND S. W. BENSON, *J. Chem. Phys.*, 34 (1961) 514.
[e] Estimated from the values for the reverse process by G. C. FETTIS AND J. H. KNOX in *Progress in Reaction Kinetics*, Vol. 2, G. Porter (Ed.), Pergamon, Oxford, 1964, p.1.
[f] E_{37}-E_{38} is estimated from *d* as 1 kcal. mole^{-1}; with the choice of $E_{38} = 0$, E_{37} is 1 kcal. mole^{-1}.

Arrhenius parameters and standard heats of reaction for (37) and (38) are given in Table 15.

These values give a very good idea of the relative efficiencies of the halogens and halogen acids. For the halogens, the exothermicity increases from $I_2 \rightarrow F_2$, whereas the reverse is true for the halogen acids. Of these, it is obvious that HF is useless as a radical trap since the hydrogen abstraction reaction has such a high activation energy. Conversely, from the point of view of efficiency and the formation of the most stable product, F_2 is the best radical trap. However, fluorine is very reactive, and, according to Semenov[235], reacts very rapidly with hydrocarbons even below 0° C, the initiation process being reaction (39)

$$F_2 + RH \rightarrow \dot{R} + HF + \dot{F} \quad (39)$$

Also the exothermicity of the trapping reaction is such that special precautions will have to be taken to prevent the formation of thermal gradients. Thus the use of F_2 is probably restricted to special systems.

The methyl reactions give a good idea of the gradation in properties of HX and X_2, but a far better and more interesting use of them may be made in systems where the radical R· has the chance to decompose in competition with trapping[209,233,234]. Below 100° C the atom X usually terminates by dimerising but at higher temperatures, its abstraction reaction will play an increasing role causing a chain process. The exception to this is for F· where this is limited due to the formation of the very stable HF, *viz.*

$$F_2 + R\cdot \rightarrow RF + F\cdot \quad (40)$$
$$F\cdot + RH \rightarrow HF + R\cdot \quad (41)$$

There is some evidence that surface processes[239,236] and complex formation[234] may obscure the trapping reactions. Hydrogen sulphide may also be classed with these types of radical traps[237,238]. Aldehydes have also been used as radical traps[239]. For acetaldehyde, the major reaction should be abstraction of the aldehydic hydrogen atom, *viz.* so that the CO at least, should provide a material balance for RH.

$$R\cdot + MeCHO \rightarrow RH + \dot{M}e + CO \quad (42)$$

Besides these, ammonia[240] and amines[241] have been used, particularly in combustion studies.

4.2.3 Chemical sensitisation

As indicated by the title of Ashmore's book[217], it is difficult to separate chemical

References pp. 104–111

sensitisation from radical trapping, since in many cases the two phenomena occur together. The catalytic action of NO in the decomposition of organic compounds is well established. Its catalytic effect in the H_2 and Cl_2 system is quite spectacular. Addition of 0.5 torr NO to 50 torr of an equimolar mixture of H_2 and Cl_2 lowers the explosion limit from 400° to 270° C[242]. An increase in the Cl atom concentration occurs via the replacement of

$$Cl_2 + M \rightleftarrows 2Cl\cdot + M \qquad \text{by} \qquad NO + Cl_2 \rightleftarrows NOCl + Cl$$

the latter process having a much lower activation energy. A thermal explosion is due to the exothermicity of the process, but autocatalysis may also occur via chain branching, which may give, therefore, a "chemically sensitised" explosion, for example in the H_2/O_2 system[243].

However, sensitisation can have an independent role. For example, the methyl-radical-sensitised decomposition of acetaldehyde allows the Arrhenius parameters for the process

$$Me\cdot + MeCHO \rightarrow CH_4 + Ac\cdot$$

to be determined[244], confirming that this is the major route to product formation in the pyrolysis. The use of HCl as a catalyst for the decomposition of dimethylether[245] showed that the slow step in the uncatalysed reaction was reaction (43)

$$Me\cdot + Me_2O \rightarrow CH_4 + \cdot CH_2OMe \tag{43}$$

Various halogen compounds have a pronounced effect on the oxidation of hydrocarbons[245,246]. Since they are usually converted to halogen acids these are the most useful catalysts, particularly HBr[247]. The addition of HBr sensitises the oxidation of most hydrocarbons. The mechanism probably involves the reactions

$$HBr + O_2 \rightleftarrows HO_2\cdot + Br\cdot \text{ (surface)} \tag{44}$$
$$Br\cdot + RH \rightarrow HBr + R\cdot \tag{45}$$
$$R\cdot + O_2 \rightarrow RO_2\cdot \tag{46}$$
$$RO_2\cdot + HBr \rightarrow RO_2H + Br\cdot \tag{47}$$

It is interesting to note that with HCl the last process would play a very minor role since it would be endothermic by some 25 kcal. mole^{-1}.

Perhaps the most striking use of chemical sensitisation has been by Benson and his co-workers for the determination of various thermodynamic quantities[248]. Chemical sensitisation may also be used to generate a particular species that may be capable of further decomposition. For example, Lin and Laidler[249] have produced the *sec*-butyl radical by the azomethane-sensitised decomposition of *n*-butane, and CH_2 insertion reactions lead to the production of hot molecules and hot radicals[250].

4.2.4 *Photosensitisation*

Another way of producing hot species is by photosensitisation techniques. Extensive studies have been made of mercury-photosensitised reactions. The subject has been reviewed by Cvetanović[251] and by Gunning and Strausz[121], where particular experimental techniques are referred to. Physical and chemical quenching may be treated mechanistically according to the Stern–Volmer scheme[252] in terms of quenching cross-sections. This involves physical quenching from $Hg(6^3P_1)$ to $Hg(6^3P_0)$ and physical and chemical quenching of $Hg(6^1P_1)$, $Hg(6^3P_1)$ and $Hg(6^3P_0)$ to the ground state $Hg(6^1S_0)$. This may take one of three forms as depicted in reactions (48)–(50); the overall spin angular momentum should remain unaltered in the process[253]

$$Hg^* + AB \rightarrow Hg + AB^* \quad (48)$$
$$\rightarrow Hg + A + B \quad (49)$$
$$\rightarrow HgA + B \quad (50)$$

In the sensitised decomposition of *t*-butanol[254], the excess energy remains almost entirely in the *t*-BuO· radical in the form of vibrational energy.

Mercury photosensitisation also provides a means of initiating a chemical reaction where species such as hydrocarbons do not absorb the wavelength available. It is also a means of placing a precise amount of energy into a molecule or radical, particularly when using isotopic photosensitisation[121]. Photosensitised oxidation of hydrocarbons is important for the light thrown on the mechanism of combustion[255]. Other atomic sensitisers that have been used are Na, Cd, Zn and inert gases, the latter particularly in radiochemical studies. Laidler[131] has shown how the multiple use of Hg, Zn and Cd can be put to very good use in the estimation of bond energies and for mechanistic studies. Molecular photosensitisers in the gas phase include benzene, SO_2[256], biacetyl, toluene, fluorobenzene, pyridine, pyrrole, styrene[257] and presumably azomethane, certainly in solution[258], and therefore also in the gas phase.

4.2.5 *Isotopes*

An important chemical method is that of isotopic labelling by both stable as well as radioactive isotopes. The fate of different atoms of the same element in a compound may be decided by isotopic substitution. Mechanistic problems are also solved this way. For example, in the mixed pyrolysis and photolysis of CH_3CHO and CD_3CDO, the relative amounts of CD_4, CH_4CD_3H and CH_3D proved that the major part of the reaction occurs via a free radical mechanism[259]. Extensive use has been made of this technique in vacuum UV photolyses and radiolyses, particularly for producing evidence for "snap-out" primary processes[260]. Monoisotopic mercury sensitisation of the decomposition of HCl has produced evidence of the extent of HgCl formation in the primary process (51)[121]

$$Hg(^3P_1)+HCl \rightarrow HgCl+H\cdot \tag{51}$$

An isotope effect can also occur in the rate of a chemical reaction[261]. Thus, isotopic substitution has a pronounced effect on the falling-off of the rate coefficient for the decomposition of ethyl chloride and also produces evidence for a four-centred transition state[262].

4.2.6 Gaseous titrations

Conventional titrations may be used for various gases[263]. The apparatus consists of a copper RV fitted with an automatic stirrer and pressure transducer. NH_3/HCl has been titrated this way with a precision of 3 %. Ethylene may be used to titrate F_2, and Cl_2/F_2, ClF/F_2 and ClF_3/F_2 mixtures. CH_4/C_2H_6 mixtures can be titrated with F_2 and alkane/alkene mixtures with Cl_2.

A more spectacular gaseous titration technique is used for the estimation of the concentrations of various atomic species. This involves the use of a fast flow discharge apparatus[264], and the spectroscopic estimation of the extent of various chemiluminescent reactions. Hydrogen[265] and nitrogen atoms[266] may both be titrated with NO, and oxygen atoms with NO_2[264] or NO[264,267] the reactions being

$$\dot{H}+NO+M \rightarrow HNO^*+M \tag{52}$$

$$\dot{N}+NO \rightarrow N_2^*+O \tag{53}$$

$$\dot{O}+NO_2 \rightarrow NO+O_2^* \tag{54}$$

$$\dot{O}+NO \rightarrow NO_2^* \tag{55}$$

In reaction (55), the NO concentration is constant since it is regenerated via reaction (54). These titrations may be used to measure the concentration of these atoms in other reactions such as (56)[265]

$$H\cdot+NO_2 \rightarrow \cdot OH+NO \tag{56}$$

4.3 GAS CHROMATOGRAPHY

4.3.1 Introduction

The advent of gas chromatography[268,269] has given a tremendous stimulus to the study of kinetics, particularly of gas-phase reactions, since it allows precision product analysis to be achieved and the purification and analysis of the starting materials to be made. The theory and application of the techniques involved have been dealt with in detail in several books[270]. The most recent of these is by Giddings; this consists of three volumes of which the first deals with the theory[271]. Apart from these, two series are devoted to the advances in gas chromatography[272] and chromatography[273]. We shall be concerned here with Gas Liquid Chromatography (GLC) and Gas Solid Chromatography (GSC). A schematic diagram of a typical set-up is shown in Fig. 51.

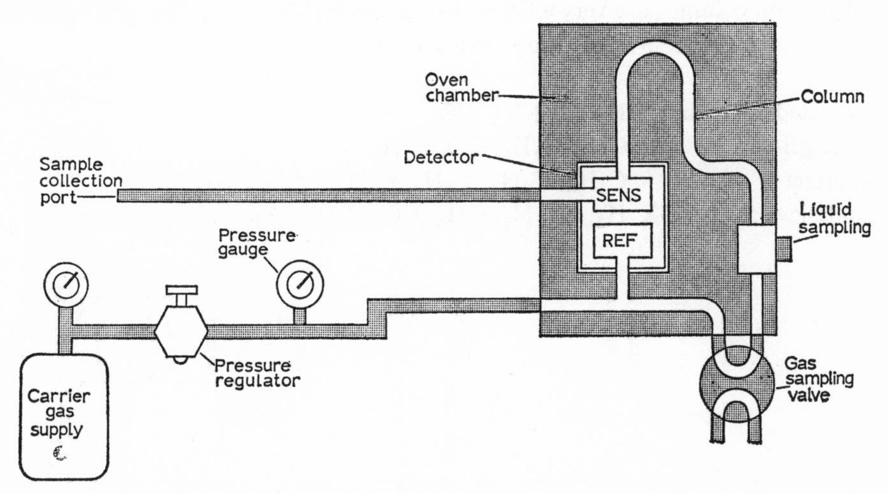

Fig. 51. Schematic diagram of a flow system for gas chromatography. From Perkin Elmer Inc.

4.3.2 Carrier gases

Carrier gases are usually hydrogen, helium or nitrogen which are dried by passing through a molecular sieve column. This is regenerated periodically with a positive flow of gas and heating to 200 °C.

4.3.3 Columns

Columns are made from glass, copper or stainless steel and are generally U-shaped or coiled. The development of capillary columns has been described by

Desty[274]. The preparation of liquid–solid and solid types of columns is described quite clearly by Purnell[275]. The inert solid support for GLC is usually kieselguhr. The quantity of the stationary phase is usually 5 or 10 % by weight but can be as high as 30 %. Since the liquid is inevitably removed from the column to some extent, the lower the proportion of liquid, the shorter the column life. Liquid is deposited on the solid by dissolving the stationary phase in a volatile solvent, mixing with the solid to form a slurry, and removing the solvent by means of a rotary evaporator. The choice of a suitable liquid for a particular analysis remains a difficult problem. Rough guides are given in the various books on chromatography and by suppliers of commercial gas chromatographs, whereas particular information can be obtained from past literature on similar kinetic studies. Two very useful stationary phases for the analysis of liquids are β–β′-oxydipropionitrile and polypropylene glycol.

The usual solids for GSC are charcoal, silica gel, alumina and molecular sieves. Adsorption columns are very efficient for the analysis of gases. The gases that can be separated on these solids (unpoisoned) are[276]

Activated charcoal	H_2, O_2+N_2, CO, NO, CH_4
Silica gel	CH_4, C_2H_6, CO, C_2H_4
Alumina	C_2H_6, C_2H_4, C_3H_8, C_3H_6, cyclo-C_3H_6
Lindesieve 5A or 13X	H_2, O_2, N_2, CH_4, CO, C_3H_6

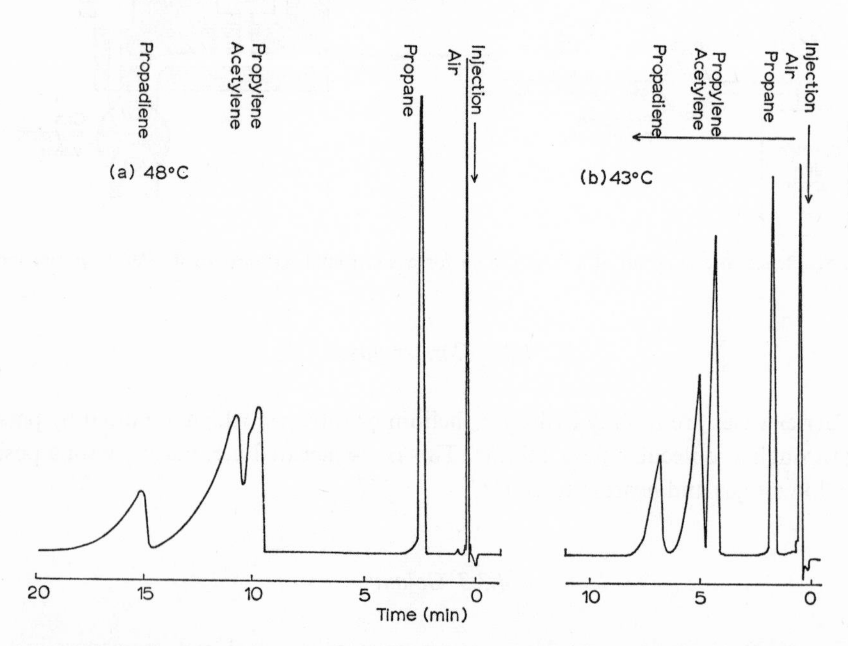

Fig. 52. The effect of a tailing reducer on alumina; (a) on untreated activated alumina, (b) on activated alumina poisoned with 2 % of dinonyl phthalate. From ref. 276.

Alumina is also very useful for the analysis of fluorocarbons. At elevated temperatures, compounds of high molecular weight are eluted but the peaks often tail. Coating the solid with 0.5–2 % of an involatile liquid such as squalene or dinonyl phthalate reduces tailing (Fig. 52)[276]. Poisoned alumina (3 % dinonyl phthalate) will separate C_4-C_6 hydrocarbons at 100° C. Purnell[277] has discussed the optimum operating conditions for these columns. For preparative scale gas chromatography, in order to prevent overloading, the diameter of the column is increased to 1–6 in. corresponding to sample injections of 0.1–100 ml. These systems are described in the references cited (275, 276 etc.).

4.3.4 Programmed temperature gas chromatography (PTGC)

A technique that has been developing rapidly is programmed temperature gas chromatography. This has been treated comprehensively by Harris and Habgood[278] and Mikkelsen[279]. It frequently happens that the early peaks in a chromatogram are very tall and narrow, and overlap one another badly, whereas the late peaks are so flat that they are hardly distinguishable from the baseline. Figure 53 shows how these early peaks can be separated and the late peaks sharpened up, even when the sample cannot be injected rapidly, by temperature programming[278]. Temperature programming becomes a working possibility when the range of the boiling points of the substances comprising the mixture under investigation is 50° or more. This would avoid multiple column techniques[280]. The range of liquid phases available is more limited than for isothermal gas chromatography since they need to have high boiling points and wide liquid ranges. Columns have to be conditioned for some time at the highest temperature to be used. It is especially important that the carrier gas should be free from water and any other condensable

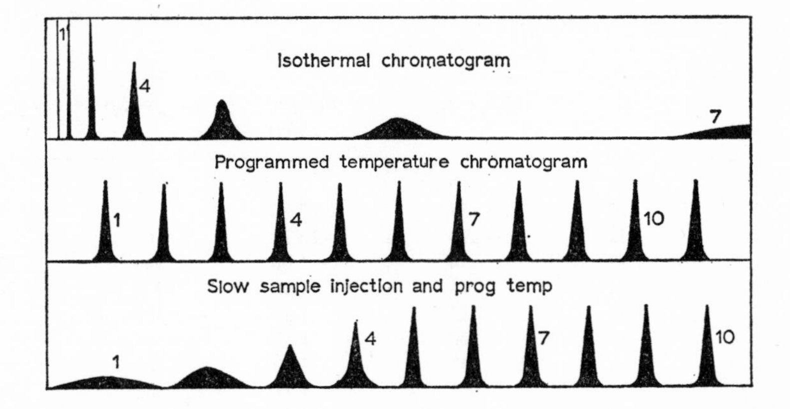

Fig. 53. Idealised improvement obtained by temperature programming. From ref. 278

References pp. 104–111

impurities since these are likely to be retained on the cold column and eluted as flat peaks when it is hot. With katharometer detectors, it is necessary to use a dual-column technique because they are so sensitive to flow-rate fluctuations. The ionisation and flame detectors are more satisfactory when only a single column is required. PTGC has a considerable application in preparative gas chromatography on a small scale.

4.3.5 Detectors

Of the many detectors that are discussed in the literature, we shall consider three which are the most important for gas-phase kinetic studies. These are the katharometer, the hydrogen flame ionisation detector (FID), and the gas density balance.

(a) Katharometers

Katharometers or thermal conductivity detectors are usually incorporated in a bridge circuit as matched pairs, such that the flow over the reference element is always pure carrier gas. There are two types of elements, the thermistor (low temperatures) and metal filament elements (high temperatures). The metal filaments are constructed from platinum or tungsten. The sensing device depends upon the change in thermal conductivity of the carrier gas stream due to the presence of the unknown gas. Table 16 lists the values of thermal conductivity coefficients for several gases at 0° C. Clearly, hydrogen and helium are the best choice for carrier gases, of which the former is the most practicable in the United Kingdom. Notice that H_2 may be estimated with a reasonable sensitivity relative to He or N_2 but

TABLE 16

THERMAL CONDUCTIVITY COEFFICIENTS (K_x) OF GASES AT 0° C

(cal. cm^{-1}. sec^{-1}. deg$^{-1}\times 10^6$)[275]

x	K_x	$(K_{N_2}-K_x)$	*Relative sensitivity*	$(K_{H_2}-K_x)$	$(K_{He}-K_x)$
N_2	58	0	0	358	290
H_2	416	−358	−25.5	0	−68
He	348	−290	−20.7	68	0
CH_4	72	−14	−1.00	344	276
C_2H_6	44	14	1.00	372	304
C_3H_8	36	22	1.57	380	312
C_4H_{10}	32	26	1.85	384	316
$CHCl_3$	16	42	3.00	400	332
CH_3OH	34	24	1.71	382	314
$(CH_3)_2CO$	24	34	2.43	392	324

gives a negative peak. Depending on the conditions, the sensitivity of a katharometer is such that 1 μmole may be estimated with a precision of 1 %. Micro-katharometers have been manufactured, the sensitivity of which approaches that of the FID.

(b) *Flame ionisation detectors*

A typical FID set-up is shown in Fig. 54. A split flow device allows a percentage of hydrogen or hydrogen + carrier gas to pass through a jet where it is burnt. Above this is placed a cylindrical metal electrode. The concentration of ions in a clean H_2/O_2 flame is extremely low. When organic compounds are added the ion concentration increases by many orders of magnitude and an ionisation current is produced, even in the absence of an applied voltage, which may be fed into an

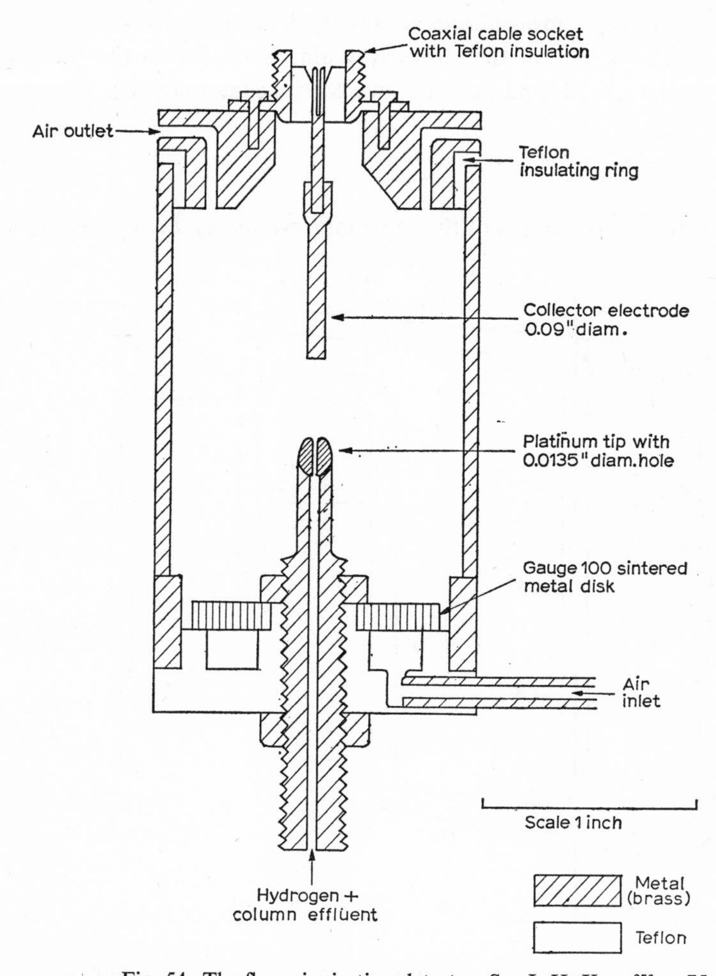

Fig. 54. The flame ionisation detector. See J. H. Knox[276] p. 75.

amplifier. The detector is relatively temperature-insensitive and therefore ideal for PTGC. However, it is also insensitive to a number of substances, notably the inert gases, CO, CO_2 and formic acid. The sensitivity of this detector is such that 10^{-2}–10^{-3} µmole may be estimated with a precision of 1 % under optimum conditions.

(c) Gas density balance (Fig. 55)

Column effluent enters the detector at A and splits at B and B′, and leaves at C and C′. Pure carrier gas splits at F and F′ and also leaves at C and C′. For balance there is zero flow across the detector element H. Addition of vapour at A changes the density of the gas mixture, causing an unbalance and a flow across H. In commercial instruments, the detector elements are usually matched thermistors or hot wires which can be incorporated into a Wheatstone bridge network. The detector is unique in that it measures an accurately defined property of the eluted vapour. The peak area, A_x, in the chromatogram, is related to the molecular weights of the carrier gas M_c and the unknown M_x by equation (M′). The carrier

$$A_x = M_x/M_c - M_x \qquad \text{(M')}$$

gas is usually nitrogen, unless x is of the same mol. wt. as N_2, *i.e.* CO, C_2H_4 etc.,

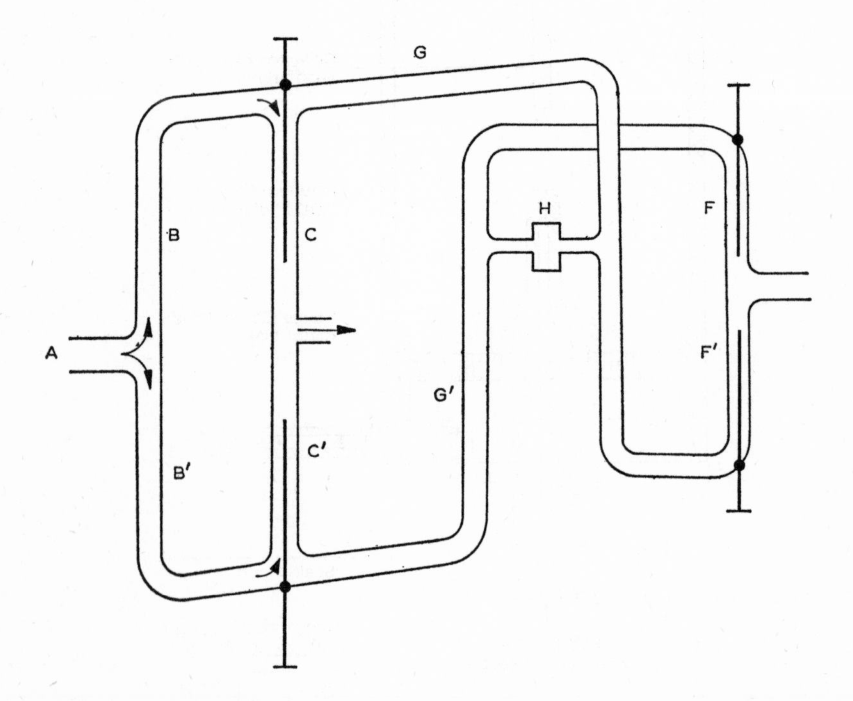

Fig. 55. Schematic view of the gas density balance detector. See J. H. Purnell[275] p. 271.

when hydrogen is used[275]. The detector block has to be thermostatted very accurately. The sensitivity of the detector is as good or better than the katharometer.

4.3.6 Product identification and trapping

An initial product identification can be carried out by comparing the retention times, under given conditions, of the unknown compound with that of suspected known compounds. To counteract minor variations in flow rate and temperature, coincidence of peaks can be checked with consecutive doping or marking of the mixture with known compounds. Coincidence should also be checked by using one or more different columns. In this context the gas density balance detector provides particular confidence in identification.

Conclusive evidence is produced by infrared or mass spectroscopic methods. This involves the selective trapping of each product as it is eluted from the column. Efficient trapping is obtained by quenching in two or more U-tubes containing glass wool followed by trapping in a breakseal trap for subsequent analysis (Fig. 56). Flow may be diverted to this trapping system by a 3-way solenoid valve operated by a micro-switch. The latter is activated by the pen of the recorder incorporated in the GLC when the peak due to the required material reaches 80% of its maximum. This system has also been used on a small preparative scale.

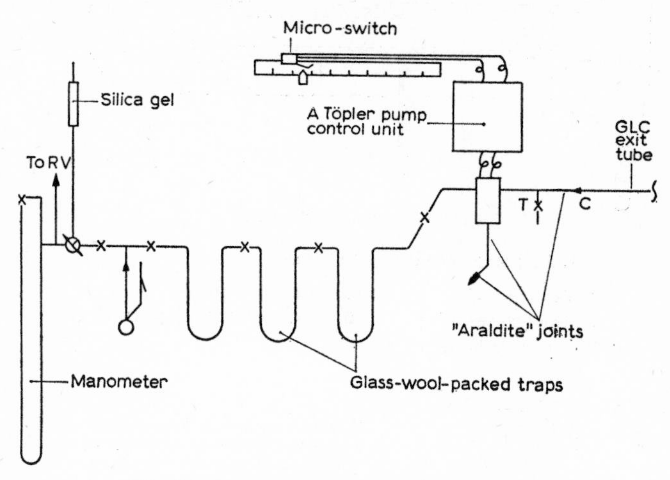

Fig. 56. Product trapping from a GLC exit tube.

4.3.7 Sampling and quantitative estimation

Samples may be taken for analysis at once, or after fractionation of the products and reactants. A set-up due to Porter and Norrish[280], is shown in Fig. 57. A sample

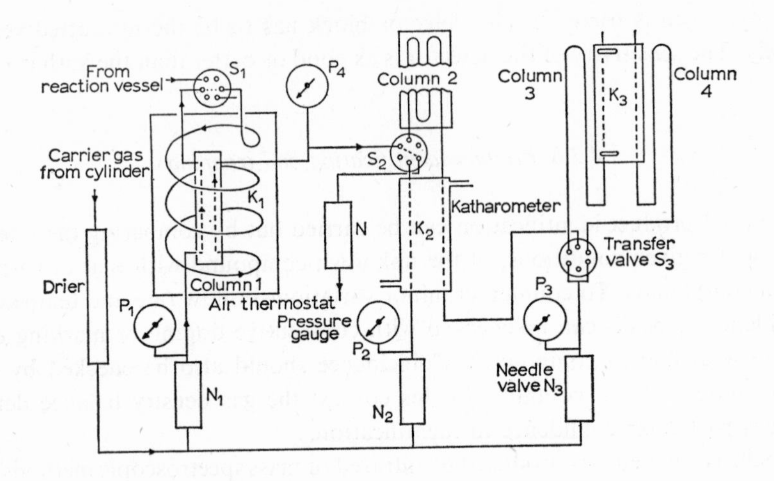

Fig. 57. Multi-column technique for product analysis. From ref. 280.

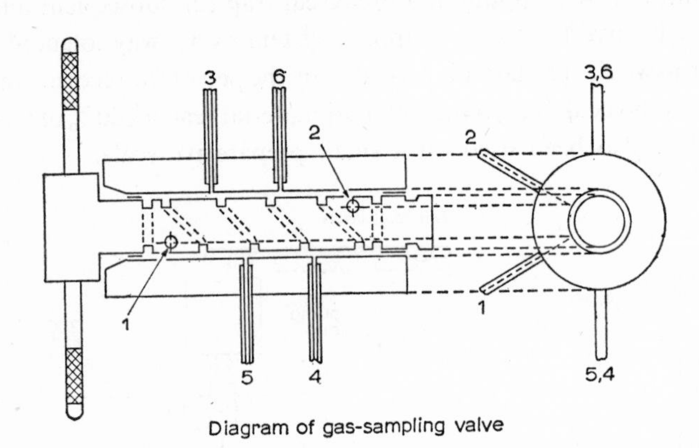

The 6 ports were connected to:

1 Vacuum
2 Reaction vessel
3, 4 Sample volume
5 Carrier gas supply
6 Chromatographic column

I X X S E X

Division of valve action by arrangement of ports

X All ports are isolated except those connecting carrier gas supply to column
E Injection of sample
I Evacuation of sample volume
S Filling of sample volume from from reaction vessel

Fig. 58. Gas-sampling valve. From ref. 282.

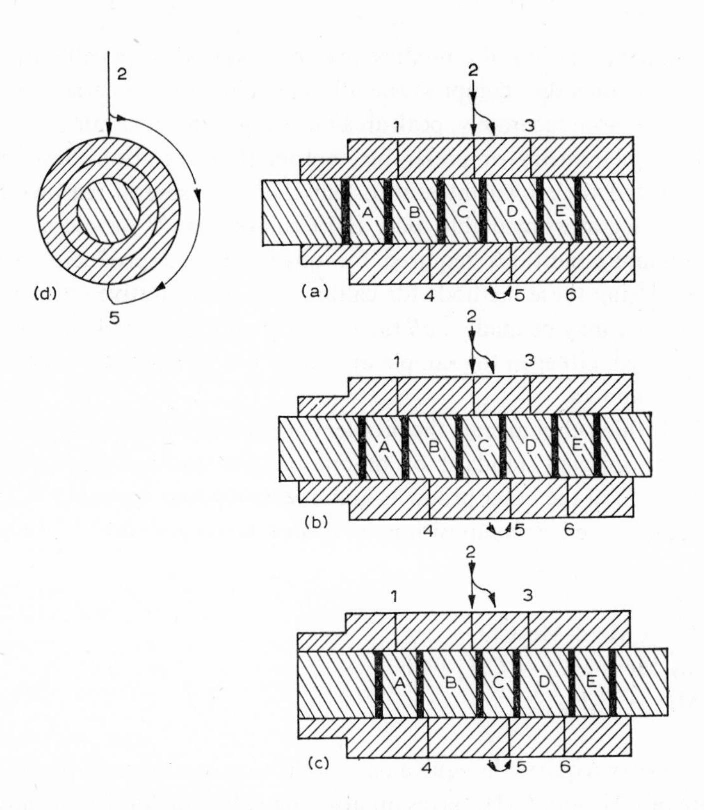

Fig. 59. Push–pull gas-sampling valve. The connections are: 1, vacuum or sample; 2 and 3, carrier gas; 4, sample volume; 5, connecting bypass 2 to 3; 6, sample volume. (a) evacuation and introduction of sample; (b) isolation of sample; (c) injection of sample on to column; (d) bypass connection. From ref. 283.

is withdrawn from the RV and injected through a series of columns such that those used for the slower eluting materials are placed first, when the faster are eluted as a single peak but are subsequently analysed in detail by subsequent columns. Suitable switching of the transfer valves allows secondary injection and the sweeping out from the system of the slower eluting materials. Alternatively, the faster eluents may be held in the transfer valve sample volume for later analysis. The use of PTGC with adsorption columns, poisoned to prevent both tailing and allow the analysis of high mol. wt. compounds, may reduce the necessity for this technique[281].

A successful sampling valve is that due to Pratt and Purnell (Fig. 58)[282]. A dead-space sample is first withdrawn, followed by a reacted sample. Cundall *et al.*[283] claim a better valve is the "push–pull" type (Fig. 59), a variation of the valve used on some commercial GLC instruments. For sample injections after fractionation, a constant volume and manometer is employed for gases, and precision microlitre syringes for liquids.

Quantitative estimation of a mixture may be obtained from calibrations, using the pure compounds that comprise the mixture. This may be carried out by peak height or often more accurately, peak area measurements, determined by weighing, triangulation, planimetry or preferably, automatic integration. Several workers have constructed their own integrators but there are several reasonably priced commercial integrators available. Batt and Cruickshank[284] have drawn attention to the optimum conditions required for quantitative, reproducible calibration using a FID. Using these methods for calibrations, quantitative estimation of the reaction mixture may be made if *all* the products are eluted and the total number of moles injected, either in the sample or from a PVT measurement of the contents of the RV, is known.

A method which circumvents the problem of total elution is to have an absolute measure of an injected standard or of one product and use this for the quantitative estimation of the other products and undecomposed reactants. For instance, in the HCl-catalysed decomposition of di-*tert*.-butyl peroxide[233], involving the reactions

$$t\text{-BuO}\cdot \rightarrow \text{Me}\cdot + \text{Me}_2\text{CO}$$

followed by $\text{Me}\cdot + \text{HCl} \rightarrow \text{CH}_4 + \text{Cl}\cdot$

or $2\dot{\text{M}}\text{e}\cdot \rightarrow \text{C}_2\text{H}_6$

the stoichiometry requires the equivalence of the total methyl and acetone amounts. A measure of CH_4 and C_2H_6 gives an absolute value for Me_2CO from which the other products may be estimated. This is particularly useful when the inevitable unknown compounds are present, and avoids the assumption that all products are eluted. For competitive reactions, which happens here to some extent, this problem virtually disappears.

4.4 SPECTROSCOPIC METHODS

4.4.1 Introduction

One may divide the use of spectroscopic methods into the three categories of product identification, quantitative estimation of reactants or products at the end of a run, and *in situ* measurement of the concentrations of reactants or products during a reaction. Of these the last is the most important, because it presents an opportunity to follow the production and disappearance of transient species as well as those already mentioned. This is particularly true for very fast reaction techniques such as flash photolysis where the concentrations of the very reactive intermediates are likely to be high.

Of the different regions of the spectrum involved, that concerning nuclear magnetic resonance has least direct application to reactions in the gas phase; although its potential has been discussed[285], no *in situ* measurements have been made. In solution kinetics, where it has been used extensively, the situation is very different. However, it should be helpful for product identification, particularly from the point of view of site attack. It appears to have some potential use in combustion studies[286]. The other regions of the spectrum are discussed in three sections below.

4.4.2 UV, *visible and* IR

McDowell and Thomas[287] have used the following set-up (Fig. 60) for the spectrophotometric analysis for nitrogen dioxide. Light from a sodium lamp is focussed by lenses onto EEL selenium photocells either through the RV onto P_1 or through an iris diaphragm onto P_2. R_1 and R_2 are variable resistances and G a galvanometer. When the external resistance is 1000 Ω, the response curves of the photocells become logarithmic. Since R_1 is proportional to the optical density ($= \varepsilon cd$, where ε is the decadic molar extinction coefficient, c is the concentration and d is the path length), R_1 is also proportional to the pressure of NO_2. Golden *et al.*[288] have modified a Cary Model 15 spectrophotometer for the direct measurement of methyl iodide, hydrogen iodide and iodine in the equilibrium system

$$I_2 + CH_4 \rightleftarrows MeI + HI$$

Measurements were made at 500, 350, 257, 245 and 235 mμ. 350 mμ is an "optical window" for this system and therefore a reference point. Calibrations were made with the substances concerned. Beer's law was obeyed up to 30 torr. I_2 was directly determined from the measurement at 500 mμ. The absorbance for I_2 was subtracted at the other wavelengths, leaving values from which the pressures of MeI and HI could be determined. In this way, 10^{-2} μmole of each could be detected.

Other spectrophotometric estimations are cited in ref. 22c. Morse and Kaufman[289] have measured the concentrations of oxygen, nitrogen and hydrogen

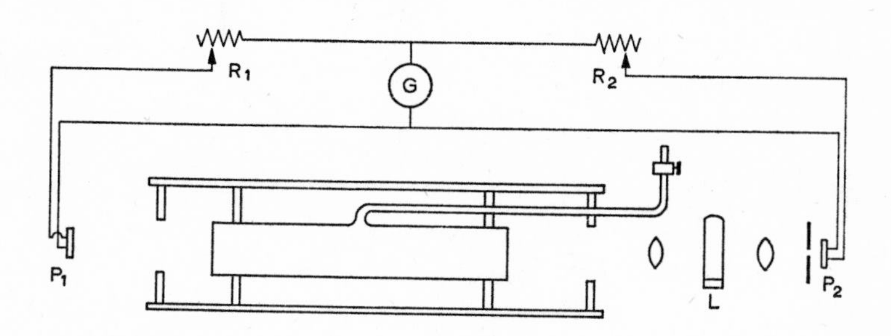

Fig. 60. Spectrophotometric determination of NO_2. From ref. 287.

References pp. 104–111

atoms by the vacuum UV line absorption of their resonance lines. Metastable states of O and N have also been detected. Increasing emphasis is being placed on direct IR absorption measurements. Cashion and Polanyi[290] have measured the formation of vibrationally excited HCl in the $H+Cl_2$ and $H+HCl$ followed by $H+Cl+M$ reactions. Ketene and acetaldehyde have been detected this way in the pyrolysis of ethylene oxide[291]. The heat of dissociation of N_2O_4 has been determined from measurements of the temperature dependence of the integrated absorption of NO_2 and N_2O_4 vibration–rotation bands[292]. Using long path cells, elusive intermediates like CH_3NO[293] and NO_3[294] have been identified and estimated. In the latter case, it has been shown that the NO_3, produced from $O_2+NO \rightleftarrows NO_3$ is the unsymmetrical OONO rather than O—N$\langle^{O}_{O}$·

IR spectroscopy appears to be a very promising tool for oxidation studies. The rate of formation of carbon monoxide, carbon dioxide and formic acid and the disappearance of ozone was measured this way with the O_3/O_2+CH_4 system[295]. The system used by Burt and Minkoff[296] for the combustion studies is shown in Fig. 61. Light from a Nernst filament is split in two and passed alternately through two heated cells F_1 and F_2 containing either fuel + N_2 or fuel+O_2. The beams are rejoined and fed into a Wadsworth monochromator containing a CaF_2 prism and finally focussed onto a thermopile, from which a particular signal may be amplified and recorded.

All these measurements for product analysis are limited by peak overlap, the possible presence of unknown compounds and the magnitude of the molar extinction coefficients, particularly in the infra-red. Here, careful calibrations have to be made as a function of pressure, using say N_2 as the diluent[22c].

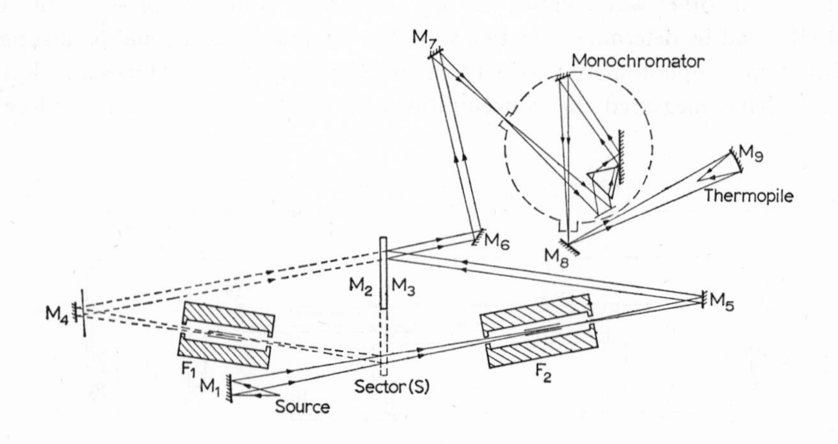

Fig. 61. *In situ* IR studies of combustion. From ref. 296.

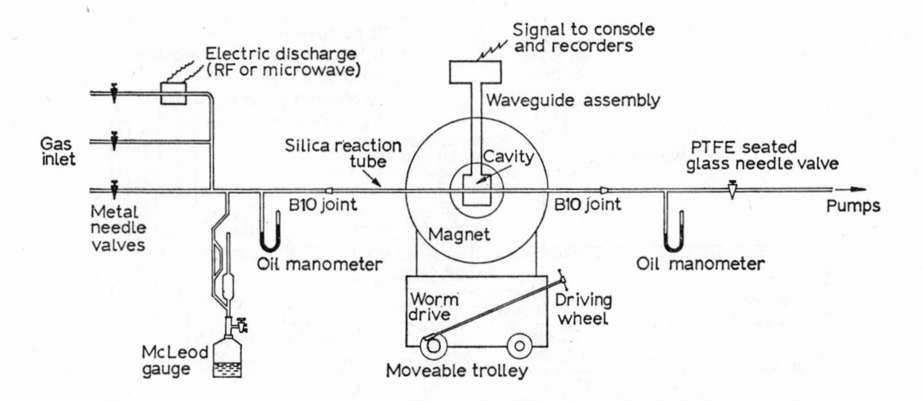

Fig. 62. Flow system with moveable cavity at room temperature. From ref. 299.

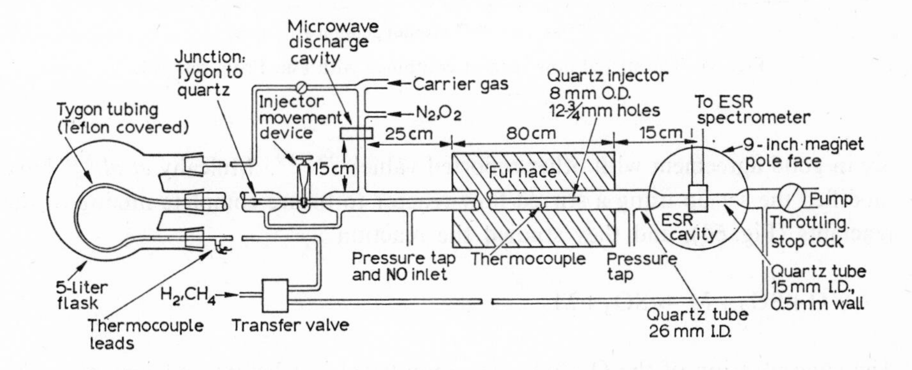

Fig. 63. Flow system with a fixed cavity at elevated temperatures. From ref. 300.

4.4.3 Electron spin resonance

Carrington and Levy[297] and Westenberg and de Haas[298] have reviewed the early work on gas-phase reactions in this field. The apparatus involves a fast-flow system where a gas, at a relatively low pressure is passed through a microwave discharge to the resonance cavity of the spectrometer. At room temperature, it is possible to have the cavity in variable positions[299] but at higher temperatures it is fixed[300] (Figs. 62 and 63, respectively). The rate coefficients found for reactions (57)–(60)

$$O + NO_2 \rightarrow NO + O_2 \quad (57)$$
$$O + NO \rightarrow NO_2 + h\nu \quad (58)$$
$$N + NO \rightarrow N_2 + O \quad (59)$$
$$2H + H_2 \rightarrow 2H_2 \quad (60)$$

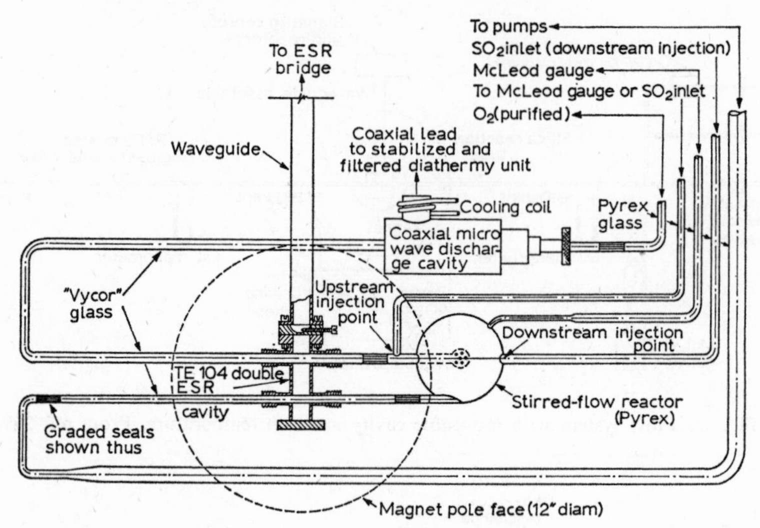

Fig. 64. The stirred flow reactor combined with ESR. From ref. 301.

are in good agreement with other reported values[300,299]. Mulcahy *et al.*[301] have modified the set-up using a stirred-flow reactor to insure complete mixing of the reactants (Fig. 64), and they studied the reaction

$$O+SO_2+M \rightarrow SO_3+M$$

The concentration of the O atoms was monitored just before and just after the RV. Nalbandyan[302] has reviewed the use of this technique in combustion studies. Apart from investigation of reactions of hydrogen and oxygen atoms and OH radicals, it has been used to establish the site at which O attacks the complex molecule. The calibration methods and the sources of error involved in estimating the concentrations of these atomic and diatomic species are discussed by several authors[298,299,303].

4.4.4 *Mass spectrometry*

Another direct, and in this case, well established, method of following the course of a reaction is by mass spectrometry. The application of mass spectrometry to analysis of organic compounds has been dealt with by Beynon[304] and Reed[305] and two volumes are concerned with the advances in this field[306,307]. The mass spectrometry of free radicals has been reviewed by Cuthbert[308].

Apart from the direct following of the rate of a reaction, identification and quantitative estimation of the transient species, the mass spectrometer may also be

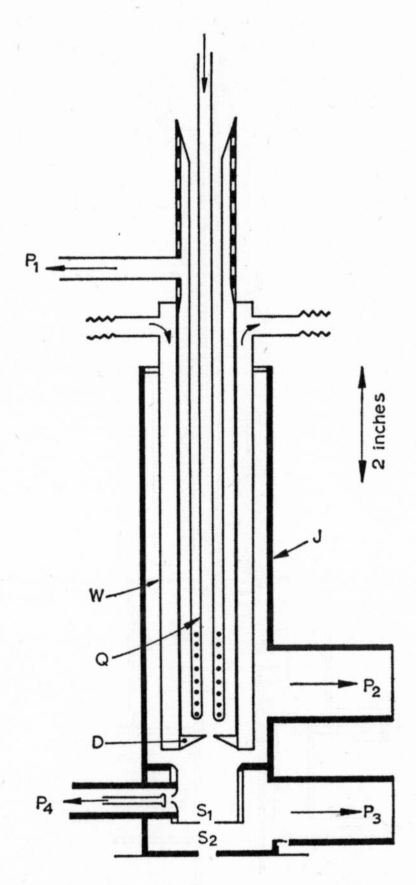

Fig. 65. Eltenton's RV and ion source. From ref. 311.

used from a mechanistic point of view. Thus, it has been shown that the formation of CH_2[309] via reaction (61) may be discounted[310].

$$\cdot CH_2CH_2O\cdot \rightarrow CH_2{:} + CH_2O \qquad (61)$$

One method for the detection of free radicals depends upon the principle that the electron energy required to ionise the free radical is less than that required to produce the ionised radical from the parent or other compound. Eltenton[311] used this method to detect free radicals in pyrolytic and combustion reactions up to pressures, in the RV, of 160 torr. His RV and ion source are shown in Fig. 65. The ionisation and ion accelerating chambers were evacuated by two, separate, large diffusion pumps (P_2 and P_3), while the filament and the analysing chambers were separately evacuated by smaller pumps (P_4). The RV consisted of a spirally-wound double quartz tube Q down the centre of which the reactant was introduced. The reaction zone

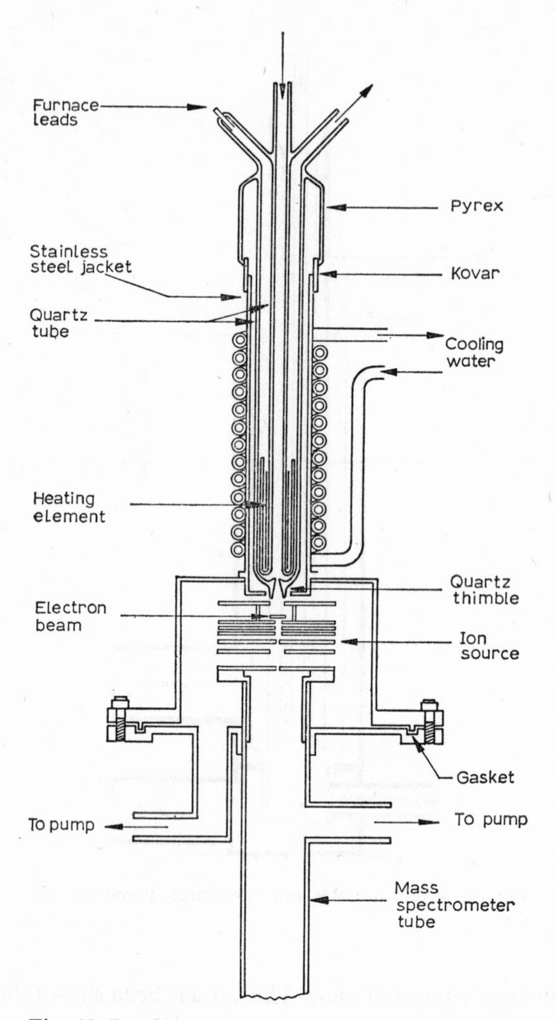

Fig. 66. Lossing's RV and ion source. From ref. 310.

was sampled through a gold leaf diaphragm D, the remaining gases being pumped out at P_1 through the annular space between Q and the water jacket W. (The chances of catalysed combination of free radicals at D is very high.) This method tends to be unreliable since the ionisation efficiency is low and varies with the electron energy. An alternative which, in many cases, allows quantitative measurements to be made, has been developed by Lossing *et al.*[310]. The net peak height and hence the concentration of the radical may be determined provided that *all* the reaction products that contribute to a radical peak are identified. The apparatus used by these authors is shown in Fig. 66. The sample leak consists of a quartz thimble, through the thin top of which a pinhole is produced by a Tesla discharge.

The errors involved in estimating net peak heights depend on the relative size of contributing peaks, the identification of all peaks and the assumption of 100% material balance. The problems involved in the detection of free radicals in mass spectrometry have been discussed by Lossing[312].

The oxidation of acetaldehyde has been studied by means of a continuous scanning technique[313]. The system was capable of producing 60 spectra per second, scanning from 12 to 80 mass units. The RV and ionisation chamber are shown in Fig. 67. Samples flow through a Lossing type quartz leak from the cylindrical RV into the ionisation chamber, from where they are evacuated by a high-speed

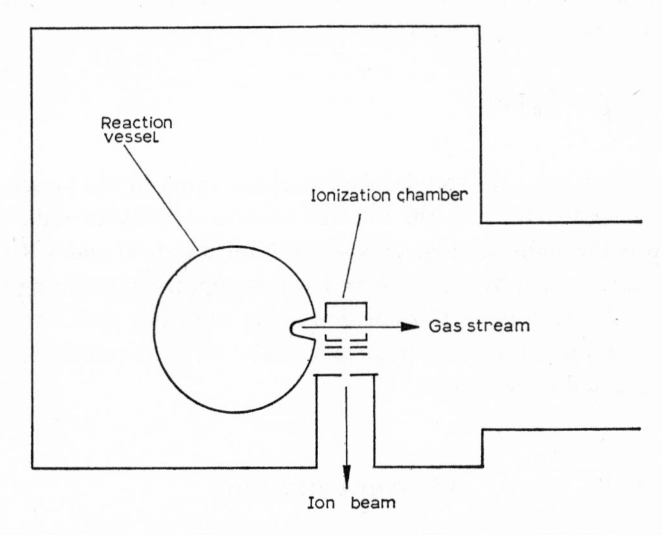

Fig. 67. Continuous sampling mass spectrometer (diagrammatic). From ref. 313.

pumping system. The distance from the leak to the electron beam is 15 mm, which is longer than the mean free path of the molecules in this region. To preserve molecular flow, the maximum pressure in the RV was limited to 100 torr; in this way, gas-phase reactions between species leaving the RV were avoided.

The mass spectrometer is the ideal instrument for studying ion-molecule reactions, and as such has been reviewed by Lampe *et al.*[314]. In the conventional analytical mass spectrometer, the pressure in the ionisation chamber is 10^{-5}–10^{-6} torr. Ion-molecule reactions are possible only at pressures $> 10^{-5}$ torr. Improvements in technique have now increased the range from here to relatively very high pressures (2 torr)[315]. The use of the mass spectrometer in isotopic substitution studies covers the whole range of gas-phase kinetics, but seems to be particularly prolific in radiochemical studies.

4.4.5 Optical pumping

Crosley and Bersohn[316] have made preliminary investigations on the use of optical pumping for the detection of hydrogen atoms and free radicals in the vacuum UV photolysis of ethane. The spin polarisation of optically pumped Rb atoms is decreased by electron exchange with unpaired electrons, and the spin-lattice relaxation time also decreases. In particular, atoms generated in S states acquire polarisation by electron exchange with the pumped atoms. Under these conditions, a cell, the walls of which contain a thin coating of rubidium, containing N_2 and C_2H_6, is irradiated with circularly polarised Rb resonance radiation at 7947A and Ar resonance radiation at 1048 and 1067A. The radical concentration, N_R, is given by equation (N′)

$$\left(\frac{1}{T_1'} - \frac{1}{T_1}\right) = N_R V \sigma \qquad \text{(N')}$$

where T_1' and T_1 are the Rb spin-lattice relaxation times in the presence and absence of UV respectively, V is the average relative velocity of radicals and Rb atoms and σ is the spin-exchange cross-section for a radical and a Rb atom. An H atom resonance was observed at $g = 1$ which appears only during photolysis. By comparing the decrease in transmission of the cell when the UV is on with the further decrease when the radio-frequency radiation is at resonance, the net H concentration could be estimated.

4.5 OTHER METHODS

4.5.1 Polarography

Polarographic analysis has been used mainly in the field of combustion for the analysis of peroxides and aldehydes. The technique has been developed for this purpose by Minkoff *et al.*[317]. The limiting diffusion currents are proportional to the concentrations of individual peroxides and are unaffected by the presence of aldehydes. Calibration is effected by determining the diffusion current for a known concentration of H_2O_2 in neutral solution. Amounts as low as 1–10 μmoles have been measured this way. However, the polarograms of organic peroxides are not easily distinguishable[318]. The method has also been used to determine aldehydes[319] and residual $MeNO_2$[320] in pyrolytic reactions.

4.5.2 Polarimetry

This technique has been used mainly in solution studies, *e.g.* of sugar hydrolysis. However, Maccoll *et al.*[321] have applied it in the investigation of the gas-phase racemisation of 2-chloro-octane during HCl elimination, to check the idea of a polar transition state.

$$\text{D-(+)-}C_8H_{17}Cl \underset{2}{\overset{2}{\rightleftarrows}} \text{L-(-)-}C_8H_{17}$$

$$1 \searrow \qquad \swarrow 1$$

$$C_8H_{16}+HCl$$

If α_t and α_0 are the rotations at a time t and zero, respectively, then k_2 is determined from equation (O′)

$$\alpha_t/\alpha_0 = \exp -(k_1+2k_2)_t \qquad (O')$$

k_1 is separately determined from pressure measurements.

4.5.3 *Conductance measurements*

These have been used for a few studies of ionic reactions in solution, for example, the classical work on the saponification of esters.

Some gases may be determined by dissolution in an appropriate solution followed by conductance measurements. Carbon dioxide[322] and hydrogen chloride[323] have been determined this way be dissolving in barium hydroxide solution and water respectively. HCl may be estimated with precision in amounts as low as 0.1–1 μmole. This method has been used to determine HCl in molecular beam reactions of hydrogen atoms with CCl_4, Cl_2, etc.

4.5.4 *Thermal conductivity*

In situ thermal conductivity measurements have been used to determine[324] the extent of the equilibrium in reaction (62)

$$N_2O_4 \rightleftarrows 2NO_2 \qquad (62)$$

over the temperature range 32–90° C. The conversion of hydrogen and carbon monoxide to methanol has also been followed in this way[325]. The use of thermal conductivity in GLC and GSC analysis has already been referred to.

4.5.5 *Calorimetry*

The rate of very exothermic reactions, such as oxidation and combustion pro-

References pp. 104–111

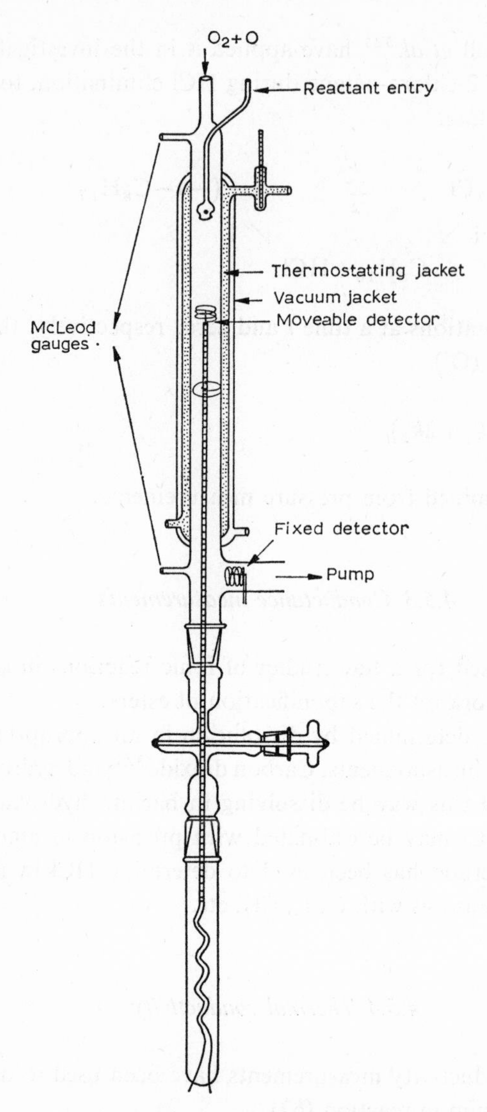

Fig. 68. Isothermal calorimeter. From ref. 328.

cesses, may be followed by measuring the heat evolved. For gas-phase reactions the change in resistance of a thin metal wire may be measured as a function of time. The wire is enclosed in a thin Pyrex or silical sleeve and suspended in the RV. Vanpeé[326] has shown that the change in temperature may be related to the change in pressure for several oxidation reactions.

The combination of atoms or radicals liberates a considerable amount of heat due simply to the energy of bond formation. At low pressures this will take place heterogeneously. Hence this presents a method of detecting free radicals or atoms by measuring the heat liberated at a coated thermometer bulb, a glowing wire or a thermocouple junction[327]. More fundamentally, Elias *et al.*[328] have used the method of isothermal calorimetry, to measure the concentrations of several atomic species. The apparatus is shown in Fig. 68. It consists of a conventional flow system which incorporates a discharge tube for the production of atoms. Two resistance wire detectors D_1 and D_2, are used, the former being moveable. The platinum resistance wires of the detectors, are electroplated with several metals which are, or their oxides, capable of catalysing the recombination of the atoms. The detector forms one arm of a Wheatstone bridge. The surface of the RV is coated with phosphoric acid or boric oxide, which are very efficient poisons for surface atom recombination. The thermal conductivity of the system must be the same in the presence and absence of atoms, while combination at the detector must be complete. Also, no other species capable of heating the detector must be present.

A known current i_0 is passed through the wire which has a measured resistance R. To counteract the heat liberated by atom combination, the current is reduced to i, such that the resistance of the wire again has the value of R. The difference in energy W required to keep the wire isothermal is given by equation (P′)

$$\Delta W = R(i_0^2 - i^2) \tag{P′}$$

and the rate of flow of atoms $F = \Delta W/\Delta H$, where ΔH is the heat of atomic combination. Callear and Robb[329] measured the extent of heating caused by Hg ($^3P_1$) atoms. The heating effect was modified according to the nature of any quench-

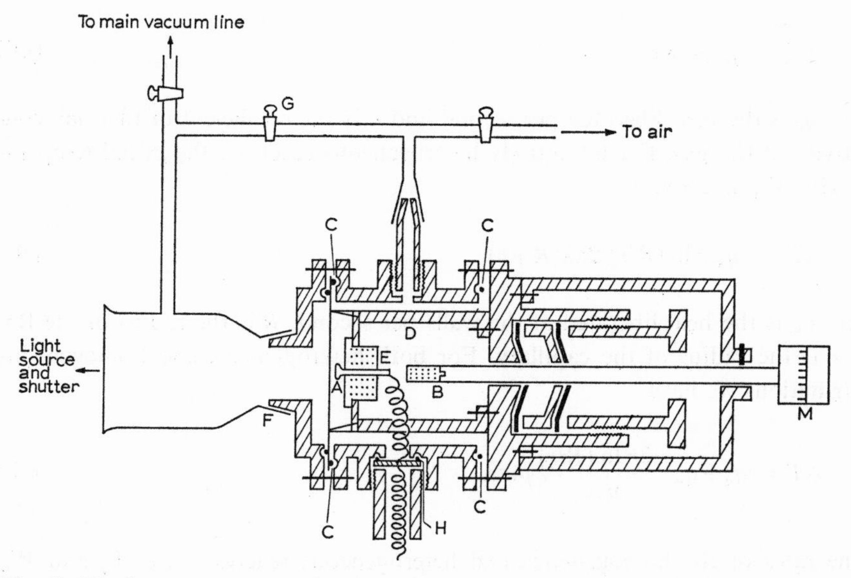

Fig. 69. Diaphragm manometer for measuring very small temperature changes. From ref. 330.

ing gas that was present. Moseley and Robb[330] have made use of the pressure change occurring in a gas resulting from the adiabatic temperature rise in the initial stages of a photochemical reaction The technique was used to determine the rate constant for methyl radical combination[331]. The small pressure changes were measured by means of a very sensitive diaphragm manometer as a change in capacity between the diaphragm and a small brass probe A (Fig. 69).

An important method (already mentioned, p. 11) for determining the extent of a heterogeneous contribution to a homogeneous reaction is that of differential calorimetry[47]. Two fine-gauge wire thermocouples enclosed in thin Pyrex or silica capillaries are mounted at the centre and close to the wall of the RV (Fig. 70).

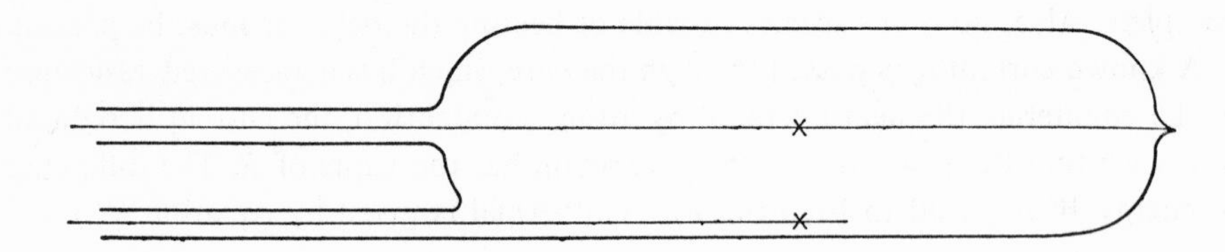

Fig. 70. Differential calorimetry. From ref. 47a.

The difference in readings of these thermocouples gives a measure of the temperature rise of the gas at the centre of the RV relative to the walls. For an entirely homogeneous reaction, the initial temperature rise for a cylindrical RV is given by

$$\Delta T_g = q_g/(4\pi\lambda) \qquad (\mathrm{Q}')$$

where q_g is the heat liberated per second and λ is the coefficient of thermal conductivity of the gas. For an entirely heterogeneous reaction, the initial temperature rise ΔT_w is given by

$$\Delta T_w = q_w r \ln (R/r)/2\pi\lambda(R+r) \qquad (\mathrm{R}')$$

where q_w is the heat liberated at the walls per second, R is the radius of the RV and r is the radius of the capillary. For both heterogeneous and homogeneous contributions we have

$$\Delta T = q_g + q_w \cdot \frac{2r \ln (R/r)}{R+r} /4\pi\lambda \qquad (\mathrm{S}')$$

If the rates of the homogeneous and heterogeneous reactions are W_g and W_w respectively, the overall rate $W = W_g + W_w$.

If Q is the molar heat of reaction,

$$q_g = W_g Q \quad \text{and} \quad q_w = W_w Q.$$

If the heterogeneous contribution is denoted by α,

$$W_g = (1-\alpha)W \quad \text{and} \quad W_w = \alpha W.$$

Substituting for q_g and q_w in terms of Q, W and we have

$$\Delta T = \frac{WQ}{4}\left\{1-\alpha+\alpha\frac{2r}{r+R}\cdot\ln(R/r)\right\} \tag{T'}$$

Hence from experimental observations of ΔT and W, α may be calculated. Markevitch[332] found that, in the range 8–100 torr and 270–370° C using stoichiometric mixtures of hydrogen and chlorine for reaction (63)

$$H_2+Cl_2 = 2HCl \tag{63}$$

$\alpha = 0$, indicating that the process is completely homogeneous. The effect of oxygen on $\Delta T/W$ produced direct evidence that the chain reaction in this case is initiated at the RV walls.

4.5.6 *Ultrasonic absorption measurements*

This technique has also been used for the well-known equilibrium (62). Sound absorption results in a perturbation of the equilibrium, such that the dissociation constant can be measured[333]. The magnitude of the absorption coefficient per wavelength at the frequency of maximum absorption can be related to the extent of dissociation of N_2O_4.

4.5.7 *Gas density balance*

A gas density balance is shown in Fig. 71. A sealed bulb G at one end of a beam A is counterpoised by C. This is pivoted at F and enclosed in an envelope H. Gases of known (ρ_a) and unknown (ρ_b) densities are run separately into H until the pointer is zeroed at D in each case. Then, for pressures p_a and p_b required,

$$\rho_b = p_a \rho_a / p_b \tag{U'}$$

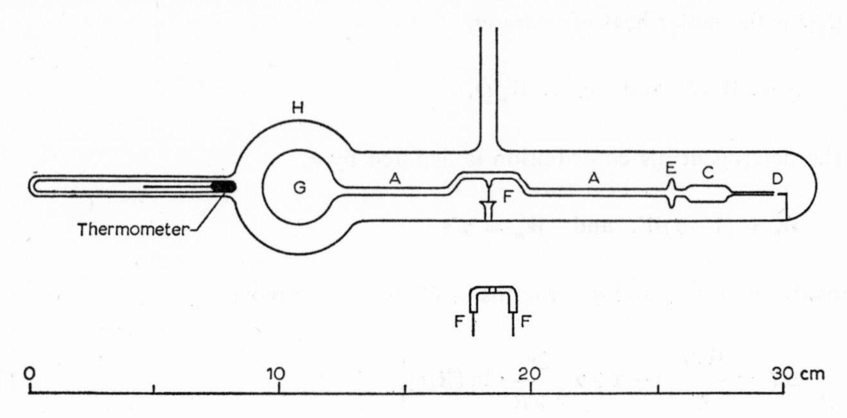

Fig. 71. Gas density balance. From ref. 22c.

Further details are given in ref. 22c. The instrument has long been superseded for conventional analysis by the mass spectrometer. However, it remains as an absolute control for gas chromatographic analysis.

4.5.8 Quartz fibre manometer

The damping of the oscillation of a quartz fibre depends on the pressure (p) and molecular weight (M) of the gas present. The time for the amplitude to decrease to one half of its original value $t_{\frac{1}{2}}$ is given by equation (V′)

$$t_{\frac{1}{2}} = A + BM^{\frac{1}{2}}p \tag{V′}$$

A and B are constants (B being characteristic of the gas present). A simple model is shown in Fig. 72. A gentle tap is required to set the fibre in vibration, the amplitude of which may be observed with a microscope carrying a scale. As the equation implies, the gauge may be used to measure pressure or molecular weight. It might be used, for example, to determine the extent of dimerisation of NO_2.

4.5.9 Interferometry and refractometry

Green and Robinson[334] have estimated small amounts of germane in hydrogen by interferometry. A parallel beam is split in two by a double slit and passed through two identical tubes, containing the gases to be compared, and then through two identical glass plates. The emergent beams are combined to form interference

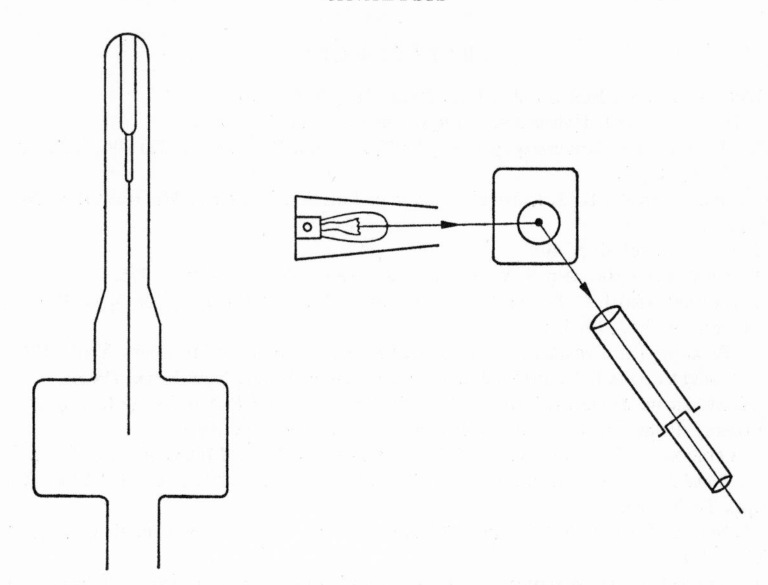

Fig. 72. Quartz fibre manometer. From ref. 22c.

fringes. For gases of different refractive index the band system is displaced sideways. The displacement can be compensated for by changing the inclination of one of the glass plates, which is attached to a micrometer screw with a measuring drum. The principles of operation of the interferometer have been described in refs. 22c and 335. A calibration curve has to be constructed. In the range of 0–15 % germane, the precision is 0.059 mole %. The method has also been used to analyse mixtures of acetylene and dimethyl ether in a flow system[336].

A photoelectric recording interferometer has been designed by Burnett *et al.*[337]. In this particular case, the temperature rise during a liquid-phase polymerisation caused a change in refractive index which could be related to the rate of the reaction. It was possible to measure a temperature change of 10^{-4} degrees at room temperature. This technique might well be adapted to study adiabatic reactions in the gas phase. Both refractive index measurement and interferometry are used extensively in combustion studies to determine temperature gradients and final temperatures[338].

4.5.10 Dielectric constant

Dielectric constants of substances vary widely and can be measured with considerable accuracy. Despite the potentialities of this technique for kinetic studies, it has been little used. (As an example, see Melville *et al.*[339], who followed vinyl polymerisation in the initial stages in this way.)

References pp. 104–111

REFERENCES

1 L. Batt and S. W. Benson, *J. Chem. Phys.*, 36 (1962) 895.
2 J. L. Hudson and J. Heicklen, *J. Phys. Chem.*, 71 (1967) 1518.
3 M. Letort, *Thesis*, University of Paris, 1937; *J. Chim. Phys.*, 34 (1937) 206; *Bull. Soc. Chim. France*, 9 (1942) 1.
4 For a discussion see K. J. Laidler, *Chemical Kinetics*, 2nd edn., McGraw-Hill, New York, 1965, p. 17.
5 K. J. Laidler, ref. 4, p. 408.
6 A. E. Axworthy, Jr. and S. W. Benson, *J. Chem. Phys.*, 27 (1957) 976.
7 K. J. Laidler and J. C. Polanyi, in *Progress in Reaction Kinetics*, Vol. 3, G. Porter (Ed.), Pergamon, Oxford, 1965, p. 18.
8 S. W. Benson, *The Foundation of Chemical Kinetics*, McGraw-Hill, New York, 1960.
9 J. G. Calvert and J. N. Pitts, Jr., *Photochemistry*, Wiley, New York, 1965.
10 N. N. Semenov, *Some Problems in Chemical Kinetics and Reactivity*, Vol. 2, (transl. by M. Boudart), Princeton Univ. Press, Princeton, N. J., 1958, p. 10.
11 V. I. Vedeneyev, L. V. Gurvich, V. N. Kondratiev, V. A. Medvedev and Y. L. Frankevich, *Bond Energies, Ionisation Potentials and Electron Affinities*, Arnold, 1966, (transl. by Scripta Technica).
12 D. R. Stull (Ed.), *JANAF Interim Thermochemical Tables*, Dow Chemical Co., Midland, Mich., 1961-66.
13 F. D. Rossini, D. D. Wagman, W. H. Evans, S. Levine and I. Jaffe, Selected Values of Chemical and Thermodynamic Properties, *N.B.S. Circular* 500, 1952.
14 K. K. Verma and L. K. Doraiswamy, *Ind. Eng. Chem. Fundamentals*, 4 (1965) 389.
15 B. H. Eckstein, H. A. Scheraga and E. R. Van Artsdalen, *J. Chem. Phys.*, 22 (1954) 28.
16 E. I. Hormats and E. R. Van Artsdalen, *J. Chem. Phys.*, 19 (1951) 778; F. E. Schweitzer and E. R. Van Artsdalen, *J. Chem. Phys.*, 19 (1951) 1028.
17 E. W. Swegler, H. A. Scheraga and E. R. Van Artsdalen, *J. Chem. Phys.*, 19 (1951) 135; H. R. Anderson, Jr., H. A. Scheraga and E. R. Van Artsdalen, *J. Chem. Phys.*, 21 (1953) 1258.
18 S. W. Benson and J. H. Buss, *J. Chem. Phys.*, 28 (1958) 301.
19 R. E. Rebbert and P. Ausloos, in ref. 9, p. 601.
20 C. W. Larson and H. E. O'Neal, *J. Phys. Chem.*, 70 (1966) 2475.
21 L. W. Siek, N. K. Blocher and J. H. Futrell, *J. Phys. Chem.*, 69 (1965) 888.
22 (a) R. S. Sanderson, *Vacuum Manipulation of Volatile Compounds*, Wiley, New York, 1948;
22 (b) R. E. Dodd and P. L. Robinson, *Experimental Inorganic Chemistry*, Elsevier, Amsterdam, 1957;
22 (c) H. W. Melville and B. G. Gowenlock, *Experimental Methods in Gas Reactions*, Macmillan, London, 1963;
22 (d) J. Yarwood, *High Vacuum Technique*, 4th edn., Chapman and Hall, London, 1967.
23 A. Maccoll, in *Technique of Organic Chemistry*, S. L. Friess, E. S. Lewis and A. Weissberger (Eds.), Vol. VIII, Part I, Interscience, New York, 1961, p. 429.
24 E. Raats, J. Harley and V. Pretorius, *J. Sci. Instr.*, 34 (1957) 510.
25 D. Verdin, *J. Sci. Instr.*, 43 (1966) 605.
26 M. Bodenstein, *Z. Physik. Chem.*, *B*, 7 (1930) 387.
27 H. C. Ramsperger, *Rev. Sci. Instr.*, 2 (1931) 738.
28 K. Judge and A. W. Luckey, *Rev. Sci. Instr.*, 24 (1953) 1150.
29 *e.g.* J. H. Knox and J. Riddick, *Trans. Faraday Soc.*, 62 (1966) 1190. For details see ref. 23, p. 430.
30 G. H. Gelb, B. D. Marcus and D. Dropkin, *Rev. Sci. Instr.*, 35 (1964) 80.
31 A. Kovács and R. B. Mesler, *Rev. Sci. Instr.*, 35 (1964) 485.
32 R. C. La Force, S. F. Ravitz and W. B. Kendall, *Rev. Sci. Instr.*, 35 (1964) 729.
33 W. D. Urry, *J. Am. Chem. Soc.*, 54 (1932) 3887.
34 R. M. Barrer, *J. Chem. Soc.*, (1934) 378.
35 J. E. Boggs, L. L. Ryan and L. L. Pak, *J. Phys Chem.*, 61 (1957) 825.
36 E. Heidemann, *Z. Physik. Chem.*, *A*, 164 (1933) 20.

37 H. J. Schumacher and G. Stieger, *Z. Physik. Chem.*, *B*, 7 (1930) 363.
38 G. Bauer, *Z. Physik. Chem.*, *A*, 174 (1935) 435.
39 G. J. Minkoff and C. F. H. Tipper, *Chemistry of Combustion Reactions*, Butterworths, London, 1962, p. 52.
40 W. Jost, *Explosion and Combustion Processes in Gases*, McGraw-Hill, New York, transl. by H. O. Croft, 1946.
41 For more details see ref. 22c, p. 362.
42 E. Daniels and F. Veltman, *J. Chem. Phys.*, 7 (1939) 756; D. Brearley, G. B. Kistiakowsky and C. H. Stauffer, *J. Am. Chem. Soc.*, 58 (1936) 43; A. Maccoll, *J. Chem. Soc.*, (1955) 965; D. H. R. Barton and K. E. Howlett, *J. Chem. Soc.*, (1949) 155; E. Swinbourne, *Australian J. Chem.*, 11 (1958) 314.
43 L. Batt and B. G. Gowenlock, *Trans. Faraday Soc.*, 56 (1960) 682.
44 K. W. Egger and S. W. Benson, *J. Am. Chem. Soc.*, 87 (1965) 3315.
45 F. O. Rice and K. F. Herzfeld, *J. Phys. & Colloid Chem.*, 55 (1951) 975.
46 W. L. Garstang and C. N. Hinshelwood, *Proc. Roy. Soc.* (*London*), *A*, 134 (1931) 1.
47 (a) M. L. Bagoyavlenskaya and A. A. Koval'skii, *Zhur. Fiz. Khim.*, 20 (1946) 1325; details are given in ref. 10, Vol. I.
47 (b) V. N. Kondrat'ev, *Chemical Kinetics of Gas Reactions*, transl. by J. M. Crabtree and S. N. Carruthers, translation edited by N. B. Slater, Pergamon, Oxford, 1964.
48 E. J. Harris and A. C. Egerton, *Proc. Roy. Soc.* (*London*), *A*, 168 (1938) 7; decomposition of $(EtO)_2$.
49 H. S. Johnston, ref. 8, p. 431; decomposition of HNO_3 at high temperatures.
50 S. W. Benson, *J. Chem. Phys.*, 22 (1954) 46 and ref. 8, p. 427.
51 F. Daniels and P. L. Veltman, *J. Chem. Phys.*, 7 (1939) 756; E. Swinbourne, *Australian J. Chem.*, 11 (1958) 314.
52 S. G. Foord, *J. Sci. Instr.*, 11 (1953) 126; D. H. R. Barton and K. E. Howlett, *J. Chem. Soc.*, (1949) 155.
53 M. Bodenstein and W. Dux, *Z. Physik. Chem.*, 85 (1913) 300.
54 E. J. Harris and A. C. Egerton, *Proc. Roy. Soc.* (*London*), *A*, 168 (1938) 1; R. J. L. Andon, J. D. Cox, E. F. G. Herington and J. F. Martin, *Trans. Faraday Soc.*, 53 (1957) 1074; J. D. Cox and R. J. L. Andon, 54 (1958) 1622.
55 Ref. 22c, p. 82; K. J. Laidler and B. J. Wojeiechowski, *Proc. Roy. Soc* (*London*), *A*, 259 (1960) 257.
56 (a) H. Kamphauson, *Glas und Instrumenten Technik*, 7 (1959) 423;
56 (b) R. Cooper and D. R. Stranks, *Technique of Inorganic Chemistry*, Vol. VI, H. B. Jonassen and A. Weissberger (Eds.), Wiley, New York, 1966.
57 A. R. Olsen and L. L. Hirst, *J. Am. Chem. Soc.*, 51 (1929) 2378; S. C. Lind and R. Livingston, *J. Am. Chem. Soc.*, 54 (1932) 94; J. C. Lilly, V. Legallais and R. Cherry, *J. Appl. Phys.*, 18 (1947) 613.
58 G. L. Pratt and J. H. Purnell, *Chem. Ind.* (*London*), (1960) 1080.
59 P. Dugleux and E. Frehling, *5th Intern. Symp. on Combustion, Pittsburgh, 1954*, Reinhold, New York, 1955, p. 491.
60 R. C. Tolman, *Phys. Rev.*, 23 (1924) 699; *The Principles of Statistical Mechanics*, Clarendon Press, Oxford, p. 163; see also ref. 4, p. 110.
61 F. A. Lindemann, *Trans. Faraday Soc.*, 17 (1922) 598.
62 S. W. Benson, *J. Chem. Phys.*, 40 (1964) 105; S. W. Benson and G. Haugen, *J. Phys. Chem.*, 69 (1965) 3898; J. N. Bradley and A. Ledwith, *J. Chem. Soc.*, *B*, (1967) 96.
63 Ref. 8, pp. 218–220.
64 R. A. Marcus and O. K. Rice, *J. Phys. & Colloid Chem.*, 55 (1951) 894.
65 T. L. Cottrell and J. C. McCoubrey, *Molecular Energy Transfer in Gases*, Butterworths, London, 1961.
66 A. F. Trotman-Dickenson, *Gas Kinetics*, Butterworths, London, 1955.
67 L. Batt and F. R. Cruickshank, *J. Phys. Chem.*, 70 (1966) 723.
68 J. H. Raley, F. R. Rust and W. E. Vaughan, *J. Am. Chem. Soc.*, 70 (1948) 88.
69 Ref. 39, pp. 1–57.
70 Ref. 8, p. 434.

71 O. K. RICE, *J. Chem. Phys.*, 8 (1940) 727.
72 A. D. FRANK-KAMENETSKII, *Diffusion and Heat Exchange in Chemical Kinetics*, Princeton Univ. Press, Princeton, N.J., 1955, translated from the Russian version of 1947.
73 L. BATT AND S. W. BENSON, *J. Chem. Phys.*, 38 (1963) 3031.
74 Ref. 4, p. 417.
75 (a) G. HADMAN, H. W. THOMPSON AND C. N. HINSHELWOOD, *Proc. Roy. Soc. (London), A*, 138 (1932) 297;
75 (b) C. N. HINSHELWOOD AND E. A. MOELWYN-HUGHES, *Proc. Roy. Soc (London), A*, 138 (1932) 311.
76 (a) H. W. THOMPSON AND C. N. HINSHELWOOD, *Proc. Roy. Soc. (London), A*, 122 (1929) 610;
76 (b) H. W. MELVILLE AND E. B. LUDLAM, *Proc. Roy. Soc. (London), A*, 132 (1931) 108.
77 O. OLDENBERG AND H. S. SOMMERS, JR., *J. Chem. Phys.*, 7 (1939) 279; 9 (1941) 114; A. H. WILLBOURN AND C. N. HINSHELWOOD, *Proc. Roy. Soc. (London), A*, 185 (1946) 353.
78 R. R. BALDWIN AND R. F. SIMMONS, *Trans. Faradoy Soc.*, 51 (1955) 680.
79 M. EUSUF AND K. J. LAIDLER, *Can. J. Chem.*, 42 (1964) 1861.
80 L. BATT AND F. R. CRUICKSHANK, *J. Chem. Soc., A*, (1967) 261.
81 S. CHAPMAN, *Phil. Trans., A*, 217 (1918) 115.
81 (a) B. K. MORSE, ref. 23, p. 548;
81 (b) R. LIVINGTON, ref. 23, p. 69.
82 M. SZWARC, *J. Chem. Phys.*, 16 (1948) 128.
83 D. KOPP, A. KOWALSKY, A. SAGOLIN AND N. SEMENOV, *Z. Physik. Chem., B*, 6 (1930) 307.
84 L. BATT, *Thesis*, Birmingham, 1959.
85 This and other types of circulating pumps are described in ref. 22c, pp. 45–51; see also D. KALLÓ, I. PRESZLER AND K. PAYER, *J. Sci. Instr.*, 41 (1964) 338.
86 E. WARHURST, *Trans. Faraday Soc.*, 35 (1939) 674.
87 S. W. BENSON AND G. N. SPOKES, *J. Am. Chem. Soc.*, 89 (1967) 2525.
88 J. D. MCKINLEY, *J. Chem. Phys.*, 40 (1964) 120.
89 M. GILBERT, *Combustion and Flame*, 2 (1958) 137, 149.
90 M. SZWARC, *Proc. Roy. Soc. (London), A*, 198 (1949) 267.
91 M. BODENSTEIN AND K. WOHLGAST, *Z. Physik. Chem.*, 14 (1908) 534.
92 V. G. GERVART AND D. A. FRANK-KAMENETSKII, *Bull. Acad. Sci. U.R.S.S., Classe sci. chim.*, (1942) 210.
93 T. J. HOUSER AND R. B. BERNSTEIN, *J. Am. Chem. Soc.*, 80 (1958) 4439.
94 E. S. LEWIS AND W. C. HERNDON, *J. Am. Chem. Soc.*, 83 (1961) 1955.
95 M. F. R. MULCAHY AND D. J. WILLIAMS, *Australian J. Chem.*, 14 (1961) 534.
96 J. DE GRAAF AND H. KWART, *J. Phys. Chem.*, 67 (1963) 1458.
97 W. C. HERNDON, M. B. HENLY AND J. M. SULLIVAN, *J. Phys. Chem.*, 67 (1963) 2842.
98 K. G. DENBIGH, *Chemical Reactor Theory*, Cambridge Univ. Press, 1966.
99 C. B. SHEPHERD AND C. E. LAPPE, *Ind. Eng. Chem.*, 31 (1939) 972.
100 J. P. LONGWELL AND M. A. WEISS, *Ind. Eng. Chem.*, 47 (1955) 1634; for specific details see ref. 95.
101 J. M. SULLIVAN AND T. S. HOUSER, *Chem. Ind. (London)*, (1965) 1057.
102 C. DRUCKER, E. JIMENO AND W. KANGRO, *Z. Physik. Chem.*, 90 (1915) 513; see ref. 22c, p. 74.
103 A. O. ALLEN, *J. Am. Chem. Soc.*, 56 (1934) 2053.
104 M. F. R. MULCAHY AND M. R. PETHARD, *Australian J. Chem.*, 16 (1963) 527.
105 *e.g.* L. BATT, B. G. GOWENLOCK AND J. TROTMAN, *J. Chem. Soc.*, (1960) 2222.
106 M. SZWARC, *Chem. Rev.*, 47 (1950) 75.
107 R. J. AKERS AND J. J. THROSSELL, *Trans. Faraday Soc.*, 63 (1967) 124.
108 K. J. LAIDLER AND B. W. WOJEIECHOWSKI, *Proc. Roy. Soc. (London), A*, 260 (1961) 91.
109 O. J. KLEJNOT AND J. E. WREEDE, *J. Vacuum Sci. Technol.*, 3 (1966) 288.
110 D. A. RICE AND J. ROACH, *J. Sci. Instr.*, 44 (1967) 473.
111 Ref. 22c, p. 231.
112 R. G. W. NORRISH AND J. H. PURNELL, *Proc. Roy. Soc. (London), A*, 243 (1958) 435.
113 D. J. LEROY, *Can. J. Research, B*, 28 (1950) 492.
114 Ref. 9, p. 764.
115 H. W. MELVILLE, *Trans. Faraday Soc.*, 32 (1936) 1525.

116 J. R. Dacy and J. W. Hodgkins, *Can. J. Research, B*, 28 (1950) 90.
117 L. J. Heidt and H. B. Boyles, *J. Am. Chem. Soc.*, 73 (1951) 5728.
118 H. E. White, *Introduction to Atomic Spectra*, McGraw-Hill, New York, 1934.
119 W. A. Noyes, Jr. and P. A. Leighton, *The Photochemistry of Gases*, Reinhold, New York, 1941.
120 A. C. G. Mitchell and M. W. Zemanchy, *Resonance, Radiation and Excited Atoms*, Cambridge Univ. Press, 1934, reprinted 1961.
121 H. E. Gunning and O. P. Strausz, Mercury photo-sensitisation, in *Advances in Photochemistry*, Vol. I, W. A. Noyes, Jr., G. S. Hammond and J. N. Pitts, Jr. (Eds.), Wiley, New York, 1963, p. 209.
122 T. Tako, *J. Phys. Soc. Japan*, 16 (1961) 2016.
123 R. Gomer and G. B. Kistiakowsky, *J. Chem. Phys.*, 19 (1951) 85.
124 Ref. 9, p. 695.
125 Ref. 22c, p. 277.
126 Ref. 9, p. 699.
127 S. G. Whiteway and C. R. Masson, *Can. J. Chem.*, 32 (1954) 1154.
128 P. Schultz, *Ann. Physik*, 1 (1947) 95.
129 W. A. Bann and L. Dunkelman, *J. Opt. Soc. Am.*, 40 (1950) 782.
130 J. T. Anderson, *J. Opt. Soc. Am.*, 41 (1951) 385.
131 K. J. Laidler, *The Chemical Kinetics of Excited States*, Oxford Univ. Press, 1955.
132 F. Briers and D. L. Chapman, *J. Chem. Soc.*, (1928) 1802.
133 (a) G. M. Burnett and H. W. Melville, in *Technique of Organic Chemistry*, Vol. VIII, Part. II, 2nd edn., *Investigation of Rates and Mechanisms of Reactions*, S. L. Friess, E. S. Lewis and A. Weissberger (Eds.), Interscience, New York, 1963, p. 1107.
133 (b) R. E. March and J. C. Polanyi, *Proc. Roy. Soc. (London), A*, 273 (1963) 360.
134 H. Kwart, H. S. Broadbent and P. D. Bartlett, *J. Am. Chem. Soc.*, 72 (1950) 1060.
135 *e.g.* W. C. Kreyl and R. A. Marcus, *J. Chem. Phys.*, 37 (1962) 419.
136 W. E. Vaughan and W. A. Noyes, Jr., *J. Am. Chem. Soc.*, 52 (1930) 559.
137 S. Robin and B. Vodar, *J. Phys. Radium*, 13 (1952) 671.
138 M. C. Sauer, Jr. and L. M. Dorfman, *J. Chem. Phys.*, 35 (1961) 497.
139 P. Harteck and F. Oppenheimer, *Z. Physik. Chem., B*, 16 (1932) 77.
140 W. Groth, *Z. Physik. Chem., B*, 37 (1937) 307.
141 W. Groth, *Z. Physik. Chem. (Frankfurt)*, 1 (1954) 300.
142 P. L. Hartman, *J. Opt. Soc. Am.*, 51 (1961) 113.
143 P. G. Wilkinson and Y. Tanaka, *J. Opt. Soc. Am.*, 45 (1955) 344.
144 P. G. Wilkinson, *J. Opt. Soc. Am.*, 45 (1955) 1044.
145 E. W. Schlag and F. J. Comes, *J. Opt. Soc. Am.*, 50 (1960) 866.
146 H. Okabe, *J. Opt. Soc. Am.*, 54 (1964) 478.
147 J. R. McNesby and H. Okabe, in *Advances in Photochemistry*, W. A. Noyes, Jr., G. S. Hammond and J. N. Pitts, Jr. (Eds.), Vol. III, 1964, p. 157.
148 G. Porter and J. I. Steinfeld, *J. Chem. Phys.*, 45 (1966) 3457.
149 N. J. Turro, *Molecular Photochemistry*, Benjamin, New York, 1965, p. 255.
150 W. V. Smith and P. P. Sorokin, *The Laser*, McGraw-Hill, New York, 1966.
151 A. K. Levine (Ed.), *Lasers*, Edward Arnold, London, 1966.
152 J. V. V. Kasper and G. C. Pimentel, *Phys. Rev. Letters*, 14 (1965) 352.
153 K. G. Anlauf, D. H. Maylotte, P. D. Pacey and J. C. Polanyi, *Phys. Letters*, 24A (1967) 208.
154 P. A. Leighton and G. S. Forbes, *J. Am. Chem. Soc.*, 51 (1929) 3549; P. A. Leighton and F. E. Blacet, *J. Am. Chem. Soc.*, 54 (1932) 3165.
155 J. W. Nicholas and F. F. Pollak, *Analyst*, 77 (1952) 49.
156 C. Christiansen, *Ann. Physik. Chem.*, 23 (1884) 298.
157 F. Weigert and H. Staude, *Z. Physik. Chem.*, 130 (1927) 607.
158 H. T. J. Chilton, *Spectrochim. Acta*, 16 (1960) 979.
159 C. A. Parker, *Proc. Roy. Soc. (London), A*, 220 (1953) 104; C. G. Hatchard and C. A. Parker, *Proc. Roy. Soc. (London), A*, 235 (1956) 518.
160 K. Porter and D. H. Volman, *Anal. Chem.*, 34 (1962) 748.

161 F. S. JOHNSON, K. WATANABE AND R. TOUSEY, *J. Opt. Soc. Am.*, 41 (1951) 702.
162 L. J. STIEF AND P. AUSLOOS, *J. Phys. Chem.*, 65 (1961) 877.
163 D. A. BECKER, H. OKABE AND J. R. MCNESBY, *J. Phys. Chem.*, 69 (1965) 538.
164 H. ESSEX, *J. Phys. Chem.*, 58 (1954) 42.
165 P. AUSLOOS AND R. GORDON, JR., *J. Chem. Phys.*, 40 (1964) 3599; G. G. MEISELS AND T. J. SWORSKI, *J. Phys. Chem.*, 69 (1965) 2867; H. H. CARMICHAEL, R. GORDON AND P. AUSLOOS, *J. Chem. Phys.*, 42 (1965) 343.
166 A. J. SWALLOW, *Radiation Chemistry of Organic Compounds*, Pergamon, London, 1960.
167 J. W. T. SPINKS AND R. J. WOODS, *An Introduction to Radiation Chemistry*, Wiley, London, 1964.
168 A. CHARLESBY (Ed.), *Radiation Sources*, Pergamon, London, 1964.
169 S. C. LIND, *Radiation Chemistry of Gases*, A. C. S. Monograph series, Chapman and Hall, London, 1961.
170 F. WILLIAMS, *Quart. Rev. (London)*, 17 (1963) 101.
171 A. V. TOPCHIEV (Ed.), *Radiolysis of Hydrocarbons*, Engl. edition edited by R. A. HOLROYD, Elsevier, Amsterdam, 1964.
172 E. COLLINSON, *Ann. Repts. Progr. Chem. (Chem. Soc. London)*, 62 (1965) 79.
173 P. AUSLOOS, *Ann. Rev. Phys. Chem.*, 17 (1966) 205.
174 C. F. SMITH, B. G. CORMAN AND F. W. LAMPE, *J. Am. Chem. Soc.*, 83 (1961) 3559.
175 S. G. LIAS AND P. AUSLOOS, *J. Chem. Phys.*, 43 (1965) 2748.
176 M. S. B. MUNSON, J. L. FRANKLIN AND F. H. FIELD, *J. Phys. Chem.*, 68 (1964) 3098.
177 L. W. SIECK AND J. H. FUTRELL, *J. Chem. Phys.*, 45 (1966) 560.
178 F. W. LAMPE, J. L. FRANKLIN AND F. H. FIELD, in *Progress in Reaction Kinetics*, Vol. I, G. PORTER (Ed.), Pergamon, Oxford, 1961, p. 69.
179 Ref. 169, p. 212.
180 G. SAXON, ref. 168, p. 220.
181 C. D. BOPP AND W. W. PARKINSON, JR., ref. 168, p. 1.
182 For a review see R. WOLFGANG, in *Progress in Reaction Kinetics*, Vol. 3, G. PORTER (Ed.), Pergamon, Oxford, 1965, p. 97.
183 E. P. RACK AND A. A. GORDUS, *J. Phys. Chem.*, 65 (1961) 944.
184 D. E. LEA, *Actions of Radiation on Living Cells*, Cambridge Univ. Press, 1946, Chapter 1.
185 M. BURTON, *Discussions Faraday Soc.*, 12 (1952) 317.
186 R. A. BACK, T. W. WOODWARD AND K. A. MCLAUCHLAN, *Can. J. Chem.*, 40 (1962) 1380.
187 H. FRICKE AND S. MORSE, *Am. J. Roentgenol. Radium Therapy*, 18 (1927) 430.
188 J. M. RAMARADHYA AND G. R. FREEMAN, *J. Chem. Phys.*, 34 (1961) 1726.
189 S. C. LIND, D. C. BARDWELL AND J. H. PERRY, *J. Am. Chem. Soc.*, 48 (1926) 1556; L. M. DORFMAN AND F. J. SHIPKO, *J. Am. Chem. Soc.*, 77 (1955) 4723; C. ROSENBLUM, *J. Phys. Chem.*, 52 (1948) 474.
190 F. W. LAMPE, *J. Am. Chem. Soc.*, 79 (1957) 1055.
191 V. O. JANSSEN, A. HENGLEIN AND D. PERNER, *Z. Naturforsch.*, 19B (1964) 1005.
192 G. R. A. JOHNSON, *J. Inorg. Nucl. Chem.*, 24 (1962) 461.
193 R. A. LEE, R. S. DAVIDOW AND D. A. ARMSTRONG, *Can. J. Chem.*, 42 (1964) 1906.
194 G. G. MEISELS, *J. Chem. Phys.*, 41 (1964) 51.
195 L. A. SAMSON, *J. Opt. Soc. Am.*, 54 (1964) 6.
196 (a) J. N. BOAG, in *Radiation Dosimetry*, G. J. HINE AND G. L. BROWNELL (Eds.), Academic Press, New York, 1956, Chapter 4;
196 (b) J. S. LAUGHLIN, Ref. 196a, Chapter 13.
197 A. C. BIRGE, M. O. ANGER AND C. A. TOBIAS, Ref. 196a, Chapter 14.
198 G. W. REID (Ed.), *Radiation Dosimetry*, Proc. Int. School Phys. "Enrico Fermi", Varina, Italy, 1964.
199 G. R. A. JOHNSON AND J. M. WARMAN, *Trans. Faraday Soc.*, 61 (1965) 1709.
200 R. D. DACKER AND P. J. AUSLOOS, *J. Chem. Phys.*, 42 (1965) 3746.
201 S. O. THOMPSON AND O. A. SCHAEFFER, *J. Am. Chem. Soc.*, 80 (1958) 553; *Radiation Res.*, 10 (1959) 671.
202 V. AQUILANTI, *J. Phys. Chem.*, 69 (1965) 3434.
203 P. KEBARLE AND A. M. HOGG, *J. Chem. Phys.*, 42 (1965) 668.

204 H. W. MELVILLE AND B. G. GOWENLOCK, ref. 22c, p. 223.
205 For a review see G. J. MINKOFF AND C. F. H. TIPPER, ref. 39, p. 89 and sequel. For particular analyses, the reader should consult current literature on combustion.
206 C. E. BRICKER AND H. R. JOHNSON, *Ind. Eng. Chem. (Anal. Ed.)*, 17 (1945) 400.
207 M. PESEZ AND J. BARTOS, *Bull. Soc. Chim. France*, (1960) 481.
208 C. S. WISE, S. L. MEHLTRETTER AND J. W. VAN CLEVE, *Anal. Chem.*, 31 (1959) 1241.
209 H. E. O'NEAL AND S. W. BENSON, *J. Chem. Phys.*, 36 (1962) 2196.
210 L. BATT AND F. R. CRUICKSHANK, *Talanta*, 14 (1967) 245.
211 W. WEST, *J. Am. Chem. Soc.*, 57 (1935) 1931; *Ann. N.Y. Acad. Sci.*, 41 (1941) 238.
212 E. J. ROSENBAUM AND T. R. HOGNESS, *J. Chem. Phys.*, 2 (1934) 267.
213 E. PATAT, *Z. Physik. Chem., B*, 32 (1936) 274, 294.
214 L. FARKAS AND P. HARTECK, *Z. Physik. Chem., B*, 25 (1934) 257.
215 See S. W. BENSON, ref. 8, p. 109.
216 F. PANETH, see ref. 8, p. 100 for further details.
217 P. G. ASHMORE, *Catalysis and Inhibition of Chain Reactions*, Butterworths, London, 1963.
218 B. G. GOWENLOCK, in *Progress in Reaction Kinetics*, G. PORTER (Ed.), Vol. 3, 1965, p. 173.
219 L. A. K. STAVELEY AND C. N. HINSHELWOOD, *J. Chem. Soc.*, (1936) 812, 818.
220 C. S. COE AND T. F. DOUMANI, *J. Am. Chem. Soc.*, 70 (1948) 1516.
221 D. E. HOARE, *Can. J. Chem.*, 40 (1962) 2012.
222 L. BATT AND B. G. GOWENLOCK, *Trans. Faraday Soc.*, 56 (1960) 682.
223 H. A. TAYLOR AND H. BENDER, *J. Chem. Phys.*, 9 (1964) 761.
224 J. SCRIVENOR, *E.R.D.E. Report 1/R/56*, Ministry of Supply.
225 M. EUSUF AND K. J. LAIDLER, *Can. J. Chem.*, 42 (1964) 1861.
226 W. A. BRYCE AND J. CHRYSOCHOOS, *Trans. Faraday Soc.*, 59 (1962) 1842.
227 D. L. COX, R. A. LIVERMORE AND L. PHILLIPS, *J. Chem. Soc., B*, (1966) 245.
228 F. O. RICE AND O. L. POLLY, *J. Chem. Phys.*, 6 (1938) 273.
229 J. R. MCNESBY AND A. S. GORDON, *J. Am. Chem. Soc.*, 79 (1957) 825; A. S. GORDON, S. R. SMITH AND J. R. MCNESBY, *J. Am. Chem. Soc.*, 81 (1959) 5059.
230 Ref. 218, pp. 174, 198.
231 E. HORN, M. POLANYI AND D. W. G. STYLE, *Trans. Faraday Soc.*, 30 (1934) 189.
232 For a review, see E. W. R. STEACIE, *Atomic and Free Radical Reactions*, A.C.S. Monograph No. 125, 2nd edn., Reinhold, New York, 1954.
233 M. C. FLOWERS, L. BATT AND S. W. BENSON, *J. Chem. Phys.*, 37 (1962) 2662.
234 L. BATT AND F. R. CRUICKSHANK, *J. Phys. Chem.*, 71 (1967) 1836.
235 N. N. SEMENOV, *Some Problems in Kinetics and Reactivity*, Vol. 2, (transl. by M. BOUDART), Princeton Univ. Press, Princeton, N.J., 1959, p. 65.
236 A. M. HOGG AND P. KEBARLE, *J. Am. Chem. Soc.*, 86 (1964) 4559.
237 N. IMAI AND O. TOYAMA, *Bull. Chem. Soc. Japan*, 34 (1961) 328.
238 K. H. ANDERSON AND S. W. BENSON, *J. Chem. Phys.*, 39 (1963) 1677.
239 L. BATT AND F. R. CRUICKSHANK, unpublished work.
240 C. F. CULLIS AND K. FRIDAY, *Proc. Roy. Soc. (London), A*, 224 (1954) 308.
241 *e.g.* D. J. WADDINGTON, *Proc. Roy. Soc. (London), A*, 265 (1962) 436.
242 P. G. ASHMORE, *5th Intern. Symp. on Combustion, Pittsburgh, 1954*, Reinhold, New York, 1955, p. 700.
243 See ref. 8, p. 438 for a discussion.
244 A. MACCOLL, ref. 23, p. 485.
245 M. SEAKIN, *Proc. Roy. Soc.* (LONDON), *A*, 274 (1963) 413; 277 (1964) 279.
246 C. F. CULLIS, A. FISH AND R. B. WARD, *Proc. Roy. Soc. (London), A*, 276 (1963) 527.
247 E. R. ALLAN AND C. F. H. TIPPER, *Proc. Roy. Soc. (London), A*, 258 (1960) 251.
248 For details see *J. Am. Chem. Soc.*, 88 (1966) 236, 241, 650, 3194, 3196, 3480, 4570.
249 M. C. LIN AND K. J. LAIDLER, *Can. J. Chem.*, 45 (1967) 1315.
250 (a) H. M. FREY, *Progress in Reaction Kinetics*, Vol. 2, G. PORTER (Ed.), Pergamon, London, 1964, p. 131.
250 (b) B. S. RABINOVITCH AND D. W. SETSER, in *Advances in Photochemistry*, Vol. 3, W. A. NOYES, JR., G. S. HAMMOND AND J. N. PITTS, JR. (Eds.), Interscience, New York, 1964, p. 1.
251 R. J. CVETANOVIĆ, *Progress in Reaction Kinetics*, G. PORTER (Ed.), Vol. 2, 1964, p. 39.

252 O. STERN AND M. VOLMER, *Physik. Z.*, 20 (1919) 183.
253 K. J. LAIDLER, *J. Chem. Phys.*, 10 (1942) 43; 10 (1942) 34; 15 (1947) 712.
254 A. R. KNIGHT AND H. E. GUNNING, *Can. J. Chem.*, 41 (1963) 2849.
255 Ref. 39; D. E. HOARE AND G. S. PEARSON, in *Advances in Photochemistry*, Vol. 3, W. A. NOYES, JR., G. S. HAMMOND AND J. N. PITTS, JR. (Eds.), Interscience, New York, 1964, p. 83.
256 R. B. CUNDALL AND T. F. PALMER, *Trans. Faraday Soc.*, 56 (1960) 1211.
257 R. B. CUNDALL, *Progress in Reaction Kinetics*, Vol. 2, G. PORTER (Ed.), Pergamon, Oxford, 1964, p. 210.
258 J. N. BRADLEY AND A. LEDWITH, *J. Chem. Soc.*, *B*, (1967) 96.
259 Ref. 8, p. 108.
260 Ref. 9, p. 595.
261 For reviews see ref. 250 (b) and W. H. SAUNDERS JR., Kinetic isotope effects, in *Technique of Organic Chemistry*, 2nd edition, Vol. VIII, Part I, *Investigation of Rates and Mechanisms of Reactions*, S. L. FRIESS, E. S. LEWIS AND A. WEISSBERGER (Eds.), Interscience, New York, 1961, p. 389.
262 A. T. BLADES, P. W. GILDERSON AND M. G. H. WALLBRIDGE, *Can. J. Chem.*, 40 (1962) 1526.
263 S. KATZ AND J. T. BARR, *Anal. Chem.*, 25 (1953) 19.
264 (a) F. KAUFMAN, *Proc. Roy. Soc. (London)*, *A*, 247 (1958) 123; (b) a review is given by I. M. CAMPELL AND B. A. THRUSH, *Ann. Repts. Progr. Chem. (Chem. Soc. London)*, 62 (1965) 17.
265 M. A. A. CLYNE AND B. A. THRUSH, *Trans. Faraday Soc.*, 57 (1961) 1305.
266 M. A. A. CLYNE AND B. A. THRUSH, *Proc. Roy. Soc. (London)*, *A*, 261 (1961) 259.
267 E. A. OGRYZLO AND M. I. SCHIFF, *Can. J. Chem.*, 37 (1959) 1690; M. A. A. CLYNE AND B. A. THRUSH, *Proc. Roy. Soc. (London)*, *A*, 269 (1962) 404; F. KAUFMAN AND J. R. KELSO, *Symposium on Chemiluminescence, Duke University, 1965*, see ref. 264b, p. 30.
268 A. J. P. MARTIN AND R. L. M. SYNGE, *Biochem. J.*, 35 (1941) 1358.
269 A. J. P. MARTIN AND A. T. JAMES, *Biochem. J.*, 50 (1952) 679; *Analyst*, 77 (1952) 915.
270 For a list see ref. 22c, p. 256.
271 J. C. GIDDINGS, *Dynamics of Chromatography*, Vol. 1, *Principles and Theory*, Arnold, London, 1965.
272 *Gas Chromatography Abstracts*, Butterworths, 1958.
273 J. C. GIDDINGS AND R. A. KELLER (Eds.), *Advances in Chromatography*, Arnold, London, Vol. 1, 1966.
274 H. DESTY, ref. 271 p. 199.
275 J. H. PURNELL, *Gas Chromatography*, Wiley, London, 1962, p. 233.
276 J. H. KNOX, *Gas Chromatography*, Methuen, London, 1962, p. 44.
277 J. H. PURNELL, ref. 275, p. 367.
278 W. E. HARRIS AND H. W. HABGOOD, *Programmed Temperature Gas Chromatography*, Wiley, New York, 1966.
279 L. MIKKELSEN, in Vol. 2 of ref. 273.
280 K. PORTER AND R. G. W. NORRISH, *Proc. Roy. Soc. (London)*, *A*, 272 (1963) 164.
281 S. SANDLER AND J. A. BEECH, *Can. J. Chem.*, 38 (1960) 1455.
282 G. L. PRATT AND J. H. PURNELL, *Anal. Chem.*, 32 (1960) 1213.
283 R. B. CUNDALL, K. HAY AND P. W. LEMEUNIER, *J. Sci. Instr.*, 43 (1966) 652.
284 L. BATT AND F. R. CRUICKSHANK, *J. Chromatog.*, 21 (1966) 296.
285 R. A. OGG, JR., *Discussions Faraday Soc.*, 17 (1954) 99.
286 Ref. 39, p. 93.
287 C. A. MCDOWELL AND J. H. THOMAS, *Trans. Faraday Soc.*, 46 (1950) 1030.
288 D. M. GOLDEN, R. WALSH AND S. W. BENSON, *J. Am. Chem. Soc.*, 87 (1965) 4053.
289 F. A. MORSE AND F. KAUFMAN, *J. Chem. Phys.*, 42 (1965) 1785.
290 J. K. CASHION AND J. C. POLANYI, *Proc. Roy. Soc. (London)*, *A*, 258 (1960) 529.
291 G. L. SIMARD, J. STEGER, T. MARINER, D. J. SALLEY AND V. Z. WILLIAMS, *J. Chem. Phys.*, 16 (1948) 836.
292 A. GUTTMAN AND S. S. PENNER, *J. Chem. Phys.*, 36 (1962) 98.
293 G. R. MCMILLAN, J. G. CALVERT AND S. S. THOMAS, *J. Phys. Chem.*, 68 (1964) 116.
294 W. GUILLORY AND H. S. JOHNSTON, *J. Phys. Chem.*, 67 (1963) 1695.
295 F. J. DILLEMUTH, D. R. SKIDMORE AND C. C. SCHUBERT, *J. Phys. Chem.*, 64 (1960) 1496.

296 R. Burt and G. J. Minkoff, *Anal. Chim. Acta*, 16 (1957) 259.
297 A. Carrington and D. H. Levy, *Symposium on E.S.R. Spectroscopy, Michigan State Univ., Aug. 1966*; see *J. Phys. Chem.*, 71 (1967) 2.
298 A. A. Westenberg and N. de Haas, *J. Chem. Phys.*, 40 (1964) 3087.
299 J. E. Bennett and D. R. Blackmore, *Proc. Roy. Soc. (London), A*, 305 (1968) 553.
300 A. A. Westenberg and N. de Haas, *J. Chem. Phys.*, 46 (1967) 490.
301 M. F. R. Mulcahy, J. R. Steven and J. C. Ward, *J. Phys. Chem.*, 71 (1967) 2124.
302 A. B. Nalbandyan, *Russ. Chem. Rev. (English Transl.)*, (1966) 243.
303 S. Krongelb and M. W. P. Strandberg, *J. Chem. Phys.*, 31 (1959) 1196.
304 J. H. Beynon, *Mass Spectrometry and its Application to Organic Chemistry*, Elsevier, Amsterdam, 1960.
305 R. I. Reed, *Applications of Mass Spectrometry to Organic Chemistry*, Academic Press, London, 1966.
306 J. D. Waldron (Ed.), *Advances in Mass Spectrometry*, Pergamon, Oxford, 1959.
307 R. I. Reed (Ed.), *Mass Spectrometry*, Academic Press, London, 1965.
308 J. Cuthbert, *Quart. Rev. (London)*, 13 (1959) 215.
309 C. J. M. Fletcher and G. K. Rollefson, *J. Am. Chem. Soc.*, 58 (1936) 2135.
310 F. P. Lossing, K. V. Ingold and A. W. Tickner, *Discussions Faraday Soc.*, 14 (1953) 34; F. P. Lossing and A. W. Tickner, *J. Chem. Phys.*, 20 (1952) 907.
311 G. C. Eltenton, *J. Chem. Phys.*, 15 (1947) 455; *J. Phys. & Colloid Chem.*, 52 (1948) 463.
312 F. P. Lossing, in *Mass Spectrometry*, C. A. McDowell (Ed.), McGraw-Hill, New York, 1963, p. 442.
313 L. P. Blanchard, J. B. Farmer and C. Ouellet, *Can. J. Chem.*, 35 (1957) 115.
314 F. W. Lampe, J. L. Franklin and F. H. Field, *Progress in Reaction Kinetics*, G. Porter (Ed.), Vol. 1, Pergamon, Oxford, 1961, p. 67.
315 F. H. Field and M. S. B. Munson, *J. Am. Chem. Soc.*, 87 (1965) 3289.
316 D. R. Crosley and R. Bersohn, *J. Chem. Phys.*, 45 (1966) 4253; D. R. Crosley, *J. Chem. Phys.*, 47 (1967) 1361.
317 G. J. Minkoff and H. Bruschweiler, *Anal. Chim. Acta*, 12 (1955) 186; *Nature*, 172 (1953) 909.
318 D. A. Skooy and A. B. H. Lauwzecha, *Anal. Chem.*, 28 (1956) 825; P. F. Urone, *Dissertation Abstr.*, 20 (1960) 2595.
319 W. B. Guenther and W. D. Walters, *J. Am. Chem. Soc.*, 73 (1951) 2127; G. McDonald, N. M. Lodge and W. D. Walters, *J. Am. Chem. Soc.*, 73 (1951) 1757.
320 K. H. Mueller, *J. Am. Chem. Soc.*, 77 (1955) 3459.
321 C. J. Harding, A. Maccoll and R. A. Ross, *Chem. Comm.*, (1967) 289.
322 R. D. Goodwin, *Anal. Chem.*, 25 (1953) 263.
323 Ref. 22c, p. 250.
324 B. N. Srivastava and A. K. Barna, *J. Chem. Phys.*, 35 (1961) 329.
325 G. T. Morgan, R. Taylor and T. J. Hedley, *J. Soc. Chem. Ind.*, 47 (1928) 117T.
326 M. Vanpée, *Bull. Soc. Chim. Belg.*, 62 (1953) 468.
327 Ref. 8, p. 114.
328 L. Elias, E. A. Ogryzlo and H. I. Schiff, *Can. J. Chem.*, 37 (1959) 1680.
329 A. B. Callear and J. C. Robb, *Discussions Faraday Soc.*, 17 (1954) 173.
330 F. Moseley and J. C. Robb, *Proc. Roy. Soc. (London), A*, 243 (1957) 119.
331 F. Moseley and J. C. Robb, *Proc. Roy. Soc. (London), A*, 243 (1957) 130.
332 A. M. Markevitch, *Zh. Fiz. Khim.*, 22 (1948) 941.
333 M. Cher, *J. Chem. Phys.*, 37 (1962) 2564.
334 M. Green and P. H. Robinson, *Anal. Chem.*, 25 (1953) 1913.
335 C. Candler, *Modern Interferometers*, Hilger and Watts, London, 1951.
336 A. Magnus and A. Kraus, *Z. Physik. Chem., A*, 158 (1932) 183.
337 G. M. Burnett, P. J. Deas and H. W. Melville, *Discussions Faraday Soc.*, 17 (1954) 173.
338 Ref. 39, p. 264.
339 C. M. Burrell, T. G. Majury and H. W. Melville, *Proc. Roy. Soc. (London), A*, 205 (1951) 309, 323, 496.
340 *Radiation Research Reviews*, G. O. Phillips, R. B. Cundall and F. S. Dainton (Eds.), Elsevier Publishing Company, Amsterdam, Vol. 1, 1968.

Chapter 2

Experimental Methods for the Study of Fast Reactions

D. N. HAGUE

Introduction

What is a fast chemical reaction? Sometimes the answer is given as a reaction with a rate coefficient larger than a certain specified value, or one with an activation energy less than a certain amount, but probably the most convenient definition is one whose kinetics cannot be followed by "traditional" means, *e.g.*, by a line system with pressure gauges for a gas reaction or with a pipette and stopwatch for solution reactions. At any rate, that is the definition which will be used here, and in this chapter will be considered methods developed, mostly in the last twenty years, for following reactions variously described in the older literature as "instantaneous", "too fast to measure", and the like. For most reactions, the half-life is dependent on the initial concentrations of the reactants and so it is often possible to slow them down, by diluting, by cooling, or both, sufficiently to make them amenable to study by "normal" methods. Such systems have not been considered here; all the methods discussed have been developed specifically for fast reactions and most are unsuitable for following reactions with half-lives in the range of minutes or longer.

Because of the correlation between rate coefficient and activation energy, the reactions which are classified as fast tend to belong to certain categories. Thus, for example, fast gas-phase reactions usually involve free radicals or atoms—a fact which increases the problems considerably since it is very difficult to generate radicals of known concentration, mix them with the other reactant and then monitor the concentrations of radicals and products at known time intervals after the start of the reaction. In solution the upper limit to the reaction velocity is governed by the rate at which the partners can diffuse together; gas reactions are not so frequently diffusion-controlled because of the considerably higher diffusion coefficients in the gas phase. The fastest reaction studied in aqueous solution is the neutralization reaction, formally

$$H^+ + OH^- \rightleftarrows H_2O$$

This has a forward rate coefficient[1] at 25 °C of 1.4×10^{11} mole. l^{-1}. sec^{-1}. It has this high value because of the special diffusion mechanism available for the proton and defect-proton in water associated with the latter's strongly hydrogen-bonded structure. Interestingly, the rate coefficient for this reaction is larger in ice than in liquid water[2]. Other fast reactions in water are often between ions, which are heavily solvated. The nature of the solvent is very important in determining the rate coefficient of a liquid-phase reaction and it is in many ways somewhat unfortunate that the common solvent, water, is apparently one of the most complex. This situation is not likely to reduce the number of kinetic studies performed with aqueous solutions but rather to act as a spur for further investigations into its detailed nature and properties. An advantage of having a large excess of solvent molecules surrounding the reacting particles is that there is usually no difficulty in dissipating the heat of reaction; in gas-phase reactions this is often a serious problem.

The problems which must be overcome in any method for following fast reactions may be grouped under two headings. In the first place, the reactions must be initiated homogeneously and in a time which is short compared with the reaction half-life. In traditional kinetic investigations the reaction will generally be started by mixing the two reactive components. For reactions which are complete in less than a few milliseconds this technique is not appropriate because the mixing time cannot be reduced much below this value. Secondly, the course of the reaction must be followed by recording some varying parameter of the system at known times after initiation of the reaction. Most of the techniques described here avoid the problem of mixing. Rather than dividing the methods according to whether they may be used for gas or solution reactions, it is convenient to group them according to the way in which they attempt to deal with these two problems. The first few methods use what has been termed the "perturbation" method: the equilibrated system is subjected to a sudden perturbation and its rapid readjustment is followed directly on a necessarily very short time scale. The second group uses the "competition" method : the system is disturbed by some physical process (often periodic in nature) which competes with the chemical reaction. Information about the rates of the chemical processes can be deduced if the physical process is quantitatively characterized. Unfortunately, however, not all the methods lend themselves to quite such a rigid demarcation. The dividing line may be drawn at about Section 3 (Chemical relaxation methods) although others (*cf.* Section 7—Flow methods) include variants from both categories.

Flash photolysis is one of the few methods for studying the kinetics of fast reactions which have been applied with equal success for both reactions in the gas and in the liquid phase. It was originally developed by Norrish and Porter for following gas reactions[3], and was first used for solution reactions[4] in 1954; it is applicable to reactions which are initiated by light. The reactants are subjected to a light flash of very high intensity in a region of the spectrum where at least one of the species is photosensitive. The type of excitation produced by the absorbed light differs from case to case, but essentially the method involves following the fate of the excited species by means of subsequent light absorption or emission.

Pulse radiolysis is very similar in principle. It may be regarded as the "radiation" equivalent of flash photolysis, in which a pulse of ionizing radiation replaces the light flash. Its application so far has been mostly in the field of solution kinetics, principally aqueous reactions, although it has recently been applied to gaseous systems as well[5]. It has proved to be an excellent way of following the reactions of the aqueous electron, and many of the most interesting results have come in this field. In Section 1 flash photolysis and pulse radiolysis are discussed together. The application of the former method to a couple of systems and of the latter to some studies on the hydrated electron are also considered.

Shock tube experiments have been carried out to a limited extent with condensed phase systems but they have been used primarily to follow the kinetics of very fast reactions in the gas phase. The reactants are contained in a long narrow tube and subjected to a very sudden, large increase in pressure. This is accompanied by a sudden temperature rise, which is sufficient to excite the components and often cause dissociation, typically into free radicals. The subsequent behaviour of these radicals can be followed. From the point of view of trying to obtain a complete understanding of the elementary reactions, it is important to inquire into the mechanisms of energy relaxation between the various degrees of freedom of the excited species formed. Work along these lines has been done, but in Section 2 the discussion is limited to the use of shock tubes for following the actual chemical reactions.

One of the principal difficulties in following reactions with half-lives between about 10^{-3} and 10^{-12} sec, then, is to initiate the reaction in a sufficiently short time. The problem is overcome in relaxation techniques by starting with a solution in which the reactants are already thoroughly mixed and equilibrated. Such a system may be used for all the commonly accepted "equilibrium" reactions, but it may also be used for investigating a stationary state in which there is no net reaction. Under a given set of conditions, for example defined concentrations, temperature and electric field-strength, the position of the system, which for convenience will be referred to as an equilibrium, is defined. If one of the external parameters (*e.g.* temperature or electric field-strength) is now altered, a new equilibrium state will generally result. If the perturbation is performed sufficiently faster than the chemical shift can occur, the latter can be observed independently. This type of delayed response is called *relaxation*, and the term can be applied to many effects (*e.g.* the delay in reorientation of dipoles in a changing electric field). When the delay between cause and effect is associated with the finite time taken for one or more chemical components to react, the process is called *chemical relaxation*. From an analysis of this behaviour it is possible to determine a characteristic time—the relaxation time, τ—which may be related to the rate coefficients for the reactions involved in the equilibrium and to the equilibrium concentrations of the reactants. In Section 3 the two main categories of chemical elaxation techniques are discussed. They are the transient or step-function methods in which the approach to the new equilibrium position after a single small jump is followed, and the stationary methods in which the interaction is measured between the chemical system and an oscillating constraint.

A technique which is finding increasing application in solution reactions is the use of *spectral line-broadening*. Work has been done especially with NMR spectra, but the ideas used here have been also applied to ESR and Raman signals. The reactions which may be studied in this way are somewhat restricted in scope since their half-lives must be of the same order of magnitude as the relaxation times of the physical process involved. But where they can be used, these methods are often powerful; for example, NMR line-broadening can be used for reactions, such as the exchange of solvent between the coordination shell of a metal and the bulk medium, which involve no net chemical change. These reactions are very difficult to study by other methods. In Section 4 the emphasis is placed on the use of NMR because most of the work in this area has been done using this technique; it also illustrates the principles rather well.

In what is perhaps the competition method *par excellence*, the process against which the chemical reaction is pitted is the *fluorescence* of one of the reactants. Absorption of low-wavelength light by a solution of a species A may produce excited molecules A* which can return to their ground states by the emission of fluorescence. If the illumination is continuous and of constant intensity, a steady state situation will arise in which the rate of formation of A* molecules is balanced by the rate of their deactivation. Under these conditions the intensity of fluorescence is proportional to the steady-state concentration of A*. The excited state of A (the first excited singlet) is attained within about 10^{-12} sec of the interaction of A with the photon, and so the lower time limit of reactions which can be studied by this method lies near what might be loosely described as the boundary of "chemical" reactions. The upper time limit is the average lifetime of the fluorescing species, which is about 10^{-8} sec. In the absence of a chemical reaction A* loses its excess energy by various processes, the most important of which is frequently deactivation by solvent molecules. It is possible to relate the rate coefficients of the fluorescence quenching reactions with the fluorescence intensities and the concentration of the quenching species; the way in which this is done is considered in Section 5. Because of the requirement that one species fluoresce and the restrictions on the reaction half-life, this method is not one of the most generally useful for following fast reactions in solution. It has, however, yielded very important results among the fastest of the chemical reactions.

The type of reaction which can be studied by *electrochemical methods* is limited in several respects. The systems must be good electrical conductors (which has tended to limit them to aqueous solutions to which large amounts of inert electrolyte have been added), one of the species involved must be electrolytically reducible at conveniently applied potentials and the rate coefficients must fall within certain boundaries. The electrolytic reduction of a species in a reaction system where these

References pp. 176–179

conditions are obeyed will have different current-voltage characteristics from the reduction process of the same species in the absence of a reaction. The way in which these characteristics change may be used to deduce the relevant rate coefficients. Electrochemical methods are considered in Section 6.

One of the first methods for following fast chemical reactions was developed in the early 1920's to try to settle what was then a lively point of controversy: whether the human lung could pump oxygen actively into the blood from the air. Before this could be settled, the answer had to be found to another question: how fast does oxygen combine with haemoglobin? The experiment designed to measure this rate was based on the *flow method*, and the events leading up to its introduction have been described in a fascinating way by one of the pioneers, F. J. W. Roughton[6]. The use of the singular—flow method— is not, perhaps, quite valid since this technique is usually considered under three headings: continuous-flow, accelerated-flow and stopped-flow. However, all three share the same basic principles. Like so many of the physical methods which have been applied successfully to the study of chemical kinetics, the flow method is beautifully simple in concept. The two reactants are allowed to flow down separate tubes leading into a mixing chamber and the chemical behaviour of the emerging mixture is followed. The main difference between the three subdivisions concerns the nature of the time-dependence of the flow pattern; connected with this is the important question of how the reaction is monitored. Many experimental and theoretical problems, such as the design of the most efficient mixing device, are common to all three types of flow technique. Section 7 discusses flow methods; it is divided into two parts. The first part deals with reactions in solution and the second with reactions in the gas phase.

Conceptually rather similar to flow systems is the *flame*. A flame has been defined as a combustion reaction which can propagate through space. In a controlled flame, as opposed to an explosion, this space-propagation occurs subsonically. Although flames have been known, of course, for many centuries, it is only in the last one or two decades, with the significant improvements in apparatus design, that systematic investigations have been undertaken of the kinetics of chemical processes involved in combustion. The self-propagation requirement implies that chain reactions, and in turn free radical reactions, are important in flames. Generally, radical reactions have comparatively low activation energies and so it would be predicted that most steps in flame reactions are fast. In fact the rates covered lie in the range between those studied by conventional static systems and those studied by flash photolysis or in the shock tube. Besides the possibility of working at high temperatures and under conditions at which high concentrations of free radicals can be obtained, the use of flames has the added advantage that wall effects are absent. Flames may be considered in two categories. One type may be generated in a system in which two components mix in the combustion zone. This is a "diffusion" flame since the rates of reaction are governed primarily by the speed at which mixing occurs and hence by the diffusion rates. In Section 8 discussion is limited to the other main category—

pre-mixed flames—where the limiting rates are determined chemically.

It has been noted that reactions in solution tend to be influenced by such effects as the formation of solvent sheaths around the reactants. Although elementary reactions can be studied in the gas phase it is still not possible to make any except the most general *a priori* predictions about reaction rates: these predictions are of the type that certain free radical reactions will have low activation energies and be fast (which is not so far removed from the type of prediction that certain acid-base reactions in aqueous solution will be diffusion-controlled). Ideally, it should be possible to mix molecules of two types under variable and known conditions and predict the outcome—will there be a reaction and, if so, what will be the properties of the products? Such predictions would use quantum mechanical descriptions of the species and involve construction of potential energy surfaces for the reaction. Although this approach is in its infancy, one of the most exciting aspects of physical chemical research today is that work is being done in this branch of reaction kinetics—and, moreover, that the results are encouraging. The experimental technique which allows this comparison uses crossed *molecular beams* of the two reactants.

Molecular beams have been used in physical investigations for many years. Beams of a single chemical species in which the molecules move effectively without collision behave in many ways like beams of light. For example, they can be focussed and will produce well-defined shadows of objects placed in their path. In the classic experiment of Stern and Gerlach, a beam of silver atoms was split in an inhomogeneous magnetic field, a result which experimentally confirmed the theoretical prediction of space-quantization of angular momentum. One great advantage of molecular beams is that it is often possible to choose molecules with certain specified properties, for example, molecules with a certain velocity or spin state. Molecular beams have been used to investigate several aspects of the kinetic theory[7], but their use in chemical problems has come comparatively recently. In a single beam, because of the effective lack of collisions, the only reactions which could be observed would be nuclear transmutations. Thus, for following chemical reactions, it is necessary for a beam of one reactant to interact with a beam of the other. This application stems from the discovery of a way of separating the reactive collisions from the much more common elastic scattering (typically these would occur with frequencies in the ratio of about 1:10). The method was first successfully applied[8] in 1955. Several systems have been studied since then but there are still very strict limitations on the type of reaction which may be followed. Because the method is potentially so powerful and the prospective rewards to be obtained from a comparison between experimentally observed behaviour and that predicted from first principles are so great, there is no doubt that ways will be found for extending these studies. Although the macroscopic rate coefficient can be computed from it, the parameter that is measured in these experiments is the angular distribution of the products relative to the reactant beams. The method is thus very different in concept from the other techniques

References pp. 176–179

discussed in this chapter; it is considered in Section 9. Also mentioned briefly is the application of molecular beams to the study of fast gas-surface reactions.

It may be apparent from this brief introductory survey that each method has its limitations, and that for a given system probably only one or two would be appropriate. The restrictions are of two types. Some are inherent in the method itself (*e.g.*, a chemical relaxation method requires a system comparatively near to equilibrium or, rather, one in which the equilibrium or stationary state may be easily displaced) while others are practical difficulties (*e.g.*, the very limited chemical range of the detectors in molecular beam experiments). Most of the methods described here are being used in many laboratories and are consequently undergoing continuous development. It can thus be confidently predicted that many of the practical difficulties in applying them to specific chemical systems will be overcome in the near future. Because of this continual improvement in technique not much has been said about the range of rate coefficients for which a given method is appropriate, or the accuracy with which the coefficients can be determined. The maximum sensitivity of most methods is within the 10 % level, although the actual sensitivity in a given application depends on several factors associated with the specific system, such as the optical absorptivity. Several of the techniques considered here have been discussed in one or more of the recent reviews on fast kinetics[9-16].

1. Flash photolysis and pulse radiolysis

1.1 FLASH PHOTOLYSIS

A schematic diagram for flash photolysis[17,18] is shown in Fig. 1. The detailed arrangements and dimensions vary considerably, but typically the reaction mixture is contained in a quartz vessel which has a length of 20 cm and a diameter of 2 cm. It is important that the ends are optically flat and parallel, since analyzing light is passed through the vessel along its axis. Parallel to the reaction vessel is a flash-light tube containing a rare gas, often krypton; the electrodes of the flash tube are connected to a bank of condensers. The reaction is initiated photochemically by dis-

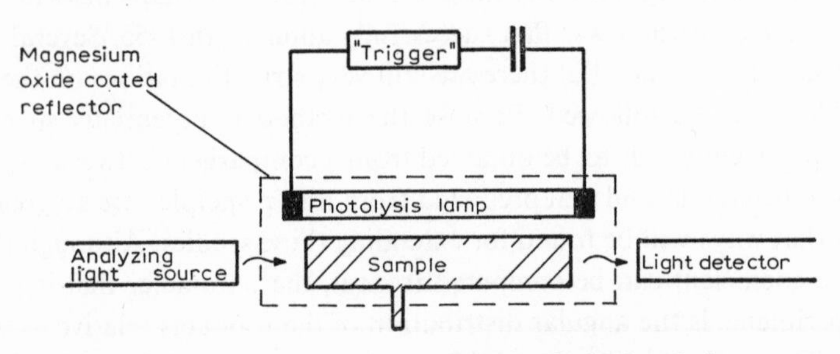

Fig. 1. Schematic arrangement for flash photolysis.

charging the condensers through the flash tube with the help of either a spark gap or a smaller "trigger" electrode placed about halfway along the flash tube. The light emitted by the flash tube is a continuum on which are superimposed the broadened atomic spectral lines of the rare gas; it extends over the whole of the visible spectrum and well into the infra-red and ultra-violet. In the arrangement shown in Fig. 1, the light will enter the reaction vessel in a highly non-uniform manner; to help increase the uniformity of illumination, the reaction tube and the flash lamp are mounted in a hollow cylindrical vessel coated on the inside with magnesium oxide, which has a very high reflectivity towards light of most wavelengths. The duration time of the photolysis flash depends on the energy to be dissipated. The two conflicting demands of a short-lived but high intensity burst have to be balanced, and the flash may have an energy of a few hundred joules spread over a few tens of microseconds.

When a new reaction is investigated its course is normally followed in the first place by means of a second light flash—the "spectroscopic" flash. Another quartz flash tube, rather smaller than the photolysis lamp and with one of its ends transparent (the electrode may be built into the side of the tube), is mounted along the axis of the reaction vessel. At the other end of the vessel is a spectrograph with plate camera which can cover the whole of the spectral range of interest. The spectroscopic flash is triggered electronically with a photocell which picks up light from the photolysis flash. The absorption spectra of the reaction mixture before and at known times after photolysis are compared, and a suitable wavelength at which to study the reaction kinetically is chosen. One of the advantages of the method is that, because of the comparatively large optical path-length, species of low extinction coefficient or present in very small concentration can be detected. However, it is sometimes very difficult to identify a transient species from its spectrum alone. In certain cases it is possible to determine the spectrum of an intermediate on an absolute scale by assuming that that fraction of a reactant which is decomposed (and this can be measured from the changes in *its* spectrum, for which extinction coefficients can be readily measured) is converted completely to the intermediate. This gives the absolute concentration of intermediate and thus, if the optical density is measured, the extinction coefficient. Such an approach is especially useful when an excited molecular state, *e.g.*, a triplet, is formed. Where such an approach is not possible, a relative concentration scale can be established and from this the reaction half-lives measured. If the concentration of the intermediate changes much more rapidly than the pressure or temperature (which is often the case), the following assumption can be made. If the same intensity is found with a given system at different times by using path lengths in the ratio a:b, then the concentrations at these times are in the ratio b:a. This is a direct consequence of Beer's Law. Provided certain restrictions regarding the practical applicability of Beer's Law[18] are borne in mind, such a method is generally useful.

When a suitable wavelength has been found at which to study the reaction kinetically, it is convenient to replace the spectroscopic flash tube with a high-intensity light source and a monochromator. The spectrograph is then replaced by a photomultiplier coupled to an amplifier and an oscilloscope. Alternatively, the analyzing light could be kept "white" until it has passed through the reaction vessel, and the photomultiplier coupled with the spectrograph. Such an arrangement increases the risk of further photochemical reaction brought about by the analyzing light. Because it is often difficult to determine the concentration of an intermediate absolutely at a given time, following any reaction which is not first order is rather difficult. From spectrograms taken before and after photolysis it is possible to estimate how much net decomposition has taken place, but many systems involve "dark" reactions which restore the original reactants. Where this is not so, it is often found that, because the extinction coefficients of the intermediates are so high, their concentrations may be followed even when they are very small compared with those of the reactants. In such cases, the extent of permanent reaction has no significant effect on the composition.

It has been estimated[18] that the temperature rise associated with the absorption of the photolyzing light may be as high as several hundreds of degrees. This temperature change (and its accompanying pressure change) may affect the kinetic investigation directly, *e.g.*, by increasing the rate coefficients, and it may also seriously interfere with the analysis by producing inhomogeneities in the reaction system. In solution reactions the heating is never significant since the solvent acts as a "thermostat". A similar result can often be achieved in gas reactions by adding a large amount of an inert gas.

1.1.1 Applications of flash photolysis

(1) An interesting application to gas-phase reactions was the study of the role of tetraethyl lead as an antiknock[19]. Combustion of many hydrocarbons proceeds in two stages: the first is a precombustion step, and among the products are aldehydes and peroxides. The second stage is the true combustion reaction. A mixture of acetylene, oxygen and amyl nitrite (the latter is known to promote knocking; it also acts as a sensitizer) was used to show that at the ignition point a very rapid build-up of OH and other radicals occurs. When tetraethyl lead was added it was found that the time lapse between the two stages was increased, the appearance of OH being correspondingly delayed, and the occurrence of detonation was greatly reduced or even eliminated. The species apparently responsible for the antiknock action was PbO. In the first place, it was formed between the two stages by the reaction of $PbEt_4$ with the oxygenated intermediates, which are known to lead to knocking. Secondly, it acted as a chain terminator by reacting preferentially with OH and the other propagating radicals to give atomic lead. From studies of the rates of appearance and disappearance of the various absorbing species, it could be

shown that the antiknock effect of tetraethyl lead in this particular system occurs entirely in the gas phase.

(2) A system which is important not only because of its intrinsic interest but also because it lies at the very heart of life is photosynthesis. Flash photolysis is an ideal technique by which to measure the effect of light on the various components of chloroplasts, both in solution and as suspensions of the chloroplasts themselves. In fact it has been necessary to pushthe technique towards its limits both as regards sensitivity of absorption and also time measurement. This topic has been discussed by Witt *et al.*[20]. The conventional technique has measured transmission changes of 0.01 % in times as short as 10^{-3} sec, but a method known as periodic flash photometry has increased the sensitivity considerably. In conventional flash photolysis the whole sequence of absorption changes following a flash is recorded. But if a measurement can be made several times its accuracy increases with the square root of the number of readings. Use can be made of this by storing many traces of the complete reaction in a computer of average transients (CAT) (optical density changes as small as 0.0001 have been measured at about 50 µsec after the flash[21]). Alternatively, the complete reaction sequence may be split up into small time-segments (*e.g.*, by the use of choppers in conjunction with the flashes), and the absorption of the system during a given segment measured many times. The absorption of the next segment is then measured, and so on, until the entire course of the reaction has been covered.

1.2 PULSE RADIOLYSIS

The main differences between pulse radiolysis and flash photolysis arise from the use in the former of ionizing radiation instead of light to initiate the reaction. Thus a pulse of electrons from a linear accelerator (or, less commonly, an X-ray pulse)

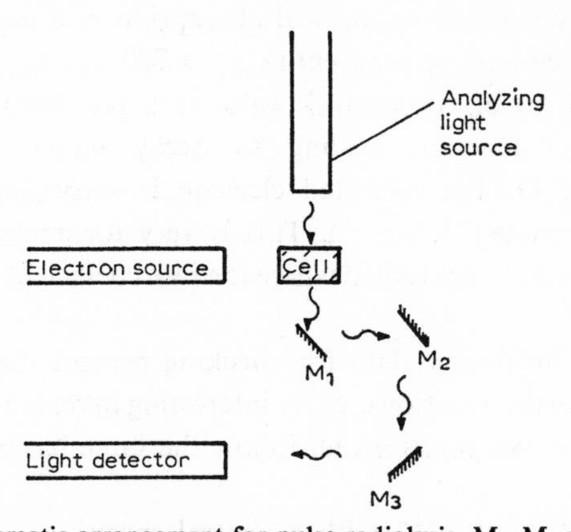

Fig. 2. Schematic arrangement for pulse radiolysis. M_1–M_3 are mirrors.

is passed through the solution and the changes produced are monitored spectrophotometrically[22], either at a known time after the electron pulse with a spectroscopic flash and a spectrograph or continuously at a given wavelength with a steady light source and photomultiplier (*cf.* Fig. 2). The same opposing demands of short duration but high intensity of the radiation pulse apply as for the principal flash in flash photolysis. Also, if the reactions studied are not first order, the absorption of radiation must be uniform throughout the reaction cell. As in flash photolysis with solutions, care must be taken to de-gas the sample properly. In the pulse radiolysis of aqueous solutions the last traces of impurity, which can interfere significantly with the reactions under study, can be removed in rather a simple way. The hydrated electrons produced on irradiation are found to disappear more slowly after successive pulses, and it has been postulated that they are reacting with the residual impurities to give non-reactive products. If this is so, then, after a sufficiently large number of pulses, the last traces of reactive impurities should have been removed, and the rate of disappearance of the hydrated electrons should reach a reproducible value. This is found to be the case, generally after about 20 pulses. Two recent reviews[23,24] include discussions on kinetic determinations in radiolysis experiments.

1.2.1 The hydrated electron

Certainly the most exciting applications of pulse radiolysis to date have been concerned with the reactions of the aqueous electron, e^-_{aq}. Although there may be disagreement as to the mechanism of their formation[24], it is now accepted that when ionizing radiation (notably an electron pulse) is passed into water, hydrated electrons are produced. They are always present in comparatively small concentrations (about 10^{-8}–10^{-6} *M*) and so in most reactions the other reactant is in excess; thus first order kinetics are nearly always observed. One property that makes e^-_{aq} especially easy to study is its strong optical absorption in a region of the spectrum where most other species are transparent (λ_{max} = 720 mμ, ε_{max} = 15,800 mole.l^{-1}.cm^{-1}). The half-life of e^-_{aq} in neutral water is a few tenths of a millisecond. The most important reaction leading to decay under these conditions is $e^-_{aq}+H_3O^+_{aq} \rightarrow H+H_2O$. The hydrated electron is surprisingly unreactive[25] towards H_2O ($k \leq 16$ mole.l^{-1}. sec^{-1}). This is very fortunate since it allows the reaction with any species to be studied where the rate coefficient is greater than about 10^5 mole.l^{-1}. sec^{-1}.

Besides providing invaluable data for checking present theories of the mechanisms of electron-transfer reactions, e^-_{aq} is interesting in reacting with an extremely wide range of species. The reactions all follow the same basic scheme

$e^-_{aq}+AB \rightarrow AB^- \rightarrow A+B^-$ with a neutral species,

$e_{aq}^- + AB^{n+} \rightarrow AB^{(n-1)+}$ with a cation, and

$e_{aq}^- + AB^{n-} \rightarrow AB^{(n+1)-}$ with an anion.

The primary product (*e.g.*, AB^-) may be stable and appear as the final product. More often it is unstable and dissociates to give a stable negative ion B^- and a radical A or a stable molecule A and a radical ion B^-. The reaction of the hydrated electron with acids illustrates this general sequence of reactions. The radical ion AH^- is formed as an intermediate, and is likely to undergo subsequent decomposition. Sometimes a hydrogen atom is formed, but in the absence of other reagents the most stable fragments are produced. In the case of oxyacids, the two possible sets of products are a hydrogen atom and the anion of the acid, or a hydroxyl ion and a free radical from the acid

$$AOH + e_{aq}^- \rightarrow AOH^- \rightarrow A + OH^-$$
$$\searrow AO^- + H$$

Generally, the hydration energy of OH^- is greater than that of the acid anion and the free energy of formation of the corresponding free radical is lower than that of a hydrogen atom. Consequently, the products of such a reaction are usually $A + OH^-$. Interestingly enough, an exception is found to this generalization when the acid is $H_2PO_4^-$

$$H_2PO_4^- + e_{aq}^- \rightarrow H_2PO_4^{2-} \rightarrow HPO_4^{2-} + H$$

This has been rationalized in terms of the fact that the dissociation energy of a P–O bond in phosphate is considerably higher than that of an O–H bond[26].

Perhaps the one outstanding result of the direct kinetic investigations of e_{aq}^- (as well as the earlier competition kinetics based on product analysis) has been that the electron as a hydrated species seems to behave more or less as expected.

2. Shock tube and adiabatic compression

One of the main reasons that shock waves form is that the speed of sound in a fluid increases with pressure. When a sound wave is generated, alternating regions of high and low pressure are formed (the differences in refractive index of such regions have been used as a basis for the analysis of relaxation in solution, as discussed below) and as the distance from the source increases there is a tendency for the wave to lose its typical sinusoidal form and become "saw-toothed"; the high-pressure regions travel faster than those of lower pressure. Eventually the crests and the troughs arrive at a given point together or, in other words, the rise in pres-

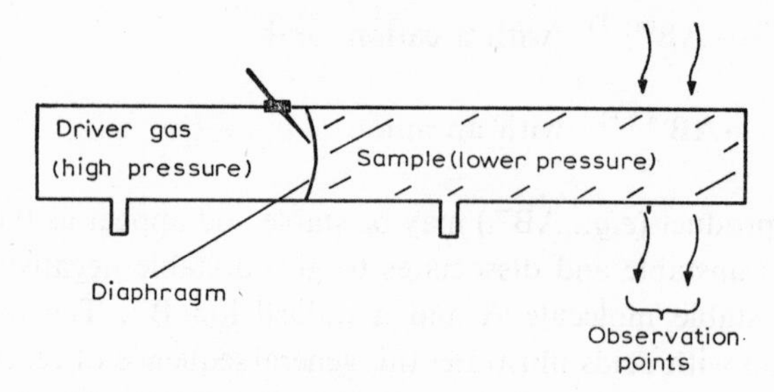

Fig. 3. Schematic arrangement for the shock tube.

sure is instantaneous. This is a shock wave, and, if their energy were not gradually dissipated to the medium, all sounds would eventually become shock waves a sufficiently large distance from the source. Shock waves are used in the study of the kinetics of gas reactions most conveniently in a shock tube. A good general account of the application of the technique to chemical kinetics has been given by Bauer[27]. Other general treatments give varying amounts of theoretical and practical detail[28,29].

In principle, the design of a shock tube is very simple; it is illustrated in Fig. 3. Typically, a smooth metal or glass tube about twenty feet long and a few inches in diameter is divided into two compartments by a thin diaphragm made of a suitable material such as cellophane or aluminum foil. About a third of the tube comprises a high-pressure region containing the "driver" gas, often hydrogen or helium. The remainder of the tube is a comparatively low-pressure region containing the gas mixture whose reactions are to be studied. The experiment is started by puncturing

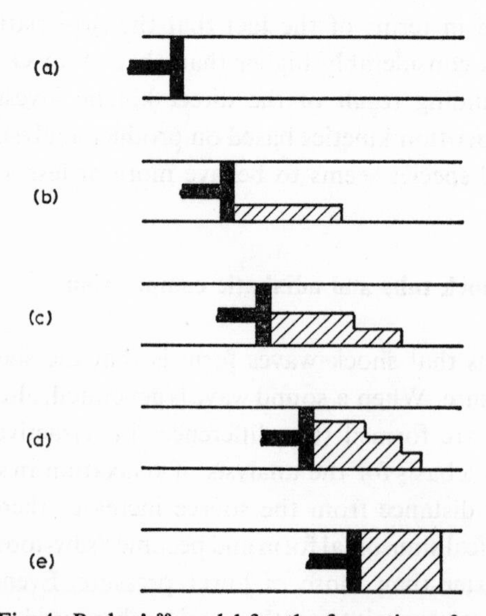

Fig. 4. Becker's[30] model for the formation of a shock wave.

the diaphragm with a sharp needle. This produces a shock wave which enters the low-pressure region and moves rapidly towards the end of the tube. Just before the end it passes several observation points located along the side of the tube, and, by following the changes in properties of the mixture as the shock wave passes, the reaction can be characterized kinetically.

What form does this "shock wave" take and how is it produced? A useful model for visualizing the formation of a shock was first proposed by Becker[30] in 1922. At the moment the diaphragm breaks, the high-pressure region may be compared with a tight-fitting piston in a tube; it is instantaneously at rest but rapidly accelerates as the boundary moves into the low-pressure area. The acceleration can be thought of as a stepwise process (*cf.* Fig. 4): each displacement generates a pressure pulse which travels into the gas ahead. Each pulse raises the gas to the piston velocity and also raises its temperature adiabatically, but because the speed of sound is higher at higher temperatures the second, third, and later pulses tend to catch up with the first. Eventually, these pulses all coalesce and form a discontinuity which moves at a constant speed along the tube. This speed is considerably greater than the speed of sound in the *undisturbed*, relatively cool, gas. The term "shock wave" is perhaps a little misleading since in the shock tube the pulse is actually a single sharp transition which is characterized by two sets of essentially uniform conditions. What is of interest, then, is to see how the chemical system readjusts to the new conditions behind the shock.

It is not proposed to deal in any detail with the hydrodynamical theory of shock waves, which has been considered thoroughly by Greene and Toennies[29]. In an ideal system (*i.e.*, one in which the medium is continuous, is non-viscous and does not conduct heat) the shock wave should be abrupt in nature and it would then be possible to treat it as a mathematical discontinuity. However, real systems are not continuous, and they are viscous and heat-conducting. Interestingly, it is found that the actual size of the transition zone between the high- and low-pressure areas is very small. It is very difficult to measure this thickness experimentally, but there are indications that it is typically of the same order of magnitude as the wavelength of visible light, about 10^{-4} cm. For convenience the coordinates of the system are referred to the moving shock front as the origin rather than the laboratory. On this system the gas enters the transition zone smoothly with the velocity of the shock wave. Within a few collision path-lengths it is heated very rapidly—*i.e.*, it gains a considerable amount of kinetic energy and entropy in a short time— and it leaves the "transition zone" with new physical properties. On this model, three conservation relationships are expressed mathematically in terms of the physical parameters of the system—the conservation of mass, of energy and of momentum. By combining these expressions with two equations of state of the gas, equations can be derived which relate any three of the energy, pressure, density and temperature of the gas on entering and leaving the transition zone (a total of six parameters). In order to define the state behind the shock front it is necessary to measure one

References pp. 176–179

parameter associated with the shock process. For practical reasons this property is generally the shock velocity, and this is one of the two functions of the observation points indicated in Fig. 3. The temperature rise at the walls of the tube is much less than in the body of the shock wave (if it were not, it would obviously be rather an unrealistic approximation to regard the gas as a non-conductor of heat), and hence it would be very difficult to estimate the temperature of the bulk of the gas directly. What is often done is to use the comparatively small temperature rise at the walls as an indication of the passage of the shock front. The time taken for the shock wave to pass between sets of sensors placed known distances apart is measured. These devices may be small areas of platinum, of which the electrical conductivity changes with temperature, deposited on plates fixed flush with the wall. The temperature, density, etc. are then calculated. (Where possible, the sample comprises at least 90 % argon, the remainder being the substance under investigation. Provided that the heat of reaction is not very large, the density and temperature in the shock front are then determined by the argon.)

A feature of shock waves not yet considered is that there is inevitably a low pressure or "rarefaction" wave produced at the diaphragm at the same time as the shock wave. This moves initially in the opposite direction from the shock wave but is reflected by the back wall of the tube, and so eventually follows the main shock wave down the tube. Relative to laboratory coordinates this rarefaction wave travels with the local velocity of sound in the gas. This is considerably less than that of the shock wave because of the substantially lower temperature, but superimposed on it is the flow motion of the driver gas towards the low-pressure region. This has the result that the rarefaction wave tends to catch up with the shock wave. Because of the simplifications it allows, it is convenient to make the measurements on the shocked gas *before* the rarefaction arrives. This consideration is an important one in deciding on the relative positions of the diaphragm and observation points, and on the relative lengths of the high- and low-pressure areas[29]. For a reason considered below, measurements are also sometimes made after the shock wave has been reflected from the front wall, but before the rarefaction wave has arrived. Such a situation is only used where absolutely necessary because it is now felt that the shock front is significantly distorted on reflection.

In addition to the shock wave velocity it is necessary, in a system where the density does not uniquely determine the composition (and this includes all but the very simplest chemical systems of interest), to measure some concentration function in order to follow the reaction. This is one of the greatest experimental difficulties associated with the method, since the changes occur so rapidly. Where possible, the concentration change is followed spectrophotometrically. This concentration monitoring is the second function of the observation points in Fig. 3. Especially for species with line spectra, the small optical density change, coupled with the fast response-time necessary, excludes the use of a conventional spectrophotometer. An example of a detection system which has been used for the hydrogen/oxygen reaction is discussed below. It is not a straightforward matter to extract reaction rate coefficients from the concentration–time data. For endo- and exothermic reactions, where the temperature rise is large, the enthalpy of reaction alters the measured parameters significantly. Another complicating factor is the contraction which occurs in the time scale. Because the heated gas is moving rapidly along the tube behind the shock, the gas which is being observed at any one instant will have been heated *not* since the shock wave passed the observation point (which would be the case if there were no net motion) but for a considerably longer time. The temperature change, which is related to the strength of the shock, is limited to a certain maximum value which can be calculated. Various devices have been used to increase this maximum value. Light driver gases, such as hydrogen and helium, produce much stronger shocks for a given pressure ratio than heavier ones. This is primarily because of their lower molecular weight and the consequent higher velocity of sound in them. The temperature change may also be increased by pre-heating the driver gas. Yet another device is to use the shock wave after reflection off the front wall; this increases the temperature rise by a factor of two. The rate data are generally obtained from the velocity and concentration measurements with the help of a computer. Equations representing the various processes are set up and trial solutions inserted; the resulting profiles should eventually fit the experimental traces.

One of the most difficult types of gas reaction to follow kinetically is the explosion. Typically, this has a radical chain mechanism and an induction period, and is therefore seriously disturbed in any apparatus in which wall effects play a significant role. Explosions may not be studied conveniently in flames (Section 8) since the constant supply of radicals diffusing back against the gas flow renders the concept of an induction period somewhat irrelevant. The shock tube is suitable for studying them in the micro- to millisecond time range, but for times of 10^{-3}–1 sec a technique which uses rapid *adiabatic compression* is being developed[31]. The apparatus consists essentially of a cylinder containing the reactant gases and a compressing piston. It is about 80 cm long and 5 cm in diameter and is mounted vertically so that the piston may be driven by a falling weight of about 100 kg. Because preliminary reaction of the gas mixture would be important just before the piston is arrested (when the gas temperature is high but not at its final value), it is important to stop the piston in as short a space as possible. The pressure in the reaction vessel at known times may be recorded, and the gas may be removed for analysis at a given time after the start of the reaction through a punctured membrane. The preliminary work using this technique has been on the combustion of hydrocarbons, but it is also suitable for following such reactions as that between hydrogen and oxygen.

2.1 THE HYDROGEN/OXYGEN REACTION

A reaction which has been studied in shock tubes as well as by means of flames

References pp. 176–179

is that between hydrogen and oxygen. The shock tube work has been done principally by Schott *et al.*[32]. They measured the course of the reaction by following the absorption due to the hydroxyl radicals. This species was chosen because it has comparatively intense absorption lines and is reasonably stable. The concentration observation point comprised two quartz windows set opposite each other into the side of the tube. On discharging a flashlamp containing water vapour, the line-spectrum of hydroxyl radicals was emitted, and a monochromator was used to pick out a suitable portion of the spectrum to pass through the shock tube. The change in the absorption of these lines due to the OH radicals produced in the H_2/O_2 reaction was then measured with a photomultiplier and oscilloscope. In the later work, on dilute gases with lean hydrogen/oxygen ratios, five piezoelectric crystals located along the wall and coupled with an amplifier and oscilloscope were used to measure the shock velocity. Reaction zone temperatures of 1150°–1850° K and pressures of a few atmospheres were achieved. The measured parameters were therefore (*a*) a trace showing the progress of the shock past the various gauge positions, and (*b*) the absorption by OH radicals of incident radiation from the flashlamp as a function of the time behind the shock front. From these data it was possible to deduce the relative importance of the possible reaction paths for the recombination of the radicals produced with different H_2/O_2 concentration ratios.

3. Chemical relaxation methods

3.1 THE RELAXATION TIME

In the chemical relaxation methods[33,34] for following fast reactions, the rate coefficients are determined from a parameter called the relaxation time, τ. The physical significance of τ will be considered in terms of the step-function model, but the same quantity is measured by the other main group of relaxation methods, the stationary methods.

Suppose that the simple equilibrium

$$\mathrm{A} \underset{k_{21}}{\overset{k_{12}}{\rightleftharpoons}} \mathrm{B}+\mathrm{C} \tag{1}$$

exists in solution, and that it is somehow "instantaneously" displaced. The concentrations of the three species will immediately move towards their new equilibrium values and, provided that the original displacement was small compared with the amount of that species present in lowest concentration (an important condition), the rate of attainment of the new equilibrium at any instant is directly proportional to the displacement from it. This behaviour may be represented by an exponential function with time constant τ.† Since the method is applicable also to a stationary

† The half-life of this first-order reaction, $t_{\frac{1}{2}} = (\ln 2)\tau = 0.693\tau$.

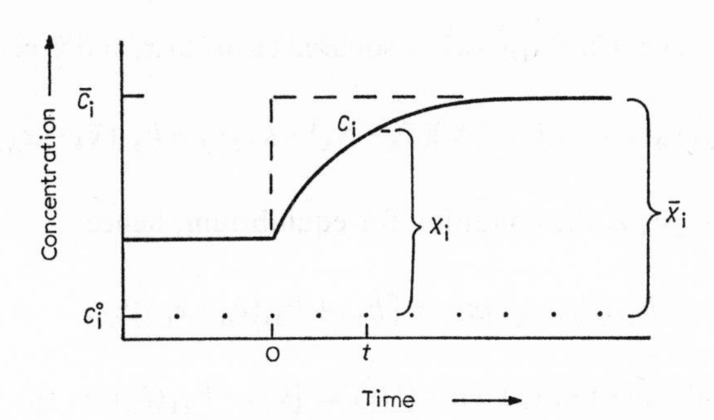

Fig. 5. Chemical relaxation in terms of the step-function model.

reaction state in which there is no net rate of transformation, such as that found at a fixed distance from the mixer in a flow-mix experiment, it is convenient to make the derivation of relaxation times as general as possible.

The full line in Fig. 5 represents the concentration of species i at times before and after the application of the step-function (dashed line). c_i^0 is a time-independent reference concentration, while $\bar{c}_i$ is the corresponding (final) equilibrium value which might be time-dependent. Then

$$x_i = c_i - c_i^0 \quad \text{and} \quad \bar{x}_i = \bar{c}_i - c_i^0$$

and the rate can be written as

$$\frac{-\mathrm{d}(\bar{x}_i - x_i)}{\mathrm{d}t} = (1/\tau)(\bar{x}_i - x_i)$$

i.e.,

$$\mathrm{d}x_i/\mathrm{d}t + (1/\tau)x_i = (1/\tau)\bar{x}_i \tag{2}$$

provided that $(\bar{x}_i - x_i) \ll c_i$.

The relationship between the reciprocal relaxation time and the rate coefficients for equilibrium (1), k_{12} and k_{21}, may be derived quite simply.

The rate expression is given by

$$-\mathrm{d}c_A/\mathrm{d}t = \mathrm{d}c_B/\mathrm{d}t = \mathrm{d}c_C/\mathrm{d}t = k_{12}c_A - k_{21}c_B c_C.$$

The concentrations c_i are expressed in terms of $\bar{c}_i$, x_i and $\bar{x}_i$: $c_A = c_A^0 + x_A$, but $c_A^0 = \bar{c}_A - \bar{x}_A$, and similarly for the species B and C. The stoichiometry gives $x_A = -x_B = -x_C$. Now,

$$\mathrm{d}c_A/\mathrm{d}t = \mathrm{d}x_A/\mathrm{d}t = k_{21}(\bar{c}_B + x_B - \bar{x}_B)(\bar{c}_C + x_C - \bar{x}_C) - k_{12}(\bar{c}_A + x_A - \bar{x}_A),$$

and for small differences $(\bar{x}_i - x_i)$—when squared terms in x_i and $\bar{x}_i$ can be neglected

$$dx_A/dt = k_{21}\bar{c}_B\bar{c}_C + k_{21}(\bar{c}_B+\bar{c}_C)(\bar{x}_A - x_A) - k_{12}\bar{c}_A + k_{12}(\bar{x}_A - x_A).$$

Now, $k_{21}\bar{c}_B\bar{c}_C = k_{12}\bar{c}_A$ is the condition for equilibrium, hence

$$dx_A/dt + [k_{12}+k_{21}(\bar{c}_B+\bar{c}_C)]x_A = [k_{12}+k_{21}(\bar{c}_B+\bar{c}_C)]\bar{x}_A,$$

which is the general equation (2) with $(1/\tau) = [k_{12}+k_{21}(\bar{c}_B+\bar{c}_C)]$.

This illustrates an unusual feature about relaxation times: their expressions always contain the sum of a contribution from the forward and the backward reaction. This has the convenient result that τ is independent of the sign of the perturbation, *i.e.*, it is immaterial for a determination of τ whether the disturbance causes the equilibrium to move to the left or to the right.

TABLE 1

RELAXATION TIMES IN TERMS OF EQUILIBRIUM CONCENTRATIONS AND RATE COEFFICIENTS FOR SINGLE-STEP REACTIONS (*cf.* Ref. 33)

Reaction	Relaxation time
1. $A \underset{k_{21}}{\overset{k_{12}}{\rightleftarrows}} B$	$\tau^{-1} = k_{12}+k_{21}$
2. $A+B \rightleftarrows C$	$\tau^{-1} = k_{12}(\bar{c}_A+\bar{c}_B)+k_{21}$
3. $A+B \rightleftarrows C+D$	$\tau^{-1} = k_{12}(\bar{c}_A+\bar{c}_B)+k_{21}(\bar{c}_C+\bar{c}_D)$
4. $A+B+C \rightleftarrows D+E$	$\tau^{-1} = k_{12}(\bar{c}_A\bar{c}_B+\bar{c}_B\bar{c}_C+\bar{c}_C\bar{c}_A)+k_{21}(\bar{c}_D+\bar{c}_E)$
5. $A+C \rightleftarrows B+C$ (C is a catalyst)	$\tau^{-1} = (k_{12}+k_{21})\bar{c}_C$
6. $A+B \rightleftarrows C$ (B is buffered)	$\tau^{-1} = k_{12}\bar{c}_B+k_{21}$

The expressions for τ for a few single-step reactions are shown in Table 1. It is interesting that in the case of a catalyzed reaction (5) (where the catalyst concentration appears on both sides of the equilibrium) or a buffered system (6) (where the concentration of the buffered species appears on one side of the equilibrium only) the only concentration term to appear in the relevant part of the expression for τ^{-1} is that of the catalyst ($\bar{c}_C$) or the buffered reactant ($\bar{c}_B$). This is because the changes in concentration of these species (x_C, and x_B, respectively) are effectively zero, hence the terms in the relaxation expression which contain x_C and x_B must also be zero. The only terms which do *not* contain x_C and x_B are those in $\bar{c}_C$ and $\bar{c}_B$, respectively[†].

So far, only reactions which involve a single step have been considered. The na-

† An important proviso for this is that the buffer equilibrium is established rapidly compared with the equilibrium under consideration.

ture of the relaxation expression for a reaction proceeding via an intermediate, *e.g.*

$$AB+C \rightleftarrows A+B+C \rightleftarrows A+BC \tag{3}$$

will depend on the nature of the intermediate, in this case B. If the concentration of B is small relative to those of the other species (*e.g.*, if B cannot be detected), then the steady state hypothesis may be applied to its formation and dissociation rates. The concentration of B will not appear in the expression for the relaxation time, and this reaction will effectively be single-step, with only one relaxation time. If, on the other hand, the intermediate B is present in significant concentrations, then the system must be regarded as two-step, and will generally be associated with two relaxation times. In general, a multi-step reaction will be characterized by a "spectrum" of relaxation times whose order depends on the number of independent relaxation steps (but not necessarily in a simple way). As the number of steps increases, the complexity of the relaxation expression also increases. Relaxation expressions have been derived for several generalized systems[33,35].

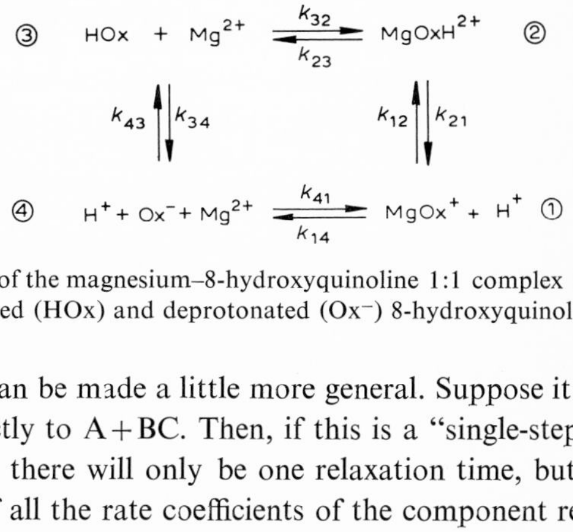

$$\begin{array}{ccc}
③\ HOx + Mg^{2+} & \underset{k_{23}}{\overset{k_{32}}{\rightleftarrows}} & MgOxH^{2+}\ ② \\
k_{43}\ \updownarrow\ k_{34} & & k_{12}\ \updownarrow\ k_{21} \\
④\ H^{+} + Ox^{-} + Mg^{2+} & \underset{k_{14}}{\overset{k_{41}}{\rightleftarrows}} & MgOx^{+} + H^{+}\ ①
\end{array}$$

Fig. 6. The formation of the magnesium–8-hydroxyquinoline 1:1 complex ($MgOx^{+}$) from protonated (HOx) and deprotonated (Ox^{-}) 8-hydroxyquinoline

The system (3) can be made a little more general. Suppose it is also possible for AB+C to go directly to A+BC. Then, if this is a "single-step" reaction (*i.e.*, $\bar{c}_B$ is relatively small), there will only be one relaxation time, but its expression will contain the sum of all the rate coefficients of the component reactions, with their corresponding concentration terms. It is sometimes a little difficult to differentiate between a single-step and a multiple-step reaction, but if more than one relaxation time is observed then the reaction is necessarily a multiple-step one. The reverse is not necessarily true.

A reaction which illustrates these principles quite well is the formation and dissociation of the 1:1 complex between magnesium and 8-hydroxyquinoline (oxine) in water [36]. The postulated mechanism is shown in Fig. 6. $MgOx^{+}$ may be regarded as being formed by two parallel paths: by the reaction of Mg^{2+} with the oxine anion (Ox^{-}) or with the neutral molecule (HOx). Under the conditions used, the concentrations of HOx and Ox^{-} were comparable.† Consequently, the reaction via path

† In this reaction scheme, H^{+} is the species present in smallest concentration, but, because it is buffered, $x_{H^{+}}$ is effectively zero. It is therefore necessary to consider the species present in *next* smallest concentration, which may be any one of $MgOx^{+}$, Ox^{-} or HOx.

③ → ④ → ① must be considered as two-step, and it will be associated with two relaxation times. Protonation reactions are generally very fast[33,37], so the relaxation time for step ③ → ④ will be short; metalation reactions of magnesium are much slower[33,38], so the relaxation time associated with step ④ → ① will be comparatively long. On the other hand, the other intermediate, $MgOxH^{2+}$, is present in much smaller concentration than $MgOx^+$, Ox^- and HOx. The steady-state approximation can be applied to its formation rate, and pathway ③ → ② → ① may be regarded as single-step and thus associated with a single relaxation time. However, these two relaxation processes are coupled, and in this scheme give rise to only two relaxation times—a short one for the protonation and a longer one for the metalations. The first of these was too fast to be measured by the technique used (temperature-jump); the expression for the second illustrates how relaxation terms can be combined.

Path ③ → ④ → ① may be written slightly differently

$$\begin{array}{l} Ox^- \quad\quad + Mg^{2+} \underset{k_{14}}{\overset{k_{41}}{\rightleftarrows}} MgOx^+ \\ +H^+ \\ \uparrow\downarrow\ K \\ HOx \end{array}$$

where K is the deprotonation constant for oxine. As above, the expressions $c_{Ox} = c^0_{Ox} + x_{Ox}$ and $c^0_{Ox} = \bar{c}_{Ox} - \bar{x}_{Ox}$, etc. are introduced. Now since c_H and c_{Mg} (effectively) are buffered, the simplification may be introduced that $x_H = x_{Mg} = 0$. On disturbing the system, *both* equilibria are displaced (the change in the position of the proton equilibrium is associated with displacement of K and a small change in the pK of the buffer-system), $(x_{Ox} + x_{HOx}) = -(x_{MgOx})$. Thus c_{Ox} must be expressed in terms of $c_{(Ox+HOx)}$, the total uncomplexed oxine concentration, *i.e.*

$$c_{Ox} = \frac{c_{(Ox+HOx)}}{1 + Kc_H}$$

This leads to the expression

$$\tau^{-1}_{3,4,1} = \frac{k_{41} \cdot c_{Mg}}{1 + Kc_H} + k_{14}$$

The relaxation time associated with path ③ → ② → ① is given simply by

$$\tau_{3,2,1}^{-1} = \frac{k_{32} \cdot k_{21} \cdot c_{\text{Mg}}}{k_{21}+k_{23}} + \frac{k_{12} \cdot k_{23} \cdot c_{\text{H}}}{k_{21}+k_{23}}$$

(Again, since c_{H} and c_{Mg} are buffered, only terms in these species will enter the relaxation expression.)

$$\tau^{-1} = c_{\text{Mg}} \left\{ \left(\frac{k_{41}}{1+Kc_{\text{H}}} \right) + \left(\frac{k_{32} \cdot k_{21}}{k_{21}+k_{23}} \right) \right\} + c_{\text{H}} \left(\frac{k_{12} \cdot k_{23}}{k_{21}+k_{23}} \right) + k_{14} \tag{4}$$

This system yields experimental values for k_{41}, k_{14} and $k_{32} \cdot k_{21}/(k_{21}+k_{23})$. It is interesting that, by changing c_{Mg} and c_{H}, it is possible to find conditions where, in turn, each of the first three terms in equation (4) is the dominant one.

3.2 STEP-FUNCTION, OR TRANSIENT METHODS

The main experimental problems are twofold: how to displace the equilibrium, *e.g.*

$$\text{A} \rightleftarrows \text{B}+\text{C} \tag{1}$$

and how to measure the accompanying changes of concentration of one of the species. The second of these is comparatively easy to solve. Concentration changes are generally measured optically, although sometimes other physical properties such as conductivity may be used, and suitable electronic equipment is generally available for measuring in the required time range. One of the more serious problems associated with measurement at very high rates is how to obtain a sufficiently large signal/noise ratio[39]. Where the signal/noise ratio is dangerously low such refinements as "multiple jumps" in conjunction with computers of average transients (CATs) are being used to increase the resolution. If the reaction under investigation is not associated with a measurable spectral change it is often possible to couple it with an indicator system (*cf.* below).

The problem of how to displace the equilibrium is more difficult to solve, and much ingenuity has gone into finding methods of doing this. Suppose reaction (1) is associated with a finite enthalpy change. At a given temperature and total reactant concentration, the position of the equilibrium will be uniquely defined. If the temperature is now raised by, say, 10°, a new set of equilibrium conditions is defined. Thus if the change in temperature can be made rapidly enough the relaxation behaviour of the system can be followed by monitoring the concentration of the optically-absorbing species (or indicator). A similar effect is found, for example, by suddenly changing the external pressure[40] on systems where the net reaction is associated with a finite volume change; the change in conductivity is followed. Although it is not, perhaps, the most sophisticated relaxation technique available, the temperature-jump has so far proved to be the most versatile and reliable, and many data have been collected on different types of system.

3.2.1 Temperature-jump

Most commonly the energy stored in a bank of condensers is discharged through the solution contained in a cell between two specially-designed electrodes (Fig. 7)[41]. The absorption change of the solution between the electrodes is followed with a beam of monochromatic light (passed through the solution perpendicular to the axis of the cell) and a photomultiplier coupled to a suitably-triggered oscilloscope. A typical oscilloscope trace, which shows the relaxation behaviour directly, is shown in Fig. 8. The relaxation time may be derived from a semilogarithmic plot or comparison with standard exponential curves. One point about this apparatus is that the reactant solution must conduct sufficiently well to carry the current efficiently and so maintain a short heating time, typically, a few microseconds. This does not necessarily mean that only ionic reactions may be studied. Often an inert electrolyte, such as 0.1 *M* KNO_3, is added to carry the current. A constraint which is generally not very serious is that the change in optical absorption must be small compared with the total absorption—for two reasons. In the first place, the condition for linearizing the rate-expression is that the relative concentration change be

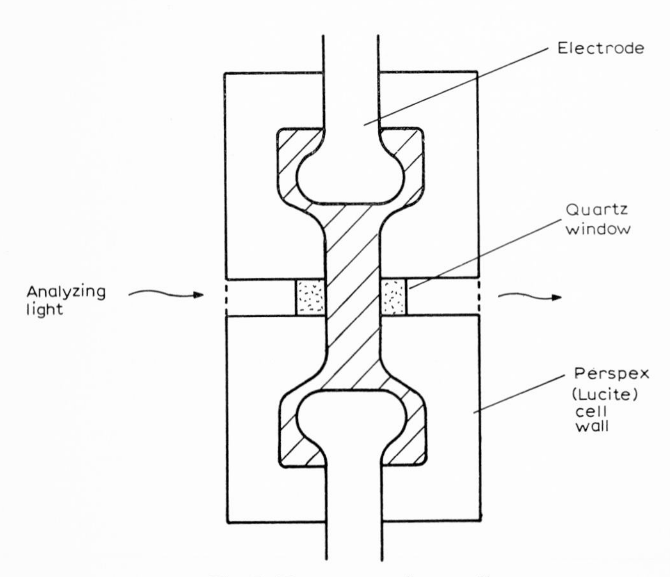

Fig. 7. Temperature-jump cell.

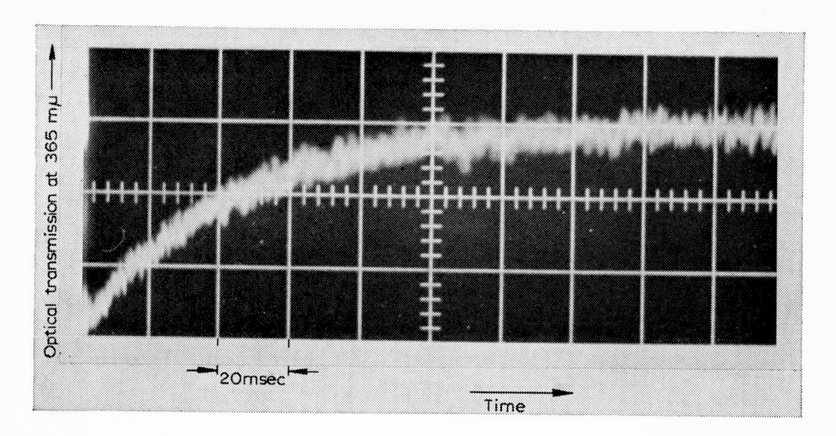

Fig. 8. Oscillogram showing relaxation behaviour for Mg^{2+}–adenosine-5′-triphosphate (ATP^{4-}) complex (5×10^{-3} *M*) and 8-hydroxyquinoline buffered at pH 8.2.

small. Secondly, the photomultiplier/oscilloscope trace measures transmittance changes whereas the concentration is proportional to the absorbance. The relation between these is approximately linear only for small total changes. Typically, the cell contains about 5 ml of solution and the optical path length is 1 cm. For most sensitive recording the optical density at the chosen wavelength should be quite high: 0.8–0.9. For a given system to be suitable for T-jump study, three properties must have suitable quantities—the overall effective equilibrium constant, the enthalpy change of at least one of the reaction steps, and the optical density change at a suitable wavelength.

An alternative approach to the problem of suddenly raising the temperature is to use a pulse of microwaves[42]. The advantage of this method is that non-conducting systems may be used. A disadvantage of the microwave T-jump is that only a relatively small temperature rise can be obtained, generally less than 1°. Recently, the heating effect of a flash lamp has been used[43]; a laser has been employed as the heat source in surface-T-jump experiments and attempts are being made to use it for heating solutions[44,45]. One difficulty is that at present the wavelengths of available lasers are not very suitable, since some species (such as Cu^{2+}_{aq}, which interferes with many reactions of potential interest) must be present to absorb the energy. The T-jump technique is also being applied to reactions of species with short life-times, an extension which will be of special interest in enzyme-chemistry. By combining the flow and T-jump techniques, reactions of species in quasi-equilibrium are being studied (Section 7)[46].

3.2.2 Applications of T-jump

It is difficult to say that a given relaxation technique has been especially useful in elucidating particular types of kinetic patterns and reaction mechanisms. In most types of system data have been collected with the help of several relaxation methods. Metal-complex formation is discussed in greater detail below, but for some metals the T-jump is a particularly suitable tool. For example, magnesium forms complexes two or three orders of magnitude more slowly than does calcium; the T-jump has been used in several cases for magnesium[47,48,36] but it is generally not possible to use this method with calcium. Metal-complex formation data have been collected by Eigen and Wilkins[49]. If possible, changes in optical properties associated with the ligand or metal are followed, but this is sometimes not practical. Such a case was

$$ML(H_2O)_x + NH_3 \rightleftarrows ML(H_2O)_{x-1}NH_3 + H_2O$$

where M is a divalent metal ion (Ni^{2+} or Cu^{2+}) and L is a chelating ligand. This reaction was coupled with the reactions

$$NH_3 + H_2O \rightleftarrows NH_4^+ + OH^-$$

and

$$OH^- + HInd \rightleftarrows Ind^- + H_2O$$

the shift in relative concentrations of HInd and Ind^- resulting in a colour change[50]. It may be possible to select conditions such that these indicator reactions are fast and do not interfere with the relaxation under investigation (as was the case here). Often, however, this is not possible, and the overall relaxation expression contains contributions from the indicator reactions[47]. The rate coefficients for these reactions must then be obtained separately.

Usually proton-transfer reactions in aqueous solution are diffusion-controlled and are, consequently, too fast to be studied in this way. Sometimes structural features such as internal hydrogen-bonding slow them down sufficiently to make their investigation with the T-jump feasible. This type of reaction has been considered fully[51]. An example of a system studied by this method is the indicator phenol red in water[52]. Because measurements must be made around pH 7 (the pK of phenol red is 7.8) in unbuffered solutions at small acid or base concentration, the error limits are quite high. A problem also arises which for most applications of the T-jump is not important. This is the long-term disturbance of the system due to polarization effects at the electrodes. In this case, at low indicator concentrations a colour gradient may be seen after the electrical discharge ranging from yellow at the "cathode" (associated with the protonated species) to red at the "anode" (the deprotonated species).

If the enthalpy change ΔH of a reaction sequence, in which one step is a protonation, is too small to permit the application of the T-jump, it might be possible to use a buffer which has a high ΔH. When the temperature is raised, the concentration of hydrogen ions will be changed slightly by virtue of the buffer's ΔH, and this "concentration jump" might be sufficient to displace the main equilibrium.

3.3 STATIONARY METHODS

Whereas in the "step-function" relaxation methods a single pulse is applied to an equilibrium system and the consequent *rapid* adjustment of the concentration is determined, in the "stationary" methods a steadily oscillating disturbance is applied and the behaviour of the system can be analyzed at leisure. This behaviour is represented in Fig. 9, in which the forcing function is given by curve ①. If the frequency of the forcing function is sufficiently low, the system can adjust itself rapidly and the concentration changes can effectively keep up with their equilibrium values, in which case the concentrations are also given by curve ①. If the forcing frequency is much too large for the system to readjust itself, there will be no observed concentration changes (case ②). The interesting situation arises when the applied frequency is of the same order of magnitude as the rate at which the changing chemical equilibrium can be set up in the system. In such a case (curve ③) there is an interaction between the system and the forcing function in which energy is absorbed by the system, and analysis of the phase-lag and amplitude loss leads to the relaxation time, and hence the reaction rate coefficients.

A time-independent "transfer function" $\mathfrak{S}$ has been introduced[33] which, like the forcing function, can be treated mathematically in terms of a complex exponential. For the practical application of stationary relaxation methods it is not necessary to consider these functions in detail; it is, however, interesting to note the connection between the measured quantities and the transfer function. It is possible to extract the relaxation time from the measured data in two general ways. One method uses the real part of the transfer function $\mathfrak{S}_{re}$, whose variation with the applied circular

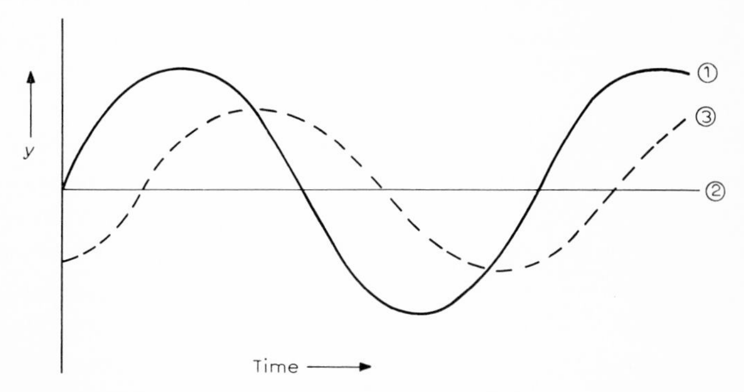

Fig. 9. Chemical relaxation on application of a periodic forcing function (y represents the forcing function and the response, as explained in the text).

References pp. 176–179

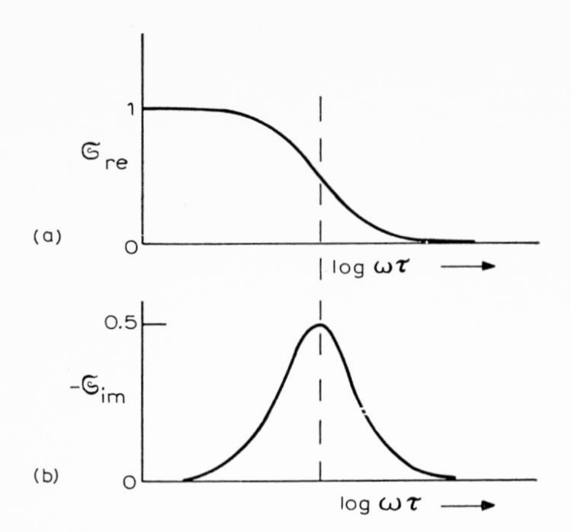

Fig. 10. Real ($\mathfrak{S}_{re}$) and imaginary ($\mathfrak{S}_{im}$) parts of the transfer function.

frequency ω is shown in Fig. 10(a). The relaxation time τ may be readily measured from a curve of this form; it corresponds to the value of ω at which $\mathfrak{S}_{re}$ has the maximum slope. A physical quantity which, for a relaxing system, shows a frequency dependence of this type is the sound velocity. This frequency dependence is called sound dispersion, and consequently the methods which follow the frequency behaviour of $\mathfrak{S}_{re}$ are called dispersion methods.

The imaginary part of the transfer function $\mathfrak{S}_{im}$ has a frequency dependence shown in Fig. 10(b). It represents the amount of energy transferred to the system per period of oscillation owing to the relaxation, and the relaxation time corresponds to the frequency at which the greatest interaction occurs. Again, the effect on sound waves has been widely used. Methods making use of $\mathfrak{S}_{im}$ all involve the measurement of energy absorption; they are called absorption methods.

The relationship between the transient and stationary approaches to the relaxation times has been considered by Eigen and de Maeyer[33]. For any chemical "equilibrium" a system of nonhomogeneous differential equations which represent the rates of concentration change may be set up. The complete solution of the system is the sum of two solutions. One of these depends on the initial conditions of the dependent variables and upon the forcing function (the "transient" solution), while the other depends on the differential equation system and on the forcing function (the "forced" solution). The latter does not depend on the initial conditions of concentration, etc. The step-function methods for studying chemical relaxation experimentally determine the transient behaviour, while the stationary methods determine the steady-state behaviour.

In principle any parameter which causes a measurable displacement in the chemical equilibrium under consideration may be used as the forcing function. Thus, for example, a high-frequency oscillating electric field might be used with solutions of weak electrolytes, or an alternating magnetic field might be used with systems where there is a difference in magnetic moment between reactants and products.

Because of the very small effects expected and, consequently, the immense experimental difficulties involved in measuring them, techniques are not yet well developed which make use of these properties. The bulk of the work has made use of the disturbing influence of high-frequency sound-waves on reacting solutions, and the remainder of this section will discuss briefly the application of acoustical methods to the study of chemical relaxation.

3.3.1 Ultrasonic absorption

When sound-waves are propagated adiabatically through a liquid, the alternating regions of high and low pressure are associated with slight increases and decreases of temperature. If a chemical system whose position of equilibrium is pressure- or temperature-sensitive is present in the liquid, the disturbance caused by the sound-waves may be used as a forcing function in the sense discussed above. It is of no consequence whether, in a given case, the system is reacting to the change in temperature or the change in pressure, since these two are completely coupled. Actually, for electrolytic reactions in water, *e.g.*,

$$NH_4^+ + OH^- \rightleftarrows NH_3 \cdot H_2O$$

the important parameter is ΔV while for reactions of nonelectrolytes, especially in non-aqueous solution (*e.g.*, the dimerization of benzoic acid in toluene), ΔH is more important.

One of the parameters which can be measured directly is α, the attenuation constant†. α may be divided into several parts, the actual number depending on the number (n) of discrete chemical relaxation phenomena in the system under consideration. Thus,

$$\alpha = \sum_n \alpha_{\text{chem}, n} + \alpha_{\text{res}}$$

where $\alpha_{\text{chem},n}$ is that part of the attenuation constant attributable to the nth chemical relaxation process, while α_{res} is the residual attenuation associated with dissipation produced by the solvent or inherent in the design of the apparatus. The plot of log α/ω^2 *vs.* log ω has the typical "step" form (*cf.* Fig. 10(a)) and is shown in Fig. 11, for $n = 2$. The residual attenuation is sometimes determined in a separate experiment on the solvent alone. Strictly, α_{res} need not be the same for the pure solvent as for the solution, for example an electrolyte may alter α_{res} by tending to break up

† The decrease in energy for a plane sound-wave propagated in the direction of the x-axis is given by a factor $e^{-2\alpha x}$, *i.e.*, the energy absorption coefficient is characterized by 2α.

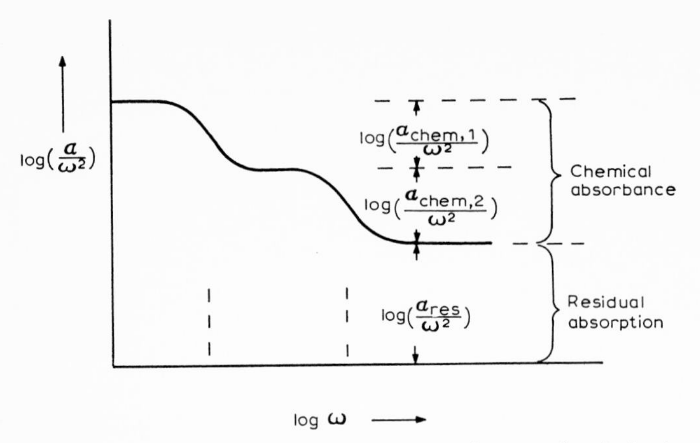

Fig. 11. Variation of (α/ω^2) with ω for a system possessing two chemical relaxation times.

the highly hydrogen-bonded water structure. Also, a relaxation effect which is too fast to be picked up as a distinct "step" will contribute to α_{res}. The most satisfactory way to separate α_{chem} from α_{res} whenever possible is to take measurements over the whole frequency range for which α_{chem} is important.

Another term which is sometimes used is the chemical absorption cross-section, Q, defined as

$$Q = \frac{2\alpha_{chem}}{(N_A c_i^0)} \text{ cm}^2 \text{. molecule}^{-1}$$

(N_A = Avogadro's Number, c_i^0 = concentration of species i in mole.cm^{-3}). A plot of log $[Q\lambda]$ (λ = sound wavelength) against log ω has the form shown in Fig. 12 (*cf.* Fig. 10(b)), where the maxima represent the relaxation times ($\omega_{max} \approx 1/\tau$). A function which gives the same type of plot against log ω is the coefficient of ab-

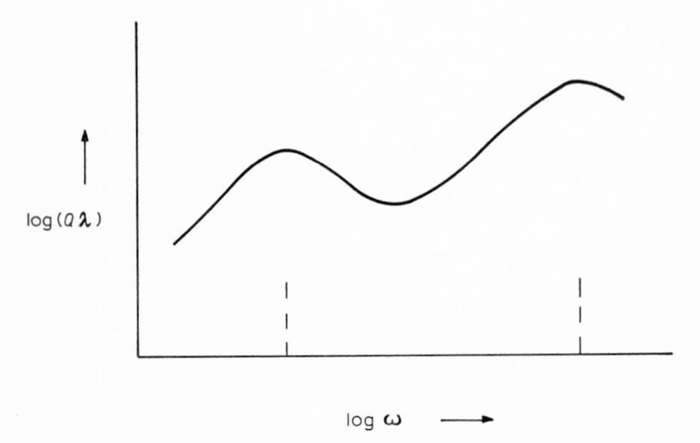

Fig. 12. Variation of $(Q\lambda)$ with ω for a system possessing two chemical relaxation times.

sorption per wavelength. When investigating chemical relaxation phenomena, measurements should be made over as wide a range of frequencies as possible. Generally, it is not possible to cover the necessary range with a single apparatus, so for a given system it might be necessary to use more than one technique. Several methods have been discussed[33]; the following one has been particularly useful for collecting data on the kinetics of protonation and metal-complex formation in aqueous solution.

3.3.2 *Optical technique*

With sound waves of frequency above about $1\,Mc.sec^{-1}$ in an aqueous solution, the Debye–Sears effect may be used to measure directly the interaction between the waves and the solution. A quartz crystal set into the bottom of a glass vessel a few cm long filled with the liquid under investigation is excited by a high-frequency generator at a known frequency, to produce a progressive sound wave in the liquid. The refractive index of a liquid is dependent on the pressure, so, to a parallel light beam passing through the vessel perpendicular to the plane of the wave-front, the alternating regions of high and low pressure behave like a diffraction grating (Fig.13).

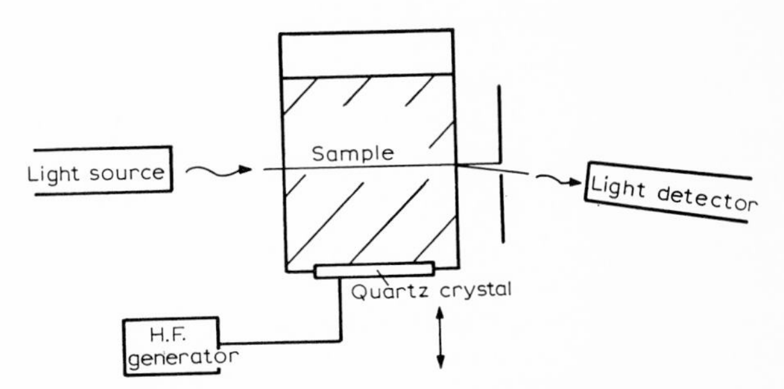

Fig. 13. Ultrasonic absorption: use of the Debye–Sears effect.

This arrangement can then be used either to determine the dispersion or the energy absorption of the solution. Thus the spacing of the diffraction orders depends on the wavelength, a dispersion characteristic. Alternatively, the light intensity in the diffraction orders can be used to determine the sound intensity. This is because an increase in the intensity of the sound wave forces more and more light from the zero-diffraction order into the higher orders (the diffraction lines are becoming "more opaque"). The first-order diffraction is generally separated by a slit and its intensity measured with a photomultiplier. It is then assumed that the intensities of the higher orders are sufficiently weak to allow the approximation that the measured intensity is proportional to the sound-pressure level[53]. The pressure level can now be determined at any distance from the sound-wave source by moving the light beam vertically. In practice, the cell is usually moved up and down through a fixed light beam and the photomultiplier signal, after amplification and rectification, is recorded by a logarithmic pen recorder. To avoid difficulties associated with the continuous illumination of a photomultiplier, it is well to have a pulsed-light signal. This can be achieved elegantly by using a periodically pulsed excitation of the quartz crystal.

3.3.3 *Ion-pair formation of metal salts*

An interesting application of ultrasonic measurements was to aqueous solutions of certain metal salts. The sound-absorption characteristics of several 2:2 electrolytes of the type MX have been measured, and found to have two maxima (*cf.* Fig. 12) from which two relaxation times can be determined[54]. It was not obvious with which relaxation processes these times should be associated, especially since the (diffusion-controlled) rate coefficients for "ion-pair" formation (which is known to occur at the concentrations used, 10^{-2}–10^{-1} *M*) give relaxation times too short to be observed. It seemed, then, that when the two ions $M^{2+}{}_{aq}$ and $X^{2-}{}_{aq}$ had diffused together to within a certain distance two other, slower, processes occurred within the ion pair. The faster of these two effects was more or less independent of the metal ion, whereas the slower was almost independent of the anion but depended strongly on the nature of the cation. The former was attributed to the loss of a water molecule from the hydration sheath of the anion, to give an ion pair in which the M^{2+} and X^{2-} are separated by a single water molecule, while the latter was attributed to the loss of this "bridging" water molecule from the coordination shell of the metal. Such a sequence is represented in Fig. 14.

4. Spectral line-broadening

4.1 NUCLEAR MAGNETIC RESONANCE

In order to appreciate its application to kinetics, a brief outline of the nuclear magnetic resonance phenomenon will be given. This treatment will be rather pic-

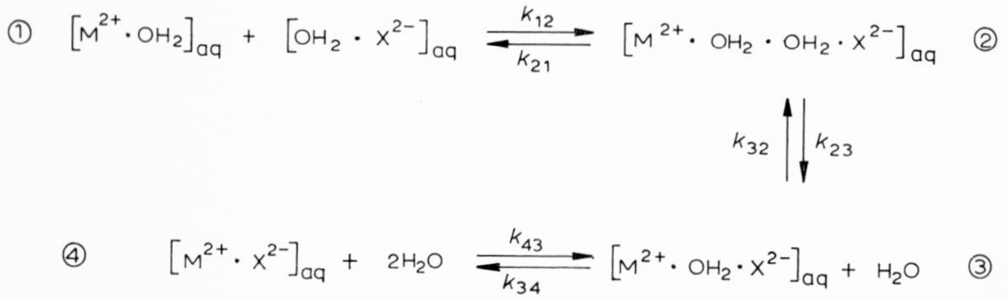

$$\text{①}\ [M^{2+}\cdot OH_2]_{aq} + [OH_2\cdot X^{2-}]_{aq} \underset{k_{21}}{\overset{k_{12}}{\rightleftharpoons}} [M^{2+}\cdot OH_2\cdot OH_2\cdot X^{2-}]_{aq}\ \text{②}$$

$$\text{②} \underset{k_{32}}{\overset{k_{23}}{\rightleftharpoons}} \text{③}$$

$$\text{④}\ [M^{2+}\cdot X^{2-}]_{aq} + 2H_2O \underset{k_{34}}{\overset{k_{43}}{\rightleftharpoons}} [M^{2+}\cdot OH_2\cdot X^{2-}]_{aq} + H_2O\ \text{③}$$

Fig. 14. Ion-pair formation between Mg^{2+}_{aq} and X^{2-}_{aq} (*cf.* Fig. 12). k_{12}, k_{21}: too fast to detect; k_{23}, k_{32}: the shorter relaxation time (+ pre-equilibrium k_{12}, k_{21}); k_{34}, k_{43}: the longer relaxation time (+ pre-equilibria k_{12}, k_{21}; k_{23}, k_{32}).

torial, and a much more complete coverage can be found elsewhere[55,56]. A nucleus whose mass number and charge are both even, which effectively means whose mass number is a multiple of four, such as ^{12}C, ^{16}O, ^{32}S, has no spin. Other nuclei may have spin values I equal to multiples of one half; they are associated with magnetic quantum numbers $m = I, (I-1), (I-2), \ldots, -I$. Of these the most valuable is the hydrogen nucleus which has a spin of $\frac{1}{2}$, but other species with finite spin have also been used in this type of kinetic investigation. A spinning nucleus has a circulating charge which generates a magnetic field and consequently has a nuclear magnetic moment μ. In the absence of an applied magnetic field a nucleus of spin $\frac{1}{2}$ has no preferred direction of alignment, but when it experiences an external magnetic field it can take up only two positions. For $m = +\frac{1}{2}$ it is aligned in the direction of H_0 and at an angle α to it, while for $m = -\frac{1}{2}$ it is aligned at an angle α against the applied field. Since $m = +\frac{1}{2}$ corresponds to a lower energy, this level is more heavily populated. The excess may be calculated by means of the Boltzmann distribution law, and at normal temperatures and moderately high fields (a few thousand gauss) it is only of the order of five parts per million. In nuclear magnetic resonance spectroscopy (*a*) the nuclei are induced to change from one state to another, and (*b*) the frequencies with which these transfers occur under given conditions are measured.

When the axis of a spinning gyroscope lies along the earth's gravitational field lines it is found that there is zero relative motion about the other two axes. If the gyroscope is tilted about one of these other axes the displaced spin-axis begins to revolve slowly about the gravitational field-line. Such "precession" is a direct consequence of the Law of Conservation of Angular Momentum. A nucleus with magnetic moment μ situated in a magnetic field H_0 behaves in a similar way. Because the magnetic moment vector is aligned automatically at an angle α to H_0, precession occurs about the field axis, the angular velocity ω_0 being directly proportional to H_0. The spinning nucleus may be induced to change its spin quantum number when it interacts with a magnetic field rotating about the same axis with the same frequency and phase. Perpendicular to the fixed magnetic field H_0 is applied an oscillating magnetic field H_1, which can be thought of as having a component moving in the same sense as the precessing nucleus. The oscillating frequency of H_1, ω, is varied until it coincides with the precession frequency ω_0†.

Two protons will precess at the same angular velocity only when the magnetic field strengths at their nuclei are the same. In any situation of chemical interest they are bonded to other species, and the applied magnetic field is shielded to some extent by the bonding electrons and further modified by the other charged parts of the molecule. In a molecule containing two or more non-equivalent protons, the NMR peaks do not come at the same place and the difference between the two positions or their positions relative to a standard—the "chemical shift" δ— is of in-

† In practice, it is easier to keep ω fixed and to vary the steady magnetic field H_0. This has the effect of varying the precession frequency ω_0.

terest. If the sensitivity of measurement is high, it is often found that the peaks have a fine structure. This arises from an interaction between the protons on neighbouring atoms which results in a splitting of the energy levels of the spin states. The magnitude of this splitting, or "spin–spin interaction" J, is *independent* of the oscillating frequency ω and the background field H_0. In the application of NMR to the study of fast kinetics the shape and width of the spectral lines are vitally important, and it will now be considered *how* the line-broadening comes about†.

So far the only mechanism considered for changing the magnetic quantum number is that in which an interaction occurs between the precessing nucleus and an oscillating external magnetic field of the same, or approximately the same, frequency. If this were the only path for the transference of energy, resonance would not be observed; energy would only be absorbed until the numbers of nuclei in the upper and lower states had been equalized. As in other forms of spectroscopy, it is necessary for the species in their excited states to be continually losing their excess energy to the surroundings by some form of radiationless transfer so that further absorption of energy can occur. The energy is transferred by "nuclear relaxation", which is exactly analogous to chemical relaxation in that it is a kinetically first-order process and the relaxation time is the time taken for the population difference to drop to $1/e$ of its original value. There are two important types of nuclear relaxation:

(*a*) *Longitudinal Relaxation* (sometimes called spin–lattice relaxation). A particular spinning nucleus is surrounded by many others of the same type which are undergoing constant thermal motion. Such thermal motion is associated with small local magnetic fields oscillating in a random manner, and it will sometimes happen that the nucleus experiences a rotating magnetic field whose angular frequency component ω is equal to its own precessional frequency ω_0. It is then possible for an interaction to occur, and a transition will take place from one spin state to the other. Such a transition can occur in either direction, *i.e.*, the upper spin state can relax to the lower or *vice versa*, corresponding to the gain or loss in translational and rotational energy by the surrounding molecules. Thus a dynamic equilibrium is maintained, and the relaxation time of this process is usually referred to as T_1. T_1 is of the order of a few seconds for most liquids.

The line-width of an NMR signal is related to the lifetime of the nucleus in a given spin-state, and in fact the order of magnitude of the broadening can be estimated from the Heisenberg Uncertainty Principle. If the uncertainty in the frequency of absorption is $\Delta\omega$, the energy difference ΔE, the mean lifetime Δt and Planck's constant h, then $\Delta E \cdot \Delta t \approx h/2\pi$. But $\Delta E = h \cdot \Delta\omega$ and so $\Delta\omega$ is about $1/2\pi\Delta t$. In other words, the line width, on a frequency scale, associated with longitudinal relaxation is of the order of T_1^{-1}.

(*b*) *Transverse Relaxation* (sometimes called spin–spin relaxation). When a group

† NMR has also been used to measure the concentration of a reactant during a chemical reaction. This application involves, in principle, the same considerations as the use of visible spectroscopy to monitor concentration and will not be considered further.

of spinning nuclei is placed in a magnetic field the nuclei will tend to precess in phase with one another. This results in a net rotating magnetic vector with a component in the plane perpendicular to the applied field. Now if the nuclei somehow lose their coherence, the resultant rotating vector decreases. Again, this decay is a first order process associated with a relaxation time which is normally given the symbol T_2. The reason for the name "spin–spin" relaxation is that one mechanism for this loss of coherence involves an interchange of spin between two identical nuclei. Another cause for loss of phase coherence is the local inhomogeneity within the sample caused by the fact that the molecules might have different neighbours. Sometimes also placed under this heading is the broadening caused by slight non-uniformities in the applied magnetic field. This is an instrumental limitation which means that the observed spectrum is broadened because it is really a superposition of spectra from molecules in different parts of the sample.

For most liquids T_1 and T_2 are of the order of a few seconds or less. They are related to the line-width in the Bloch equations and it is possible to derive an exact expression for the line shape and width in terms of T_1 and T_2 (actually, the line width is often determined primarily by T_2). In a reacting system the lines are often broadening further and the experimental line shapes are compared with the solutions of the Bloch equations to which the appropriate terms have been added allowing for the exchange[57]. The exact method of doing this is rather complicated and will not be dealt with here; the detailed treatment generally involves the use of computers, especially in exchange systems which involve more than two lines. It is also possible to measure T_2 directly by the "spin-echo" method[58], a technique which may also be used for following fast chemical reactions[56,59].

Ideally, then, the most kinetic information may be obtained from an NMR signal by comparing the observed with the computed pattern. In many cases this is not possible, or at best very tedious, yet very good approximations of the relevant rate constants may be obtained in a much more straightforward way.

Suppose there are two protons, in environments A and B, in the system under investigation, and that they have resonances at frequencies ω_A and ω_B, respectively, with no spin coupling between them; the kinetics of possible exchange between A and B are to be followed. Suppose also that the concentrations of A and B are equal (although this is not an important restriction since the treatment has been extended[60] to the more general case where $c_A \neq c_B$). The observed spectrum comprises two lines of equal intensity separated by $\delta = |\omega_A - \omega_B|$, the chemical shift. The term τ is now introduced, and this represents the mean lifetime of states A and B, or the time spent by the proton in the environments A and B†. (If the concentrations o

† In some ways the nomenclature is a little confusing. In NMR it is traditional to use T_1 and T_2 to describe the relaxation times of specific nuclear processes and τ the mean lifetime of a nucleus in a particular state. In chemical relaxation methods, however, τ refers to the relaxation time. Chemical relaxation methods and NMR are similar in that in each, *one* time function is measured and its variation with concentration is followed. The difference comes in the relationships of the respective time functions to the concentrations.

A and B are equal, $\tau_A = \tau_B$, but if they are not, the lifetimes are related by the mole fractions N_A and N_B: $\tau_A/N_A = \tau_B/N_B$.) A second case may be considered, that in which two protons in environments A and B are coupled through spin–spin interaction J. Here the spectrum comprises four lines and, if $\delta \gg J$, the two doublets are well-separated and have almost equal intensity. Now it is assumed that only one proton undergoes exchange (A, say, with the solvent). This will have a similar effect on the doublet of B as is observed when A and B exchange, but the modification of the B spectral lines will give information about τ_A. The same treatment therefore applies, to a good approximation, with J replacing δ.

The basis of kinetic measurements by NMR is a comparison of the kinetic behaviour (exchange) with nuclear properties of known frequency, δ or J. What happens as the lifetime τ changes? Fig. 15 shows the changes in two NMR lines separated by δ (or J) as the exchange rate increases. This behaviour may be classified under three headings.

(*a*) "Slow" exchange—the lines begin to broaden and to approach each other. This approximation is given by $\tau\delta \gg 1$ which, on substitution into the Bloch equations, gives

$$1/T_2' = 1/T_2 + 1/\tau$$

for each line. (T_2, T_2' are the transverse relaxation times in the absence and presence of exchange, respectively. They are related to the width of the line at half height ($\Delta\omega$, $\Delta\omega'$) by

$$\Delta\omega, \Delta\omega' = 1/\pi(T_2, T_2').)$$

Methods of evaluating τ from the separation, height and width of the lines have also been given[61].

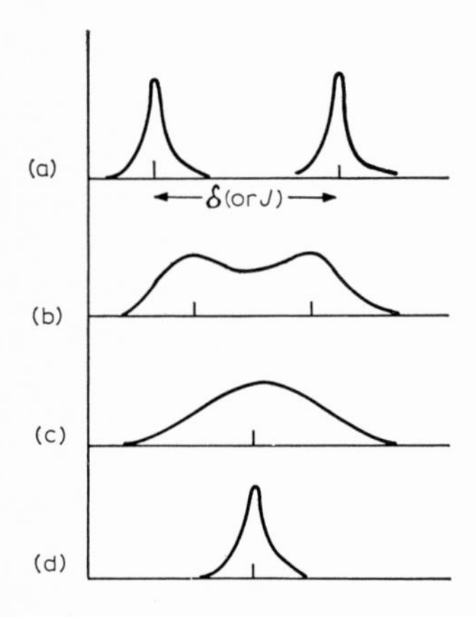

Fig. 15. Variation of NMR line shape with increasing [(a) to (d)] exchange rate (intensities not to scale).

(*b*) The intermediate region—at the coalescence point (c) the exchange broadening is generally greatest. This corresponds to the condition $\tau\delta \simeq 1$, and

$$\tau^{-1} = 2^{\frac{1}{2}}\pi(\Delta\omega' - \Delta\omega).$$

A procedure often used to evaluate τ is to estimate the temperature at which the lines coalesce[62]. The disadvantage of this method is that the coalescence point is usually not very sharply defined.

(*c*) "Fast" exchange—the single line sharpens. This approximation is given by $\tau\delta \ll 1$, and substitution into the Bloch equations gives

$$1/T_2' = 1/T_2 + \tau \sum_{i=1}^{n} N_i \cdot \delta_i^2$$

for *n* lines. Again, methods have been given for evaluating τ from the variation of peak height or width[61].

The rate coefficient for the exchange process *k* is obtained from the measured mean lifetime of species A. In a reaction involving A, B, ... with kinetic orders a, b, ..., then

$$dc_A/dt = kc_A^a c_B^b \ldots = c_A/\tau_A.$$

The first order rate coefficient (τ_A^{-1}) is derived in all cases, and it refers to the system under conditions of chemical equilibrium (*cf.* the rate coefficient derived from the chemical relaxation methods where this equilibrium is slightly displaced). In order to evaluate a, b, ..., τ_A is measured as a function of the concentrations of A, B, ...

4.1.1 Applications of NMR

1. One of the reasons that ethanol is so frequently used to exemplify the application of NMR is that it has a comparatively simple yet distinctive spectrum. It also exhibits very interesting behaviour when mixed with water and dilute acids or bases, and illustrates very well how much information can be obtained from changes in NMR line shape. Three aspects of this behaviour will be considered.

The spectrum of pure dry ethanol is shown in Fig. 16, in which attention should be focussed on the hydroxyl triplet and the methylene modified-quadruplet. If a little water is added, an extra line due to the H_2O-proton appears between the OH and CH_2 bands. As the concentration of water increases three things happen—the OH triplet sharpens to a singlet; this then broadens and coalesces with the H_2O band; at the same time the CH_2 signal becomes a simple quadruplet. These are all consistent with an increasing rate of proton exchange between the OH group and

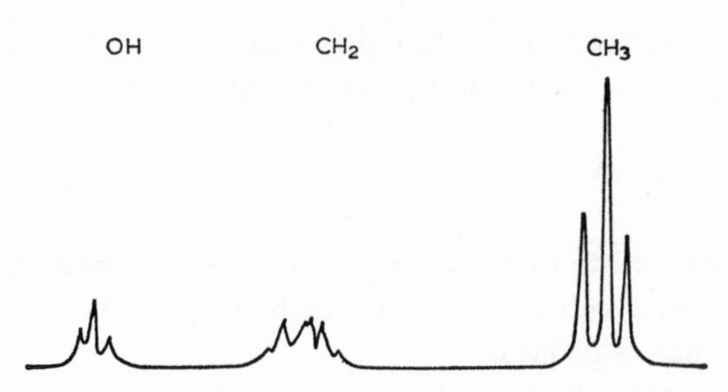

Fig. 16. Proton resonance spectrum of pure dry ethanol (*cf.* ref. 64).

species present in water (*e.g.*, OH^-, H_2O). The exchange lifetime has been determined from the temperature dependence of the coalescence[63].

The addition of a small amount of hydrochloric acid or caustic soda to pure ethanol leads also to a sharpening of the OH-peak and the corresponding simplification in the methylene band. Again, exchange of the hydroxyl proton is the cause but this time several reactions of the type

$$EtOH + A \rightleftharpoons EtO^- + AH^+$$

are involved. A comparison of the shape of the CH_2-band with the theoretical curves has led to the evaluation of the overall first order rate coefficient for the exchange by all of these reactions[64]. This is divided into two components, associated with those reactions involving water and those involving only the derivatives of ethanol. The first of these may be evaluated from the simple H_2O/EtOH exchange, and, by following the variations of the component first order rate coefficients with c_{H^+} or c_{OH^-}, the second order rate coefficients for many of the individual steps have been evaluated.

A third interesting observation is that in pure dry ethanol the chemical shift of the hydroxyl proton relative to the methylene group decreases as the temperature is raised. This has been interpreted in the following way. In ethanol it is known that the hydroxyl protons are involved in hydrogen-bonding to the oxygen atoms of other ethanol molecules and it is reasonable that an H-bonded and a non-H-bonded proton will be subject to different magnetic shielding. If the exchange between these two states is fast, the proton resonance corresponds to the average shielding for the two states. As the temperature is altered, the proportions of ethanol molecules hydrogen-bonded will also change, and so the resonance frequency will be temperature-dependent. This explanation is confirmed by the observation of a similar effect when a non-H-bonding solvent (*e.g.*, chloroform) is added. Many similar "hydrogen-bond shifts" have been observed, and it is sometimes possible to obtain the lifetime of the H-bonded state, or at least to indicate its limits[65].

2. One of the surest ways of reducing the nuclear magnetic relaxation time for a species is to introduce a paramagnetic ion into its vicinity. The NMR spectra of aqueous

and methanolic solutions of transiton metal ions with unpaired electrons have been investigated and much information has been obtained concerning the solvent in the inner coordination shell of the metal. The nuclei in this solvation sheath interact strongly with the unpaired electrons of the metal ion and thus experience a different magnetic field compared with those in molecules in the bulk of the solvent. Thus there is a chemical shift of, say, the water or methyl OH-protons of a coordinated molecule relative to the pure solvent. Most metals exchange solvent water rapidly, so only a single proton resonance line is observed for water resulting from an averaging of the resonances of the bonded and free molecules. Methanol behaves similarly, and for Co^{2+} it has been possible, by lowering the temperature, to slow down the exchange to such an extent that no averaging of the NMR resonances occurred[66]. Another approach to this problem has been to look at the ^{17}O resonance line-broadening associated with the entering of $^{17}OH_2$ molecules into the inner coordination shell of the paramagnetic cation[67]. From these NMR measurements it has been possible to evaluate the rate coefficients of the reaction

$$M(H_2O)^{n+} + \overline{H_2O} \rightarrow M(\overline{H_2O})^{n+} + H_2O$$

for many metals, among them Mn^{2+}, Fe^{2+}, Co^{2+}, Ni^{2+} and Cu^{2+}. It is interesting to note that the rate coefficients measured in this way generally agree very well with those measured by relaxation studies on ion-pair formation of metal salts in water (Section 3)[68].

4.2 ELECTRON SPIN RESONANCE AND OTHER SPECTRAL METHODS

Just as certain nuclei possess spin and associated magnetic moments, so an unpaired electron has a spin and a corresponding magnetic moment. As for the proton, the electron can have spin values of $+\frac{1}{2}$ and $-\frac{1}{2}$. When a strong magnetic field is applied the electron may take up only two positions and it may be induced to pass between the two levels by the application of an oscillating electromagnetic field perpendicular to the steady field[69]. The experimental arrangement is rather similar to that used in NMR spectroscopy. The steady magnetic field corresponding to the resonance condition is determined for an oscillating field of fixed frequency. This time the resonance occurs in the microwave region when H_0 is of the order of 10^4 gauss. The frequency is higher than for NMR because of the much larger magnetic moment of the electron compared with, for example, the proton. It is traditional in ESR spectroscopy to plot the variation of the absorption with field as the first derivative (*i.e.*, the slope of the absorption *vs.* field plot is used as the ordinate, as in Fig. 17) so that the peak corresponds to the field at which the curve cuts the abscissa. This practice makes it especially easy to determine the distance between the points of extreme slope, and this is the parameter which is usually measured in

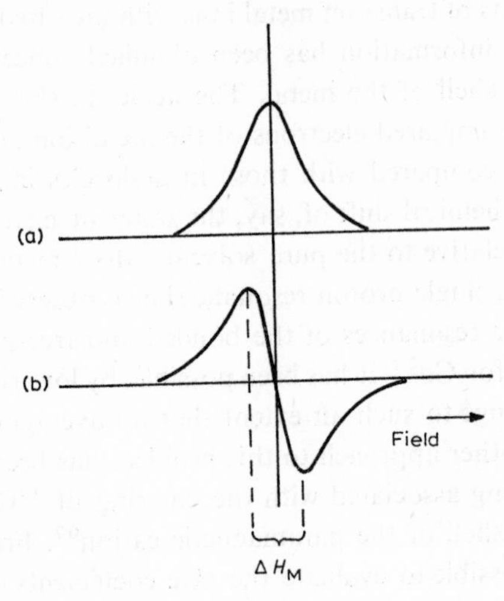

Fig. 17. ESR absorption line (a) and its first derivative (b).

an estimation of the kinetics of a reaction. Generally in a given radical the unpaired electron interacts with species of finite nuclear spin (*e.g.*, ^{1}H, ^{14}N) with the result that the absorption has a hyperfine structure whose nature depends on this interaction. This has the beneficial result that it is usually possible to identify with certainty the species whose reaction is being followed.

Again, it is possible to describe the shape of a spectral line mathematically; as has been noted, the shape of a spectroscopic line associated with a transition between a pair of states is affected by the mean lifetime of the states. Lorentz was the first to relate the lifetime and linebreadth, and more recently other authors have extended this work. Application of the Uncertainty Principle gives the result, if the line shape is assumed to be Lorentzian, that $\Delta\omega = 1/\pi\tau$ where $\Delta\omega$ is the line-broadening at half-height and τ is the average lifetime of the radical under observation. (In fact, the shape of the curve in the case of ESR is probably modified slightly by the exchange process, a factor which somewhat reduces the accuracy of the method.)

The negative ion of naphthalene can be made in a suitable solvent, such as dimethoxyethane (DME), by treatment with an alkali metal in the absence of oxygen. This stable free radical-ion has an ESR spectrum with hyperfine structure. If more naphthalene is added, a broadening of the spectral lines occurs which is attributed to the reaction

$$(\text{naphthalene})^- + (\text{naphthalene}) \rightleftarrows (\text{naphthalene}) + (\text{naphthalene})^-$$

$$(N^- + N \rightleftarrows N + N^-).$$

This reaction is of the second order with rate coefficient k, and it may be shown[70] that

$$c_{N^-}/\tau_{N^-} = c_N/\tau_N = kc_N c_{N^-}$$

(*cf.* in the NMR treatment, above), thus $k = 1/c_N\tau_{N^-}$. This may be expressed in terms of the line-width at half height, but it is a much simpler matter experimentally to measure the difference in field between the points of extreme slope (ΔH_M in Fig. 17) than the half-height width. Thus k is more simply derived from $k = 1.6 \times 10^7$ $\Delta H_M/c_N$, where the numerical constant contains the Landé g-factor, the Bohr magneton and Planck's constant.

This is one of several reactions of this type in which an organic negative radical-ion and its parent molecule react in the presence of an alkali metal. It is found, rather interestingly, that the rate coefficients depend on the nature of the metal. To account for this, it has been postulated that the metal is involved in a bridging role in the activated complex, *e.g.*, dipy . . K^+ . . $dipy^-$ for the case of 2,2′-dipyridyl (dipy)[71]. A more extreme case of this association between the radical-ion and the ion of the alkali metal used to form it occurs in the reaction of benzophenone with its negative ion. The spectrum of $(benzophenone)^-$ in DME has many hyperfine lines caused by the interaction of the free electron with the ^1H and, when the metal is sodium, the ^{23}Na nuclei. When benzophenone is added, the structure, due to the proton interaction, disappears and only the lines associated with the sodium interaction remain. To account for this, it has been suggested that the odd electron moves rapidly over all the proton positions (*too* fast for the lines characteristic of the electron in the different proton environments to be seen), but relatively slowly from one sodium nucleus to another. Seen another way, this means that the transfer of an electron from molecule to molecule is associated with the transfer of the cation[72].

Before leaving ESR techniques a couple of methods will be mentioned which measure the rate of decay of radicals during a matter of milliseconds. Normally ESR cannot be used to follow decay rates in this time range because of the large output time constants of the spectrometer. In what was effectively a combination of pulse radiolysis and ESR, ethyl radicals were produced in liquid ethane by pulses of electrons and two techniques were used to follow their rate of disappearance[73]. The first of these was a rotating sector of the type which has been used in photosynthesis experiments (*cf.* Section 1); the average radical concentration was measured directly from the height of the recorded ESR signal at a fixed field, corresponding to maximum absorption. The time function was varied by altering the speed of rotation of the disc. The second technique used sampling. As discussed in Section 1, the signal/noise ratio of a signal can be increased by averaging many measurements of the same quantity. The intensity of maximum absorption was measured many times at a fixed time-interval after irradiation and averaged; this procedure was

References pp. 176–179

repeated at different times. In the case chosen, a second order recombination rate coefficient of about 10^8 mole.l^{-1}.sec^{-1} was obtained, and the values from the two experiments agreed to within 7 %.

There have been recent attempts[74] to apply the same type of argument to the broadening of Raman lines as a result of chemical reaction. Thus, for example, the 1435 cm^{-1} band of the trifluoroacetate ion, $CF_3CO_2^-$, is broadened considerably in acid as compared with neutral solution. This has been interpreted in terms of the shortness of the lifetime of the species on account of the exchange reaction

$$H^+ + CF_3CO_2^- \rightleftarrows CF_3CO_2H$$

The broadening of other spectral bands (*e.g.*, in electronic spectra in solution[75]) has also been discussed in terms of the mean lifetimes of the molecule in the excited state.

5. Fluorescence quenching

In order to be able to estimate the rate of a reaction by the quenching of fluorescence it is necessary to measure two functions: the relative fluorescence intensities with known concentrations of quencher present and the mean lifetime of the fluorescing molecule in the absence of the quencher[71,77,78]. Because of the efficiency of oxygen as a fluorescence quencher, it is necessary to work with de-oxygenated solutions. In principle the first of these two measurements is quite straightforward. Monochromatic light of a suitable wavelength is directed at a quartz cell containing the solution under investigation and a photomultiplier is used to measure the intensity of the fluorescent light emitted by the solution normal to the irradiating light beam. By the use of suitable filters it is possible to eliminate the effects of any scattered light and diaphragms can be placed so as to make the signal effectively independent of the optical density of the solution[79]. There is one important practical constraint: the fluorescent intensities of the two solutions must be compared under the same illuminating conditions. The same cell may be used consecutively for the two solutions, or the light beam may be divided and passed in turn through the two solutions contained in two measuring cells. Commercial spectrofluorimeters are available. Since only a comparison is required, this is not a very difficult measurement to make. Far more difficult is the measurement of the fluorescent lifetime of the excited species A* in the absence of chemical quenching. Since the lifetime is typically between 10^{-9} and 10^{-7} sec, the ordinary oscilloscope cannot sweep fast enough to present the fluorescence decay curve directly. The method generally used is to determine the phase shift $\Delta\phi$ between a sinusoidally modulated light source and the resultant fluorescence (the latter has the same period as the incident light). The applied frequency ν is usually about 10^7 cycle. sec^{-1}. The lifetime τ_0

is then given by $\tau_0 = \tan \Delta\phi / 2\pi\nu$. The modulation may be brought about by passing the light through a liquid cell across which a suitable periodic disturbance has been applied, for example, ultrasonic waves can cause a suitable liquid column to behave like a diffraction grating (the Debye–Sears effect, *cf.* Section 3). Other rapid periodic-switching devices such as a Kerr cell between two crossed polarizers may be used (although the opaqueness of nitrobenzene to UV light and its tendency to photodecomposition make it a somewhat unsuitable Kerr cell filler).

How may these measured functions be related to the rate coefficients of reactions of the excited species? It is useful to consider how the excitation arises and the various ways in which the excited species may dissipate its excess energy. Suppose A is excited to A* which reacts with B to give products. There are two distinct mechanisms which may be considered for such a fluorescence quenching process. Before the transfer of energy from A* to B is possible the species must form an encounter complex in which the solvent cages of A* and B have been sufficiently modified to allow significant chemical interaction between the two reactants. By the dynamic, or diffusional, pathway A* is formed in comparative isolation from B, and the encounter complex is produced as a second step

$$A(+B) \rightarrow A^*(+B) \rightleftarrows A^* \cdot B \xrightarrow{k_2} \text{products}$$

The static path involves the excitation of a molecule of A which is already in close proximity to B

$$A + B \rightleftarrows A^* \cdot B \xrightarrow{k_2} \text{products}$$

Consider, now, how the energy is dissipated. In the absence of B, A* may lose its energy either as fluorescence emission or in some non-radiative process such as interaction with the solvent. On the convention used by Förster[76] and Weller[78] the rate coefficients or probabilities for these two processes are denoted by n_f and n_d sec^{-1}, respectively. The lifetime τ_0 of the excited species is then $(n_f + n_d)^{-1}$ sec and the quantum yield ϕ_0 of the fluorescence process is $n_f/(n_f + n_d) = n_f \cdot \tau_0$. When B is present, a third method of energy dissipation from A* is by reaction with B; for this a pseudo-first order rate coefficient $k_2 c_B$ sec^{-1} may be assigned. The lifetime of A*(τ) is now given by

$$\tau = (n_f + n_d + k_2 c_B)^{-1} \text{ sec}$$

and the quantum yield ϕ by

$$\phi = n_f/(n_f + n_d + k_2 c_B) = \phi_0/(1 + k_2 c_B \tau_0)$$

Thus the ratio of the quantum yields (which is equal to the ratio of the measured fluorescence intensities with and without B present, I/I_0) is given by

References pp. 176–179

$$\phi/\phi_0 = (1+k_2 c_B \tau_0)^{-1}$$

allowing k_2 to be determined from the variation of ϕ/ϕ_0 with c_B, provided that τ_0 is known. This equation is associated with the names of Stern and Volmer.

In one important respect, this derivation is not quite complete. Just as there are two ways in which the encounter complex A*·B can be formed, so there are two ways in which it can react. Because the average reaction time is comparable to the time taken for the steady state to be set up, only a certain fraction w of the excited molecules will obey the Stern–Volmer equation. The remaining $(1-w)$ reacts immediately after excitation and so does not contribute to the relative fluorescence yield. Put another way, if a molecule of A has a B within the "reaction distance" when it is excited, it may react "immediately" and so will not fluoresce. As may be predicted, the effect of this transient excess reactivity is more important the harder it is for A* and B to diffuse apart, *i.e.*, the greater the viscosity of the medium, and the more efficient is the reaction. Thus $\phi/\phi_0 = w/(1+k_2 c_B \tau_0)$, and the stationary rate coefficient may be evaluated if w is known. The latter can be calculated from the expression $w = \exp(-V_D c_B)$, where V_D is a characteristic reaction volume surrounding A* and w represents the probability that no B molecule will be found inside this space. V_D is a function[78] of the diffusion coefficients of A and B, the mean lifetime of A* in the absence of B(τ_0) and the effective encounter distance. In most cases approximate values of w can be calculated and then, by successive approximations, the stationary rate coefficients and encounter distances which best fit the data are computed.

It sometimes happens that one of the products of the reaction of A* with B is still excited and can itself fluoresce. Such a situation arises in "fluorescence transformation" and, if the spectrum of the fluorescing product is sufficiently different from that of A*, some very useful additional information about the main reaction may be gained. The quantum yield for the species formed in the fluorescence transformation (B*) is given by

$$\phi' = \phi_0' \left[\frac{k_2 c_B \tau_0}{1+k_2 c_B \tau_0}\right]$$

(neglecting the factor w). Using the fact that $\phi'/\phi_0' + \phi/\phi_0 = 1$, plots of the respective ratios of fluorescence intensities against concentration of B should be of the form shown in Fig. 18. I'/I_0' tends to unity as the concentration of B is increased, while I/I_0 tends to zero. Should an important side reaction not have been considered in the reaction scheme, the shape of the relative intensity *vs.* concentration plots will show this.

Before discussing reactions that have been studied fluorimetrically it should be pointed out that the excited form A* may be expected to have very different reactivity patterns from the ground state form of A. Such differences arise because of

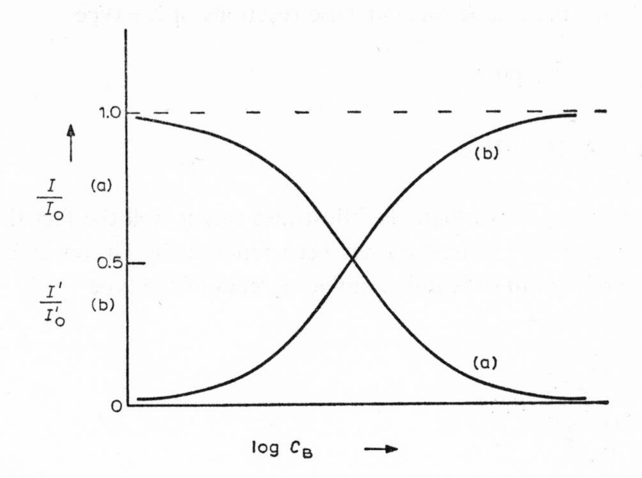

Fig. 18. Variation of relative fluorescence intensities with concentration in fluorescence transformation (see text).

the different energies and electronic wave functions of A and A* and are often sufficiently large to make it not very fruitful to consider the two as the same species. It is possible to divide the systems which have been studied kinetically into four groups.

(*i*) The "simple" fluorescence quenching process in which neither product fluoresces: $A^*+B \rightarrow A+B$. It is thought that this process involves electron transfer, but it has proved very difficult to demonstrate this unequivocally.

(*ii*) Complex formation, either with a molecule of the same species in the ground state or with another species

$$A^*+A \rightarrow A^*A; \qquad A^*+B \rightarrow A^*B \tag{5}$$

Pyrene undergoes this type of "dimerization" in solution; as the concentration increases the original fluorescence spectrum disappears and a new one appears at longer wavelengths. The latter has been shown to be associated with the excited pyrene dimer, which is formed by the diffusion-controlled reaction (5).

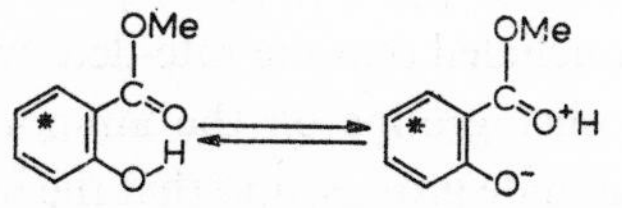

(*iii*) Isomerization: $A^* \rightarrow A^{*\prime}$, where, for example, A can be methyl salicylate and the reaction involves the transfer of a proton across the H-bond (*iv*) Much work has been done on acid–base reactions of the type

$$A^*H + B = A^{*-} + HB^+$$

and

$$A^* + HB = A^*H^+ + B^-.$$

An example of group (*iv*) which also illustrates rather well the fact that A and A* frequently behave very differently has been reported by Urban and Weller[80]. The fluorescence of certain substituted amino-pyrenes of the type

was studied in various alcohols. In the ground state these amino-pyrenes ($ArNR_2$) are weak acids but in the excited state they are very strong acids, having dissociation constants between about 10^2 and 10^6 mole.l^{-1}. Thus in strongly acidic media the unexcited species is mainly in the protonated form $ArNR_2H^+$. From restrictions of the Franck–Condon type it is to be expected that the excited form will be $Ar^*NR_2H^+$ and that its molecular environment will be that appropriate to the ground state, even though in the excited state the deprotonated form is thermodynamically the more stable. However, at room temperature the fluorescence spectrum was found to be due to Ar^*NR_2, and this was explained in the following way. The deprotonation reaction

$$Ar^*NR_2H^+ \overset{k_{12}}{\rightleftarrows} Ar^*NR_2 + H^+ \tag{6}$$

is so fast that the lifetime of $Ar^*NR_2H^+$ reacting by (6) is much less than its fluorescence lifetime. As the temperature was lowered, another emission spectrum appeared and the first decreased in intensity. The deprotonation reaction (6) was evidently being slowed down, so that k_{12} eventually became much less than the reciprocal fluorescence lifetime of $Ar^*NR_2H^+$. Thus during the lifetime of $Ar^*NR_2H^+$ no more proton loss occurred and only fluorescence due to this species was observed. The lifetimes τ_0 were not actually measured, but there was very good reason to suppose that they were about the same as for unsubstituted pyrene, of the order of 10^{-7} sec. One very interesting result of this study was that, although the dissociation constants varied over four orders of magnitude, k_{12} was found to be independent of the substituents on the ammonium–pyrene, but depended significantly on the alcohol used as solvent. It was concluded that the rate-determining step was independent of factors involving the other groups on the ammonium–pyrene, *e.g.*, changes in solvation around the sulphonate groups, and that the important step was the transfer of H^+ to the alcohol. The reaction is also significant in being the first very fast dissociation reaction of a strong acid whose kinetics have been measured.

6. Electrochemical methods

In suitable cases one of this family of techniques may be used to follow fast reactions in solution. They may be appropriate if one of the reactants is electrochemically reduced under conditions where the others are not. In the normal electrolytic reduction process a potential difference is applied between two electrodes placed in the conducting solution, resulting in the migration of ions towards the electrodes and the discharge at the cathode of that species which is most easily reduced—*i.e.*, $X+e^- \rightarrow A$ (in this section, the electroreduced species will be given the symbol X and other species will be represented by A, B, . . .). Because of the discharge of X at the electrode a concentration gradient is set up between the bulk of the solution and the surface of the cathode. The current flowing is proportional to the rate at which X is reduced. This can be expressed mathematically in terms of measurable quantities—an exercise which is greatly simplified if certain conditions are obeyed. In the first place, the species X, which may be an ion, a molecule or a free radical, must be discharged "immediately" on arriving at the cathode; in other words, the discharge process should not be the rate-limiting step. Secondly, it is a help if the cathode is of a shape whose surface area may be simply expressed mathematically—*e.g.*, a sphere or an "infinite" plane. The third simplifying condition is that the mass transfer from the bulk of the solution to the electrode surface be of one or, at most, two types. Three types may be considered: diffusion down a concentration gradient, general convectional motion and movement down or up a potential gradient. The latter is no problem for uncharged X, but it can also be effectively eliminated in the case of ions by using a large excess of an "inert" electrolyte such as potassium chloride to carry the current in the bulk of the solution. These "inert" ions are not discharged at the electrodes until most of the species with smaller electrode potential has been discharged. Typically, the conditions are chosen so that the limiting physical process in the electrolysis is diffusion down the concentration gradient. A current which is characteristic of the system is measured, called the "diffusion current".

Now, suppose that the species which is discharged (the depolarizer) is involved in an equilibrium of the type

$$X+B \rightleftarrows C \tag{7}$$

of which the equilibrium constant is considerably greater than unity. Again a concentration gradient is set up but, if the dissociation of C into X and B is slower

References pp. 176–179

than the electrode process, the discharge of X is controlled partly by diffusion and partly by the rate of attainment of equilibrium (7). In an extreme case, the overall discharge process may be controlled completely by reaction (7). In this case it is also possible to measure a characteristic current, which is less than the diffusion current. This is the "kinetic current" since it is controlled by the kinetic process (7).

Yet a third situation may be considered: where the depolarizer is regenerated by a chemical reaction—*e.g.*,

$$X+e^- \rightarrow D \quad \text{followed by} \quad D+E \rightarrow X+F \tag{8}$$

In such a situation a "catalytic current" is measured. The crux of the electrochemical methods is to relate the ratio of kinetic, or catalytic, and diffusion currents to the rate coefficients of the respective rate-limiting chemical processes. In fact, the currents are comparatively easy to measure, and simple and cheap apparatus is available for doing this. The main disadvantage is that the theory is somewhat difficult to apply quantitatively.

Four main methods which use electrode processes have been developed, although a few others are being investigated[81]; of these, by far the most work has been done on the first.

Group 1, in which a steady voltage is applied between the electrodes:

(*i*) polarography uses a dropping mercury cathode and the current is measured;

(*ii*) the rotating-disc method uses a spinning platinum disc as the cathode and the current in the steady state situation at a series of rotation speeds is measured;

(*iii*) the potentiostatic method uses fixed electrodes and the fall of current with time is measured.

Group 2, in which a steady current is applied:

(*i*) in the galvanostatic method (also called electrolysis at constant current) the rise of voltage with time is measured.

6.1 POLAROGRAPHY

The reason why most applications have used polarography is that the equipment is particularly simple. The polarographic method will be considered here in a little more detail, as well as a couple of systems studied kinetically with it. Discussions may be found elsewhere of the electrochemical methods[82,83].

The anode is nonpolarizable (*i.e.*, its potential is independent of the current flowing through the cell) and is typically a pool of mercury at the bottom of the cell. The dropping mercury cathode comprises a capillary tube of about 0.03 mm bore connected to a mercury reservoir and held with its tip below the surface of the electrolyte. The mercury head, of the order of 50 cm, is adjusted to give a drop every 3 sec or so. A platinum wire is immersed in the mercury reservoir and the circuit is

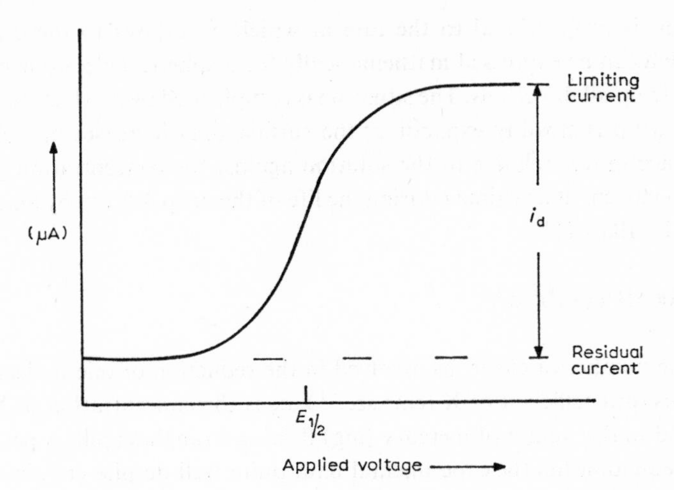

Fig. 19. Current–voltage variation associated with a polarographic wave.

completed through a potentiometer. Because the reduction process of oxygen might interfere, the latter is often removed from the solution before the experiment. The applied potential difference between the electrodes is gradually increased and the corresponding current, usually of the order of a few microamps, measured. A typical current–voltage plot is shown in Fig. 19; the "polarographic wave" represents the reduction of X. The half-wave potential, $E_{\frac{1}{2}}$, is typical of the species X, and the limiting current i_d is determined†.

$E_{\frac{1}{2}}$ is often similar to the electrode potential given by the Nernst equation

$$E = E_0 - RT/nF \ln (a_A/a_X)$$

where a_A, a_X are the activities of the reduced and oxidized species, respectively, but sometimes under polarographic conditions the Nernst equation is not obeyed. The displacement from the potential calculated by the Nernst equation, the "over-voltage", may be as much as a volt or so, and is negative for cathodic processes. The periodic change in area as each mercury drop grows and falls results in an oscillating current varying between (almost) zero and a maximum value, i_d, just before the drop falls. In practice either i_d can be measured with a device which is relatively undamped or the average current, $\bar{i}_d$, with a well-damped instrument. $\bar{i}_d$ over the complete drop-life is $\frac{6}{7}$ of i_d. Because the current is so small the complete current–voltage curve may be traced several times with the same sample without causing noticeable change in concentration.

† More strictly, the difference between the limiting and the residual currents, since if several species are electroreduced a corresponding number of steps is obtained at successively higher half-wave potentials.

The current is proportional to the rate at which X diffuses to the cathode, a function which can be expressed mathematically for a spherical electrode on which X converges from all directions. The situation is complicated by two factors. Because the mercury drop is steadily expanding, the surface area increases gradually and also the surface moves relative to the solution against the concentration gradient. The diffusion current at any time t during the life of the drop is given by an equation first derived by Ilkovič[84]

$$i_t = 706nD^{\frac{1}{2}}c_X \mathrm{m}^{\frac{2}{3}}t^{\frac{1}{6}} \ \mu\mathrm{A}$$

where n is the number of electrons involved in the reduction of one molecule of X, D is the diffusion coefficient of X ($cm^2.sec^{-1}$), c_X is the concentration of X (millimole.l^{-1}) and m the weight of mercury (mg) flowing from the capillary per second. The Ilkovič equation fits the experimental data quite well despite certain assumptions of rather doubtful validity used in its derivation—*e.g.*, it supposes no stirring and neglects the curvature of the electrode surface. Because the *average* diffusion current (*i.e.*, the difference between the average current at applied potentials considerably above and below the half-wave potential, *cf.* Fig. 19) is usually measured it is common to see the Ilkovič equation in the form

$$\bar{i}_d = 607nD^{\frac{1}{2}}c_X m^{\frac{2}{3}}t_m^{\frac{1}{6}}$$

where t_m is the drop time in seconds. Now if the depolarizer takes part in a chemical reaction, the Ilkovič equation is not obeyed, but is modified in a way depending on the type of reaction. Several such cases have been discussed[83].

A system giving a kinetic current is a well-buffered solution of pyruvic acid, $CH_3COCOOH$. The relevant reactions can be abbreviated

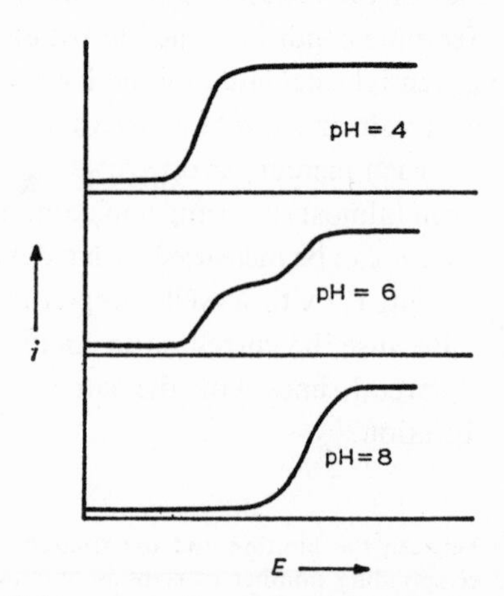

Fig. 20. Polarograms of pyruvic acid buffered at different pH.

$$A^- + H^+ \underset{k_b}{\overset{k_f}{\rightleftarrows}} HA(X) \xrightarrow{(e^-, H^+)} G \tag{9}$$

The undissociated acid HA is reduced at a more positive potential than (*i.e.*, "before") A^-. At low pH only one wave, with $E_{\frac{1}{2}}$ appropriate to the reduction of HA, is observed while at high pH again one wave is found, but at the half-wave potential associated with A^-. At intermediate pH *both* waves are observed (Fig. 20). How does this double wave arise, and how can it be used to measure the rate coefficients k_f, k_b?†

The disadvantage of the rigorous approach[85] to this type of problem is that the mathematics are very difficult for any case except the most simple (*e.g.*, this first order reaction). Instead, it is common to consider the somewhat vague concept of the "reaction layer". This is an approach which gives a physical idea of the processes involved as well as allowing rate coefficients to be derived for more complicated kinetic mechanisms. The reaction layer is a hypothetical layer surrounding the electrode within which all the HA molecules produced by reaction (9) reach the electrode and are reduced. Its thickness μ depends on the reverse rate coefficient, k_b ($\mu = D^{\frac{1}{2}} k_b^{-\frac{1}{2}}$). Suppose the applied potential is only sufficient to discharge HA. Even when the concentration of HA near the electrode is very small, it will be discharged and the equilibrium (9) will be disturbed. In this situation, the concentration of A^- near the electrode will decrease and a contribution to the current due to the diffusion of A^- and subsequent production of HA will be added to the (small) contribution due to the diffusion of HA. As the voltage is increased, the current will increase to an (average) maximum value, $\bar{i}_k$, which is less than $\bar{i}_d$ and depends on the pH and the relative values of k_f and k_b. On further increasing the voltage, A^- will be discharged and the remainder of the current–voltage curve will be like that found at high pH. The Ilkovič equation has been modified for this situation

$$\frac{\bar{i}_k}{\bar{i}_d} = \frac{0.886[(k_f K)t_m]^{\frac{1}{2}} c_{H^+}}{1 + 0.886[(k_f K)t_m]^{\frac{1}{2}} c_{H^+}}$$

where $K = k_f/k_b$. This expression can be modified slightly to take account of spherical diffusion and the different diffusion coefficients of A^- and HA. The forward rate coefficient for pyruvate plus a proton[86] is 1.3×10^{10} mole.l.$^{-1}$sec^{-1}.

The electrolytic reduction of hydrogen peroxide involves a very large overvoltage, yet H_2O_2 is a relatively good oxidizing agent. This is why it is possible to find many systems involving H_2O_2 which give a catalytic current. In reaction (8), hydrogen peroxide would be substance E. Ferric ion (X) can be reduced polarographically to

† It was originally thought that each step might arise from the simple discharge of the equilibrium concentrations of the two species. This was disproved by showing that the pH at which the double steps occurred were considerably different from the pK_a of the acid (around which pH the concentrations of A^- and HA are similar).

ferrous (D) at potentials at which H_2O_2 is virtually unaffected. In the presence of an excess of hydrogen peroxide, Fe^{II} is reoxidized to Fe^{III}; thus the current for the reduction of ferric ion is increased above its normal value. The difference between the limiting currents in the presence and absence of the oxidant is the catalytic current $\bar{\imath}_c$ and expressions have been derived[87] for the ratio $\bar{\imath}_c/\bar{\imath}_d$.

Polarography and the electrochemical methods in general have been used to study the kinetics of many systems, and the results usually agree with those obtained from other fast reaction techniques. However, there are some data from acid-base reactions in particular which do *not* agree. Sometimes the rate is too low, and this might be explained in terms of a reaction preceding the main reaction. Sometimes the rate coefficient is larger than that calculated for a diffusion-controlled reaction and so must be regarded with a great deal of suspicion. But, provided that the possible complications are taken into account[83] the electrochemical methods can be expected to give reliable data.

7. Flow methods

7.1 FLOW METHODS IN SOLUTION

The application of flow methods to the study of kinetics in solution has been discussed in detail[88,89,15]; it stems[90] from the original work by Hartridge and Roughton on the reaction of oxygen with haemoglobin. An excellent survey has been given by Roughton and Millikan[91] which discusses the general aspects of flow methods.

Fig. 21 shows the general experimental arrangement. Two solutions, A and B, are allowed to flow into a mixing chamber and the mixed solution flows with a velocity of v cm.sec.$^{-1}$ along a single tube past an observation position x cm from the mixer. A suitable function of the concentration of reactants or products, such as optical density at a given wavelength or conductivity, is monitored electronically and the extent of the reaction in a small section of the reaction mixture corresponding to a known time x/v sec after the start of the reaction thus determined. By varying certain functions, the reaction profile can be constructed, and exactly how this is done depends on whether the method is continuous-flow, accelerated-

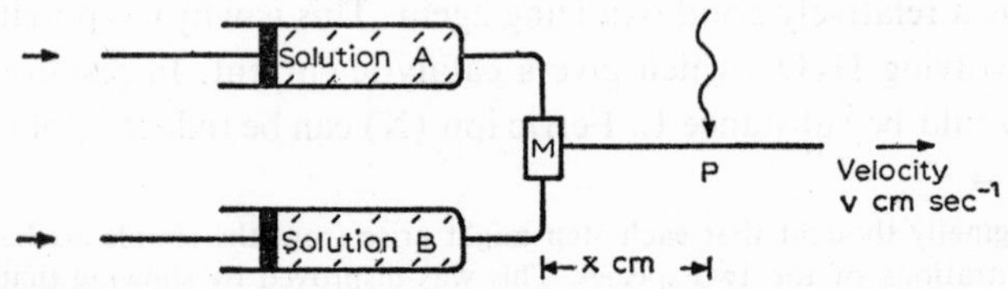

Fig. 21. Generalized scheme of flow-mix system, showing mixing chamber M and observation position P.

flow or stopped-flow. The order in which they are listed here is the order in which they were developed, which was to some extent fortuitous and depended partly on the availability of specialized apparatus such as fast-responding photocells and on the specific reactions which were first investigated. Although the techniques are being used to a greater extent to follow inorganic reactions, for which in many cases they are ideally suited, it is rather interesting that most of the earlier work was done on biological systems. This is especially so in view of the complications often associated with the latter, such as extreme sensitivity to atmospheric oxygen, or severe scattering of the analyzing light by species of large molecular weight. In general, the technique which is most useful is the stopped-flow, and it is the one used unless a feature of the reaction makes one of the others preferable. One aspect of flow methods which should be emphasized, however, is their versatility. The design of a particular apparatus will depend very largely on the type of reaction to be followed.

For stopped-flow, the reagents are kept in two hypodermic syringes of equal volume, typically between 2 and 50 ml, and the plungers are pushed together so as to deliver equal amounts of solution to the mixer. The syringes are mounted on a metal block and a bar, running along parallel guide-rods and pushed either manually or mechanically, ensures even delivery of solution. The observation position is fixed, and the distance between it and the mixer outlet is often as short as is practicable (usually a few cm). The reactant liquids are made to flow and are then abruptly stopped; the change in concentration in the small column of mixed liquid at the observation point P is followed with time. By the time the mixed liquid reaches P a certain fraction of the reaction, which depends on the half-life, on v, and on x, will be over and the lower limit on the half-lives of reactions which may be followed thus depends on v and x. If the arrest of the liquid flow occurred instantaneously, the half-life of the fastest observable reaction would be about x/v sec. Of course, instantaneous arrest is not possible and the usefulness of stopping the flow depends largely on how fast it can be done. The mixed solution passes into a third syringe arranged[92] so that the plunger is forced against a block. If the main syringes are motor-driven, the motor must evidently be disconnected immediately when the liquid flow ceases. The timing is critical, but once the apparatus is set up it is very easy to follow a complete reaction with only a fraction of a millilitre of each reagent.

In the continuous-flow technique the reaction profile is followed by keeping the flow rate constant and varying the distance between M and P. Thus, a measured time parameter is replaced by a measured distance, x. (In an alternative procedure, x can be kept constant and the flow rate v varied stepwise.) The reagents may be contained in large reservoirs and forced into the mixing chamber under a constant head; the observation position can be changed at leisure. It is especially important here to maintain a constant flow rate and also to ensure that the bore of the observation tube is uniform. The continuous-flow method is especially useful when an observation technique with a particularly long response time is being used. Thus, for example, it can be used in conjunction with an ESR spectrometer or a commercial spectrophotometer such as a Beckman DU. The main disadvantage is the vast amount of reactant solution required.

The third method, accelerated-flow, is similar to stopped-flow in many respects. It, also, has the advantage of requiring only small quantities of solution, and the reaction is followed at only one position, x. The reactants are driven from syringes, and the concentration changes at P are followed from the start of the drive until the maximum flow-velocity has been achieved. The motion of the syringe driver-bar is also followed electronically and, by making use of the fact that the time between mixing and observation is inversely proportional to the flow rate, the conventional concentration–time reaction profile can be constructed. When it was first introduced, the accelerated-flow method had the advantage over the stopped-flow in that reactions with shorter half-times could be followed. Since more efficient stopping devices are now used this is barely true, and the rather more complicated electronic circuitry required in the former has led to its eclipse by the stopped-flow method. Both techniques require detecting systems with very fast response times.

It has already been mentioned that the greatest single difficulty to be overcome in applying "conventional" techniques to the investigation of the kinetics of fast reactions is the problem of mixing the reagents in a time short compared with the half-life of the reaction. In fact, it is true to say that the crux of any fast kinetic technique is the way in which this mixing problem is either overcome or avoided. Much thought and experimentation has gone into the design of the mixing chambers for flow systems and although much of the work was done on continuous-flow, the same considerations apply to the other flow methods. A simple capillary T-junction mixer is quite adequate for reactions of half-life greater than about 10 milliseconds[88], but to increase the time range below this, special designs are necessary. It is important for the motion in the outlet tube to be as near to "mass flow" as possible—a (hypothetical) condition typified by zero relative motion along the axis of flow between liquid elements at the centre and at the periphery of the column—from the point of view of following the concentration changes at P since it is assumed that the concentrations are uniform across the tube. If a velocity greater than a certain empirically-derived critical velocity (which depends on the diameter of the tube and the viscosity and density of the liquid) is achieved, the flow will become "turbulent". Associated with turbulent flow is a rotational movement which helps to give efficient mixing. This feature is especially important for stopped-flow work[93]. Velocities of the order of 10 metre.sec.$^{-1}$ are quite common; a distance of 1 cm from the mixing chamber would then correspond to a reaction time of about a millisecond. The mixer design most commonly used in such work has at least two pairs of inlets arranged semi-tangentially with respect to the central outlet. A further important point about mixer design is that efforts must be made to avoid cavitation (the local "boiling" of the solution generally associated with a sudden decrease in pressure). Rough edges and sudden decreases in overall cross-sectional area should be avoided and wherever possible it is advantageous to de-gas the solutions. The reaction half-life limits measurable by flow techniques are a few tenths of a millisecond at one end and seconds or even minutes at the other. The upper limit is often set by the diffusion of unreacted solution from the delivery tubes towards the observation position. Such a difficulty been has overcome[94] by closing a tap between M and P after the mixture has been delivered.

A few interesting modifications to the general experimental arrangement have been made. Of particular interest for biological problems, where the amounts of the reactants available are liable to be very small, is an apparatus of the stopped-flow type which uses an optical microcell as the observation chamber. In this variant[95] the volume of each solution is as little as 15 μl. A slight modification to the continuous-flow method, in which each experiment yields just one point on the reaction profile plot, uses quenching of the reaction followed by chemical or physical analysis over an unrestricted time[96]. Thus the reagents are mixed as normal (Fig. 21), but at P they are mixed further with a substance which quenches the main reaction, and a sample is removed for analysis. A reaction which has been studied in this way[97] is the radiochloride exchange in acetone of organosilicon chlorides with Cl^-, a reaction which could be quenched by precipitating the free chloride ions as sodium chloride

$$R_3SiCl + {}^*Cl^- \rightleftarrows R_3Si^*Cl + Cl^-$$

followed by

$$Na^+ + \begin{cases} Cl^- \\ {}^*Cl^- \end{cases} \rightarrow \begin{cases} NaCl \\ Na^*Cl \end{cases}$$

Counting the washed precipitate enabled the fraction of radiochloride to be determined. The variables in such a technique are again v and x. A system which combines flow and relaxation (Section 3) techniques[46] will be of great value for studying the reactions of unstable intermediates, for example in enzyme reactions where several fast steps may precede the rate-determining step. In a flow tube it is often possible to observe these intermediates, and at a given distance from the mixer a steady state will be set up. If this steady state is perturbed by a sudden increase in temperature a relaxation process will result, and the rate coefficients of all the steps before the rate-determining one may in principle be deduced from the spectrum of relaxation times. One such technique incorporates a mixer in close proximity to one of the electrodes of an electrical-discharge temperature-jump apparatus and allows the mixture to flow towards the other electrode[98]. The change in optical absorption of the solution is observed perpendicular to the direction of flow.

One of the inorganic systems studied by flow methods[99] takes advantage of the fact that reactions with half-lives in the millisecond range may be followed but faster ones will not be seen. The apparatus used was of the conventional stopped-flow type[100] using two 50 ml reagent syringes driven *via* a magnetic slip-clutch by a variable-speed motor. The syringes were connected to the mixing chamber by 3-way taps which were used to evacuate the system (for the removal of air-bubbles) and to admit the reactant solutions. The stopping syringe activated a trigger circuit which had the dual purpose of triggering the monitoring oscilloscope and stopping the drive motor. One reactant solution contained an organic halide, RX, and a cobalt(II) solution, while the other contained sodium cyanide (of at least 5 times the concentration of the cobalt) and an equivalent amount of sodium hydroxide to suppress hydrolysis. These solutions were de-oxygenated since the first product of the reaction is air-sensitive. The first product, $Co^{II}(CN)_5^{3-}$, was formed within the mixing time, of the order of 1 msec, and it was the reaction of *this*, unstable, compound with the organic halide, RX, whose kinetics were of interest.

$$5\,CN^- + Co^{2+} = Co(CN)_5^{3-}$$

(very fast; occurred within the mixing time)

$$2\,Co(CN)_5^{3-} + RX = Co(CN)_5R^{3-} + Co(CN)_5X^{3-}$$

(fast; kinetics of this reaction followed)

The latter reaction was followed by observing the increase in light absorption due to the product $Co(CN)_5X^{3-}$. A two-step mechanism was proposed for this process; the rate-determining production of R· radicals

$$Co(CN)_5^{3-} + RX \rightarrow Co(CN)_5X^{3-} + R\cdot$$

being followed by

$$Co(CN)_5^{3-} + R\cdot \rightarrow Co(CN)_5R^{3-}.$$

7.2 FLOW METHODS IN THE GAS PHASE

Mention has already been made of one of the inherent difficulties of following the kinetics of fast gas reactions. The absence of large concentrations of inert "third bodies" means that temperature equilibration in gases tends to be comparatively slow. This is quite a serious problem in flow systems since it is difficult to counteract any thermal effects due to the heat of reaction. In addition any change in pressure caused by a possible mole number change and the pressure drop associated with the flow make it difficult to define the reaction conditions accurately. Despite this, kinetic data obtained using gas-phase flow systems have been reported[101].

The general experimental arrangement is similar to that shown in Fig. 21. One reaction component is passed along a tube inside a thermostatted compartment which is often, though not invariably, held at a comparatively high temperature. The other component is injected into the main stream and as the gases emerge from the thermostat they pass a detector. In the cases when the reaction is quenched, it is assumed that this process also occurs instantaneously—an assumption which is difficult to uphold rigorously. The distance between the injection point and the detector may be varied, and for each measurement the steady state is set up. If one of the reactants is in the atomic state, it is often made from the molecular species by passing it through a tube across which a radio-frequency discharge is applied between two external electrodes. The uncertainties in defining the reaction time and temperature tend to be minimized by using long reaction tubes and low flow rates.

An example of a gaseous process whose kinetics have been measured by the flow method is the decay of nitrogen atoms in the presence of oxygen molecules[102]. A stream of nitrogen atoms, generated from gaseous molecular nitrogen with the help of rf power, was passed along a quartz reaction tube of about 150 cm length and 2 cm diameter situated inside a furnace held between about 150° and 450° C. There were four inlet jets approximately equally spaced inside the furnace for admitting gas to the reaction system. The light emitted by the emergent gas was observed with a photomultiplier placed with its axis normal to the tube, and the pressure near the observation point was measured with a manometer. The oxygen was admitted to the atomic N stream through the first jet and then at successive downstream jets nitric oxide was added. The latter was to determine the concentration of N atoms in the gas arriving at the particular jet since active nitrogen can be titrated with NO, the end-point being a sharp transition from a blue NO-emission to a green continuum (the air afterglow). The end-point corresponds to the complete removal of N atoms by the very rapid reaction

$$N + NO \rightarrow N_2 + O$$

The primary reaction, whose kinetics were followed in this system, was

$$N + O_2 \rightarrow NO + O$$

It was found to be first order in both N atom and oxygen molecule, and the second-order rate coefficient of $8.3 \times 10^9 \exp(-7100/RT)$ l. mole^{-1}. sec^{-1} was evaluated.

8. Flames

A flame which is commonly used in the laboratory for kinetic measurements (such as that from a modified Bunsen burner) is essentially a steady-state system with the combustion zone fixed. The processes involved in flames are radical chain reactions in which the three normal stages (initiation, propagation and termination)

References pp. 176–179

are discernible, but the first of these is not very important since the reaction is started in the unburnt fuel by active species which have diffused against the flow. In principle there is nothing other than experimental convenience to distinguish such a fixed flame from one moving along a tube of fuel.

What type of system is suitable for studying in this way? Firstly[103,104,105], the situation is simplified considerably if the reactants are gases. Work has been done on suspensions of liquid droplets and even fine solid particles, but complications arise associated with the volatilization steps prior to the reaction. Secondly, and far more importantly, the system must have the rather delicate balance of properties necessary to maintain a steady-state flame. Active species and heat must be produced and transported back against the flow of fresh reactants sufficiently rapidly to be able to sustain the reaction, but not so fast as to produce an explosion. Chemical systems which have the necessary properties are quite numerous and rather varied in nature. The simplest flames are "decomposition flames", in which a single substance (*e.g.*, ozone, hydrazine) undergoes decomposition exothermally. Far more common are flames in which there are two initial components, which are termed fuel and oxidizer. This category includes such well-known examples as hydrocarbon and oxygen, hydrogen and oxygen or hydrogen and bromine, as well as some rather less likely ones such as diborane and hydrazine or atomic nitrogen and oxygen.

The size of the combustion zone is determined by the rates of the reactions and by the gas flow rate, which in turn depends on the pressure of the gas–the larger the pressure and flow rate the smaller the combustion zone. As in other flow systems, the time parameter is replaced by a distance, so it is generally true that the combustion zone should be as large as possible to give the greatest spatial resolution. A flame can be looked at theoretically either in terms of a continuous fluid or in terms of individual molecules; the theory most often used is the Chapman–Enskog Theory of dilute gases[106]. It is simplified considerably if the geometry of the flame is kept relatively simple, which is the case, especially, when it can be treated as uni-dimensional, *e.g.*, as flat and circular or as spherical. The use of such shapes also simplifies the experimental problems significantly since the flame dimensions must be measured. In practice it is usually possible to define a system completely from one independent and $(n+1)$ dependent variables, where n is the number of chemical components involved in the reaction†.

Normally the measured quantities will be the weight fractions of $(n-1)$ species, the flame velocity (for a one-dimensional flat flame this is the velocity component of the cold gas normal to the front), the temperature profile of the flame and the distance, although in certain cases other parameters may be substituted[103]. These are the quantities whose measurement is difficult and produced the problems which

† It is generally necessary to draw a dividing line between those species which are present in significant concentration and those which are not. The position of this line depends to a large extent on the nature of the system.

held up progress in flame kinetics. In addition the pressure (assumed to be constant throughout the flame), the mass flow rate and the gas composition at the inlet can be determined relatively easily. From a knowledge of these quantities, and by applying continuity equations for each species (in which the rate of formation is balanced against the mass flux across the flame) and the energy conservation equation, values of the rate coefficients involved may be deduced. In general, for a reaction of unknown mechanism it is not possible to work out the complete kinetic pattern from a single flame[103]. Thus, for example, the properties in two flames of different concentration ratios at positions of the same temperature would have to be compared.

Temperature is defined statistically in terms of the way in which the energy of the particles comprising the system is distributed, which means that the temperature is defined effectively by the Maxwell–Boltzmann distribution function. Because the species encountered in flames are generally polyatomic, they have internal degrees of freedom (rotational, vibrational or electronic), and it is possible to define separate temperatures with respect to translation and, say, vibration, provided that the energy in each degree of freedom has an equilibrium distribution of the form described by the Maxwell–Boltzmann equation. These temperatures will be the same only if the energy is equilibrated throughout the system, but in flames we are dealing with time scales which may be short compared with the relaxation times for this energy redistribution. Thus it may be that in the reaction a species is produced in which some particular energy level is populated to an unusual extent. In this case it is quite conceivable that the temperatures defined for the various degrees of freedom are not equal, and so the term "temperature" may have no precise meaning†. To circumvent this difficulty, it is usually assumed that equilibrium exists on a local scale (which is usually true in the slower and cooler flames) and that the particles have an essentially Maxwellian energy distribution and so may be assigned a single temperature. The matter has been discussed in greater detail by Broida[107].

A very schematic representation of a system for following kinetics in flames is shown in Fig. 22. The arrangement used in a given case will depend to a large extent on the nature of the reaction. The type of equipment available has been discussed very thoroughly by Fristrom and Westenberg[103]; it will be considered here briefly under four headings.

(*i*) *The burner*. One of the most important requirements of a flame used in a kinetic investigation is that it be "stabilized". This means that a stable equilibrium situation must exist such that the combustion zone maintains its fixed position relative to the observation points. Theoretically, it should be possible to stabilize a flame by balancing the propagation velocity against the gas velocity, but this cannot be done in practice since the gas pressure, and so the gas velocity, cannot be kept sufficiently constant. It has already been noted that considerable simplification arises

† Indeed, if the energy distribution in a particular degree of freedom is far from Maxwellian, no temperature may be defined for that degree of freedom.

References pp. 176–179

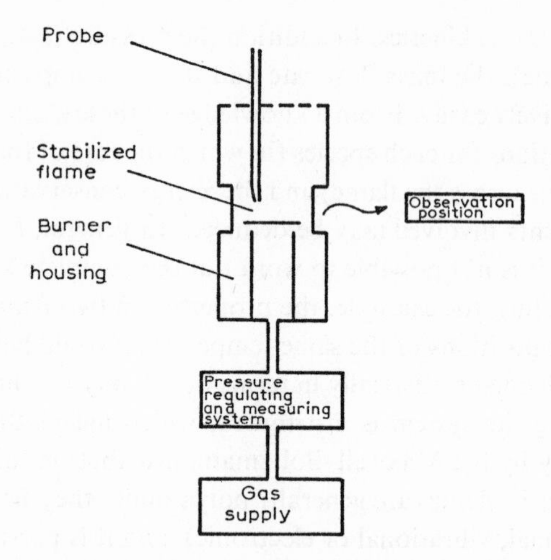

Fig. 22. Schematic arrangement for flame studies.

if a uni-dimensional flame is used. A flat, free-burning flame cannot be produced, but a convenient way of obtaining a flat flame is to support it on a circular mesh screen. By acting as a heat sink (and so removing heat from the combustion zone) such a screen also stabilizes the flame. It is usually necessary to enclose the flame in a gas-tight housing to avoid the effects of small fluctuations in atmospheric conditions and also to allow gas pressures other than atmospheric to be used.

(*ii*) *Pressure regulation and measurement*. One of the simplest and most satisfactory ways of maintaining a constant pressure on a gas flame is to locate a restriction (effectively a barrier with a small hole) between the burner and the pumping system. The dimensions of this "choking orifice" and the relative pressures above and below it are chosen so as to give the emergent gas the velocity of sound at the hole. If such a condition can be achieved the orifice acts rather like a valve; the flow becomes a function of the cross-sectional area and of the upstream pressure, but is independent of any fluctuations in the downstream pressure. The pressure, which is usually around atmospheric, can be conveniently measured with a mercury manometer.

(*iii*) *Temperature and velocity profile measurements*. There are several methods of measuring the temperature profile of a flame. As with the concentration measurements discussed below, the temperature of a flame usually changes very greatly and very rapidly with distance. Thus, the method used for monitoring it must be capable of covering a wide range, but the spatial discrimination must be good. One common method uses a microthermocouple or a micro resistance probe. There are difficulties connected with the use of probes in a flame: firstly the danger of affecting its geometry and thus the kinetics of the relevant reactions and, secondly, the possibility of catalyzing these reactions. The latter is often minimized by coating the probe with a ceramic material. A technique which is used frequently in aerodynamics for measuring the flow velocity may be used to estimate the acceleration and hence the extent of expansion and the temperature of the gases as they pass through the flame. Very small, inert particles are suspended in the reactant gases and their movement is photographed under stroboscopic light. A modification of this method traces a temperature contour by using particles of a substance which evaporates at a known temperature. Yet a third method for determining the temperature uses the variation with temperature in the vibrational or rotational fine structure of the spectrum of a species present in the flame. Probably the most common way of estimating the flow rate is to use an orifice or capillary upstream of the flame and measure the pressure above and below the restriction. The standard equation governing critical flow through the restriction is then applied[103].

(*iv*) *Concentration measurement.* The determination of the concentration profiles of $(n-1)$ species (n is defined above) completely defines the flame system chemically. It is convenient to consider the measurement of these concentrations in two categories, depending on whether the species is "stable" or not. The stable species are molecular, such as H_2, H_2O, and CO_2: species which may be kept for long periods without readily undergoing further change. They are determined most conveniently by using a probe. Thus a sample is removed from a known part of the flame through a small tube which is designed so that the reaction is frozen through rapid decompression and also, often, by water cooling. The concentrations may then be determined with a time-of-flight mass spectrometer, which can perform a complete analysis within the millisecond time range. Because molecular species tend not to have convenient, strong absorption bands it is usually necessary to employ specially modified techniques, such as multiple-pass optics, if absorption methods are to be used for analysis.

Unstable species fall into two categories: free radicals, including reactive atoms, and ions. The latter are present in such small concentrations that their measurement is not normally required. Sampling coupled with mass spectrometry may be used for free radicals as for molecules. Absorption and ESR spectrometry have also found limited application, although they are potentially very useful. The technique which has been most widely used depends on the addition to the flame of an indicator which reacts at a known rate or to a known extent with one of the radicals or atoms present. The rate of this reaction is followed either by sampling or spectroscopically. The details of this type of technique depend to a large extent on the nature of the species concerned, and the methods which have been developed naturally involve those radicals, such as H, OH and O, which tend to be present in high concentration in flames[104].

Two methods which have been used to estimate the concentration of hydrogen atoms in hydrogen–oxygen flames diluted with nitrogen will illustrate the idea. Sugden and co-workers[108] added very small amounts of lithium and sodium salts, in the form of a fine spray from a calibrated atomizer, to the reactant gases. In the flame, complete decomposition takes place and the emission spectra of Li and Na

References pp. 176–179

are observed. The latter forms no significant concentration of any compound, but lithium produces a large amount of the stable gaseous hydroxide, LiOH

$$Li + H_2O \overset{K}{\rightleftarrows} LiOH + H \tag{10}$$

This equilibrium, for which the equilibrium constant K can be estimated, is set up comparatively rapidly. If the total amounts of sodium and lithium in the flame are equal, then at a given position the free Na concentration is equal to the free Li concentration (both estimated from the intensity of emission) plus the concentration of LiOH. By using the equilibrium constant for equation (10) the concentration of hydrogen atoms may be calculated. In the second method, Fenimore and Jones[109] added small amounts of D_2O, which reacts with hydrogen atoms at a known rate

$$H + D_2O \rightarrow HD + OD.$$

Samples are withdrawn at different points and the concentration of HD measured mass spectrometrically. This method which, incidentally, only works with rich flames (where the HD is not consumed by an excess of oxygen) gave results which were in excellent agreement with those obtained by the first method. It should be noted that in both cases the indicator concentration was very much less than the concentration of the hydrogen atoms.

9. Molecular beams

The successful use of molecular beams for investigating physical properties of gas molecules has been discussed by several authors[110,111]; more recently, chemical applications have also been discussed[112,113]. A molecular beam differs from a gas

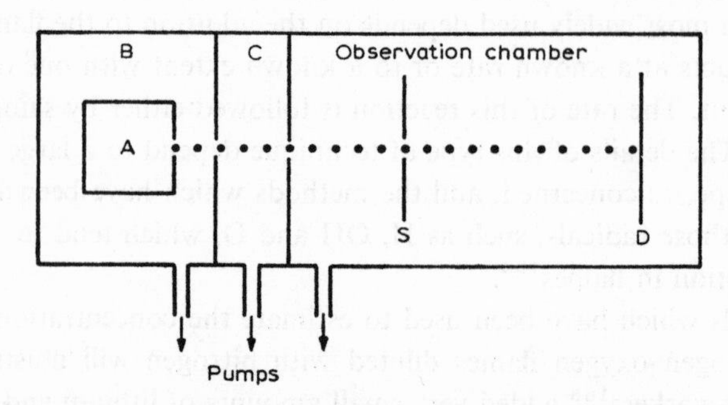

Fig. 23. Schematic arrangement for the production of a single molecular beam. A is the source of beam material; B the oven chamber; C a short isolating chamber; S a collimating slit; D the detector.

jet in that the mean free path of molecules, λ_s, in the former is very much longer than the width of the beam. This means that when a beam passes through a low-pressure region it retains its identity whereas a jet disintegrates within a very short distance. The details of a molecular beam apparatus depend to a large extent on the nature of the experiment, and Fig. 23 shows schematically how a single molecular beam may be produced. A source of beam material A has a slit through which the molecules enter the oven chamber B. The beam then passes through a short isolating chamber C and into the observation chamber which contains a collimating slit S. The position and intensity of the beam are detected at D. Traditionally, the three chambers are connected to independent pumping systems. The requirement that λ_s be large compared with the beam width restricts the width of the slits and also, because λ_s is inversely proportional to it, the pressure. This is why the intensity of a molecular beam cannot be raised above what is really a very low level (and one which requires a detector of extreme sensitivity); if it were, the condition for effusive flow would no longer be satisfied.

The critical feature of any apparatus for following chemical reactions in crossed molecular beams[114] is the detector. The surface-ionization detector, which provides a very sensitive way of estimating alkali metals in the presence of their salts, comprises parallel filaments of tungsten and a platinum–tungsten alloy which can be heated in turn and are partly surrounded by a metal sheath held at a negative potential relative to the wires. Suppose, for example, that the reaction being studied is $MeI + K \rightarrow Me + KI$. The ionization potential of potassium in K and KI and the work function of tungsten are such that, if a K atom or a KI molecule strikes the heated tungsten wire, the alkali metal may lose an electron to the wire. (In fact, the ionization potential of the beam molecules must be less than the work function of the wire material.) This produces a positive ion which is attracted to the sheath, and the positive ion current is a direct measure of the rate at which K and KI are striking the wire. Now the ionization potential of potassium in its salts is higher than that of the free metal and the work function of the Pt–W alloy is (under certain conditions) less than that of W. The second wire gives a positive current with free potassium but not with KI[8]. Thus the concentrations of K and KI can be determined at any position from the two positive ion currents. The surface-ionization detector is restricted to reactions involving the alkali metals and a few others. In the original experiments the angle between the beams was kept constant at 90° and the detector moved only in the plane described by the two beams. It is now apparent that additional useful information may be had if the beam angle can be varied and the density of the products outside the beam-plane can be measured. Other variables are the temperatures of the beams, which need not be the same. The source of the alkali metal is a double-oven; the inner oven contains the molten metal and is kept at a constant temperature, while the temperature of the outer oven may be varied. By this method the beam temperature can be varied by about 300° without affecting the vapour pressure. It was found that the normal triple-pumping arrangement could be dispensed with by enclosing the whole apparatus in a copper box attached to a large liquid-nitrogen trap. The background pressure was maintained at about 10^{-7} mm Hg. Typically the beams were about 1 mm wide at the scattering centre.

How the measured intensities of the products at different angles relative to the reactant beams may be correlated with the projected reaction mechanism has been discussed admirably by Herschbach[112,114]. By applying various conservation laws it is possible to derive relationships between the velocity vectors of reactant and product particles. Although some of the finer points are modified slightly in a full quantum mechanical treatment, Newtonian mechanics still apply and it is instructive to discuss the collisions in terms of classical behaviour. A so-called "Newton Diagram" is constructed from which it is possible to compute the product distribution for different final kinetic energies. In a collision between particles (1) and (2) to give (3) and (4) it is useful to think in terms of the movement of the centre of mass. Its velocity is given by

$$\boldsymbol{v}_c = \left(\frac{m_1 \boldsymbol{v}_1 + m_2 \boldsymbol{v}_2}{m_1 + m_2}\right)$$

where m_1 and m_2 are the masses of the reactants and $\boldsymbol{v}_1$, $\boldsymbol{v}_2$ their velocities. If the relative velocity vector of the two particles is $\boldsymbol{v}\ (= \boldsymbol{v}_1 - \boldsymbol{v}_2)$, it follows that

$$(\boldsymbol{v}_1 - \boldsymbol{v}_c) = \left(\frac{m_2}{m_1 + m_2}\right) \boldsymbol{v} \quad \text{and} \quad (\boldsymbol{v}_2 - \boldsymbol{v}_c) = -\left(\frac{m_1}{m_1 + m_2}\right) \boldsymbol{v}.$$

These follow from the linear momentum conservation relationship. Similar expressions relate the recoil velocities $\boldsymbol{v}_3$ and $\boldsymbol{v}_4$

$$(\boldsymbol{v}_3 - \boldsymbol{v}_c) = \left(\frac{m_4}{m_1 + m_2}\right) \boldsymbol{v}' \quad \text{and} \quad (\boldsymbol{v}_4 - \boldsymbol{v}_c) = -\left(\frac{m_3}{m_1 + m_2}\right) \boldsymbol{v}'.$$

The magnitude of the final relative velocity vector $\boldsymbol{v}'$ is determined by the energy conservation restriction

$$\boldsymbol{v}' = \left(\frac{2E'}{\mu'}\right)^{\frac{1}{2}}$$

where μ' is the reduced mass of the products. E', the final relative translational kinetic energy, is restricted by the relationship

$$E' + W' = E + W + \Delta D_0$$

in which E is the initial relative translational kinetic energy, W and W' are the initial and final internal excitation energies (rotational, vibrational or electronic), respectively, and ΔD_0 is the difference in dissociation energies of the new and old bonds, measured from the zero-point vibrational levels. (Note that the translational kinetic energy of the centre of mass remains constant during the collision. E and E' refer to the *relative* translational kinetic energies.) The direction of the final relative velocity vector is determined by the conservation of *angular* momentum. Thus if $\boldsymbol{L}$, $\boldsymbol{L}'$ are the orbital angular momenta associated with the relative motion of the collision partners and $\boldsymbol{J}$, $\boldsymbol{J}'$ ($= \boldsymbol{J}_1+\boldsymbol{J}_2$; $\boldsymbol{J}_3+\boldsymbol{J}_4$, respectively) the sums of the momenta of the individual reactant and product molecules, then $\boldsymbol{L}'+\boldsymbol{J}' = \boldsymbol{L}+\boldsymbol{J}$. The Newton diagram shows the recoil spectrum of products as a set of spheres centred on the tip of the centre of mass velocity vector $\boldsymbol{v}_c$ and with radii depending on the values of E'. The next steps in the comparison of the observed product density pattern with a theoretical model are to project these distributions on to the laboratory coordinate system and to average over the initial velocity distributions in the incident beams. It is possible to obtain additional information about the reaction through a velocity analysis of the reaction products.

In a recent review[112] reference is made to all of the reactions of the type $RX+M \rightarrow R+MX$ (where X is a halogen and M an alkali metal) which have been studied in crossed molecular beams. The relationship between the rate coefficient k and the parameters measured in molecular beam experiments has been considered[115]. The ratio of cross-sections of reactive (σ_r) and elastic (σ_e) scattering was shown to be proportional to the ratio of the rate coefficients for reactive and elastic scattering, k_r/k_e (the collision yield). Several very interesting generalizations have been made[112]. It seems that most of the chemical energy released by the reaction goes into internal degrees of freedom of the products rather than appearing as increased translational kinetic energy. Also, the angular distribution of the products is usually quite anisotropic, and the preferred recoil direction is strongly correlated with the magnitude of the reaction cross-section, σ_r. For example, in the reaction between methyl iodide and potassium the KI produced recoils backwards with respect to the potassium beam while the CH_3 is scattered forwards. Although it is much too early to speak of complete agreement between experimental observations and theoretical predictions, some recent theoretical work on the MeI–K reaction in which the methyl group was treated as a point mass[116] has yielded very encouraging results.

9.1 GAS–SOLID REACTIONS

A conceptually rather simple application of molecular beams to kinetics has been used to measure the life-time of adsorption of hydrogen molecules on a nickel surface. The apparatus is shown in Fig. 24[117]; the molecular beam strikes a ro-

References pp. 176–179

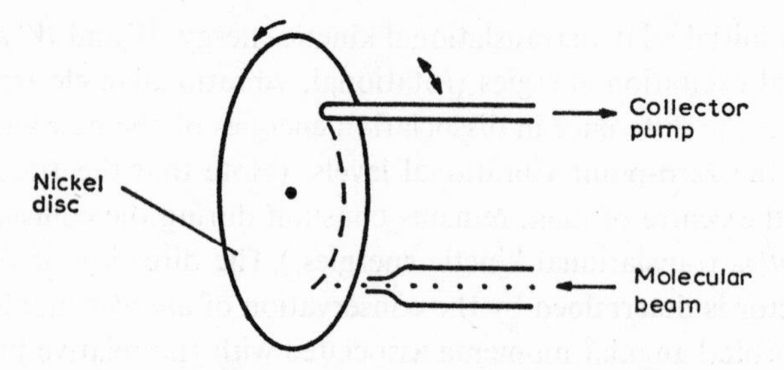

Fig. 24. Use of a single molecular beam for following gas–solid reactions.

tating disc and the molecules are adsorbed on to the surface. A small collector tube is located directly above the disc such that a molecule evaporating from the surface when it is under the tube is pumped off and passes down the tube to a mass spectrometer to be counted. The position of the tube can be varied relative to the impact point as shown by the dashed line. In the system chosen all the hydrogen was adsorbed, and the rate of desorption (associated with an average retention time of about 10^{-4} sec) followed a simple and reproducible first order kinetic law from which it was possible to derive a frequency factor and an activation energy. By using a mixed molecular beam of hydrogen and deuterium it was found that no exchange occurred, suggesting an adsorption process in which the bonds of the original molecules are not completely broken. It would seem that this method has considerable potential for measuring the kinetics of fast gas–solid reactions, including the study of single steps in heterogeneous catalytic reactions, although there are bound to be difficulties in obtaining surfaces of known composition.

ACKNOWLEDGEMENT

It is a pleasure to acknowledge stimulating discussions with Dr. D. B. Rorabacher.

REFERENCES

1 M. Eigen and L. de Maeyer, *Z. Elektrochem.*, 59 (1955) 986; M. Eigen, *Angew. Chem. (Intern. Ed. Engl.)*, 3 (1964) 1.
2 M. Eigen, L. de Maeyer and H.-Ch. Spatz, *Ber. Bunsenges. Physik. Chem.*, 68 (1964) 19.
3 R. G. W. Norrish and G. Porter, *Nature*, 164 (1949) 658; G. Porter, *Proc. Roy. Soc. (London), A*, 200 (1950) 284.
4 G. Porter and M. W. Windsor, *Discussions Faraday Soc.*, 17 (1954) 178.
5 M. C. Sauer and L. M. Dorfman, *J. Am. Chem. Soc.*, 86 (1964) 4218.
6 F. J. W. Roughton, *Z. Elektrochem.*, 64 (1960) 3; also in ref. 15, p. 5.
7 *cf.* N. F. Ramsey, *Molecular Beams*, Oxford University Press, Oxford, 1956, Chapter 2.
8 E. H. Taylor and S. Datz, *J. Chem. Phys.*, 23 (1955) 1711.
9 The study of fast reactions, *Discussions Faraday Soc.*, 17 (1954).

10 Schnelle Reaktionen in Lösungen, *Z. Elektrochem.*, 64 (1960) 1.
11 M. BOUDART, *Ann. Rev. Phys. Chem.*, 13 (1962) 241; S. W. BENSON AND W. B. DE MORE, *ibid.*, 16 (1965) 397; B. H. MAHAN, *ibid.*, 17 (1966) 173.
12 L. DE MAEYER AND K. KUSTIN, *Ann. Rev. Phys. Chem.*, 14 (1963) 5; G. G. HAMMES, *ibid.*, 15 (1964) 13; H. STREHLOW, *ibid.*, 16 (1965) 167.
13 S. L. FRIESS, E. S. LEWIS AND A. WEISSBERGER (Eds.), *Technique of Organic Chemistry*, Vol. VIII, Part II, *Investigation of Rates and Mechanisms of Reactions*, Interscience, New York, 1963.
14 E. F. CALDIN, *Fast Reactions in Solution*, Blackwell, Oxford, 1964.
15 B. CHANCE, R. H. EISENHARDT, Q. H. GIBSON AND K. K. LONBERG-HOLM (Eds.), *Rapid Mixing and Sampling Techniques in Biochemistry*, Academic Press, New York, 1964.
16 I. AMDUR AND G. G. HAMMES, *Chemical Kinetics: Principles and Selected Topics*, McGraw-Hill, New York, 1966.
17 R. G. W. NORRISH AND G. PORTER, in ref. 9, p. 40; R. G. W. NORRISH AND B. A. THRUSH, *Quart. Rev. (London)*, 10 (1956) 149; also ref. 14, p. 104.
18 G. PORTER, in ref. 13, p. 1055.
19 K. H. L. ERHARD AND R. G. W. NORRISH, *Proc. Roy. Soc. (London), A*, 234 (1956) 178.
20 H. T. WITT, B. RUMBERG, P. SCHMIDT-MENDE, U. SIGGEL, B. SKERRA, J. VATER AND J. WEIKARD, *Angew. Chem. (Intern. Ed. Engl.)*, 4 (1965) 799.
21 B. KE, R. W. TREHARNE AND C. MCKIBBEN, *Rev. Sci. Instr.*, 35 (1964) 296.
22 L. M. DORFMAN AND M. S. MATHESON, in G. PORTER (Ed.), *Progress in Reaction Kinetics*, Vol. III, Pergamon, Oxford, 1965, p. 237.
23 *Radiation Res.*, Suppl. 4 (1964).
24 *Solvated Electron*, Advances in Chemistry Series, Vol. 50, American Chemical Society, 1965.
25 E. J. HART, S. GORDON AND E. M. FIELDEN, *J. Phys. Chem.*, 70 (1966) 150.
26 M. ANBAR, in ref. 24, p. 55.
27 S. H. BAUER, *Science*, 141 (1963) 867.
28 J. N. BRADLEY, *Shock Waves in Chemistry and Physics*, Methuen, London, 1962; J. N. BRADLEY, Chemical application of the shock tube, *Roy. Inst. Chem. (London), Lecture Ser. No. 6* (1963) 1; S. H. BAUER, *Ann. Rev. Phys. Chem.*, 16 (1965) 245; also ref. 16, p. 193.
29 E. F. GREENE AND J. P. TOENNIES, *Chemical Reactions in Shock Waves*, Academic Press, New York, 1964; A. G. GAYDON AND I. R. HURLE, *The Shock Tube in High-Temperature Chemical Physics*, Reinhold, New York, 1963.
30 R. BECKER, *Z. Physik*, 8 (1922) 321.
31 W. JOST, *Symp. Combustion, 3rd*, (1949) 424; W. JOST, *Z. Naturforsch.*, 18a (1963) 160; W. JOST, *Symp. Combustion, 9th*, (1963) 1013.
32 *e.g.* R. W. GETZINGER AND G. L. SCHOTT, *J. Chem. Phys.*, 43 (1965) 3237; *cf.* S. H. BAUER, G. L. SCHOTT AND R. E. DUFF, *J. Chem. Phys.*, 28 (1958) 1089.
33 M. EIGEN AND L. DE MAEYER, in ref. 13, p. 895.
34 Ref. 14, Chapters 4 and 5; ref. 16, Chapter 6.
35 G. W. CASTELLAN, *Ber. Bunsenges. Physik. Chem.*, 67 (1963) 898; R. A. ALBERTY, G. YAGIL, W. F. DIVEN AND M. TAKAHASHI, *Acta Chem. Scand.*, 17 (1963) S34.
36 D. N. HAGUE AND M. EIGEN, *Trans. Faraday Soc.*, 62 (1966) 1236.
37 M. EIGEN, W. KRUSE, G. MAASS AND L. DE MAEYER, in G. PORTER (Ed.), *Progress in Reaction Kinetics*, Vol. II, Pergamon, Oxford, 1964, p. 287.
38 M. EIGEN, *Pure Appl. Chem.*, 6 (1963) 97.
39 *e.g.* Ref. 33; H. RÜPPEL, V. BÜLTEMANN AND H. T. WITT, *Ber. Bunsenges. Physik. Chem.*, 68 (1964) 340.
40 H. STREHLOW AND M. BECKER, *Z. Elektrochem.*, 63 (1959) 457; Ref. 33; H. HOFFMAN, J. STUEHR AND E. YEAGER, in B. E. CONWAY AND R. E. BARRADAS (Eds.), *Chemical Physics of Ionic Solutions*, Wiley, New York, 1966, p. 255.
41 *e.g.* G. CZERLINSKI AND M. EIGEN, *Z. Elektrochem.*, 63 (1959) 652; G. G. HAMMES AND P. FASELLA, *J. Am. Chem. Soc.*, 84 (1962) 4644; also ref. 33.
42 G. ERTL AND H. GERISCHER, *Z. Elektrochem.*, 65 (1961) 629.
43 H. STREHLOW AND S. KALARICKAL, *Ber. Bunsenges. Physik. Chem.*, 70 (1966) 139.
44 D. DE VAULT, in ref. 15, p. 165.
45 G. CZERLINSKI, in ref. 15, p. 183.

46 M. EIGEN AND L. DE MAEYER, in ref. 15, p. 175; M. EIGEN, L. DE MAEYER, B. H. HAVSTEEN AND L. D. FALLER, to be published; also ref. 45.
47 H. DIEBLER, M. EIGEN AND G. G. HAMMES, *Z. Naturforsch.*, 15b (1960) 554.
48 G. CZERLINSKI, H. DIEBLER AND M. EIGEN, *Z. Physik. Chem. (Frankfurt)*, 19 (1959) 246; M. EIGEN AND G. G. HAMMES, *J. Am. Chem. Soc.*, 82 (1960) 5951 and 83 (1961) 2786.
49 M. EIGEN AND R. G. WILKINS, in *Mechanisms of Inorganic Reactions*, Advances in Chemistry Series, Vol. 49, American Chemical Society, 1965, p. 55.
50 D. W. MARGERUM AND M. EIGEN, in V. GUTMANN (Ed.), *Proc. 8th Intern. Conf. on Coordination Chemistry, Vienna, 1964*, Springer-Verlag, Berlin, p. 289.
51 M. EIGEN, *Angew. Chem. (Intern. Ed. Engl.)*, 3 (1960) 1; also ref. 37.
52 A. BEWICK, M. FLEISCHMANN, J. N. HIDDLESTON AND WYNNE-JONES, *Discussions Faraday Soc.*, 39 (1965) 149, 163; M. EIGEN AND G. ILGENFRITZ, *ibid.*, 39 (1965) 163.
53 C. V. RAMAN AND N. S. NATH, *Proc. Indian Acad. Sci. Sect. A*, 2 (1935) 406.
54 *e.g.* M. EIGEN, *Discussions Faraday Soc.*, 24 (1957) 25; M. EIGEN, *Z. Elektrochem.*, 64 (1960) 115; M. EIGEN AND K. TAMM, *ibid.*, 66 (1962) 93, 107.
55 *e.g.* J. A. POPLE, W. G. SCHNEIDER AND H. J. BERNSTEIN, *High-resolution Nuclear Magnetic Resonance*, McGraw-Hill, New York, 1959; J. D. ROBERTS, *Nuclear Magnetic Resonance*, McGraw-Hill, New York, 1959; H. S. GUTOWSKY, in A. WEISSBERGER (Ed.), *Technique of Organic Chemistry*, Vol. I, Part IV, *Physical Methods of Organic Chemistry*, Interscience, New York, 1960, Chapter 41; H. STREHLOW, in ref. 13, Chapter 17.
56 C. S. JOHNSON, in J. S. WAUGH (Ed.), *Advances in Magnetic Resonance*, Vol. I, Academic Press, New York, 1965, p. 33.
57 H. S. GUTOWSKY, D. W. MCCALL AND C. P. SLICHTER, *J. Chem. Phys.*, 21 (1953) 279; H. M. MCCONNELL, *ibid.*, 28 (1958) 430; S. ALEXANDER, *ibid.*, 37 (1962) 974.
58 H. Y. CARR AND E. M. PURCELL, *Phys. Rev.*, 94 (1954) 630.
59 A. ALLERHAND AND H. S. GUTOWSKY, *J. Chem. Phys.*, 41 (1964) 2115.
60 H. S. GUTOWSKY AND A. SAIKA, *J. Chem. Phys.*, 21 (1953) 1688.
61 A. LOEWENSTEIN AND T. M. CONNOR, *Ber. Bunsenges. Physik. Chem.*, 67 (1963) 280.
62 W. D. PHILLIPS, C. E. LOONEY AND C. P. SPAETH, *J. Mol. Spectry.*, 1 (1957) 35; L. W. REEVES AND K. O. STRØMME, *Can. J. Chem.*, 38 (1960) 1241, and *Trans. Faraday Soc.*, 57 (1961) 390.
63 I. WEINBERG AND J. R. ZIMMERMAN, *J. Chem. Phys.*, 23 (1955) 748; W. G. SCHNEIDER AND L. W. REEVES, *Ann. N. Y. Acad. Sci.*, 70 (1958) 858.
64 J. T. ARNOLD, *Phys. Rev.*, 102 (1956) 136; Z. LUZ, D. GILL AND S. MEIBOOM, *J. Chem. Phys.*, 30 (1959) 1540.
65 J. T. ARNOLD AND M. E. PACKARD, *J. Chem. Phys.*, 19 (1951) 1608; U. LIDDEL AND N. F. RAMSEY, *ibid.*, 19 (1951) 1608.
66 Z. LUZ AND S. MEIBOOM, *J. Chem. Phys.*, 40 (1964) 2686.
67 T. J. SWIFT AND R. E. CONNICK, *J. Chem. Phys.*, 37 (1962) 301.
68 M. EIGEN, in S. KIRSCHNER (Ed.), *Advances in the Chemistry of the Coordination Compounds*, Macmillan, New York, 1961, p. 371.
69 D. J. E. INGRAM, *Free Radicals as Studied by Electron Spin Resonance*, Butterworths, London, 1958; D. H. WHIFFEN, *Quart. Rev. (London)*, 12 (1958) 250; A. CARRINGTON, *ibid.*, 17 (1963) 67.
70 R. L. WARD AND S. I. WEISSMAN, *J. Am. Chem. Soc.*, 79 (1957) 2086.
71 W. L. REYNOLDS, *J. Phys. Chem.*, 67 (1963) 2866.
72 F. C. ADAM AND S. I. WEISSMAN, *J. Am. Chem. Soc.*, 80 (1958) 1518.
73 R. W. FESSENDEN, *J. Phys. Chem.*, 68 (1964) 1508.
74 M. M. KREEVOY AND C. A. MEAD, *J. Am. Chem. Soc.*, 84 (1962) 4596, and *Discussions Faraday Soc.*, 39 (1965) 166; A. K. COVINGTON, M. J. TAIT AND WYNNE-JONES, *ibid.*, 39 (1965) 172.
75 F. A. MATSEN, in A. WEISSBERGER (Ed.), *Technique of Organic Chemistry*, Vol. IX, *Chemical Applications of Spectroscopy*, Interscience, New York, 1956, p. 694.
76 T. FÖRSTER, *Fluoreszenz Organischer Verbindungen*, Vandenhoeck and Ruprecht, Göttingen, 1951.
77 E. J. BOWEN AND F. WOKES, *Fluorescence of Solution*, Longmans, London, 1953; N. WOTHERSPOON AND G. ORSTER, in A. WEISSBERGER (Ed.), *Technique of Organic Chemistry*, Vol. I, Part III, *Physical Methods of Organic Chemistry*, Interscience, New York, 1960, Chapter 31.

78 A. WELLER, in G. PORTER (Ed.), *Progress in Reaction Kinetics*, Vol. 1, Pergamon, Oxford, 1961, p. 189; A. WELLER, in ref. 13, p. 845.
79 A. WELLER, *Z. Physik. Chem. (Frankfurt)*, 3 (1955) 238.
80 W. URBAN AND A. WELLER, *Ber. Bunsenges. Physik. Chem.*, 67 (1963) 787.
81 H. STREHLOW, *Ann. Rev. Phys. Chem.*, 16 (1965) 167.
82 (a) I. M. KOLTHOFF AND J. J. LINGANE, *Polarography*, 2nd edn., Interscience, New York, 1952;
(b) P. DELAHAY, *New Instrumental Methods in Electrochemistry*, Interscience, New York, 1954;
(c) G. W. C. MILNER, *Principles and Applications of Polarography and Other Electroanalytical Processes*, Longmans, London, 1957;
(d) P. DELAHAY, *Double Layer and Electrode Kinetics*, Interscience, New York, 1965;
(e) *Talanta* (Heyrovsky Honour Issue), 12 (1965) 1061–1379; also ref. 14, Chapter 9.
83 H. STREHLOW, in ref. 13, Chapter 15.
84 D. ILKOVIČ, *Collection Czech. Chem. Commun.*, 6 (1934) 498; *cf.* D. MACGILLAVRY AND E. K. RIDEAL, *Rec. Trav. Chim.*, 56 (1937) 1013, and ref. 82(a), Chapter 2.
85 J. KOUTECKÝ AND R. BRDIČKA, *Collection Czech. Chem. Commun.*, 12 (1947) 337.
86 M. BECKER AND H. STREHLOW, *Z. Elektrochem.*, 64 (1960) 129, 813.
87 P. DELAHAY AND G. L. STIEHL, *J. Am. Chem. Soc.*, 74 (1952) 3500; *cf.* ref. 83.
88 F. J. W. ROUGHTON AND B. CHANCE, in ref. 13, Chapter 14.
89 Ref. 14, Chapter 3.
90 F. J. W. ROUGHTON, *Proc. Roy. Soc. (London), A*, 104 (1923) 376.
91 F. J. W. ROUGHTON AND G. A. MILLIKAN, *Proc. Roy. Soc. (London), A*, 155 (1936) 258; F. J. W. ROUGHTON, *ibid.*, 155 (1936) 269; G. A. MILLIKAN, *ibid.*, 155 (1936) 277.
92 Q. H. GIBSON AND L. MILNES, *Biochem. J.*, 91 (1962) 161.
93 R. L. BERGER, in ref. 15, p. 33.
94 R. H. PRINCE, in ref. 10, p. 13.
95 P. STRITTMATTER, in ref. 15, p. 71.
96 *e.g.* W. R. RUBY, *Rev. Sci. Instr.*, 26 (1955) 460.
97 J. T. HAYSOM AND R. H. PRINCE, private communication.
98 M. EIGEN, L. DE MAEYER AND B. H. HAVSTEEN, to be published.
99 J. HALPERN AND J. P. MAHER, *J. Am. Chem. Soc.*, 87 (1965) 5361.
100 G. DULZ AND M. SUTIN, *Inorg. Chem.*, 2 (1963) 917.
101 *cf.* S. W. BENSON AND W. B. DE MORE, *Ann. Rev. Phys. Chem.*, 16 (1965) 397.
102 M. A. A. CLYNE AND B. A. THRUSH, *Proc. Roy. Soc. (London), A*, 261 (1961) 259.
103 R. M. FRISTROM AND A. A. WESTENBERG, *Flame Structure*, McGraw-Hill, New York, 1965.
104 G. DIXON-LEWIS AND A. WILLIAMS, *Quart. Rev. (London)*, 17 (1963) 243.
105 A. G. GAYDON AND H. G. WOLFHARD, *Flames: Their Structure, Radiation and Temperature*, Chapman and Hall, London, 1953; G. J. MINKOFF AND C. F. H. TIPPER, *Chemistry of Combustion Reactions*, Butterworths, London, 1962.
106 J. O. HIRSCHFELDER, C. F. CURTISS AND R. B. BIRD, *Molecular Theory of Gases and Liquids*, Wiley, New York, 1954; corrected, 1964.
107 H. P. BROIDA, in H. C. WOLFE (Ed.), *Temperature: its Measurement and Control in Science and Industry*, Vol. II, Reinhold, New York, 1955, Chapter 17.
108 E. M. BULEWICZ, C. A. JAMES AND T. M. SUGDEN, *Proc. Roy. Soc. (London), A*, 235 (1956) 89.
109 C. P. FENIMORE AND G. W. JONES, *J. Phys. Chem.*, 62 (1958) 693.
110 R. G. J. FRASER, *Molecular Rays*, Cambridge University Press, Cambridge, 1931; N. F. RAMSEY, *Molecular Beams*, Oxford University Press, Oxford, 1956; K. F. SMITH, *Molecular Beams*, Methuen, London, 1955.
111 J. ROSS (Ed.), *Advances in Chemical Physics*, Vol. X, *Molecular Beams*, Interscience, New York, 1966.
112 D. R. HERSCHBACH, in ref. 111, Chapter 9.
113 Ref. 16, Chapter 9.
114 D. R. HERSCHBACH, *Discussions Faraday Soc.*, 33 (1962) 149.
115 S. DATZ, D. R. HERSCHBACH AND E. H. TAYLOR, *J. Chem. Phys.*, 35 (1961) 1549; also ref. 8.
116 M. KARPLUS AND L. M. RAFF, *J. Chem. Phys.*, 41 (1964) 1267; L. M. RAFF AND M. KARPLUS, *ibid.*, 44 (1966) 1212.
117 J. DEWING AND A. J. B. ROBERTSON, *Proc. Roy. Soc. (London), A*, 240 (1957) 423.

Chapter 3

Experimental Methods for the Study of Heterogeneous Reactions

D. SHOOTER

1. General introduction

A wide variety of methods have been used to study the kinetics of heterogeneous reactions. The emphasis in this chapter is on those methods which have been particularly developed for the study of such reactions, and as a consequence, the study of kinetics by conventional measurements, *e.g.* chemical analysis, pressure changes etc., is only briefly mentioned.

The subject matter really covers several fields of work and it is not possible to include more than a fraction of the available references. Where good reviews of the subject exist, references to individual publications have been deliberately limited. More direct references have been given where suitable reviews have not been found.

2. The solid–gas interface

2.1 ADSORPTION

There are two usually distinguishable types of adsorption at the gas–solid interface, these are physical adsorption and chemisorption. Physical adsorption, or Van der Waals adsorption, occurs when the adsorbed molecule is held to the surface by weak Van der Waals or dispersion forces of the type responsible for cohesion in liquids. Chemisorption is a true chemical reaction involving electron transfer between the solid and the gas. The difference between the two types of adsorption is sometimes blurred but a number of criteria can be used to distinguish between them[1–3]. Physical adsorption is non-specific and generally takes place at low temperatures through forces of attraction similar to those causing liquefaction. Multilayer adsorption can take place and the heat of adsorption is less than 5 kcal.mole^{-1}. Chemisorption is much more specific for solid–gas pairs and can occur at temperatures well above the boiling point of the gas. A single adsorbed layer is formed, with a heat of chemisorption which can be as high as 150 kcal.mole^{-1}. The data of physical adsorption are beyond the scope of this article, although the process is widely used for the measurement of the surface area and pore structure of solids[1,4,5]. Further information can be obtained from the many reference sources on this subject[1,4,6].

Rates of chemisorption vary over a very wide range and many different experimental methods have been evolved to study specific gas–solid interactions. However, before proceeding to a description of the experimental methods, certain criteria must be defined to ensure that the correct phenomenon will be observed. Before any experiment can begin, it is necessary that all contaminating layers are removed in order to provide a 'clean' surface which is characteristic of the bulk material (or as near as the degree of unsaturation and lack of symmetry at the surface will allow). In practice such 'clean' surfaces are quite difficult to obtain and only in a relatively small number of chemisorption experiments can it be said that this criterion has been reasonably satisfied. Secondly, a 'cleaned' surface must be maintained free of contamination for the duration of the kinetic measurements. This usually requires the use of high vacuum (10^{-6} torr) or ultra high vacuum (10^{-9} torr) techniques, because of the high reactivity of most clean solid surfaces.

2.1.1 Preparation of clean surfaces[2,3,12]

(a) Chemical techniques

Reduction with hydrogen has been applied to metals and semiconductors in order to remove oxide layers. Purity of the hydrogen is very important and small amounts of oxygen or water vapour (0.01 %) prevent successful reduction[7]. Frankenburg[8] used a vacuum furnace to clean tungsten powder by reduction at temperatures up to 750° C. The temperature required to reduce the oxide depends on its thermodynamic stability. This temperature is about 20° C for palladium and platinum, above 300° C for nickel and cobalt and above 400° C for iron.

A variant of this method is reduction with atomic hydrogen[9] produced in the reaction vessel by an electrodeless discharge. This is accomplished by winding the reaction tube with a few turns of wire connected to a 15 Mc/sec oscillator. A current is then passed through the coil whilst hydrogen is admitted to a sufficient pressure (10^{-1} torr) to give a luminous discharge. The disadvantages are that the atomic hydrogen might remove contaminants from the walls into the gas phase. Hydrogen will be absorbed by the adsorbent and must be removed by stringent outgassing. Although surfaces produced in the above manner are active for chemisorption, most of the examples quoted in the literature have not been conclusively shown to give a completely clean surface. Cobalt is perhaps an exception where Rudham and Stone[10] obtained adsorption results with reduced cobalt very similar to those obtained for evaporated films.

Carbon can be removed from tungsten, molybdenum and nickel field emission tips by annealing in oxygen or hydrogen[17]. Becker *et al.*[11] have made a detailed study of the removal of carbon from tungsten ribbons and conclude that it can be substantially removed by heating in oxygen (10^{-6} torr) for 24 hours. The efficiency of carbon removal is dependent on the initial purity of the tungsten and it may be necessary to continue the heating for much longer periods[105].

References pp. 270–278

The techniques are limited to reactions which effect the removal of impurities as volatile products and are probably best used in conjunction with other techniques. It is sometimes necessary to exercise great care *e.g.* in oxidation reactions to prevent complete reaction of the cleaning agent with the surface being studied.

(b) Thermal desorption

Clean surfaces of a number of metals and non-metals can be obtained by heating a ribbon or wire *in vacuo*, close to the melting point. This method, pioneered by Langmuir[13] and Roberts[14], can be applied to those materials whose oxide layers or other surface impurities possess a higher vapour pressure than the bulk material. By the use of techniques such as Auger electron emission[15] and field emission[16,17], the metals tungsten, molybdenum, rhenium, tantalum, niobium, platinum and nickel and the semiconductors silicon[18,19] and germanium[20] have been shown to give clean surfaces after high temperature heating. Carbon can be cleaned by heating to 1000° C[21]. The generation of clean tungsten surfaces has received the greatest attention, and the evidence that high temperature heating produces a clean surface has been summarized by Hagstrum and D'Amico[22]. It is concluded that an atomically clean surface can be obtained by heating above 2000° C. A clean silicon surface is obtained by heating for several minutes at 1400° C[18,19].

Heating of the specimens can be accomplished by direct resistance heating or by electron bombardment. The latter technique is often used in conjunction with ion bombardment techniques (see Section 2.1.1e). Resistance heating is normally used for metals, but electron bombardment is more useful for semiconductors and for samples which cannot be readily attached to current leads because of the experimental set-up[23,24].

However, there are pitfalls in this technique which are not at first apparent. One disadvantage is that surface impurities may diffuse into the bulk on heating instead of leaving the surface as gaseous species. Silicon, in a system containing borosilicate glass, is usually contaminated[25,26] by a surface layer containing 10^{12} to 10^{15} boron atoms per cm^2. Conversely, species in the bulk may diffuse to the surface on heating, renewing the contamination. This has been clearly demonstrated for commercial samples of molybdenum, rhenium, tantalum and tungsten containing trace amounts of carbon[11,27,28]. Diffusion of carbon to the surface of silicon and nickel has also been observed[30,31]. This problem may be common to a wider range of surfaces, which were previously thought to be clean, and such impurities can only be removed with difficulty. Thermal etching may occur at elevated temperatures[29].

(c) Evaporation of films

One of the most versatile methods for producing a clean surface is by evaporation of a film onto a suitable substrate, such as glass, metal, mica or the cleavage face of a crystal. Evaporation can be accomplished by resistance heating, r.f. induction heating, radiation heating, electron bombardment or by sputtering caused by ion bombardment. The general techniques have been reviewed by a number of authors[32–35] and vapour pressure *versus* temperature data[36] give a useful guide to the conditions necessary for evaporating a particular material.

The simplest method of evaporation is by direct heating of a wire or ribbon. If the vapour pressure at the melting point is too low for successful evaporation, the material can be wound or electroplated on a higher melting point material, usually tungsten. For materials not readily available in wire form, films can be evaporated from pieces lodged in a tungsten spiral or placed in an inert crucible. Low melting point metals such as sodium have been prepared as films by distillation from a side tube into a larger vessel. Electron bombardment heating of the material or the crucible can be used instead of direct heating. Sputtering of metals and semiconductors by bombardment with inert gas ions[55] is also widely applicable (see Section 2.1.1e). A list of films prepared by the various techniques is given in Table 1.

One of the disadvantages of evaporated films is that their crystallinity is not usually well defined. The BET surface area is often much larger than the geometrical area and may be proportional to the weight of evaporated material[40,42,50,51] indicating a very porous structure. The physical and chemical properties may thus be quite different from those of the bulk material. This problem can be overcome to some extent by sintering the film before use and it is preferable to heat the film to a higher temperature than that to be used in carrying out the experiments. Low melting point metals and Group IB metals sinter above ~ 80° C. Metals melting between 1400 and 3000° C show an increasing surface area with melting point because they become increasingly difficult to sinter.

TABLE 1

PREPARATION OF EVAPORATED METAL FILMS

Technique	*Material*	*Reference*
(1) Direct evaporation of a wire	Ti, V, Cr, Fe, Co, Ni, W, Mo, Ta, Nb, Ir, Pd, Pt, Rh	33, 37, 38, 42
(2) Overwinding on tungsten	Cu, Ag, Au, Zn, Co	37, 38
(3) Evaporation from tungsten spiral or crucible	Ge, Si	40, 43
(4) Electroplating on tungsten	Mn	38, 39
(5) Electron bombardment heating	C	41
(6) Sputtering with inert gas ions	Ge, Si, Ga, Sb, Ti, V, Cr, Fe, Co, Ni, Cu, Zr, Nb, Mo, Ru, Rh, Pd, Ag, Hf, Ta, W, Re, Os, Ir, Pt, Au, Th, U, Al, Be	43, 44, 45
(7) Distillation	Na, K, Ca, Ba, As, Sb, Bi	46, 47, 48, 49

Although most of the work has been done on polycrystalline porous films, more recent research has shown ways of producing single crystal films[52] which are well defined and exhibit diffraction patterns characteristic of particular faces. Careful choice of substrate, substrate temperature and evaporation rate are required. In general single crystal films are most easily formed on substrates which are themselves well defined crystals, *e.g.* silver on mica[53], and germanium on calcium fluoride[54].

(*d*) *Cleavage and crushing of crystals*

The generation of a clean surface *in vacuo* by crushing large crystals is a relatively simple technique, provided that the material fractures easily. Impurity layers are not removed from the original surface, but this can be neglected by comparison with the new surface exposed. Clean surfaces of germanium have been produced in this manner[56], but the technique has not been widely applied. The avoidance of high temperature heating reduces the likelihood that impurity will be sufficiently mobile to diffuse to the surface.

Crushing of a crystal produces a great many crystal planes. A single crystal plane can more readily be produced by cleavage, although the total clean surface produced is normally very small, of the order of a few square millimeters. Cleavage to obtain a specific crystal plane requires the use of a single crystal whose orientation is accurately known. Most materials will only cleave readily along certain crystal planes, thus limiting the number of different faces which can be studied. The cleavage can be obtained by forcing a wedge into a premachined slot in the crystal[59,60] or by applying a bending force[58]. Surfaces produced in this way are not always smooth and often contain a large number of steps[58]. Examples of materials which have been cleaved to generate clean surfaces are given in Table 2.

(*e*) *Ion bombardment*

The method of ion bombardment to remove surface layers was introduced by

TABLE 2

GENERATION OF CLEAN SURFACES BY CLEAVAGE

Material	*Faces produced*	*References*
InSb	(110)	59
GaAs	(110)	58, 59
BiTe	(0001)	59
Ge	(111)	59, 60
Si	(111)	58
NaCl, KCl	—	57
Mica	basal plane	61
Fe	(100)	62
MgO	—	63
Pyrolytic graphite	basal plane	64

Langmuir and Kindon[65] many years ago and has since been used intermittently by other workers[66–68]. However, Farnsworth *et al.*[30,69,70] made the technique completely acceptable by showing that surfaces cleaned in this way gave a low energy diffraction pattern which was characteristic of a clean surface. Further confirmation was obtained by Hagstrum and D'Amico[22] who showed that tungsten surfaces produced by ion bombardment did not differ from those produced by high temperature heating.

The technique requires a combination of high temperature outgassing, ion bombardment and subsequent annealing by further heating. It is necessary to repeat the ion bombardment several times to remove contaminant which has diffused from the bulk during the heating. Annealing is necessary to remove defects produced by ion bombardment and inert gas occluded near the surface. The total heating time may be several hundred hours at 500–1000° C in order to obtain a surface which remains substantially clean for a number of hours.

A thermionically maintained discharge, with an argon pressure of 10^{-3} torr, is used to bombard the crystal face. Discharge voltages in the range 200–600 volts are used, giving a positive ion current density of 10–100 microamp. cm^{-2} on the crystal face. An ion bombardment of a few minutes is normally sufficient to remove the surface layers but bombardment times and annealing temperatures must be adjusted to the material being studied. Low current densities are used to prevent pitting of the crystal face. The apparatus must be designed to shield the crystal from material which might evaporate from the hot filament, or from atoms sputtered from other parts of the apparatus[24,70,71]. This method is useful in obtaining clean surfaces on both single crystals and polycrystalline materials. These include metals, semiconductors and insulators[72]. Only a relatively small area can be produced and the electrodes necessary for the ion bombardment technique may interfere with the experiments to be carried out on the clean surface[23,24].

A circuit diagram suitable for combined electron bombardment heating and ion bombardment is shown in Fig. 1. Electron and ion energies up to 1000 electron volts can be obtained at current densities of 5 microamp. cm^{-2} to several milliamp. cm^{-2}.

(*f*) *Field desorption*

This is a specialised technique which has been applied in field emission and field ion microscopy (see Section 2.1.5c). It is achieved by giving the tip a positive potential. Tungsten can then be removed at liquid helium temperatures with an applied field of 5.7×10^3 V.cm^{-1}. Perfectly regular surface structures are exposed containing many different lattice planes. Clean surfaces have been produced on tungsten, nickel, iron, platinum, copper, silicon and germanium. It is potentially applicable to a wide range of materials, but the area of clean surface exposed is only about 10^{-10} cm^2.

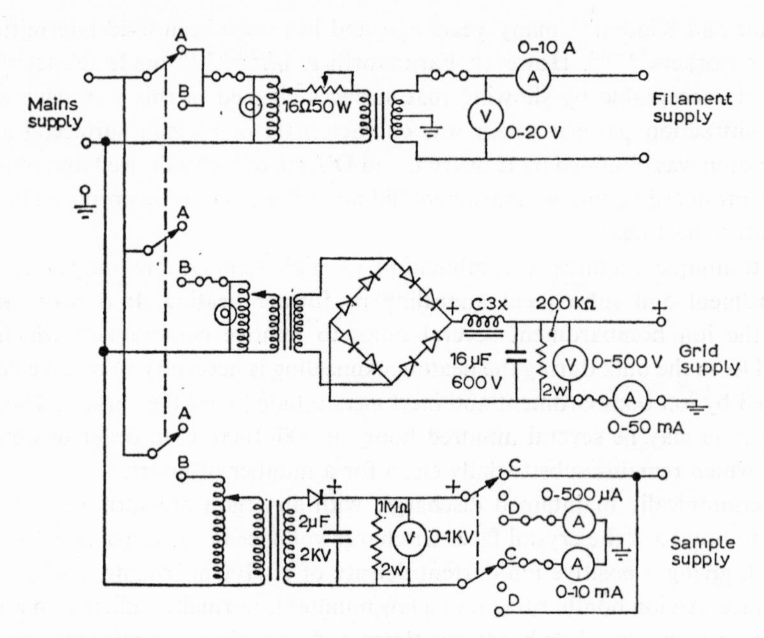

Fig. 1. Circuit diagram for electron and ion bombardment. Position A, filament only; B, filament and grid; C, electron bombardment (sample positive); D, ion bombardment (sample negative).

2.1.2 High vacuum technique

Conventionally, the term "high vacuum" is applied to pressures around 10^{-6} torr, and the term "ultra high vacuum" to pressures around 10^{-10} torr. In fact there is no great difference between the techniques required for producing "high vacuum" and "ultra high vacuum", but the requirements become more stringent at the lower pressures. Prior to 1948, the lower limit of pressure observed in the conventional ion gauge was about 10^{-8} torr. However, it was found that this reading was caused by a residual photoelectric current and was not a true ion current proportional to the pressure[73]. A consideration of this problem led Bayard and Alpert[74] to develop the 'inverted' ionisation gauge which is capable of measuring pressures down to 10^{-11} torr. From this point, the development of ultra high vacuum technique was very rapid. The appreciation of these experimental techniques is, in many instances, a necessary adjunct to the kinetic study of adsorption and catalysis. However, the subject matter is so extensive that it cannot be adequately reviewed in this chapter, and it is therefore necessary to provide reference to the many excellent publications on this subject[75–79,88].

For most studies of adsorption kinetics it is necessary to employ vacuum techniques. The question of whether high vacuum or ultra high vacuum conditions are required depends on the gas, the surface being studied and the time available to complete the experiment. Consider a Maxwellian gas of molecular weight M, at temperature $T°$ K and pressure p torr. The number of molecular collisions N with the surface per cm^2 per second is given by

$$N = \frac{3.5 \times 10^{22} p}{\sqrt{TM}}$$

For nitrogen at 298° K and 1 torr pressure, $N = 3.84 \times 10^{20}$ molecules. cm^{-2}. sec^{-1}. Not every molecule which strikes the surface will stick, so the rate of adsorption will be given by

$$\frac{dN}{dt} = \frac{3.5 \times 10^{22} ps}{\sqrt{TM}}$$

where s is a sticking probability or sticking coefficient. s is dependent on temperature and the fraction of the surface already covered with adsorbed species. For the more common gases (N_2, CO, O_2), on clean metal surfaces $s \sim 1$ to 10^{-1}, but for less reactive materials, *e.g.* O_2 on germanium[69], $s \sim 10^{-3}$. If the rate of surface contamination from residual gases is to remain low, say $< 1\%$, for the duration of the experiment, then the experiment must be completed in a time t seconds where

$$tsp \sim 2.5 \times 10^{-8}$$

For nitrogen adsorbing on a metal surface at 10^{-9} torr about 100 seconds are available for experiment. For oxygen on a germanium surface, a pressure of 10^{-7} torr will allow a similar experimental time. It can be seen that very good vacuum conditions are necessary for many experiments, and for those extending to hours or days ultra high vacuum is essential. In certain circumstances these conditions can be relaxed if the composition of the residual ambient is known. For example, a residual pressure of 10^{-6} torr, which is mainly nitrogen and carbon monoxide, will not effect experiments on a germanium surface because these gases are not adsorbed.

It can be concluded that the experimental conditions must be specified for each adsorbate–adsorbent pair with regard to the time taken for the experiment if contamination is to be avoided and meaningful results obtained.

2.1.3 *Measurement of adsorbed volume*

Measuring the rate of adsorption by noting changes in the adsorbed volume is in principle quite simple. A known mass of gas is admitted to the system and the quantity adsorbed at any instant can be calculated from the residual pressure. The adsorption rate is then found from the rate of pressure change with time. Apparatus and vacuum conditions must be arranged to correspond with the surface area available for adsorption and the rate.

(a) Conventional technique

The essential parts of a volumetric apparatus are a closed system of known volume, a source of adsorbate and a pressure measuring device. This type of apparatus is more commonly used for the determination of adsorption isotherms, but kinetics can be measured provided that the adsorbent surface is large and the response of the pressure measuring device is much faster than the rate of adsorption. These criteria are usually only satisfied by powders or films which show 'activated' adsorptions. Bond[3] has discussed the phenomenon of activated adsorption. Although the results may often be ascribed to contamination or incorporation into the bulk of the material, in other instances a genuine activated chemisorption is found.

McLeod gauges were used as the pressure measuring device in the older work[80,81] but thermal conductivity gauges or other instruments with an electrical output and a fast response have obvious advantages[78]. A recent publication by Marcus and Syverson[82] describes an apparatus equipped for high temperature studies and high speed pressure measurement (Fig. 2). Liquid adsorbates can be drawn into a capillary (V_2) before vaporisation into the measuring volume (V_1). The adsorption cell is surrounded by a vapour jacket and is also attached to a circulating pump which gives a circulation rate of about 1 l. min^{-1}. A Bourdon tube with the free end attached to the centre plate of a differential condenser forms the pressure sensing element. Changes in capacitance are converted into an output voltage which can be recorded. The response characteristics allow a pressure change from

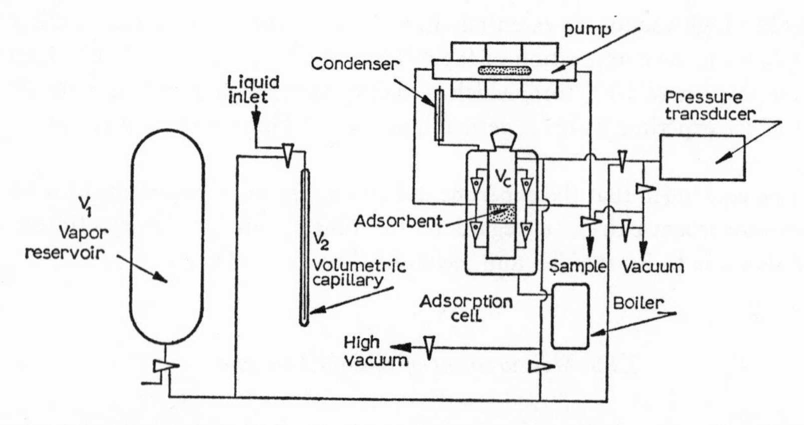

Fig. 2. Constant volume apparatus for the study of adsorption kinetics, equipped for high temperature studies and high speed pressure measurements. From ref. 82.

vacuum to atmospheric in less than 0.05 seconds with an error of ± 1 torr. The response characteristics of the complete system allow the adsorption cell to attain 95 % of its final pressure in 0.3 seconds.

Table 3 gives results which have been obtained with evaporated metal films, indicating the type of adsorption found for various gas–metal combinations. Chemisorption on oxides also shows fast and slow effects. The situation is more complicated than for metals because both the metal ions and the oxide ions are potential chemisorption sites. Usually, oxides are prepared as powders and the concept of a clean surface is necessarily different from that for metals. It is difficult to obtain a stoichiometric oxide surface and diffusion effects are often important. The data of chemisorption on oxides is given in Table 4.

The kinetics of adsorption can be categorised as follows:

(1) *Non-activated adsorption.* This is characterised by (a) zero or negative temperature coefficient of velocity, (b) initial rate independent of coverage, and (c) rate proportional to pressure.

(2) *Activated adsorption without a precursor.* This is characterised by (a) exponential increase in rate with increasing temperature, (b) continuous fall in rate with increasing coverage, and (c) rate directly proportional to pressure.

(3) *Activated adsorption with a precursor.* This is characterised by (a) exponential increase in rate with increasing temperature, (b) continuous fall in rate with increasing coverage, and (c) no simple dependence of rate on pressure.

Although part of the data for activated adsorption, particularly on metal surfaces, can be ascribed to contamination or diffusion effects, there are some experimental results which cannot be explained in this way. The theoretical reasons for this type of behaviour have been discussed by Bond[3] and by Hayward and Trapnell[2]. In general, such chemisorptions obey the Elovich[83] equation

$$\frac{\mathrm{d}N}{\mathrm{d}t} = a\mathrm{e}^{-bN}$$

where N is the amount adsorbed and a and b are constants. This equation can be derived for a uniform and a non uniform surface on the basis of a variation of activation energy with amount adsorbed.

(b) Ultra high vacuum technique

As noted in Section 2.1.2, measurement of the kinetics of chemisorption on clean metal surfaces generally requires ultra high vacuum techniques, in order to accomplish the experiment in a reasonable period of time. The variant of the classical adsorption method known as the flash-filament technique has been developed by several groups of workers[84–87] and recently summarised by Ehrlich[88].

The kinetic data obtained from the flash filament technique depend on three parameters, the absolute rate of adsorption ($A\mathrm{d}N/\mathrm{d}t$) occurring on a sample of

TABLE 3

CHEMISORPTION ON METAL FILMS

Gas	*Very fast chemisorption*	*Slow chemisorption*	*No chemisorption up to 0° C*
H_2	Ti, Zr, Nb, Ta, Cr, Mo, W, Fe, Co, Ni, Rh, Pd, Pt, Ba	(W, Fe, Ni, Pd) Mn? Ca, Ge	K, Cu, Ag, Au, Zn, Cd, Al, In, Pb, Sn
O_2	All metals except Au	(most metals)	Au
N_2	La, Ti, Zr, Nb, Ta, Mo, W	(Ta, W, Cr, Fe) Fe?, Ca, Ba	As for H_2; plus Ni, Rh, Pd, Pt
CO	As for H_2; plus La, Mn?, Cu, Ag, Au	(W, Fe, Ni) Al	K, Zn, Cd, In, Pb, Sn
CO_2	As for H_2; less Rh, Pd, Pt	Al	Rh, Pd, Pt, Cu, Zn, Cd
CH_4	Ti, Ta, Cr, Mo, W, Rh	Fe, Co, Ni?, Pd	—
C_2H_6	As for CH_4; plus Ni?, Pd	Fe, Co	—
C_2H_4	As for H_2; plus Cu, Au	Al	As for CO
C_2H_2	As for H_2; plus Cu, Au, K	Al	As for CO; less K
NH_3	(W, Ni, Fe)?	—	—
H_2S	W, Ni	—	—

Reproduced from *Chemisorption*, D. O. HAYWARD AND B. M. W. TRAPNELL, Butterworths, London, 1964. Metals which show slow chemisorption effects after an initial rapid chemisorption are indicated in brackets in the second column.

TABLE 4

CHEMISORPTION ON OXIDES

Gas	*Fast chemisorption*	*Slow chemisorption*
H_2	—	ZnO, Cr_2O_3, Al_2O_3, V_2O_3
O_2	NiO, Cr_2O_3, Fe_2O_3, MnO	Cu_2O, CoO, ZnO, MgO
CO	NiO, CoO, ZnO, Cu_2O, MnO, Cr_2O_3	(MnO, Cr_2O_3)
CO_2	NiO, CoO	—
N_2O	NiO, CoO	NiO, CoO
SO_2	Cu_2O	
C_2H_4	CoO, ZnO, Cu_2O, Ag_2O	Cr_2O_3

surface area A, the pressure in the system p, and the instantaneous surface concentration N. These quantities can be measured in a static system or a flow system.

(*i*) *Static system*

Gas is introduced into a system of volume V, containing the filament and a pressure gauge, at time t_0. The rate of change of pressure, or number of molecules in the gas phase, is given by

$$V\frac{\mathrm{d}N}{\mathrm{d}t} = F_\mathrm{F} - NS_\mathrm{F}$$

where S_F is the pumping speed of the filament and F_F the rate of evolution from the filament. The process is indicated in Fig. 3. On admission of the gas, the pressure drops due to adsorption and continues to do so until the rate of evaporation is equal to the rate of adsorption. S_F is related to the sticking coefficient s by

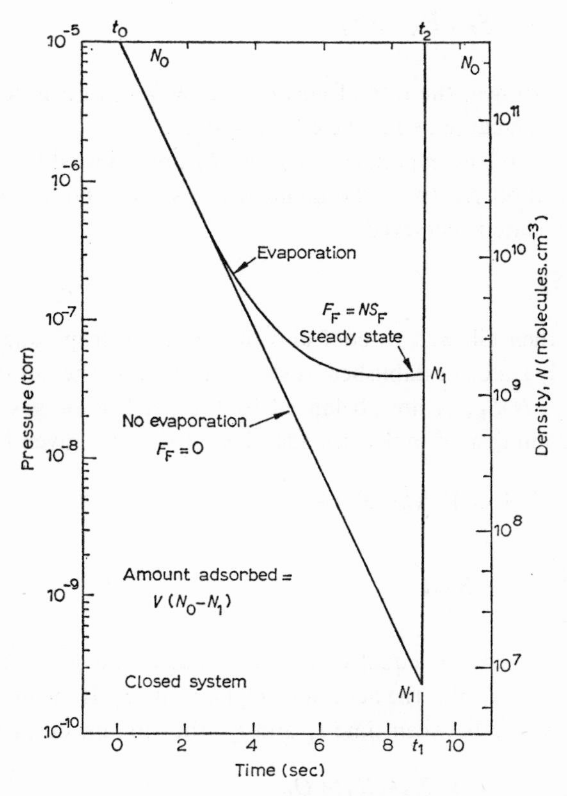

Fig. 3. Pressure cycle for adsorption experiments in a closed system on sample of unit area. t_0, start of adsorption; t_1, start of heating; t_2, end of adsorption. Amount adsorbed = $V(N_0 - N_1)$ = amount desorbed. From ref. 88.

the following equation

$$S_F = \frac{\bar{v}}{4} As$$

where $\bar{v}$ is the mean velocity of molecules in the gas phase. s can therefore be determined from the slope of the density *vs.* time curve, provided desorption (F_F) is negligible. Where this is not true, the rate coefficients may still be found by measuring dN/dt at different initial values of N. A plot of the net rate of adsorption ($NS_F - F_F$) against N for a constant amount adsorbed (see Fig. 4) gives a line of slope S_F and an intercept F_F. On heating the filament, the pressure ideally should return to the starting pressure.

(ii) Flow system

In a flow system, the pressure is controlled by the rate of gas admission to the cell (F_A) and the rates at which it is adsorbed by the surface and removed by pumping (NS_E). The mass balance equation is therefore

$$V \frac{dN}{dt} = F_F - NS_F + F_A - NS_E$$

If F_F and S_E are known, the rate of adsorption ($NS_F - F_F$) can be deduced from the pressure *vs.* time curve as for the closed system.

At the beginning of the experiment, F_A and S_E are adjusted to give the desired starting concentration N_0 (with the filament hot so that no adsorption occurs). When the steady state is achieved

$$F_A = N_0 S_E$$

The filament is then allowed to cool and the pressure drops due to adsorption until a new steady state is established as shown in Fig. 5. The net flow of gas into the system, $(N_0 - N)S_E$, is now balanced by the net loss of gas to the sample ($F_F - NS_F$). The number of molecules adsorbed at time t is given by

$$[n(t_1) - n(t_0)]A = V(N_0 - N_1) + Q_i$$

where

$$Q_i = S_E \int_{t_0}^{t_1} (N_0 - N)dt$$

i.e. the amount adsorbed is equal to the difference in the amounts of gas in the system at time t_0 and t_1 plus the net inflow Q_i (see Fig. 5). If the filament is flashed to a high temperature between time t_1 and t_2, the amount desorbed is given by

$$[n(t_1) - n(t_2)]A = V(N_2 - N_1) + Q_0$$

$$Q_0 = S_E \int_{t_1}^{t_2} (N - N_0)dt$$

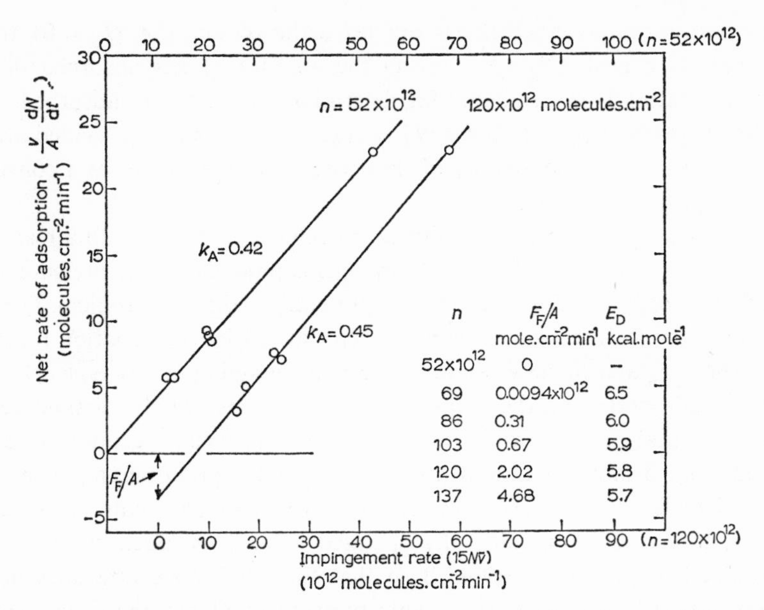

Fig. 4. Net rate of adsorption as a function of impingement rate. Xe at $T_G = 300°$ K on tungsten $\sim 80°$ K. Slope = sticking coefficient k_A; intercept = evaporation rate = F_F/A. From ref. 88.

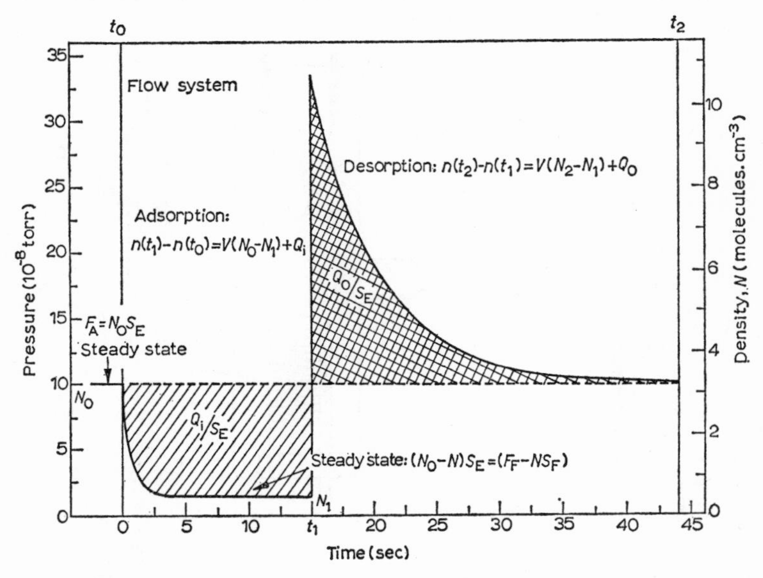

Fig. 5. Pressure cycle in flow system during adsorption on sample of unit area. Q_1, amount of gas added to cell during adsorption interval $t_1 - t_0$; Q_0, amount lost during desorption interval $t_2 - t_1$; S_E, exit speed; F_A, rate of gas entry from reservoir. From ref. 88.

If desorption is complete before gas can leave the system, (*i.e.* $Q_0 = 0$), the gas density rises to a peak $(N_2 - N_1)$ giving the amount of gas adsorbed directly. Such rapid heating is not essential for determining the amount adsorbed. When the original steady state is restored $N_2 = N_0$, $Q_i = Q_0$ and the shaded areas in Fig. 5 must be equal. The identity of these two quantities serves as a convenient check on experiment.

The ideal systems described are difficult to achieve in practice. In a real closed system adsorption occurs on the walls and gas is pumped by the pressure gauge. The walls may degass as the pressure falls, presenting a difficult problem. Therefore the equation for a closed system is really represented by the equation for a flow system, where F_A and S_E now denote gas flow or pumping of any type other than that due to adsorption on the filament. Closed systems are also limited because for a given initial pressure, only a small range of surface concentrations can be measured. Fig. 6 shows a typical apparatus for adsorption measurements. The flow into the adsorption cell is controlled by valve V_A and the pumping rate can be altered by adjusting the magnetically operated valve E. A much higher pressure is maintained in the gas reservoir B so that the admittance rate does not fall during the experiment. In order to avoid pressure gradients, the gauge and filament must be brought close together, the limit being set by the need to shield the filament from heat generated by the gauge and from stray electrons.

The type of apparatus described can also be used for making desorption studies. Because the heating of the sample, to effect desorption, changes the temperature of the filament, processes which occur at different temperatures can be recognised.

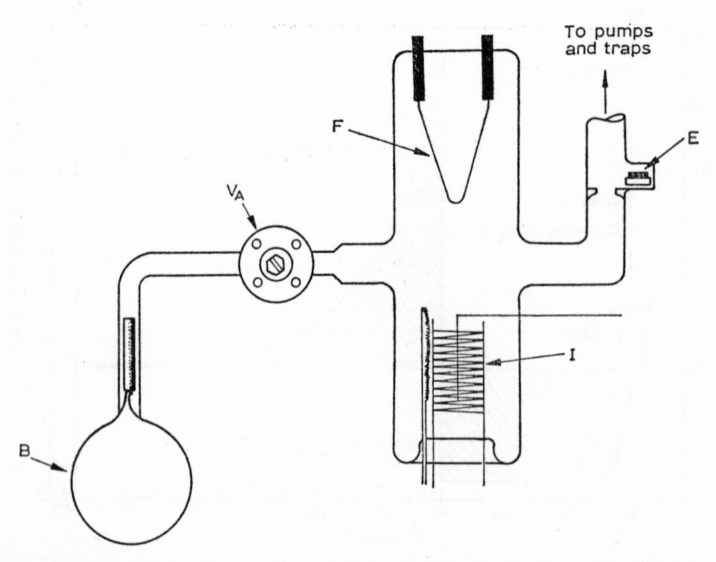

Fig. 6. Flow system for adsorption studies, with independent control of entrance and exit speed. F, adsorption sample; I, ionisation gauge. Flow out of gas reservoir (B) and into cell controlled by valve V_A; gas flow out of cell regulated by adjustable exit port E. From ref. 88.

The reaction order (x) for desorption can readily be determined from the pressure–time curve. This is best demonstrated by considering that the rate of loss of gas from the cell is negligible during the desorption, then

$$N = \int_{t_1}^{t_2} \frac{F_F}{V} \, dt$$

$$-\frac{dn}{dt} = \frac{F_F}{A} = n^x v_x \exp\left[\frac{-E_D}{RT}\right]$$

where

$$x = 1, \quad \frac{\Delta N}{\Delta N_\infty} = 1 - \exp[-X]$$

$$x = 2, \quad \frac{\Delta N}{\Delta N_\infty} = \frac{n(t_1)X}{1+n(t_1)X}$$

where

$$X = \int_{t_1}^{t_2} v_x \exp\left[\frac{-E_D}{RT}\right] dt$$

For a hyperbolic heating curve *i.e.* $1/T = a+bt$

$$X = \frac{Rv_x}{-bE_D} \exp\left[\frac{-E_D a}{R}\right]\left(\exp\left[\frac{-E_D bt}{R}\right] - 1\right)$$
$$= C(\exp[Bt]-1)$$

Evolution curves for first and second order desorption are shown in Fig. 7. They differ qualitatively in shape and their dependence on the initial concentration of

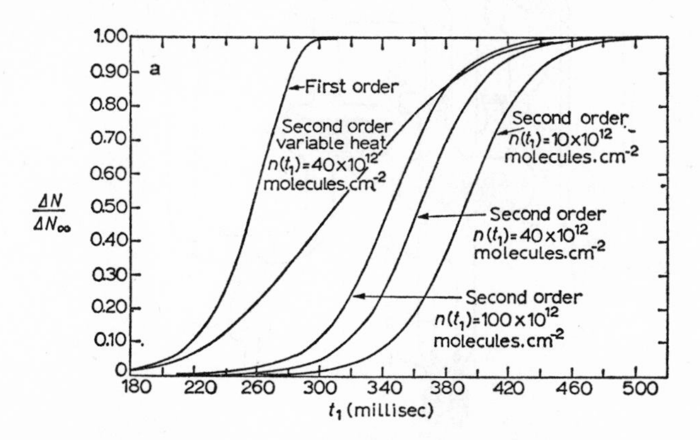

Fig. 7. Dependence of evolution curve on rate law for desorption. First order *versus* second order desorption; $E_D = 80$ kcal. mole^{-1}; $\nu_1 = 3.54\times10^{13}$ sec^{-1}; $\nu_2 = 8\times10^{-3}$ molecules^{-1}. sec^{-1}.cm^2. For variable heat, $E_D = E_D^0 - \eta n$; $E_D^0 = 80$ kcal. mole^{-1}; $\eta = 0.3$ kcal. mole^{-1} per 10^{12} molecules. cm^{-2}. From ref. 88.

adsorbed gas. First order desorption is independent of initial concentration and second order desorption gives a characteristic *s* curve. If the pumping rate cannot be neglected by comparison with the desorption rate, the analysis becomes more complicated and is best followed using a heating curve such that $1/T$ remains linear with time. Ehrlich[88] gives an analysis of the kinetics under these conditions and illustrates the technique by reference to the desorption of nitrogen from tungsten.

A suitable apparatus for flash desorption studies is illustrated in Fig. 8. The Bayard–Alpert gauge is coupled with a high speed ion current detector to provide an instantaneous record of gas density. The filament and gauge need to be mounted in separate containers to minimise electrical interference, but connected by wide bore tubing to prevent pressure gradients. Control and measurement of temperature is of prime importance. This can be accomplished by means of a simple bridge circuit[89] or preferably by means of the 'desorption spectrometer'[90] shown in Fig. 9. The heating and measuring functions of the filament are separate, direct current is used for heating, but resistance is measured by a 10 KC impedance bridge. Flashing of the filament results in an unbalance signal proportional to the resistance. This is amplified, detected, filtered and displayed on one axis of an oscilloscope. In converting filament resistance to sample temperature, it should be remembered that the temperature is not uniform over the whole sample. Corrections for end losses become smaller as the rate of filament heating is increased and the filament diameter decreased.

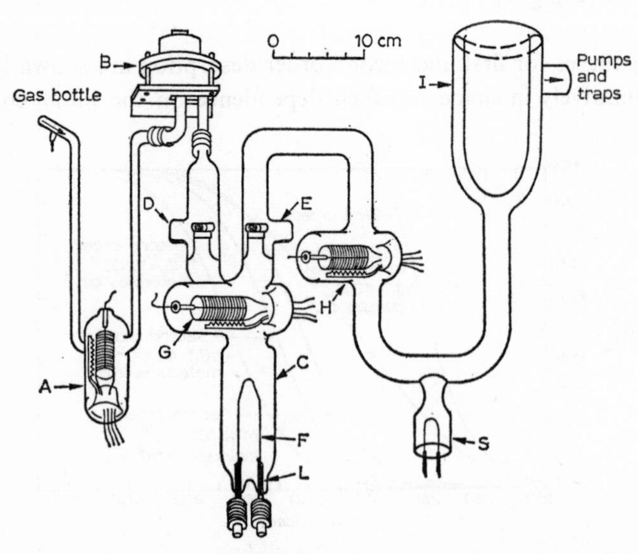

Fig. 8. Flash desorption cell and ultra high vacuum system. C, desorption cell with sample F, mounted on glass covered tungsten leads L, and gauge G. Volume of cell is defined by ground glass port D and E. A and H, ion gauges for leak checking and pressure monitoring; B, Granville–Phillips valve; I, liquid nitrogen trap; S, selective getter bulb. From ref. 88.

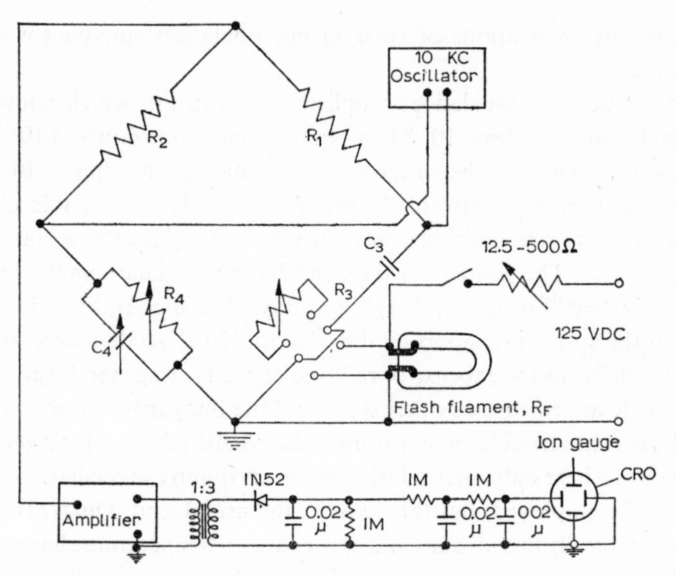

Fig. 9. Desorption spectrometer using 10 kc bridge. R_1, R_2 and R_3 are part of standard impedance bridge. C_3 = 30 μf, C_4 = 0–100 μμf. From ref. 88.

Apparatus suitable for the study of evaporated metal films under ultra high vacuum conditions has been described by Roberts[91] and by Hayward *et al.*[92]. In addition to an ion gauge for pressure measurement, both authors employ a mass spectrometer to characterise the species present in the gas phase. This is particularly important when studying hydrocarbons[91] which may decompose in the system. However, there are also problems associated with the measurement of hydrogen and oxygen. The standard tungsten filament found in an ion gauge reacts with both gases. Hydrogen is dissociated into atoms at the emission temperature, and this must be avoided by the use of a coated filament. Lanthanum boride coated on tantalum, tungsten or rhenium is suitable for this purpose and gives a satisfactory electron emission at a temperature of 1300° C[93–95]. Oxygen reacts with the carbon present in a tungsten filament to form CO and CO_2[27,90]. The carbon can be depleted by operation of the filament for many hours in a low partial pressure of oxygen[11,111]. An alternative is the use of an iridium filament.

2.1.4 *Measurement of adsorbed weight*

Measurement of weight changes is a very direct method of obtaining adsorption kinetics, and is also used to follow the course of other heterogeneous reactions (see Section 3). It is particularly suitable for adsorption measurements with small surface areas at relatively high pressures, where the volumetric method is insensitive. Several comprehensive reviews are available[96–98] which discuss the theory, construction and applications of vacuum microbalances suitable for adsorption measurements.

The term vacuum microbalance is applied to instruments which can work under vacuum conditions of at least 10^{-6} torr, and in some cases down to 10^{-9} torr, and follow weight changes of the order of a few micrograms ($\mu g = 10^{-6}$ g). For adsorption measurements with small surface areas, balances capable of detecting changes of 10^{-2} μg are desirable. Several types of balance have been used for adsorption studies. The simplest is the helical spring balance employing a spring of silica[99], tungsten[100] or copper–beryllium alloy[101], as used by McBain *et al.*[99,102]. Extension of the spring is proportional to the total load, so that increased sensitivity is only available at the expense of reduced capacity. In general, spring balances cannot be made as sensitive as other types and the buoyancy correction cannot be eliminated. Another troublesome feature is their sensitivity to vibration, especially in high vacuum where only internal friction effects damp the oscillations. Magnetic damping can be employed, but this reduces the useful load. Quartz is very fragile and must be carefully thermostatted if errors due to temperature changes are to be eliminated.

Beam microbalances are of two types, pivotal balances and torsion microbalances. Pivotal microbalances have been used by Gregg[103] and by Czanderna and Honig[104–106]. The balance described by Czanderna[106] will operate under ultra high vacuum conditions with a load of up to 20 g, and can measure weight changes of 2×10^{-7} g. Torsion microbalances are suspended by a fine wire, usually tungsten, which acts as the primary fulcrum. This type of microbalance has been most widely used and is described by Gulbransen[97] and Rhodin[96]. By using special care in ensuring uniformity of cross section of the beam and equality of arm length, balances which can detect weight changes of 10^{-7} to 10^{-8} g can be constructed. The sensitivity can be made dependent on, or independent of load by suitable design. Wolsky and Zdanuck[107] used a thin silica beam which sagged when loaded, thus enabling the sensitivity to increase with increasing load. One disadvantage of torsion microbalances is their tendency to undergo sideways yawing. This is often caused by shock or vibration which must therefore be avoided.

Buoyancy corrections can be minimised with beam microbalances by trying to make both limbs of the balance identical and using a counterweight with the same density as the sample. A more serious problem is the spurious mass changes produced by thermomolecular flow. This effect has been treated theoretically by Thomas and Poulis[108], and occurs at pressures between 10^{-3} and 1 torr, with a maximum effect between 10^{-2} and 10^{-1} torr. This is the Knudsen pressure range, where the mean free path of the gas molecules is comparable with the dimensions of the sample and the annular space between sample and balance tube. In order to eliminate these spurious mass changes, very elaborate thermostatting is necessary. One way to avoid this problem is perform the experiments in the presence of an inert gas in order to increase the total pressure above the Knudsen range.

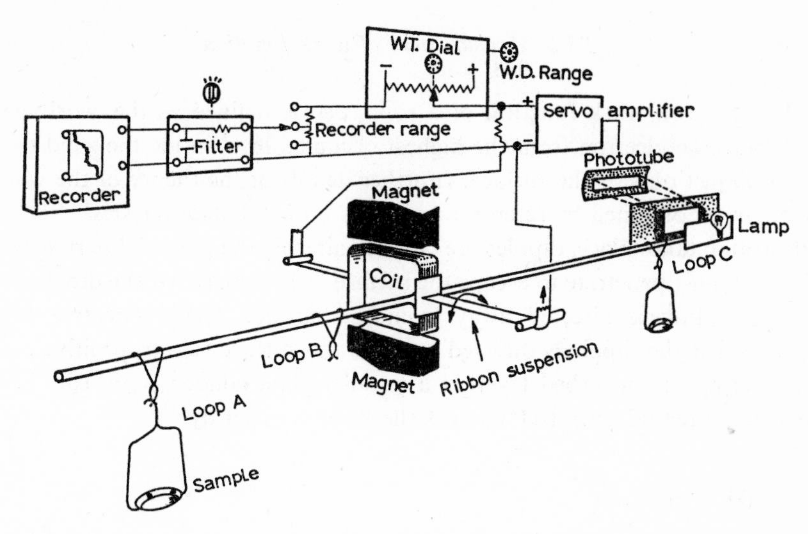

Fig. 10. Schematic diagram of a recording microbalance. From L. Cahn and H. R. Schultz, in *Vacuum Microbalance Techniques*, Ed. K. H. Behrndt, 3 (1963) 30.

The simplest method of beam microbalance operation is to measure the deflection of the ends of the beam by optical magnification. This method has the disadvantage that the maximum weight change that can be observed is limited and the vacuum system must usually be opened to reset the zero point. However, a beam microbalance can be readily converted to a null point instrument by electromagnetic adjustment. A commercially available instrument which features electromagnetic compensation and automatic recording is shown schematically in Fig. 10.

The beam is constructed from aluminium and suspended from an elastic metal ribbon. As the beam rotates, the flag on the right hand end uncovers the photocell, increasing the photocurrent. This current is then amplified and applied to the coil to rebalance the beam. Coil voltage is proportional to sample weight change, and this is recorded by means of an accurate potentiometer and a chart recorder. A different arrangement has been described by Cochran[109]. He used a variable permeance transducer to detect the movement of a steel rod attached to the balance. The signal from the transducer was amplified and used to control the current to the compensating solenoid, which restored the beam position. Langer[112] has described a more recent construction using a photocell and electromagnetic compensation.

Microbalances have been used to measure adsorption kinetics under a wide variety of conditions. Several authors[110,111] have studied adsorption on single crystal surfaces of semiconductors under ultra high vacuum conditions and utilising ion bombardment to obtain a clean surface. Gulbransen *et al.*[113] and Walker *et al.*[114,115] have made a special study of microbalance operation at high temperatures (up to 2500° C).

2.1.5 Measurement of work function[2,116]

The electron work function of a solid, $e\phi$, is defined as the work necessary to remove an electron from the highest occupied level inside the solid to a point *in vacuo* just outside the surface. (Further details of the theory of the work function can be obtained by reference to works on solid state physics[117,118].) When adsorption takes place, dipoles are created, altering the potential barrier which the electron must penetrate to escape the surface. ϕ is increased if the dipole produced on adsorption is directed away from the surface (electronegative layer), or decreased if the dipole is directed towards the surface (electropositive layer). By considering the adsorbed layer as a parallel plate condenser, it can be shown that the potential drop (ΔV) across the layer is given by

$$\Delta V = 4\pi N_s \theta \mu$$

where N_s = number of surface sites per cm^2, θ is the fraction of surface covered and μ is the dipole moment of the adsorbed molecule. $e\Delta V = e\Delta\phi$, the change in energy required to remove an electron from the solid after forming the adsorbed layer. Therefore

$$\Delta\phi = 4\pi N_s \theta \mu$$

The kinetics of adsorption can be studied from the rate of change of ϕ provided μ is constant. However, this is not necessarily the case except over a limited range. It must also be remembered that each crystal plane will have a different work function in the clean state. If more than one crystal plane is exposed, an average value will be obtained. Added to this is the fact that many gases adsorb differently on different crystal planes which makes the interpretation of results rather complicated.

Methods of work function measurement are of two types, electron emission methods and condenser methods. In the former method heating, irradiation with light of a suitable wavelength or application of sufficiently strong electrical fields is used to cause electrons to tunnel through the surface potential barrier. The latter method consists of measuring the contact potential difference between the surface under study and a reference electrode.

(a) Thermionic emission

For thermionic emission from a uniform surface, the saturation current density j extrapolated to zero applied field is given by

$$j = A(1-\bar{r})T^2 \exp(-e\phi/kT)$$

where A is a universal constant (120 amp. cm^{-2}. °K^{-2}), $\bar{\tau}$ is the average reflection coefficient for electrons at the surface, T is the absolute temperature, and k is the Boltzmann constant. j can be measured as a function of temperature and a plot of $\ln(j/T^2)$ against $1/T$ should give a straight line of slope $-e\phi/k$. The method is limited, because the high temperatures needed to obtain suitable emission currents cause the complete desorption of most adsorbed layers. One exception is that of adsorbed layers of alkali metals on tungsten[119], where the low work function makes emission possible at quite low temperatures (~ 150° C). A suitable electrode assembly has been described by Becker[120].

(b) Photoelectric emission

When a surface is irradiated by light of variable frequency ν, electrons will be emitted if the incident photons impart sufficient energy to enable them to overcome the potential barrier at the surface. At the threshold frequency ν_0,

$$h\nu_0 = e\phi$$

However, at normal temperatures the threshold is blurred due to the emission of thermally excited electrons with energies above the Fermi level. To surmount this difficulty, it is necessary to use the theory developed by Fowler[121]. This theory shows that the quantum yield, I (number of electrons liberated per incident photon), is given by

$$I = AT^2 \cdot F\left(\frac{h\nu - h\nu_0}{kT}\right)$$

where A is a constant, T is the absolute temperature and the expressions for the function $F[(h\nu - h\nu_0)/kT]$ are as given by Fowler. A series of curves of the function against $h\nu/kT$ can be plotted and the experimental plot of I/T^2 against $h\nu/kT$ fitted to them. The horizontal shift required to superimpose the calculated and experimental curves is $h\nu_0/kT$ and the work function can be obtained. If the curves do not fit well, some features of the experiment are usually suspect.

There are two practical difficulties. One is the measurement of the very small photocurrent, which may be as low as 10^{-14} amp, requiring the use of a vibrating reed electrometer or similar instrument. The second is that, for work functions above 5 eV, ν_0 lies in the far ultraviolet. This makes the study of some adsorptions very difficult and 6 eV is about the practical limit of such measurements. A suitable light source is the quartz mercury arc. The energy of the incident beam can be measured with a calibrated photocell, a vacuum thermopile or a radiometer. A suitable cell for adsorption studies[122] is shown in Fig. 11. The sample being studied forms the cathode B. It can be a metal foil or a film formed by evaporation from the filament E. A wire C is fused through the glass to make contact with the

References pp. 270–278

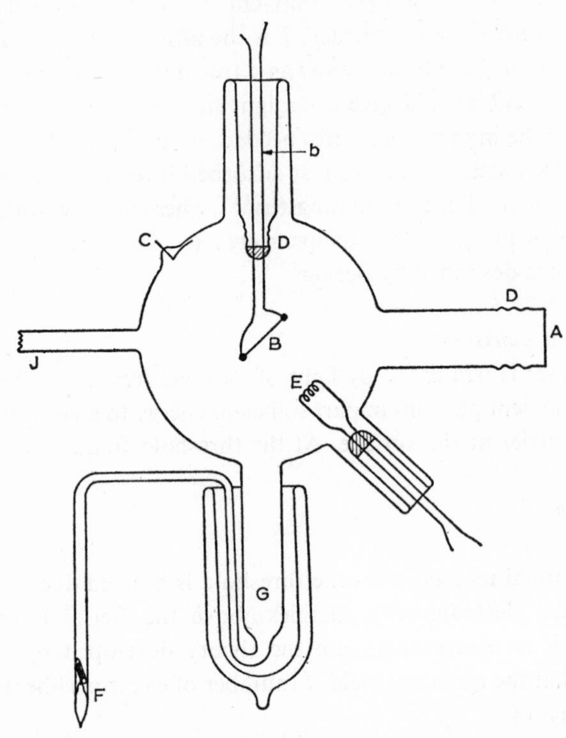

Fig. 11. Cell for the measurement of work function by the photoelectric method. A, quartz window; B, cathode; C, wire to anode film; D, graded glass seals; E, filament; F, substance to be adsorbed; G, cold trap; J, outlet to pumps. From R. Suhrman, *Advances in Catalysis*, 7 (1955) 308.

metal coating on the walls which forms the anode. The cathode can be electrically heated, or cooled by liquid air poured into the inverted seal. Monochromatic light passes through the quartz window A and strikes the cathode.

After the sample has been cleaned by direct heating or electron bombardment, the photoelectric yield as a function of wavelength is measured and the work function calculated. The substance in F is then admitted by breaking the seal and the work function remeasured. Because of ionisation by collision the photoelectric method cannot be employed above 10^{-3} torr.

(*c*) *Field emission*

The presence of a high external electric field ($\sim 10^7$ V. cm^{-1}) reduces the width of the potential barrier at a metal surface, allowing electrons to tunnel *through* the barrier and be emitted. Electrons which tunnel through the barrier have little kinetic energy and will follow lines of force which radiate from the emission tip. If the tip is surrounded by a conducting fluorescent screen, the electrons are collected and form a magnified image of the tip. Bright areas on the screen represent regions of low work function, and dark areas, regions of high work function. This is the principle of the field emission microscope which was invented by Müller[123] and has formed the subject of many publications by Gomer[124,125] and others. Work in this area was recently reviewed by Ehrlich[88].

The electron omission from the tip is described approximately by the Fowler–Nordheim equation[126], which can be expressed in the form

$$i/V^2 = \alpha \exp(-\beta\phi^{\frac{3}{2}}V)$$

where i is the total current, V is the applied voltage and α and β are constants. A plot of ln (i/V^2) against $1/V$ gives a slope $\beta\phi^{\frac{3}{2}}$. β is not normally known, but if the work function of the clean surface is known, the change in work function on adsorption can be calculated from the *change* in slope on adsorption. The Fowler–Nordheim equation does not accurately describe the emission process. Work functions obtained by this method are in general slightly smaller than the true values. A more accurate evaluation is discussed by Ehrlich[88].

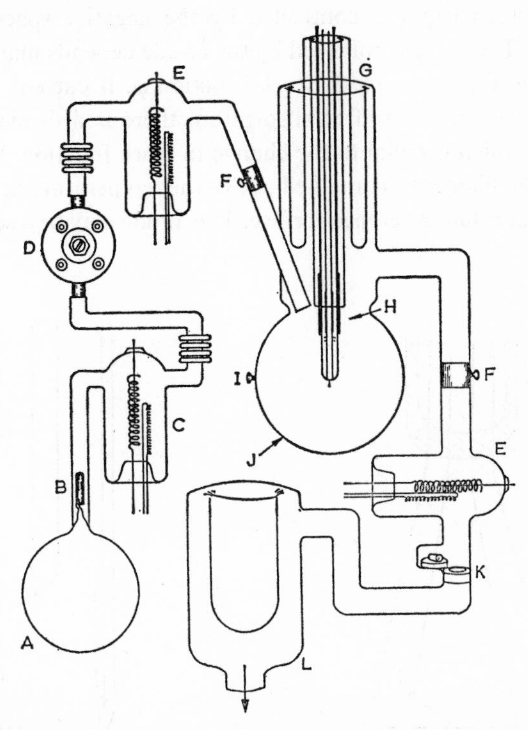

Fig. 12. Field emission microscope for adsorption studies. A, gas bottle; B, break-off seal; C, inverted ionisation gauge (also serves as a selective getter); D, Granville–Phillips valve; E, ionisation gauge; F, grounding rings; G, double Dewar; H, emitter assembly (tip mounted on hairpin support wire, equipped with potential leads for measuring resistance); I, anode terminal; J, Willemite screen settled onto tin oxide conductive coating; K, ground glass port; L, trap. From ref. 88.

An apparatus suitable for adsorption studies is shown in Fig. 12. The sample bulb is connected to a source of gas and to a pumping system. A dewar, which can be filled with liquid hydrogen, surrounds the emitter assembly H supporting the tip. In order to attain the high fields ($\sim$ 0.3 volts. A^{-1}) necessary for appreciable emission, the sample is shaped into a needle-like point with a radius of less than 2000 A. Electrons emitted from the tip are accelerated towards the Willemite screen and there produce a magnified image. The resolution is limited to $\sim$ 20 A by the tangential velocity of electrons in the free electron gas, therefore only aggregates of atoms can be detected. The technique can be used to establish the existence of an electropositive or an electronegative gas film and to study the relative rates of adsorption, surface diffusion and desorption with various crystal faces. Kinetics of hydrocarbon adsorption on iridium have been measured with this method[127].

(d) The space charge limited diode[116,128]

If a small accelerating potential is applied between the anode and cathode of a diode the emission current is controlled by the negative space charge near the cathode surface. The current collected by the anode depends mainly on the applied voltage V and on the mean anode work function ϕ. If current voltage characteristics are plotted before and after adsorption, there is a displacement along the voltage axis ΔV which is equal to the change in work function $\Delta\phi$. The cathode is usually a tungsten filament maintained above the temperature at which adsorption occurs, giving a constant reference surface. The anode may be a second filament[129],

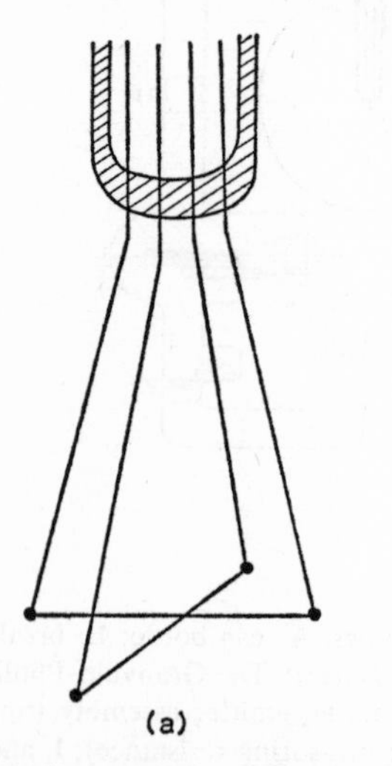

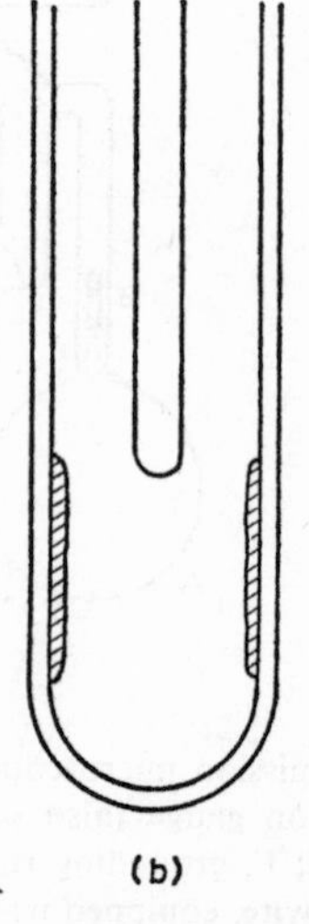

Fig. 13. Electrode assemblies for the determination of work functions with the space charge limited diode; (a) crossed filaments[129]; (b) filament cathode, evaporated metal film anode[128].

or an evaporated metal film[128] on the wall of the cell. The electrode arrangement used in these experiments is shown diagrammatically in Fig. 13.

(e) The contact potential

If two metals of differing work function are connected electrically, at the same temperature and without a source of emf, the electrostatic potentials just outside the two surfaces are different. This potential difference V_{12} is known as the contact potential difference and is equal to the difference in the work functions of the two metals ϕ_1 and ϕ_2. If a compensating potential, equal and opposite, is applied across the plates, the field becomes zero.

The two metals act as the plates of a condenser, one of the plates being used for adsorption of the gas and the other being a reference electrode. Adsorption of gas causes a change in contact potential equal to the change in work function. Disappearance of the electrical field between the two plates may be detected either by the steady condenser method[130] or the vibrating condenser method[131,132].

A glass adsorption cell used for the static condenser method is shown in Fig. 14. The condenser is formed from two concentric cylinders separated by a 2 mm gap.

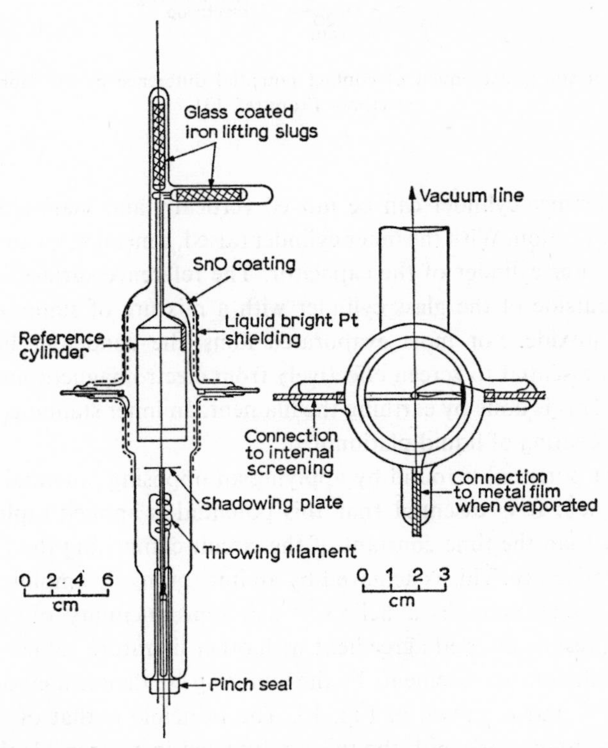

Fig. 14. Cell for the measurement of contact potential difference by the static condenser method. From ref. 130.

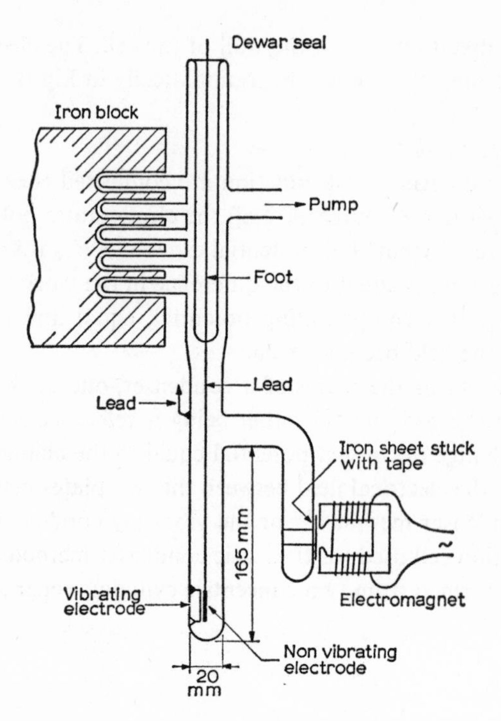

Fig. 15. Cell for the measurement of contact potential difference by the vibrating condenser method. From ref. 131.

The inner reference cylinder can be moved vertically and maintained concentric with the outer section. With the inner cylinder raised, a metal film can be evaporated to form the outer cylinder of the capacitor. The reference surface is prepared by coating the outside of the glass cylinder with a mixture of stannous oxide + 1% antimony pentoxide. For most evaporated films, the condenser has a capacity of 30 pf. It is essential to screen effectively from electromagnetic and electrostatic interference. This is done by earthing the filament, an inner stannous oxide coating and an outer coating of liquid platinum.

The contact potential is found by applying an opposing potential until the null point is reached. It is essential that this potential is applied rapidly, in a time much smaller than the time constant of the circuit comprising the condenser and the external resistance. This is achieved by an integrating dc amplifier coupled to a servo motor which actuates a helipot[130]. A reproducibility of ±0.03 volts is claimed, with results in good agreement with other literature values.

A cell suitable for measurements by the vibrating condenser method was devised by Mignolet[131] and is shown in Fig. 15. The principle is that of an electrically driven hollow tuning fork with the tail rigidly fixed in an iron block. The fundamental vibration frequency is about 450 c. sec^{-1} and it will stand an amplitude of

0.7 mm. This is greater than the normal separation of the condenser plates, 0.1 to 0.3 mm. The vibrating electrode is attached to the glass. The fixed electrode is flexibly supported so that it can be moved to one side to allow evaporation of a metal film onto the vibrating electrode. Adequate screening against stray electromagnetic or electrostatic currents is essential, as for the steady condenser method.

A major difficulty with contact potential work is the provision of a completely inert reference electrode. Poisoned nickel has been used[131], but aged gold or platinum is more widely applicable. Bewig and Zisman[133] have described reference electrodes of gold and platinum coated with Teflon resin. These electrodes were more stable than the bare metals in wet and dry oxygen, nitrogen and in carbon dioxide, hydrogen and helium.

2.1.6 Measurement of magnetic changes

Chemisorption often involves the unpaired electrons in solids, causing a change in the magnetic properties due to a change in the number of unpaired electrons per adsorbed atom or molecule. Because the magnetic properties are bulk properties, the surface-to-volume ratio must be high in order to make the changes detectable. Powder samples are often the most suitable for this reason. The classical method of following magnetic changes is by measurement of magnetic susceptibility, but more recently the techniques of nuclear magnetic resonance (NMR) and electron spin resonance (ESR) have been employed. At the present time, magnetic methods have been more widely used to study finite changes in the magnetic properties rather than the kinetics of the adsorption process.

(*a*) *Measurement of magnetic susceptibility*

If a substance is placed in a magnetic field, the intensity of magnetization may be slightly smaller (diamagnetic), or larger (paramagnetic) than that produced in a vacuum by the same field. A few materials are ferromagnetic, *i.e.* the intensity of magnetisation is very much larger than for diamagnetic and paramagnetic substances, and depends upon field strength in a complicated way. There are a number of methods for measuring magnetic susceptibility[134]. One which has been modified for making chemisorption measurements is the Faraday Method[135,136].

If the sample is placed in an inhomogeneous magnetic field with an axis of symmetry, such as that formed by two appropriately shaped pole pieces, a force f will be exerted on the sample where

$$f = mxH\partial H/\partial S$$

Here, m is the mass of the sample, x is the mass susceptibility, H the field strength and $\partial H/\partial S$ the field gradient. Selwood *et al.*[135,136] measured the force by the

References pp. 270–278

extension of a spiral spring. However, any of the other vacuum microbalance techniques described in Section 2.1.4 could be applied. Only a small sample is required and the apparatus may be operated at any temperature from 20° K to 1200° K. Selwood[134] suggests calibration of the field with a substance of known susceptibility such as water, nickel chloride solution, hydrated ferrous ammonium sulphate or cane sugar.

In paramagnetic substances, the relationship between the atomic or molecular moment (μ) and the magnetic susceptibility per mole (X_m) is given by the Curie–Weiss law[137]

$$\mu = 2.84\sqrt{X_m(T+\Delta)}$$

where T is the absolute temperature and Δ is the 'Weiss' constant. For the transition metals, $\mu = \sqrt{n(n+2)}$ where n is the number of unpaired electrons per atom. From the changes in susceptibility, the number of unpaired spins destroyed or created per adsorbed molecule can be estimated.

With ferromagnetic materials, the magnetic susceptibility is large and dependent on field strength. For this reason the specific magnetisation (XH), or more desirably the saturation magnetism, is measured. The saturation moment per atom μ (at zero temperature and infinite field) is equal to the number of unpaired electrons n. Temperatures down to 4.2° K and fields of 10,000 gauss can be obtained, the results are then extrapolated to find the saturation magnetism. The Faraday method, or an induction method[138], can be used to measure the specific magnetisation.

Although the methods have been mainly applied to the change in magnetic properties of the substrate on adsorption of gases, they are also applicable to the study of changes in the magnetic properties of absorbed gases. Oxygen adsorbed on chabasite, charcoal, silica and platinum has been studied[139]. Magnetic measurements are able to distinguish between molecular oxygen which is paramagnetic and combined oxygen which is diamagnetic. The rates of incorporation and oxide formation can also be measured. Nitrogen dioxide on silica, charcoal and alumina gel has also been investigated.

(*b*) *Nuclear magnetic resonance* (NMR)[140]

A sample possessing a resultant nuclear magnetic moment, when placed within a steady magnetic field of several thousand oersted, will exhibit resonance adsorption by the application of radiation of the correct polarisation and frequency. The latter can be produced by a circuit oscillating at several megacycles per second. Nuclei possessing magnetic moments favourable for NMR studies are given in Table 5. Those nuclei in heavy type have nuclear spin quantum number equal to $\frac{1}{2}$, and are generally the most favourable for observation in amorphous solids.

There are three widely used methods for observation of continuously excited nuclear magnetic resonance. Two of the methods use variable frequency rf os-

TABLE 5

NUCLEI POSSESSING A MAGNETIC MOMENT FAVOURABLE TO NMR STUDIES

^{1}H ^{2}H
^{7}Li ^{9}Be ^{11}B ^{13}C ^{19}F
^{23}Na ^{27}Al ^{29}Si ^{31}P ^{35}Cl
^{45}Sc ^{51}V ^{55}Mn ^{59}Co ^{63}Cu ^{65}Cu ^{71}Ga ^{75}As ^{77}Se ^{79}Br ^{81}Be
^{87}Rb ^{93}Nb ^{111}Cd ^{113}Cd ^{117}Sn ^{119}Sn ^{121}Sb ^{125}Te ^{127}I ^{129}Xe
^{133}Cs ^{141}Pr ^{199}Hg ^{203}Tl ^{205}Tl ^{207}Pb ^{209}Bi

cillators to supply the rotating magnetic field and are known as the Bloch[141] (crossed-coil) and the Pound–Knight[142] NMR Spectrometer respectively. A third method, due to Bloembergen *et al.*[143] utilises a radiofrequency bridge. The Bloch spectrometer is most useful for the study of solids, but can only be used with sample temperatures up to 300° C, whereas the Pound–Knight spectrometer can be used up to 600° C.

A block diagram of the crossed coil NMR spectrometer is shown in Fig. 16. The sample, contained in a tube, is placed in a uniform field of several thousand oersteds. This field is modulated, at 20 to 60 cycles per second, by coils the axes of which are parallel to each other and to the steady field. This modulation causes a recurrent scanning of the resonance condition. A high frequency radio transmitter supplies energy to the system through a coil placed at right angles to the steady field. The effect of this alternating field is to change the tilt of the nuclei already oriented and precessing in the steady field. Choice of frequency is governed by the relation

$$v = \mu_n H_0 / Ih$$

where v is the angular frequency of precession, μ_n is the nuclear magnetic moment, H_0 is the uniform field, I is the nuclear spin quantum number and h is Planck's constant.

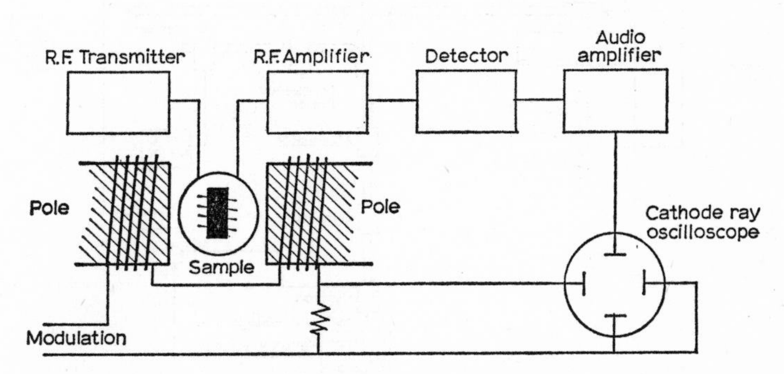

Fig. 16. Block diagram of the crossed coil nuclear magnetic resonance spectrometer due to Bloch, Hansen and Packard. From ref. 134.

Voltages are induced, by the precessing nuclear magnetic moment, in a receiving coil which is placed at right angles to the steady field (H_0) and to the axis of the coil providing the driving field. The induced voltages are then amplified and detected on an oscilloscope in a resonant circuit. A maximum in the voltage will be detected when the frequency of the rf current is equal to v, the angular frequency of precession, *i.e.* when the rf field is in resonance with the precessing nuclei. A detailed description for an advanced design of apparatus is given by Gutowsky *et al.*[144]. Commercial apparatus is now widely used.

(*c*) *Electron paramagnetic resonance* (EPR)[140]

EPR can be used to observe the unpaired electrons in paramagnetic substances and the technique is in many respects similar to NMR. The main differences are (*i*) that electron magnetic moments are 10^3–10^4 times larger than nuclear moments and (*ii*) they generally possess a contribution to their magnetic moment from orbital angular momentum, in addition to intrinsic or spin angular momentum. Because of their larger magnetic moment, resonance appears at rf frequencies of several thousand megacycles. The rf frequencies at which resonance occurs are given by

$$v = gu_B H_0/h$$

where v is the frequency, g is the Landé splitting factor, u_B is the electron spin magnetic moment, H_0 is the steady magnetic field and h is Planck's constant.

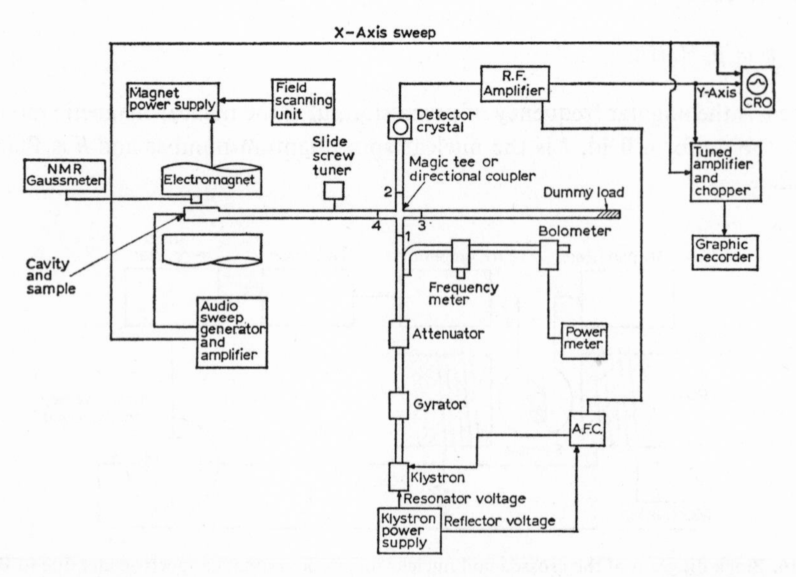

Fig. 17. Diagram of EPR spectrometer with automatic frequency control (AFC). From ref. 140.

For a free electron $g = 2$, but various interactions cause some deviations.

A commonly used EPR spectrometer is shown diagrammatically in Fig. 17. It consists of a four arm microwave bridge, which acts in a manner analogous to a Wheatstone bridge. On arm (1) is the Klystron power supply which acts as a source of microwaves. The microwave cavity containing the sample is connected to arm 4. Arm 3 contains a dummy load which absorbs all the microwave power incident on it. Arm 2 is connected to a silicon–tungsten detector crystal. The microwave bridge has the property that when arms 3 and 4 are matched to the bridge, *i.e.* when all power incident on the cavity and the dummy load is absorbed, no power is detected in arm 2. The microwave cavity can only be matched at certain frequencies and the sample is placed in that part of the cavity where the microwave magnetic field is maximal and the microwave electric field is zero. As in NMR, the microwave magnetic field is at right angles to the steady field produced by the electromagnet. Coupling of the microwave cavity to the bridge is accomplished by a small hole or iris, which is adjusted in size to give a zero reflection coefficient, *i.e.* a 'matched' cavity. Alteration of the slide screw tuner will then slightly unbalance the bridge, the EPR absorption is observed and can be displayed on a cathode ray oscilloscope or a graphic recorder.

2.1.7 Measurement of changes in electrical conductivity[122,146]

(a) Metals

When a gas is adsorbed on a metal, electrons may be abstracted from the conduction band or added to it. In either case, the conductivity will change. However, since conductivity is a bulk property, the change will only be measurable if the surface-to-volume ratio is high. More precise interpretation can be given to measurements on single crystals, but in practice this is difficult and thin evaporated films are more often used. The metal film can be evaporated by any of the usual methods (see Section 2.1.1) onto the walls of a vessel containing two contacts and the film resistance between the contacts measured[122]. It is often necessary to heat the film above the adsorption temperature to anneal out defects, otherwise the resistance may change slowly with time. Unless the films are prepared taking extreme precautions to avoid contamination, variable results may be obtained. This point has been well demonstrated by Sachtler[145], who found that the adsorption of hydrogen on nickel caused the conductivity to change in opposite directions for 'clean' and 'moderately clean' films. It must be assumed that the film is uniform and that grain boundary resistance is negligible. Since metal films can be very porous, this may not be true. However, the recent advances in technique which have enabled the formation of single crystal metal films[52] should remove some of these objections. The second assumption is that the only effect of chemisorption is to change the number of the conduction electrons and it ignores the effect on the structure of the film, such as changes in the lattice constant due to penetration of the adsorbate or local sintering due to a high heat of adsorption.

The simplest method of measuring conductivity is to determine the voltage drop across the sample and the current through the sample. The conductivity (σ ohm^{-1}. cm^{-1}) is then given by

$$\sigma = IL/VA$$

where I is the current in amperes, V is the voltage drop, and L and A are the length and the cross sectional area of the specimen respectively. A refinement of this technique is the potential probe measurement. This technique uses a potentiometer which draws negligible current to measure the potential drop across part of the specimen. L in the above equation then becomes the distance between the potential probes. Under these conditions, the effect of contact resistances is minimised. An electrometer can be used where the circuit has a high impedance[146].

Bridge methods are also a standard technique for conductivity measurements. The basic Wheatstone bridge circuit is illustrated in Fig. 18. When the bridge is balanced, *i.e.* the detector D indicates zero current, then the unknown resistance R_x is given by

$$R_x = R_s(R_1/R_2)$$

One important deficiency of the Wheatstone bridge is that it measures both sample and contact resistance. Thus it is particularly inadequate for measurements below 10 ohms. For measurements on metallic specimens with low resistances down to

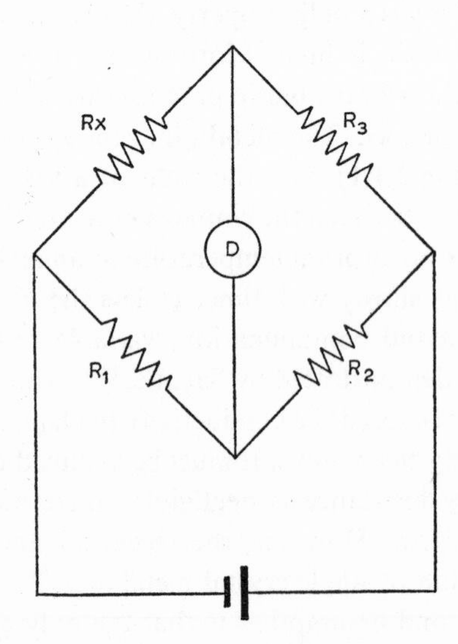

Fig. 18. Basic Wheatstone bridge circuit for the measurement of electrical conductivity.

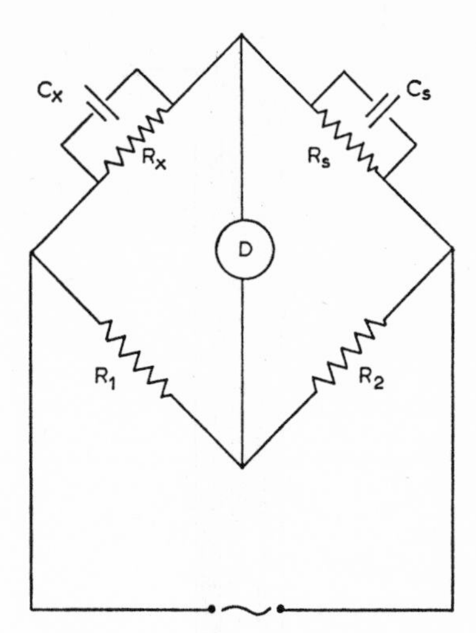

Fig. 19. Alternating current bridge circuit for the measurement of electrical conductivity.

10^{-2} ohms the Kelvin double bridge can be employed[146]. Alternating current methods can be used to overcome difficulties associated with direct current measurements. Amongst these are barriers at internal surfaces and contact resistances, which occur particularly in powders and evaporated films. The bridge used requires two balances, one for the resistive and one for the capacitative elements of the impedance being studied (Fig. 19). Current is supplied to the bridge from an ac oscillator with a frequency range of 20 cycles per sec to 10 Mc per sec. The detector D is normally a pair of earphones connected directly, or in the case of radiofrequencies, through a radio receiver. Since most adsorption studies have been conducted on powders or evaporated films in order to obtain a sufficiently high surface-to-volume ratio, the ac bridge method has a particular advantage. Other high frequency techniques which have been used are the Q-meter[147] (which measures the breadth of a resonance curve) and microwave adsorption[148].

(b) Semiconductors

The term "semiconductor" covers a wide range of materials, including the elements silicon and germanium, and compounds such as oxides, sulphides, selenides, etc. A survey of the physics and chemistry of semiconductors is given in the book edited by Hannay[149] and particular reference to semiconductivity and adsorption on oxide surfaces is to be found in an article by Gray[150].

In semiconductors, the surface electrical double layer penetrates the crystal to a distance of about 10^{-4} cm, and the mobility, as well as the number, of current carriers may be altered by the adsorption process. For this reason, conductivity measurements alone are insufficient to quantitatively determine the extent of charge transfer across the surface. For a simple *n*-type semiconductor, adsorption

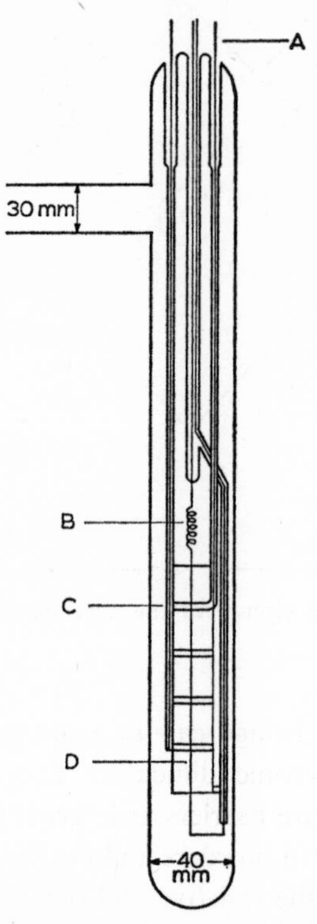

Fig. 20. Reaction tube for semi-conductivity measurements on thin films. A, Pt conductance leads and Pt/Pt–Rh thermocouples; B, tungsten tensioning spring; C, four Pt foil conductance electrodes fused onto the inner wall; D, plated tungsten evaporation filament. From ref. 150.

of a gas which donates electrons to the surface will cause an increase in conductivity. Adsorption of a gas which accepts electrons will cause a decrease in conductivity. Donation of electrons into the almost empty conduction band of an *n*-type semiconductor proceeds readily to the extent of a monolayer, this is known as cumulative chemisorption. However, abstraction of electrons from an *n*-type semiconductor is much more difficult. Once the electrons in the nearly empty conduction band have been abstracted, further electrons can only come from levels deep in the crystal. Generally, the potential barrier is too high for the latter process to occur and chemisorption stops after a fraction of a monolayer has been formed. This is called depletive chemisorption. On a *p*-type semiconductor, the above processes are reversed. Donation of electrons from the adsorbate causes a decrease in conductivity and depletive chemisorption. Abstraction of electrons by the adsorbate causes an increase in conductivity and cumulative chemisorption. A theoretical treatment of chemisorption as a boundary layer problem has been given independently by Aigrain and Dugas[151], Hauffe and Engell[152] and Weisz[153]. This and more recent work has been summarised by Law[154].

A further problem arising with semiconductivity measurements is that the surface conductivity may be of a different type to the bulk conductivity. Measurements of work function or field effect mobility are then necessary to fully define the process.

Both the potential probe and the bridge methods are applicable to semiconductors, but the dc methods are limited with films and powdered or sintered materials because of the high contact resistance between particles. An apparatus employing a four potential probe has been described by Anderson [155]. A method for studying the rates of adsorption on evaporated films has been developed by Gray *et al.*[156] using the apparatus shown in Fig. 20. Four platinum electrodes are fused to the inside of a hard glass or silica tube. A filament serves to evaporate a film of the material to be studied using the techniques described in Section 2.1.1c. The film produced can be from a few hundred to a few thousand angstroms in thickness. Conductivity changes as a function of gas adsorption can be studied either with a potentiometer, or ac bridges covering the frequency ranges 0.5 $c \cdot sec^{-1}$ to 150 $Kc \cdot sec^{-1}$. If the original material evaporated is a metal, its rate of oxidation can be measured and further adsorption of oxygen or other gases studied.

Handler has used the apparatus shown in Fig. 21 to study adsorption on thin

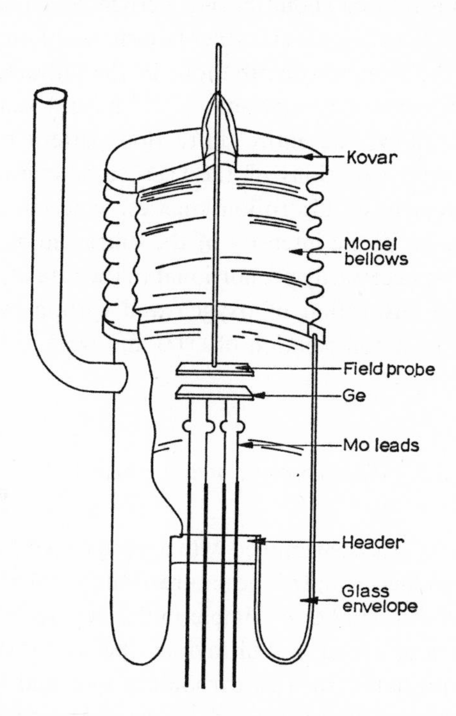

Fig. 21. Diagram of the experimental tube used for measurement of the surface conductance and field effect mobility. From P. Handler, *Semiconductor Surface Physics*, Ed. R. H. Kingston, (1957) p. 29.

slices of germanium and silicon. The surface conductance is measured by a potential probe. The field effect mobility is measured at the same time; this is defined as the change in surface conductivity produced by a field applied normal to the surface on the sample. The applied field induces a charge on the semiconductor surface. If the charge is positive the conductivity of a *p*-type surface increases, whereas that of an *n*-type surface decreases. By comparison with Hall-effect measurements[149] on the bulk material, it is possible to decide if the surface and bulk show the same type of semiconductivity.

2.1.8 Electron diffraction

The technique of Low Energy Electron Diffraction (LEED) is well suited to the observation of surface monolayers and the structure of clean surfaces. Farnsworth *et al.*[30,31,60,69,70] have done a great deal of work in this field, obtaining measurements on single crystals cleaned in ultra high vacuum by ion bombardment. One disadvantage of the method used is that the diffracted electrons are collected by a Faraday cage which is rotated about the axis perpendicular to the incident beam. Even with automatic recording this is rather laborious although it gives a relatively accurate picture of the intensity distributions in the diffraction pattern.

More recently Germer and co-workers[157–159] have developed a different type of apparatus which allows the entire diffraction pattern to be displayed on a fluorescent screen. Both variations of the LEED technique are rather difficult to apply to the measurement of adsorption kinetics, which in effect requires a measurement of the change in the intensity of the diffracted beam(s) with time. This intensity change is not necessarily proportional to the rate of adsorption; however, detailed data on the adsorption of oxygen and carbon monoxide on nickel[70] have been obtained by a combination of LEED and work function measurements.

2.1.9 Spectroscopic measurements[160–162]

Optical spectroscopy has been applied with a good deal of success to the identification of chemisorbed species and of the nature of the surface bond. Infrared spectra have been most useful in studies of simple molecules, such as carbon monoxide adsorbed on platinum or nickel, and ultraviolet spectra for the characterisation of more complex intermediates, such as carbonium ions and ion radicals. The frequency of the adsorption band (or bands) often serves to identify the adsorbed species by comparison with spectra of known compounds. Quantitative information may then in principle be obtained by measuring the area under the adsorption curve, or the maximum intensity, both of which are a function of the concentration of the adsorbing species. This is usually expressed by Beer's law,

$$\log (I_0/I) = C\varepsilon L$$

where I_0, I are the intensities of the incident and transmitted light respectively, C is the concentration in mole . l^{-1}, ε is the molar extinction coefficient and L is the path length. Log I_0/I is called the optical density or adsorbance.

In general, chemisorption will produce new spectral bands which are not characteristic of the adsorbate or the adsorbent. However, absence of such bands cannot be taken as evidence of an absence of chemisorption. A difficulty present in any attempt to make kinetic measurements is that extinction coefficients are often significantly altered as a result of adsorption. These changes, which cannot as yet be interpreted theoretically, make it difficult to correlate the observed absorbance with the coverage of adsorbed molecules. The change in extinction coefficient is dependent on both the adsorbate and the adsorbent. For example, an increase of ε was observed with increasing coverage for ethylene adsorbed on copper oxide, whereas the reverse occurred with nickel oxide[163].

Since chemisorption results only in the formation of a monolayer, the radiation must traverse many layers for the adsorption bands to be detectable. This may be achieved in three ways (a) by transmission through a highly dispersed medium, (b) by multiple reflections, or (c) by a series of total internal reflections. Applications of the latter two methods to the study of adsorption has been limited up to the present time[164,165]. The first method requires the preparation of the adsorbate in the form of a powder or a porous glass. In order to reduce light scattering to a minimum, the maximum size of the particles must be smaller than the wavelength being used. In the infrared region, this implies particles smaller than one micron for non-absorbing species. For species which absorb in the infrared, such as metals, even smaller particles are required. On moving towards the ultraviolet, the scattering losses increase rapidly, thus making it even more important to obtain very small particles.

Eischens and Pliskin[160] prepared samples by spreading a slurry of the powder on a calcium fluoride disc and evaporating the volatile liquid. Fahrenfort and Hazebroek[166] supported the powder on a 200 mesh screen, claiming that sample treatment was possible over a wider temperature range, with no wavelength limitations due to strong absorption by the catalyst. Some powders can be pressed into wafers and used directly[167]. Metals are generally supported on silica, alumina or titania. Cabosil (SiO_2) and Alon C (γ-Al_2O_3) with particle sizes less than 200 A are suitable. The metal is deposited by impregnation with a metal salt solution, drying and reduction in hydrogen. It is preferable that the size of the final metal particle be less than 100 A. Transparent massive samples of silica gel and silica–alumina have been prepared by Leftin and Hall[168] and transparent γ-alumina

References pp. 270–278

by Peri and Hannan[169]. These samples have the advantages of greater ease of manipulation and smaller scattering losses than powder and can be readily impregnated with metals. However, this type of preparation is not yet fully reproducible.

Commercial spectrophotometers are suitable for the measurement of adsorption because it is the sample rather than the instrument which limits the resolution. Special types of sample cell are required and it is desirable that the cell is attached to a vacuum line and a gas handling system to effectively manipulate the gaseous ambient. Eischens and Pliskin[160] used a pyrex cell with calcium fluoride windows at both ends for measurements in the infrared region. A calcium fluoride plate held the powder sample and coula be heated by a tungsten heating element wound on a quartz tube. This type of cell has two disadvantages. It must be mounted vertically to prevent loss of powder, which requires modification in the normal spectrometer arrangement, and it cannot be baked out at high temperatures because of the glyptal resin used to seal the calcium fluoride windows to the pyrex body. For compressed discs and massive plates, the cell need not be vertically mounted. Cells which allow the sample to be outgassed in one section and then moved to the measuring section containing the salt windows have been described[169-172]. A design used by Peri and Hannan[169] is illustrated in Fig. 22. The infrared cell is constructed of Vycor and quartz. Calcium fluoride windows are sealed to thin silver flanges and to the Vycor cell with silver chloride. The aerogel plate can be reproducibly moved between the furnace section and the infrared beam by an enclosed magnet. In this way, spectra of the gas phase alone can be obtained. Disadvantages are the use of wax to seal the ground joint and

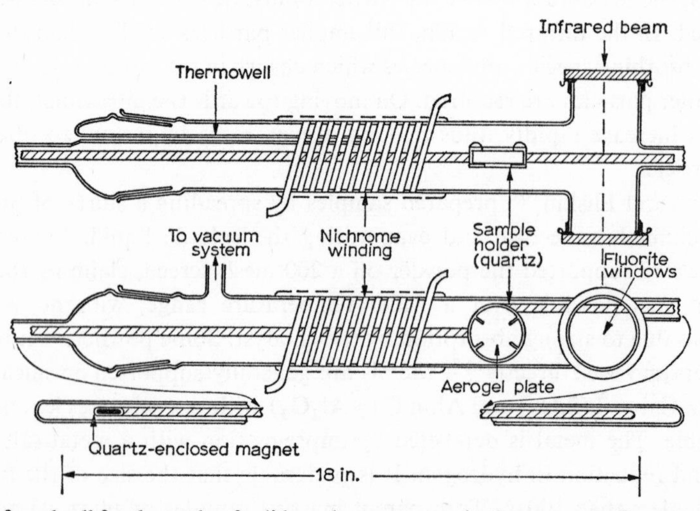

Fig. 22. Infrared cell for the study of solids under vacuum or in the presence of gaseous ambients. From ref. 169.

glyptal to seal pin holes round the windows, although the design permits bake-out of the sample to 600° C. O'Neill and Yates[171] have described a similar design, with magnesium oxide windows directly sealed to soda glass, permitting the complete cell to be heated to 400° C and eliminating the use of ground joints.

A cell for combined infrared–ultraviolet spectra devised by Leftin[173] is shown in Fig. 23. The sample is mounted with platinum wire in a rectangular quartz cage which is a close fit in the quartz adsorption cell used to observe the ultraviolet spectrum. The UV cell is sealed directly to a pyrex tube which connects with the IR cell at the other end. Sodium chloride windows are used for measurement of the infrared spectrum. Provision is made for gas admission and evacuation and for heating of the sample by an external furnace.

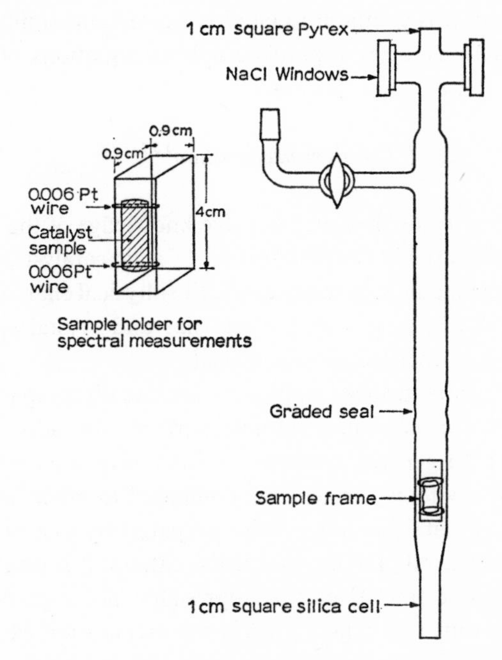

Fig. 23. Cell for catalyst spectral studies in the infrared, visible and ultraviolet regions. From ref. 173.

Potassium bromide windows have been used by Boreskova *et al.*[174] and by Liengme and Hall[175]. The latter cell is a modification of the design by Little *et al.*[176]. Sample discs can be moved between the IR windows and a furnace, with the cell connected by a flexible metal tube to a vacuum and gas handling system. An identical IR cell is also placed in the reference beam. Because the gaseous atmosphere is present in both cells, it is automatically subtracted by the instrument and only the spectrum of adsorbed species is recorded.

2.2 CATALYSIS

The literature of catalysis is so vast that it is impossible to do full justice in one part of a chapter, even to the experimental methods that are in use. Most catalytic reactions are quite complex and a complete reaction mechanism can rarely be determined from a measurement of kinetics alone. It is safe to say that no catalytic gas–solid reaction has yet been fully characterised. Because of the many difficulties, work on catalytic reactions tends to veer towards two opposite poles. One of these can be described as the "clean surface approach", which uses the simplest reactions under high vacuum conditions and attempts to define the catalyst surface completely. The other approach is to study much more complex reactions in order to obtain overall kinetic information which describes the process. Where the mechanism is not understood, it is common to employ expressions similar to those used for homogeneous reactions, or even purely empirical equations.

2.2.1 Catalyst preparation[177,178]

The practice of catalysis involves an extensive knowledge of the procedures and experimental techniques used in the preparation of solid catalysts. Chemical composition is not a sufficient guide. In many cases, the physical characteristics such as surface area, pore volume, pore size, particle size and crystal structure modify or control the catalytic properties of the material.

Catalysts for the "clean surface" approach are usually prepared by methods similar to those used for obtaining suitable surfaces for adsorption measurements (Section 2.1.1). In general, evaporated films have been most widely used because of their relatively large surface area compared to other forms of catalyst. The more conventional catalysts are usually prepared by one of the techniques outlined below. For preparation of any particular catalyst it is usually necessary to refer to the original publication. Even this approach is not a guarantee of success because there is rarely sufficient detail to define the preparation exactly.

(*a*) *Ignited catalysts*

One of the simplest methods of preparation is by decomposition of a thermally unstable compound. The nitrate or chloride is often preferred, sulphates tend to decompose at higher temperatures. Where the presence of residual traces of anion is to be avoided, the metal salts of organic acids are particularly useful. Formates, oxalates, acetates *etc.*, decompose at low temperatures and often reduce the metal at the same time. For the preparation of catalysts from anions, the ammonium salt is frequently used. Metallic salts of complex acids can be used as a source of metal oxide mixtures. Decomposition of the appropriate chromate, tungstate, molybdate or vanadate will produce the mixed oxide.

(b) Impregnated catalysts

Impregnation can be carried out on an inert porous support. This procedure often serves to greatly increase the effective surface area and provide mechanical and chemical stability. If the support is not porous, it is still possible to coat the surface, although the adhesion of the catalyst may be less satisfactory. The general procedure is to immerse the support in a solution of the catalyst component, remove the excess solution after a suitable impregnation time, dry and ignite. Nitrates are again preferred, because of their solubility and ease of decomposition. The other premises mentioned under *(a)* also apply. If necessary, the support can be evacuated, prior to immersion, to remove air which might otherwise prevent a uniform distribution of active material.

(c) Precipitated and gelled catalysts

This method can be applied to the preparation of single catalysts or, by co-precipitation, to multi-component catalysts. Hydrous oxides, sulphides, carbonates and phosphates can be used. After precipitation, the catalyst must be washed free of impurities which might have an adverse effect on the final catalytic properties. The presence of these impurities can be minimised by the use of dilute solutions. An alternative is the mixing of a metal salt with ammonia or an ammonium salt. Any ammonium nitrate remaining in the precipitate is readily removed by washing because of its high water solubility, and the last traces can be eliminated by calcination.

Where two or more components are co-precipitated, special steps must be taken to ensure homogeneity of the final catalyst. This can be achieved by adding a solution of both components to an excess of the precipitating agent, rather than the other way round. The physical properties of precipitated catalysts will often depend on the conditions of precipitation (*e.g.* concentration of solutions, order and rate of mixing, temperature of precipitation, washing, drying and calcination) all of which must be carefully studied.

Gel formation is a special case of precipitation and is particularly suitable for catalysts whose major components are hydrous oxides such as silica or alumina. The production of silica–alumina catalysts by this technique has been studied in great detail because of their commercial applications[179]. Crystalline alumino-silicates or zeolites more commonly known as 'molecular sieves' have recently been the subject of a great deal of research. Preparative details are still mainly in the patent literature[180].

(d) Skeletal catalysts

A special type of catalyst which is typified by "Raney Nickel"[181] is prepared by leaching out one component from a binary alloy leaving a skeletal structure of the desired catalyst. Raney Nickel itself is made by leaching out aluminium from an aluminium–nickel alloy with sodium hydroxide. Cobalt and iron catalysts have also been prepared in this manner.

References pp. 270–278

(e) Carriers and promotors[182]

The main functions of a carrier or support are usually to lend mechanical strength, increase stability to sintering and provide a larger active surface area than would otherwise be available. There is evidence that, in many instances, compound or complex formation takes place between the catalyst and the support, with a consequent effect on the catalytic properties. The most commonly used support materials are silica, alumina, silica–alumina, titania, silicon carbide, diatomaceous earths, magnesia, zinc oxide, iron oxide and activated carbon.

A promotor is generally defined as a substance added to a catalyst, in small concentration, which gives a marked improvement in performance, but which does not possess appreciable activity by itself. Promotors may modify the electronic properties of the catalyst, retard sintering or increase the number of lattice defects. In most cases satisfactory promotors have been found by trial and error, without their function being fully understood.

2.2.2 *Diffusion*

Any catalytic reaction can be thought of as occurring by the following processes:
(1) Diffusion of the reactants to the catalyst surface.
(2) Adsorption of the reactants on the surface.
(3) Reaction on the surface.
(4) Desorption of the products from the surface.
(5) Diffusion of the products away from the surface.

It is generally assumed, for reasons of mathematical simplification, that one of these processes is slower than the rest and that this is the rate determining step. Because the reaction mechanism is rarely known in sufficient detail, steps 2, 3 and 4 are usually considered together. Where the rates of adsorption/reaction/desorption are slower than the rate of diffusion, the true 'chemical' kinetics will be observed. If, however, the reverse is true, 'diffusion' kinetics will be observed and the results will not be characteristic of the chemical reactivity of the system.

The influence of diffusional limitations in gas phase reactions has been extensively treated by Wheeler[183,184] and from a chemical engineering viewpoint by Hougen and Watson[185,186]. More recently a monograph by Satterfield and Sherwood[187] has appeared. The problem of diffusion can be separated into two parts, the first is diffusion or mass transfer to the external surface of the catalyst and second, for those catalysts which are porous, diffusion within the catalyst pores. When diffusion is the rate limiting process, reaction rate, selectivity and activation energy are affected.

(a) Mass transfer

For very fast reactions, only the external surface of the catalyst particle is accessible to the reactants and the overall rate of reaction becomes limited by the rate of mass transfer to the surface. Wheeler[184] has derived an equation which gives the approximate first order rate coefficient, k_∞, for mass transfer to the catalyst surface

$$k_\infty = 10\sqrt{\frac{V_L}{\bar{M}a^3P_t}}\ \text{sec}^{-1}$$

where V_L is the linear velocity (cm . sec^{-1}) computed on the basis of an empty reactor, $\bar{M}$ is the average molecular weight of the diffusing species, a is the catalyst pellet size (cm) and P_t is the total pressure (atmospheres) in the reactor. The assumption made in deriving the equation is that all gases have the same viscosity and Schmidt number. The latter is a dimensionless function of linear velocity, viscosity, density and tube diameter. One of the inaccuracies of the above equation is that for a static system ($V = 0$), it predicts zero diffusion rate (and therefore zero reaction rate). Levenspiel[188] gives a criterion for the absence of mass transfer limitations as

$$kV_p \leqq K_g S_{ex}$$

where k is the rate coefficient per unit volume of pellet, V_p is the volume of one pellet and K_g is the mass transfer coefficient (moles . h^{-1} . ft^{-2} . atmosphere^{-1}). The mass transfer coefficient can be found from the dimensionless equation

$$\frac{K_g d_p Y}{D} = 2.0+0.6\left(\frac{\mu}{D}\right)^{\frac{1}{3}}\left(\frac{d_p U_p}{\mu}\right)^{\frac{1}{2}}$$

where d_p is the particle diameter, Y the mole fraction of diffusing species, D the diffusion coefficient, μ the viscosity and U the linear velocity of the gas. In this equation, the term 2.0 represents the diffusion process in a static system. Tholnese and Kramers[189] have given a similar equation, but the authors do not include a term to allow for the static system.

The above equations are quite complicated to apply but the calculation can sometimes be avoided. Satterfield and Sherwood[187] have made the following generalisation. Where a catalyst bed of small pellets requires to be at least 1 ft long in order to reach 90 % of the equilibrium product concentration, the rate of mass transfer to the catalyst surface cannot be important. An experimental procedure to determine the influence of mass transfer effects in a flow system is illustrated by means of Fig. 24. Experimental runs are made at constant space velocity (ratio of feed rate to catalyst volume) but with differing feed rates and catalyst volumes. Where the conversion of reactant is independent of feed rate, mass transfer effects are small, but where the conversion starts to fall, mass transfer

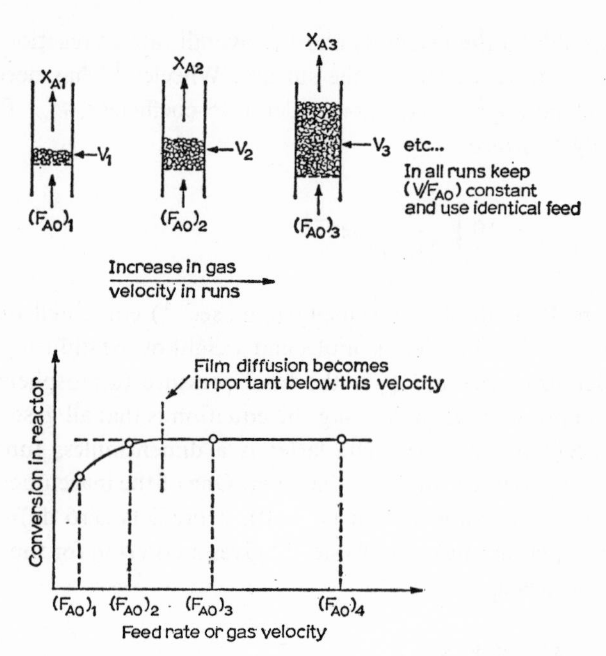

Fig. 24. Method to determine the influence of mass transfer on the rate of reaction. F_{AO}, feed rate; V, catalyst volume; V/F_{AO}, space velocity; X_A, extent of conversion. From ref. 188.

effects are becoming important. A second method, illustrated in Fig. 25, is to compare two series of experimental runs using different volumes of catalyst and a series of feed rates. At high space velocities, where mass transfer effects are small, the two sets of data will coincide. When film diffusion becomes important, the two curves will diverge. The first method is generally more accurate because it can cover a wider range of conditions and allows easier detection of mass transfer limitations. These can be removed or reduced by any means which increases gas mixing. In practice this may be done by using high linear gas velocities which produce a high degree of turbulence. This point is discussed further in the section on reactor design (Sec. 2.2.3). The temperature coefficient of the reaction also indicates the presence of mass transfer effects. Diffusion coefficients have a small temperature coefficient, so that apparent activation energies of 1–2 kcal . mole^{-1} will be observed when mass transfer is limiting the reaction rate.

(b) Pore diffusion

When mass transfer to the catalyst surface is fast compared to chemical reaction, the reaction rate may still be limited by the rate of diffusion in the catalyst pores. Bulk diffusion (*i.e.* the same process as for mass transfer to the external surface) occurs when the mean free path between molecular collisions is small compared to the pore radius. Knudsen diffusion occurs when the mean free path is large compared to the pore radius. Bulk diffusion will predominate with all catalysts at very high pressures, *e.g.* 100 atmospheres, or at low pressures with catalysts having pores larger than 5000 A radius. The bulk diffusion coefficient (D_B) varies with pressure (P) and absolute temperature (T) according to the equation

$$D_B = \frac{D_0(T/273)^{1.75}}{P} \text{ cm}^2 \text{ . sec}^{-1}$$

where D_0 is the bulk diffusion coefficient at NTP and can be obtained by experiment or from International Critical Tables.

Knudsen diffusion will occur in small pores, *i.e.* when the pore radius is less than 200 A. The coefficient for this type of diffusion (D_k) is given by

$$D_k = 9.7 \times 10^3 r\sqrt{T/M} \text{ cm}^2 \text{ . sec}^{-1}$$

where r is the mean pore radius in cm and M is the molecular weight of the gas.

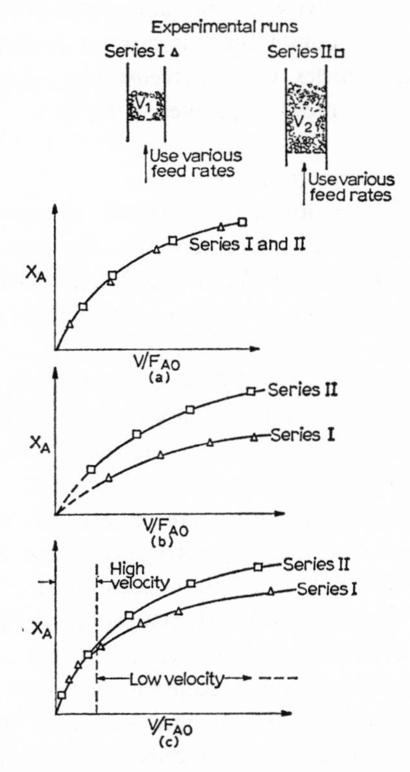

Fig. 25. Second method to determine the influence of mass transfer on the rate of reaction. Curve (a), mass transfer effects unimportant; curve (b), mass transfer effects important at all gas velocities covered; curve (c), mass transfer effects important only at low gas velocities. From ref. 188.

Wheeler[183,184] has suggested an empirical equation to determine the effective diffusion coefficient per catalyst pore in intermediate cases

$$D = D_B[1-\exp(-D_k/D_B)]$$

The diffusion coefficient per catalyst pore is related to D_E, the effective diffusion coefficient per catalyst particle, by the equation

$$D_E = \tfrac{1}{2}\theta D = \tfrac{1}{2}\rho_p V_g D$$

where θ is the porosity, ρ_p is the particle density ($g \cdot cm^{-3}$) and V_g is the pore volume ($cm^3 \cdot g^{-1}$).

Many catalysts are pellets or extrusions formed from porous particles. Such catalysts usually give a bimodal or wide pore size distribution, and measurements of effective diffusion coefficients based on mean pore radius can be very misleading. Smith *et al.* [190–192] have developed and tested experimentally some models which represent the effective diffusivity as the sum of diffusion in micropores and in macropores. The former occurs within the primary particles and the latter in the interstices between the particles. Good agreement between theory and experiment has been obtained for a number of pelleted catalysts.

(c) *Measurement of effective diffusivity*

Calculations of effective diffusivity have the disadvantage that simplifying assumptions about pore structure are necessary. Experimental determinations do not suffer from this disadvantage. Weisz and Prater[193] have described an experimental method which has been used with some elaboration by Robertson and Smith[191]. The apparatus used is illustrated in Fig. 26a. A controlled flow of pure dry helium is fed through calibrated flowmeters at a rate of 1–2 l. min^{-1} to one face of a catalyst pellet. Nitrogen is similarly fed to the opposite face of the pellet. The two faces are isolated from each other and the flow rates adjusted so that the pressure drop across the pellet is zero. After passing through the cell, or diffusing through the pellet, the respective gases are passed through the sample side of a thermal conductivity cell. Pure helium or nitrogen is passed through the reference side of the cell. With valves 1, 2, 5, 6 closed, the concentration of helium in nitrogen (and *vice versa*) is obtained by noting the steady state voltage reading for the appropriate thermal conductivity cell. Closing valves 3, 4, 7, 8 and opening 1, 2, 5, 6 allows nitrogen to be bled into the helium stream and *vice versa.* The rates are adjusted until the same voltage readings are obtained as during the diffusion experiment. This procedure avoids calibration of the cells and gives an independent measure of the diffusion rates of each gas.

Fig. 26b shows details of the diffusion cell. The pellets are mounted inside 'tygon' tubing and clamped in place. Leakage is normally very low, but can usually be detected by fluctuating rates of gas transfer across the pellet. Some improvement can be obtained by heating the specimen before placing it in the tubing. The gas flow is directed tangentially near the pellet face to eliminate kinetic energy effects

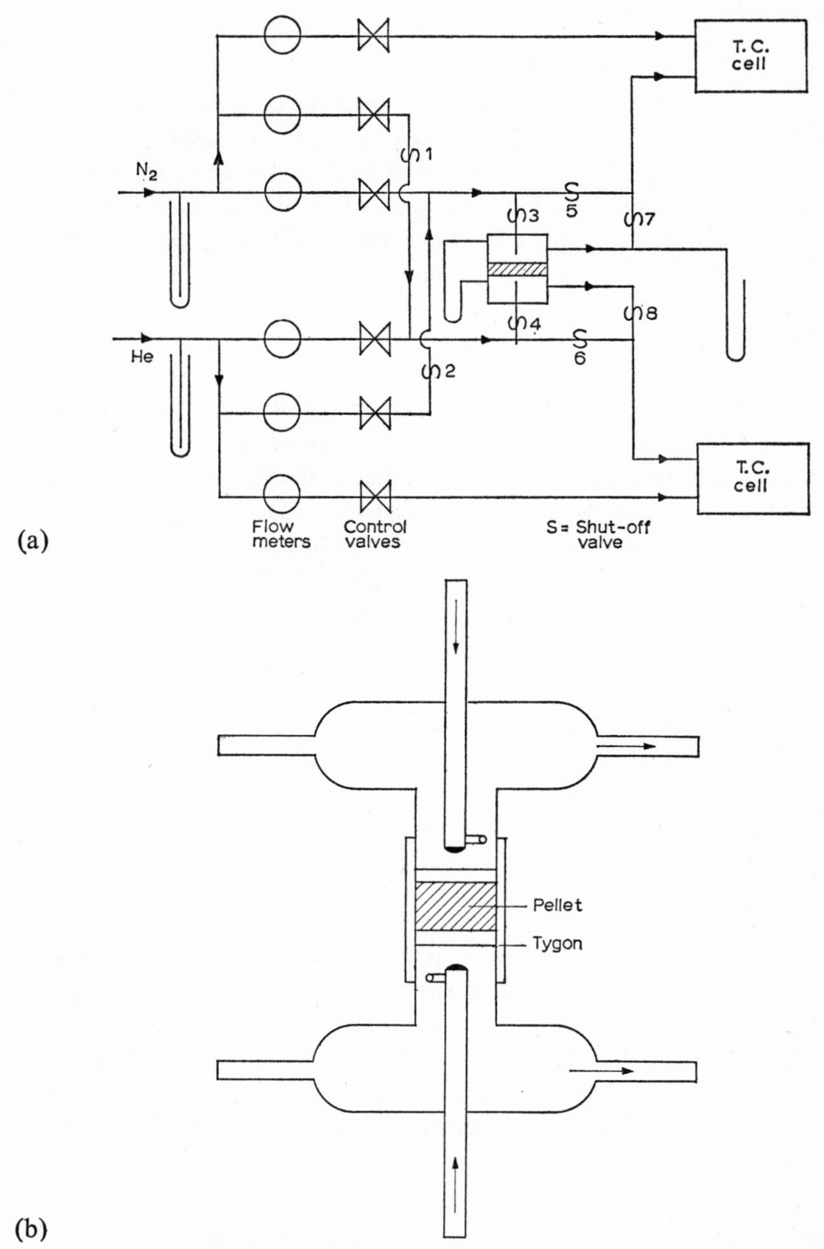

Fig. 26. (a) Schematic diagram of diffusion apparatus. (b) Details of the diffusion cell. From ref. 191.

and the build up of a stagnant layer. The effective diffusion coefficient per catalyst particle D_E is given by

$$D_E = \frac{l}{a} F C_d$$

where l is the length of the pellet, a is the cross sectional area over which diffusion takes place, F is the gas flow rate and C_d is the concentration of diffusing gas in the reference gas. Thus the effective diffusivities of both helium and nitrogen can be measured. Variation of the total pressure will show whether bulk diffusion or Knudsen diffusion is occurring, since the former is inversely proportional to pressure and the latter independent of pressure.

(*d*) *Effectiveness factors*

Mass transfer inside a porous catalyst particle was first analysed mathematically by Thiele[194] and the analysis further developed by Wheeler[183,184] and by Weisz and Prater[193]. More recent work is reviewed by Satterfield and Sherwood[187]. These analyses provide a quantitative description of the factors which determine the effectiveness of a porous pellet. The effectiveness factor, denoted by η†, is defined as the ratio of the measured reaction rate to the intrinsic chemical reaction rate, *i.e.* the reaction rate obtained on a plane surface of the same area.

For a first order reaction in a spherical particle

$$\eta = \frac{1}{\phi_L}\left[\frac{1}{\tanh 3\phi_L} - \frac{1}{3\phi_L}\right]$$

where ϕ_L is the Thiele diffusion modulus‡

$$\phi_L = \frac{R}{3}\sqrt{\frac{k_V}{D_E}}$$

where R is the radius of the sphere and k_V is the first order reaction rate coefficient per unit volume of catalyst particles. For particles of arbitrary shape, $R/3$ is replaced by the ratio of particle volume to external surface. For reactions which are not first order, k_V is replaced by $k_V C^{n-1}$, where C is the reactant concentration and n is the order of reaction. The effectiveness factor η for different values of ϕ_L is plotted in Fig. 27. The curves for different reaction orders and particle shapes lie very close together. The above equation for the Thiele modulus has important

† Wheeler[183,184] uses the symbol f and calls it "fraction of surface available".

‡ This definition is somewhat different from that originally proposed by Thiele[194]. It is equivalent to h as used by Wheeler[183,184] and to $\phi/3$ as used by Weisz and Prater[193].

implications. The effectiveness factor will approach unity as the pellet size or intrinsic reaction rate coefficient decrease, or as the effective diffusivity increases.

It is not always possible to determine the modulus ϕ_L directly, because the intrinsic rate constant k_V is not known. Weisz and Prater[193] have suggested a method for the determination of η which does not require a knowledge of k_V. The observed reaction rate $-\mathrm{d}n/\mathrm{d}t$ can be expressed as follows

$$-\mathrm{d}n/\mathrm{d}t = k_V V_c C^M \eta \text{ mole . sec}^{-1}$$

where V_c is the actual volume of catalyst [also $V_c = V_R(1-\varepsilon)$ where V_R is the reactor volume and ε is the void fraction]. Eliminating k_V by using the equation for the diffusion modulus and rearranging gives

$$\frac{R^2}{D_E}\left(-\frac{1}{V_c}\frac{\mathrm{d}n}{\mathrm{d}t}\right)\frac{1}{C} = 9\phi_L^2\eta$$

The terms on the left hand side can all be measured or calculated, and the effectiveness factor obtained from a prepared graph of η against $\phi_L^2\eta$. For $\eta < 0.5$, $\eta = 1/\phi_L$ and the effectiveness factor can be determined directly. Other methods for the determination of η are given in references 187 and 193.

So far only the case of an isothermal reaction without volume change has been considered. A volume change during reaction (isothermal case) has no effect on η if only Knudsen diffusion is operative but has a small effect if bulk diffusion is

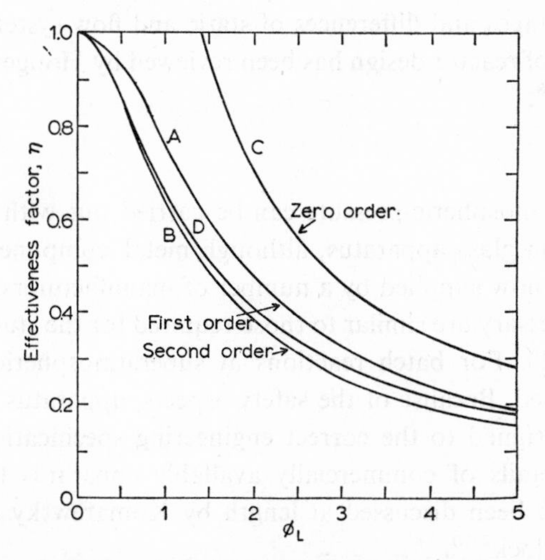

Fig. 27. Dependence of effectiveness factor on the parameter ϕ_L. Curves A, B, C are for diffusion into a flat plate of thickness L and sealed on one side and on the ends. Curve D is for diffusion into a porous sphere, first order reaction. From ref. 184.

operative. For the non isothermal case, effectiveness factors greater than unity are possible. Weisz and Hicks[195] have considered this problem in some detail and constructed a number of graphs for various heats of reaction and activation energies. When a reaction is limited by pore diffusion, the reaction rate is proportional to $\sqrt{k_v}$. If the temperature effects can be expressed as a simple Arrhenius relationship $k_v = A \exp(-E/RT)$, then the measured activation energy E will be about half the true activation energy. Very low values of the activation energy, *i.e.* 1–2 kcal . mole^{-1} are only observed when mass transfer to the external catalyst surface is limiting the rate.

Determination of the effects of pore diffusion, where the catalyst is being poisoned, have been discussed by Wheeler[183,184]. Complex kinetics involving parallel or consecutive reactions have also been studied by Wheeler[183,184] and more recently by Jottrand[196] and by Roberts and Satterfield[197]. The effects of volume change and heat transfer have been considered by Weekman[198] and by Wurzbacher[199].

2.2.3 *Reactor design*

The principles of correct reactor design for heterogeneous reactions are the same as those for homogeneous reactions, but with additional restrictions caused by the presence of a separate phase which constitutes the catalyst. The kinetic information obtained should accurately represent the steady state chemical activity and selectivity of the catalyst. This is true whether the information is required to determine the mechanism of the process, to evaluate different catalysts, or to provide design information for a full scale chemical process. As in the case of homogeneous reactions, the advantages and differences of static and flow systems must be considered. The subject of reactor design has been reviewed by Hougen and Watson[185] and by Levenspiel[188].

(*a*) *Static systems*

Reactions at subatmospheric pressure can be carried out with the aid of high vacuum techniques in glass apparatus, although metal components to fabricate vacuum systems are now supplied by a number of manufacturers. The techniques and precautions necessary are similar to those required for the study of adsorption kinetics (Section 2.1). For batch reactions at superatmospheric pressure autoclaves are usually used. Because of the safety aspects, apparatus and particularly reactors must be designed to the correct engineering specifications. The design, construction and details of commercially available apparatus for the study of batch reactions have been discussed at length by Komarewsky *et al.*[178] and by Melville and Gowenlock[200].

It is important to ensure that gaseous diffusion is not influencing the kinetic measurements. This is generally true if the rate coefficient per unit *geometrical* area is much less than unity[193]. True first order reactions which are pore diffusion limited, still show first order kinetics if Knudsen diffusion is operative, but show an apparent order between 1 and $\frac{1}{2}$ if bulk diffusion is rate controlling. If the reaction is not first order and diffusion effects are present, changes in gas pressure which may result during reaction will cause a change in effectiveness factor and hence the apparent order.

Further possible disadvantages with static systems are that impurities may poison the catalyst, changing the activity with time, or the reaction products may act as inhibitors.

(b) Flow systems

The design and construction of flow systems is more complex than similar batch systems[178,200] and requires larger volumes of reactant. It is often desirable to have automatic control of pressure, temperature and flow rates to increase the accuracy of measurement as well as the safety of operation.

Kinetic measurements in flow systems are usually made with the catalyst packed in a tubular reactor. Three methods of operation are possible: (*i*) as an integral reactor; (*ii*) as a differential reactor, and (*iii*) as a recirculation or recycle reactor.

The integral reactor operates at a large degree of reactant conversion, contact time being varied by a change in flow rate or catalyst volume. Graphical differentiation of the results gives the rate of reaction which allows the construction of a kinetic equation. However, the procedure is not very accurate, especially for complex reactions. Since the heat of reaction is released or adsorbed, it is difficult to maintain isothermal conditions in the converter. In a differential converter, the contact time is small so that the gas composition is approximately constant throughout the catalyst bed. The rate coefficient can be measured directly, but the main difficulty lies in obtaining sufficient analytical accuracy to measure the difference between inlet and exit gas compositions. An advantage of the differential reactor is that mass transfer to the catalyst surface can be neglected, because of the lack of concentration gradients.

The analytical problems associated with differential reactors can be overcome by the use of the recirculation reactor. A simplified form, called a "Schwab" reactor, is described by Weisz and Prater[193]. Boreskov and other Russian workers have described a number of other modifications[201–205]. The recirculation reactor is equivalent kinetically to the well-stirred continuous reactor[206] or backmix reactor[188], which is widely used for homogeneous liquid phase reactions. Fig. 28 illustrates the principle of this system. The reactor consists of a loop containing a volume of catalyst V and a circulating pump which can recycle gas at a much higher rate, G, than the constant feed and withdrawal rates F.

The conversion rate across the catalyst bed is

$$R = (G+F)(X_e - X_i)/V$$

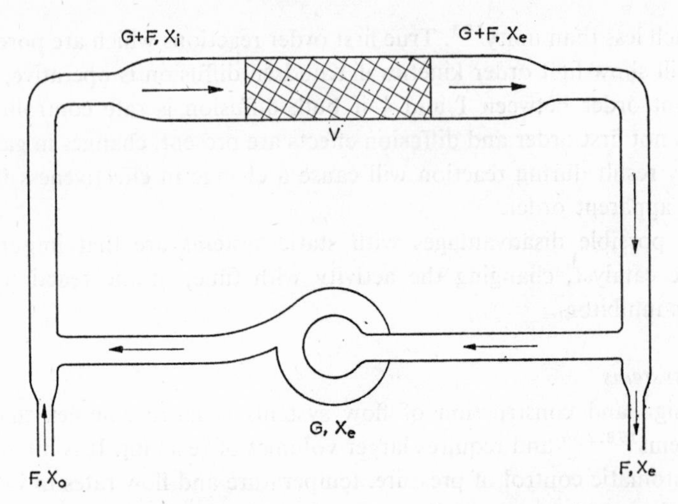

Fig. 28. Principle of the recycle reactor for kinetic measurements in flow systems.

where X_e and X_i are the mole fractions of product. The average overall rate based on the feed and product is

$$R_0 = F(X_e - X_0)/V$$

A mass balance across the reactor shows that

$$GX_e + FX_0 = (G+F)X_i$$

Differential conversion is achieved when $X_i \to X_e$, which occurs when $G \gg F$. Thus a net conversion level, $X_e - X_0$, is obtained whilst the conversion per cycle is differential. This type of reactor is not subject to the analytical errors of the differential reactor except at low overall conversions. A second advantage is that heat transfer is improved, due to the large recycle gas volume, and isothermal operation is more easily achieved. Much of the work associated with developing this type of reactor has been concerned with the kinetics of simple reactions. However, recent work has shown the effects of more complex kinetics involving consecutive and simultaneous reactions[207]. One of the major difficulties with the recirculation reactor is the design of a suitable recycle pump, especially for high temperatures and pressures. Various designs are discussed by Melville and Gowenlock[200]. The thermal syphon principle[204] and reciprocating flow with a single piston[203] have been used by Russian workers. Land[208] has described a diaphragm type pump which operates *via* the vibration of a loudspeaker coil. Boudart *et al.*[209] have constructed a glass pump with a teflon coated piston, actuated by alternately energising a pair of solenoids. The latter two pumps work at subatmospheric pressures and have pumping rates of about one liter per minute. If all the exit gas is liquifiable, an alternative solution is to cool and recycle the liquid. However, a relatively large preheater is required to vaporise the reactants again.

The gas flow through tubular reactors is of particular importance because the composition at any point is influenced by the linear velocity of the gas, the size of the reactor and the size of the catalyst particles. When gas flows through a pipe at low linear velocity (low Reynolds number), the radial velocity is not uniform. As the linear velocity increases, turbulence increases and the velocity profile approaches what is called "plug" flow. However, in a packed bed, plug flow can never be completely attained because of the high voidage near the reactor wall.

Certain criteria have been calculated which allow mixing effects to be minimised. These can be summarised as follows:

(1) Inert packing equivalent to a length of six particle diameters shall be present in front of the catalyst bed[210]. This allows transition to steady flow conditions characteristic of the packed bed.

(2) To eliminate wall effects, the ratio of converter diameter to pellet diameter should be greater than 30. The velocity profile then deviates from uniformity by about 20%. For a differential reactor, ratios greater than 5 are probably acceptable.

(3) An aspect ratio (bed length to pellet diameter) greater than 20 and preferably greater than 100 should be used to minimise axial mixing[211].

The criteria listed above are more difficult to attain in a laboratory reactor than on the full scale. In particular, the opposing requirements of uniform flow and good heat transfer make (2) difficult to achieve.

(c) *Special types of reactor*

One type of reactor which can be useful for kinetic measurements is the continuous stirred tank reactor (CSTR). The kinetic model is identical with that for the recirculation reactor[206,212], and the designs are based on the reactors used for homogeneous reactions. Carberry *et al.*[206,214] have described a CSTR which contains catalyst inside the blades of the impellor. Ford and Perlmutter[213] coated the catalyst on the walls of the tank. The disadvantages of the CSTR compared to the recirculation reactor are that the free volume in the reactor is greater than that of a packed bed, only a limited range of particle sizes can be accommodated and the extent of gas–solid contacting is less well defined. However, a circulation pump is not required.

A fluidised bed reactor is often used on the large scale in exothermic reactions, because of its very good heat transfer properties. The catalyst bed is said to be fluidised when the particle weight is just equal to the viscous drag of the flowing gas and the particle remains suspended in the gas stream. The bed then resembles a dense fluid. Because of their industrial importance, chemical engineers have done a good deal of work aimed at characterising their properties[188,212]. However, the behaviour of fluidised beds is not yet well understood and recent studies have

References pp. 270–278

tended to confirm the view that the evaluation of kinetic data and scale up of fluidised beds is not easy. The two main disadvantages are firstly, that gas bubbles appear, giving a variable bed density, and secondly, that the low gas velocity and turbulent motion of the catalyst particles permit back mixing of the products. The catalyst particles are generally of a small size and catalyst attrition due to abrasion and impact with the walls is quite high.

A third type of reactor system has been developed by Emmett *et al.*[215,216] known as the microreactor. This apparatus has the advantage of speed and simplicity and is particularly useful for evaluating the product distribution from complex reactions. Fig. 29 illustrates the principle of this type of reactor. It consists of a small catalytic reactor placed directly above a gas chromatograph. An injection of reactant is made through the top of the reactor into a stream of carrier gas, usually hydrogen, helium or nitrogen. The slug of reactant passes through the reactor, the products and reactants are separated by the chromatographic column and subsequently recorded by the detector. Systems for working at superatmospheric pressure have been devised[217], but most of the work has been of a qualitative nature. The problem of obtaining good kinetic data from a microreactor system is receiving attention and Schwab and Watson[218] have shown that kinetic data can be as reliable as that from a continuous flow system. Hall *et al.*[219] have recently measured the kinetics of butenes isomerisation in a microcatalytic reactor[220].

The advantages of the pulse technique are that only small quantities of reactant are required and the results are obtained rapidly. A major difficulty in obtaining kinetic measurements is the absence of a steady state, causing the pulse to change shape as the products are formed and as it passes through the reactor. Some

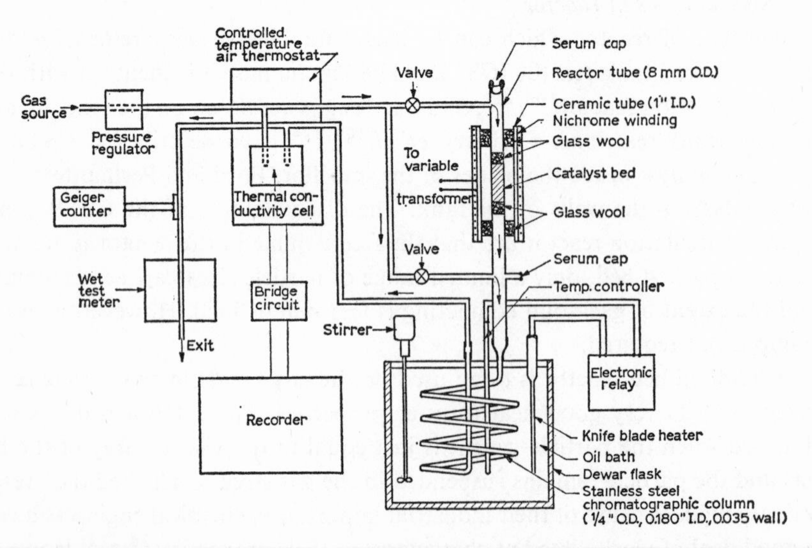

Fig. 29. Micro-catalytic reactor used in conjunction with a gas chromatograph. From P. H. Emmett, *New Approaches to the Study of Catalysis*, Pennsylvania University Press, 1962, p. 149.

caution must be observed in using the pulse technique for the determination of activation energies. The adsorption characteristics change as the temperature is increased, giving a smaller contact time and an activation energy which is lower than the true figure.

2.2.4 *The use of isotopes*

Radioactive and non-radioactive isotopes have proved to be of great value in providing insight into the mechanism of catalytic reactions. They have been used less frequently to provide information about reaction kinetics. Deuterium, tritium, carbon-14, nitrogen-15 and oxygen-18 are the most commonly used isotopes.

(a) Hydrogen isotopes

The use of deuterium has been reviewed by Taylor[221] and more recently by Emmett *et al.*[222]. It can be obtained commercially in a sealed flask or prepared from heavy water by reaction with sodium or magnesium or electrolysis of deuterium solutions[223]. The impurities are usually H_2 and HD with a trace of nitrogen. Deuterium has been used to study (*i*) the rate of exchange with gaseous hydrogen[24,224,225], (*ii*) the rate of exchange with hydrogen bound to the catalyst surface[226] and (*iii*) the rate of exchange with the hydrogen in organic molecules[227,228].

(*i*) $H_2 + D_2 \rightleftharpoons 2HD$

(*ii*) $D_2 + H\text{–}(S) \rightleftharpoons HD + D\text{–}(S)$

(*iii*) $C_xH_y + nD_2 \rightleftharpoons C_xH_{y-n}D_n + nHD$

Deuterium and its exchange products are normally detected by mass spectrometry (the use of mass spectrometers is discussed in the next section).

Tritium can be used for the same type of reaction as deuterium, but its radioactivity in most cases makes it less convenient to handle. It also appears to have a larger isotope effect compared to deuterium. Tritium is a weak β-emitter with a half life of 12.5 years and is more readily analysed with a radioactivity counter. Normal glass apparatus with 1 mm walls will completely adsorb the radiation, but it is necessary to take precautions against accidental breakage. As the tritium will exchange with moisture and stopcock grease, special care must be taken in cleaning and glassblowing used apparatus.

The other modifications of hydrogen which have been employed in kinetic measurements are *para*-hydrogen and *ortho*-hydrogen. These modifications are not isotopes in the normal sense, differing only in the direction of their nuclear spins. At room temperature, the equilibrium mixture of *ortho* and *para* hydrogen (normal hydrogen) contains 25% *para*-hydrogen. As the temperature is lowered, the proportion of *para*-hydrogen increases, until it reaches 100% at liquid hydrogen temperature. The preparation of hydrogen rich in *para*-hydrogen is accomplished by contact with a paramagnetic catalyst such as charcoal at liquid hydrogen or liquid nitrogen temperature[229]. Once prepared, the enriched "*para*-hydrogen" can be stored for several months at room temperature in the presence of glass, mercury or tap grease. An apparatus for the continuous production of *para*-hydrogen has been described by Farkas and Sachsse[230]. The early work on preparation properties and reactions of hydrogen modifications has been summarised by Farkas[231]. Many kinetic studies have been made by Eley and co-workers[37,38,232,233]. A flow system used to measure the *para*-hydrogen conversion over nickel oxides is shown in Fig. 30[234]. Hydrogen is purified, dried and split into reference and sample streams. The latter stream is then passed over an excess of nickel oxide/alumina catalyst at liquid nitrogen temperature which enriches the *para*-hydrogen content to 50%. After being partially converted to normal hydrogen by the catalyst, the stream then passes through the thermal conductivity cell for comparison with the reference stream.

Analysis of hydrogen mixtures with a thermal conductivity cell is well established. The most accurate measurements are obtained by use of a thermal conductivity gauge with the walls immersed in liquid nitrogen and the wire heated to 160° K. This is the temperature when the difference in the rotational specific heats of *ortho*- and *para*-hydrogen is a maximum[235]. Various modifications of thermal conductivity gauges have been made to improve their convenience in use[236]. A room temperature flow analyser based on a thermal conductivity cell has been developed by Weitzel and White[237] which is claimed to be as sensitive as low temperature units. Bridge current and temperature must be controlled very carefully, but the unit is relatively insensitive to changes in pressure and flow rate.

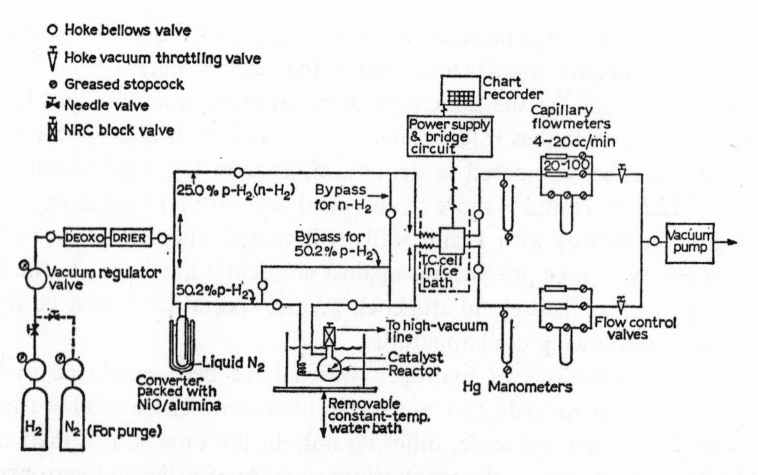

Fig. 30. Diagram of flow system for the measurement of the *para*-hydrogen conversion over catalysts. From ref. 234.

Recently gas chromatographic techniques have been developed which allow the separation of the various hydrogen modifications and their subsequent detection by a thermal conductivity cell. Hydrogen, deuterium and hydrogen deuteride have been separated on alumina columns at 77° K using helium[238,240] or neon[239] carrier gas. Coating with chromia[239] or ferric oxide[238] was used to achieve complete equilibration between the *ortho* and *para* isomers of hydrogen and deuterium. Separation of *ortho*- and *para*-hydrogen has been achieved on alumina[241] and 13X molecular sieve[242] at 77–90° K using a helium carrier gas.

(b) Oxygen isotopes

The use of oxygen-18 for the study of catalytic reactions was pioneered by Winter[243] who obtained kinetic results with a number of oxide systems. More recently, Russian workers have used oxygen exchange to assist in the study of catalytic oxidation and the subject has been reviewed by Boreskov[244].

Two types of oxygen exchange are possible: homolytic exchange, *i.e.* exchange between two adsorbed molecules on the surface, and heterolytic exchange, *i.e.* exchange between lattice oxygen and adsorbed oxygen

$$^{18}O_2 + {}^{16}O_2 \rightleftharpoons 2\,{}^{18}O(ads) + 2\,{}^{16}O(ads) \rightleftharpoons 2\,{}^{18}O{}^{16}O$$
$$^{16}O(lattice) + 2\,{}^{18}O(ads) \rightleftharpoons {}^{18}O(lattice) + {}^{16}O{}^{18}O(ads)$$

The rate of reaction is followed by measuring the changes in intensity of the peaks corresponding to masses 32($^{16}O_2$), 34($^{16}O^{18}O$) and 36($^{18}O_2$) in a mass spectrometer. Winter[243] used a static system with a capillary leak to the mass spectrometer. If the rate of exchange is high, it is more satisfactory to carry out the exchange in a circulation system to eliminate errors due to diffusion limitations. The circulation system used by Dzisjak *et al.*[245] is shown in Fig. 31. It circulates gas at the rate of half to one litre per hr with a circulation volume of about one litre. 10 ml samples are removed for analysis and the oxygen pressure can be kept constant during the experiment.

In the absence of exchange with the catalyst (heterolytic exchange), the rate of homolytic exchange is given by

$$K = \frac{1}{\tau} \ln \left\{ \frac{C_{34}^{*} - C_{34}^{0}}{C_{34}^{*} - C_{34}} \right\}$$

where τ is the reaction time (sec), C_{34}^{*}, C_{34}^{0}, C_{34} are the fractional concentrations of $^{16}O^{18}O$ at equilibrium, zero time and time τ respectively, and $K = ZS/N_g$. Z is the overall number of molecules which have already exchanged (molecules . $cm^{-2}.sec^{-1}$), S is the surface area (cm^2) and N_g is the overall number of oxygen molecules in the system. Further evidence about the rate limiting step can be obtained from the oxygen pressure dependence. If molecular oxygen is involved,

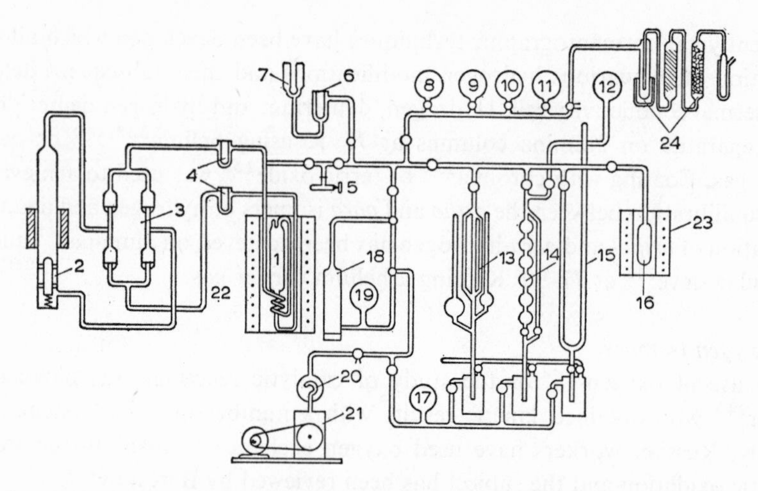

Fig. 31. Apparatus for the study of oxygen exchange with a circulation system. 1, reaction vessel; 2, circulation pump; 3, valve box; 4, cold traps; 5, gas sampling ampoule; 6, cold trap; 7, manometric valves; 8 to 12, gas storage bulbs; 14, gas burette; 16, adsorption vessel; 24, oxygen purification system. From ref. 245.

the order with respect to oxygen will vary from zero (at high coverage) to 1 (at low coverage). Similarly if atomic oxygen is involved, the order will vary from zero to 0.5.

When exchange with lattice oxygen takes place (heterolytic exchange), the kinetic equations become more complicated. The overall fraction of ^{18}O in the gas phase (α) decreases, *i.e.* $d\alpha/d\tau$ is negative. The kinetic data can be represented by the following equation

$$\frac{dy}{d\tau} = -K_y + \phi \left(\frac{d\alpha}{d\tau}\right)^2$$

where $\phi = 8K_3/(K_2+2K_3)^2$. $K = K_1+K_2+K_3$ are the exchange coefficients for homolytic exchange (K_1), heterolytic exchange involving a single oxygen atom from the catalyst (K_2) and heterolytic exchange involving two oxygen atoms from the catalyst (K_3). $y = C_{34}^* - C_{34}$. Heterolytic exchange involving one or two oxygen atoms can be distinguished by the fact that C_{34} increases with time, ($dC_{34}/d\tau > 0$), if one oxygen atom is involved and decreases with time ($dC_{34}/d\tau < 0$) if two oxygen atoms are involved.

(c) *Radioactive isotopes*

Carbon-14, the radioactive isotope of carbon, has been widely used in mechanistic studies with organic hydrocarbons but not in kinetic studies. ^{14}C decays to ^{14}N with the emission of a beta particle and has an extremely long half life (5568 ± 30 years)[246]. The emitted radiation can be detected by an ionisation chamber,

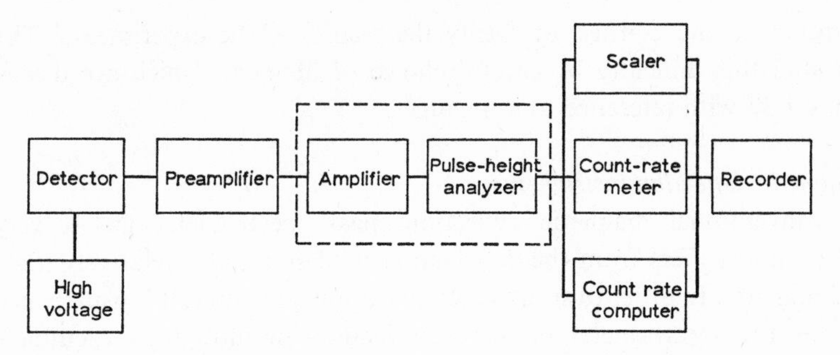

Fig. 32. Block diagram of a pulse counter for radio-isotope measurements. From ref. 246.

a Geiger–Müller counter or a scintillation counter, particularly the liquid scintillation detector[246]. It is necessary to connect the detector to an electronic circuit which is able to count the radioactive pulses. Fig. 32 shows a block diagram of typical components. The output from the detector must normally undergo one or more stages of amplification before being counted and recorded. Isotope effects from ^{14}C are known, *i.e.* the rate of reaction using ^{14}C compounds is not the same as that using ^{12}C compounds. This must be remembered when making rate measurements[247,248].

None of the other radioactive isotopes have been extensively used in catalytic reactions. However, sufficient data exists on the preparation of suitable compounds to make more extensive use of other isotopes possible, where other methods are not satisfactory. The problems and hazards associated with the handling of radioactive materials must be carefully considered and reference to the specialist publications dealing with radioisotopes is necessary[246,249,250].

2.2.5 *Mass spectrometry*

Mass spectrometers have been widely used to study adsorption phenomena and catalytic reactions. Several excellent reference books on mass spectrometry[251-253] are available and advances in the subject are periodically reviewed[254]. The magnetic deflection spectrometer is the most popular type of instrument and many commercial versions are available. Attention is concentrated in this section on types of mass spectrometer which have found particular application in the study of heterogeneous reactions. Much of the relevant work can be found in the *Transactions of the American Vacuum Society*[255].

One major difficulty is associated with the use of all mass spectrometers employing a hot filament. Interaction of gases with the filament by adsorption or chemical reaction tends to change its emission characteristics and thus alter the sensitivity. New chemical species may be produced which appear as part of the mass spectrum and confuse or falsify the results of the experiments. There are ways round this difficulty by careful choice of filament, which are discussed in Section 2.1.3b with reference to ion gauges.

(a) Magnetic deflection instruments

The conventional magnetic deflection mass spectrometer has a very high resolution, but suffers from the disadvantages of size and a relatively high background pressure. In order to study reactions at low pressures, it is usually necessary to operate the spectrometer in the high vacuum or ultra high vacuum region. This requirement demands a small spectrometer which can be baked out to reduce background contaminants. Davis and Vanderslice[256] have developed a suitable instrument for ultra high vacuum work and this design is now sold commercially by the General Electric Corporation, Schenectady.

The instrument is a 90° sector, 5 cm radius analyser and can resolve adjacent masses up to about 140. It can detect partial pressures of 10^{-10} torr and when used with a suitable electron multiplier, the mass scan can be displayed on an oscilloscope. The spectrum can be scanned in a few milliseconds thus enabling rapid reactions to be followed.

(b) The omegatron

Sommer *et al.*[257] were the first to construct this type of mass analyser, which was later simplified by Alpert and Buritz[258]. The principle of operation is similar to that of the cyclotron (Fig. 33). Ions, formed by an electron beam passing through

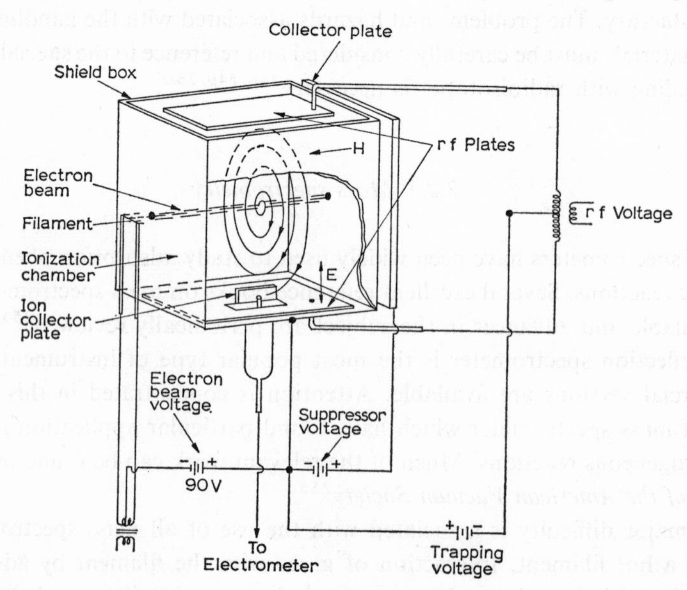

Fig. 33. Diagram of the simplified type of omegatron developed by Alpert and Buritz[258]. From ref. 79.

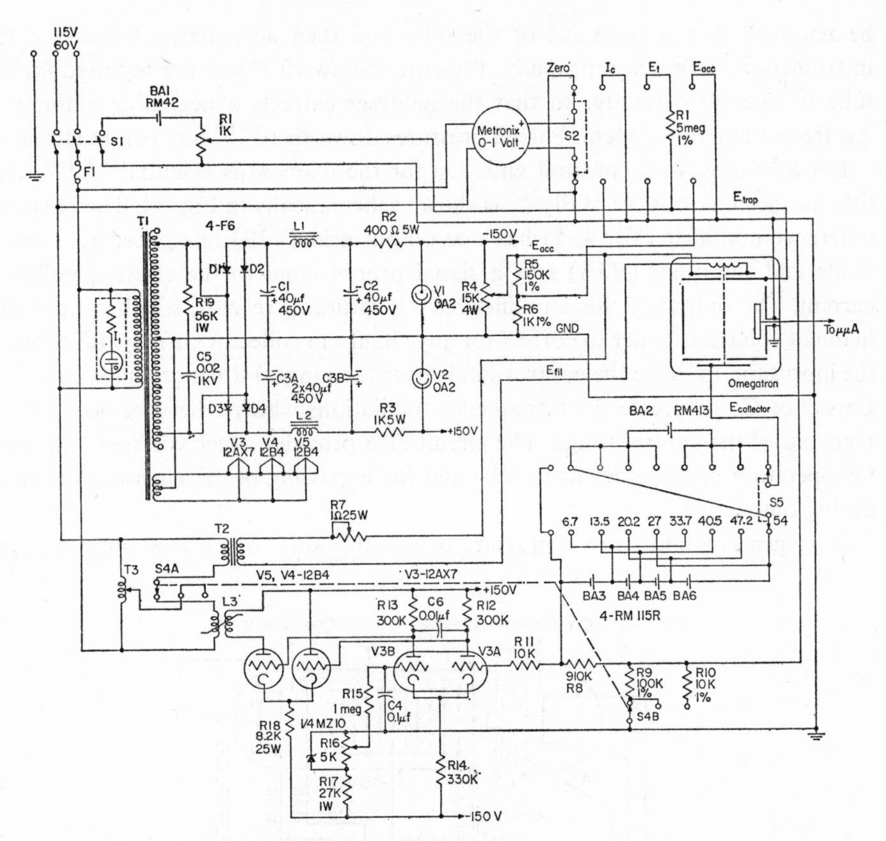

Fig. 34. Circuit diagram of an omegatron power supply with constant emission regulator.

a shielded box, are subject to crossed magnetic and high frequency electric fields. For a given strength of the magnetic and electric fields, ions of a certain e/M ratio will spiral along the magnetic lines of force and be trapped by the ion collector. Ions of mass M will be collected when

$$M = 1.525\ Be/F_c$$

where B is the magnetic field strength in kilogauss, e is the charge on the ion and F_c is the rf frequency in megacycles. The mass range can be scanned by varying the magnetic field, or more usually, by varying the rf field. Many other workers have constructed and used omegatrons, mainly to monitor vacuum conditions[259–262]. The kinetics of hydrogen–deuterium exchange[24] and oxygen adsorption[111] have been studied with this instrument. A disadvantage of the omegatron is that the resolution decreases with increasing mass. Although it is extremely sensitive for the study of hydrogen reactions, adjacent mass numbers higher than 40 are not well resolved. The large magnet required is a second disadvantage, this must be removed during bake-out of the tube and then accurately realigned if the instrument is to function properly. Pressures below 10^{-6} torr are required for the tube to operate correctly, so that the analyser extracts a negligible amount of gas from a reaction system. Partial pressures down to 10^{-10} torr can be detected.

For kinetic work a constant emission for the filament is essential[263,264]. One that has been used successfully[265] is shown schematically in Fig. 34. The balanced difference amplifier (V3, 4, 5) has one output grid (V3B) referenced to a zener diode and the other (V3A) sees a signal proportional to the electron collector current. The output of this amplifier feeds a saturable reactor, which controls the filament voltage in order to correct for any change in collector current. R7 controls the loop gain, to make the control stable over the range 0–10 microamps emission. Capacitor C_6 suppresses high frequency oscillations which otherwise occur at the high end of the control range. The circuit also provides other voltages necessary for operation of the omegatron tube and for bypassing of the emission regulator during outgassing.

If oxygen is admitted to a pressure of 5×10^{-6} torr, which is about the upper

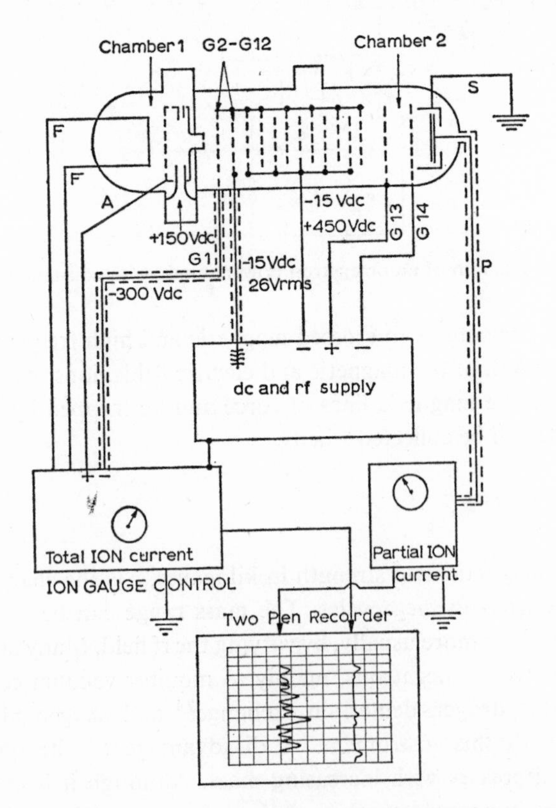

Fig. 35. Diagram of the rf mass spectrometer tube (topatron) and its electrical control system (after Váradi).

limit for omegatron operation, the collector current only decreases by 0.5% with the emission control in operation. Without the emission regulator, the same increase in oxygen pressure causes the collector current to decrease to 10% of its original value, making quantitative measurements impossible.

(*c*) *Topatron*

The theory of this type of spectrometer was discussed by Bennett[266] and by Redhead and Crowell[267]. It is a resonance spectrometer operating on the principle of a linear accelerator. The construction and operation of the device has been described by Váradi *et al.*[268,269]† and by Ehlbeck *et al.*[270,271].

It has the advantage of measuring total pressure as well as partial pressure and requiring no magnet. A schematic diagram of the device is shown in Fig. 35. The tube consists of two compartments. One contains an ion source consisting of a thoria-coated tungsten filament and a box shaped anode (A). Ions are produced in the box and are then attracted by the first negatively charged grid (G1). A proportion of the ions are collected at this grid and the measured ion current is proportional to the total pressure. Ions not trapped at G1 travel through the analysing grid system (G2–G12) which has an rf voltage between each alternate grid. If the ions have the correct velocity V to clear a grid distance S during a half period of the rf cycle, they gain energy from the rf field and are able to reach the collector S. From the relationship

$$\frac{S}{V} = \frac{1}{2f} = \frac{S}{\sqrt{V_0 2e/M}}$$

where f is the radio frequency in Mc. sec^{-1} and V_0 is the ion accelerating voltage, e/M can be found.

After bake-out at temperatures of up to 450° C, the tube will operate over the pressure range 10^{-3}–10^{-8} torr. Mass numbers up to 100 can be resolved, or up to 250 at the expense of sensitivity.

(*d*) *Farvitron*

The farvitron, which is also a resonance spectrometer, was developed by Reich and Bächler[272,273] from work by Tretner[274]. A schematic diagram of this instrument is shown in Fig. 36‡. Electrons emitted from the filament (K) are accelerated by the grid W and ionize the gas inside electrode A. The ions formed are

† A commercial instrument is marketed by Leybold in Europe and by the Raytheon Co. in the U.S.A.

‡ The instrument developed by Reich and Bächler is available commercially from E. Leybold Nachfolger, Germany.

References pp. 270–278

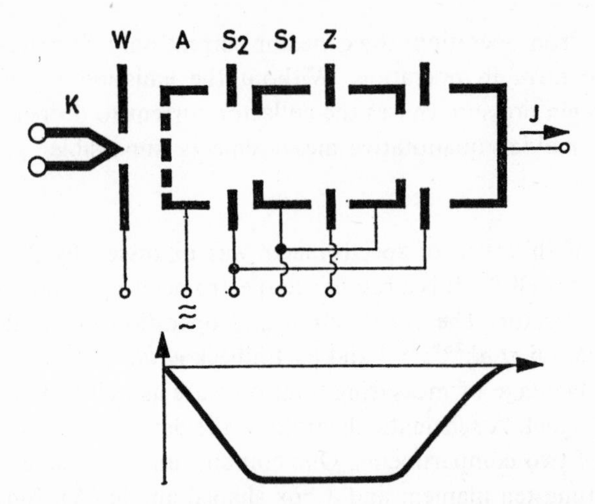

Fig. 36. Schematic construction of the farvitron gauge head and general indication of the potential distribution in the longitudinal axis. The potential trough is obtained by the different negative voltages applied at the accelerating grids S_1, S_2 and Z.

forced by the potential distribution in the instrument to oscillate along the tube axis between A and S. The oscillating frequency (Mc . sec^{-1}) is given by

$$f = C\sqrt{Ue/M}$$

where C is an instrument constant and U is the maximum potential drop through which the ions pass. Those ions whose oscillating frequency is in phase with the applied field will be gradually accelerated towards the signal electrode J. This signal is amplified, demodulated and transmitted to the screen of an oscilloscope. If the applied r.f. voltage is oscillated in phase with the mains frequency the mass range can be scanned 50 times per sec, allowing rapidly changing phenomena to be studied. The instrument operates over the range 10^{-4} torr to 2×10^{-9} torr although the resolution is less than that of the omegatron or the topatron.

(e) Other types

A cycloidal mass spectrometer was devised by Bleakney and Hipple[275] which operates with crossed magnetic and electrostatic fields. Measurements by other workers[276] show that with a magnetic field of 5000 gauss, masses 1–500 can be detected. Adjacent mass numbers are separable up to 100, but the resolution decreases for higher mass numbers. Partial pressure detection down to 10^{-12} torr is claimed.

The quadruple mass filter was described by Paul and Raether[277]. It consists of four parallel cylindrical rods to which is applied a sinusoidal voltage. Under given conditions of voltage and frequency, only ions of certain e/M ratio reach the collector, the rest lose their charge by collision with the rods. A hot filament and a Penning type source have both been used as a means of producing ions. The instrument operates over the pressure range 5×10^{-3} torr to 10^{-10} torr. Resolution is practically constant over the mass range 1–200 and adjacent mass numbers can be resolved.

2.3 NON CATALYTIC REACTIONS

2.3.1 Oxidation of metals

The first stage in the interaction of oxygen with a metal must be chemisorption of oxygen. Since the standard free energy of all metals (except gold) is negative, the chemisorbed layer should be formed very rapidly. A study of the initial stages therefore normally requires ultra high vacuum conditions to eliminate contaminants and to allow a sufficiently low partial pressure of oxygen to be used. The formation of the chemisorbed layer can be followed by the methods described in Section 2.1.

Diffusion of oxygen and/or metal atoms near the surface then occurs and the initial oxide layer is formed. The further growth of this layer involves two steps: (a) reaction at the metal/oxide and oxide/oxygen interfaces, and (b) transport of material through the oxide layer. As the oxide layer gets thicker, the rate of growth is controlled by (b) which becomes the slower of the two steps. If the volume of the oxide produced is less than that of the parent metal, the simple rate law for an unimpeded reaction is observed

$$y = K_1 + K_2 t$$

where y is the film thickness at time t. The oxidation of light metals, sodium, calcium *etc.* follows the above equation. With most metals, the volume of oxide is greater than that of the metal and a protective layer is formed, resulting in a logarithmic or parabolic rate law as indicated below

$$y = K_1 \log (K_2 + K_3 t)$$

or

$$y^2 = K_1 + K_2 t$$

Theories of oxidation have been developed by Wagner[278,279] and by Mott[280,281]. In general the logarithmic rate law applies to very thin oxide layers which form protective coatings and the parabolic rate law to thick oxide layers. More recent reviews of the subject have been given by Grimley[282], Kubaschewski and Hopkins[283] and by Wyn Roberts[284].

Weight changes, measured with a microbalance, are a convenient way of following the kinetics of oxidation. The use of this technique is discussed in Section 2.1.4. Many metals have been studied[285,287–289] and differing behaviour is often observed in differing temperature ranges. Very high temperatures are required for the more noble metals, which demand the construction of special furnaces and apparatus[113–115,286].

Condit and Holt[290] have reviewed the use of radioactive tracers in the study of oxidation. Radioactive platinum has been used to study the oxidation of cobalt[291] and radioactive silver to study the oxidation of molybdenum[292]. If the tracer element remains at the surface it indicates that the oxygen is diffusing into the metal, and if it is located some distance from the surface shows diffusion of the metal through the oxide. The tracer can be detected by autoradiography and its distribution studied by gradual removal of the oxide layer. Its distance from the surface can also be estimated by measurement of the energy of the emitted radiation at some point outside the surface[293,294].

An important characteristic of radioactive decay is that the momentum of the emitted particle must be balanced by the momentum of the product nucleus. In suitable cases it should be possible to detect the recoil nucleus. Thorium-228, which decays to radium-224 having an energy of 97 keV, has been used to study the oxidation of a number of metals[290]. Its advantages are its low volatility (as thorium oxide) and its relatively low rate of diffusion in lighter metals. The maximum range of recoils in solids is of the order of 300–500 A and for thinner oxide layers, its distance from the surface can be measured. One difficulty with quantitative work is that the radium undergoes a sequence of further decay, which complicates calculation of recoil ranges, and calibration may be necessary.

2.3.2 Dissociation reactions

These reactions are mainly of the type

$$\text{A(solid)} \rightarrow \text{B(solid)} + \text{C(gas)}$$

but also a few of the type

$$\text{A(solid)} \rightarrow \text{B(gas)} + \text{C(gas)}$$

The extent of reaction *versus* time can usually be represented by a sigmoidal curve. There is an induction period when the initial nuclei of phase B are being formed, followed by an accelerating rate which eventually decays due to the depletion of phase A.

A survey of the kinetics of dissociation reactions has been made by Garner[295].

Much of the experimental data has been obtained by observing weight losses using microbalance techniques (Section 2.1.4). This technique is almost a necessity for studying hydrate decompositions, because of the difficulties of handling easily condensable water vapour. It is preferable to use single crystals of known surface area, rather than numbers of small crystals. Reaction may also be started by rubbing the crystal surfaces with decomposition product. This technique ensures that the interface is formed parallel to the surfaces of the crystal and reduces the induction period for the reaction. Britton *et al.*[296] have studied the decomposition of calcium and magnesium carbonates under conditions where the carbon dioxide was continuously removed. Bircumshaw *et al.*[297–299] have studied the decomposition of nickel formate, ammonium permanganate and potassium perchlorate.

The other commonly used method in kinetic work is to measure the pressure rise during decomposition. Bircumshaw and Newman[300] studied the decomposition of ammonium perchlorate using a McLeod gauge as the measuring device. They also studied the decomposition in a stream of nitrogen, analysing the products chemically by adsorption of Cl_2 in potassium iodide. For measurements in a stream of inert gas, a katharometer could be used as a detection device. Other methods of pressure measurement[78] are also applicable, especially where a rapid reaction demands that a record be obtained.

2.3.3 Transport reactions[301]

Chemical transport reactions are those in which a solid (or liquid) substance A reacts with a gas to form a vapour phase product. The reverse reaction then occurs in a different part of the system with the reformation of substance A

$$\text{A(solid)} + \text{B(gas)} \rightarrow \text{C(gas)} \xrightarrow{\text{Transport}} \text{A(solid)} + \text{B(gas)}$$

The process appears to be one of sublimation or distillation. Substance A does not possess an appreciable vapour pressure however and is transported chemically due to the existence of a concentration gradient.

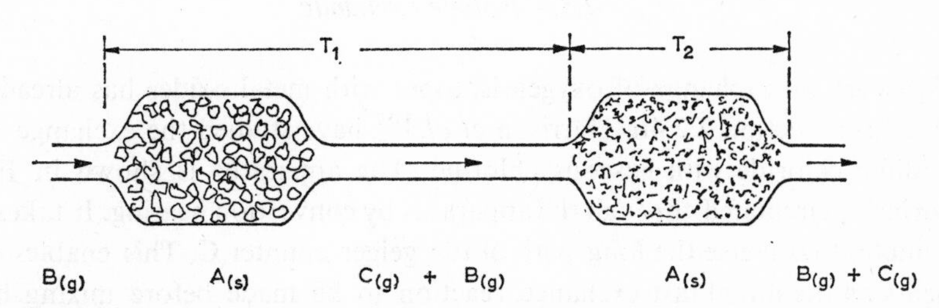

Fig. 37. Apparatus for the measurement of transport reactions using a temperature difference.

If the reverse reaction proceeds readily, a closed tube containing the materials will probably suffice. The concentration gradient can be established by a temperature difference across the tube. An improved apparatus is shown in Fig. 37. The enlargements in the tube are to reduce the linear velocity of the transporting gas. Typical examples are the transport of iridium[302] and ruthenium[303] oxides.

A concentration gradient can also be established by means of a pressure differential. Fig. 38 illustrates this principle. An inert gas is mixed with the substance being transported, lowering its partial pressure and causing it to be deposited.

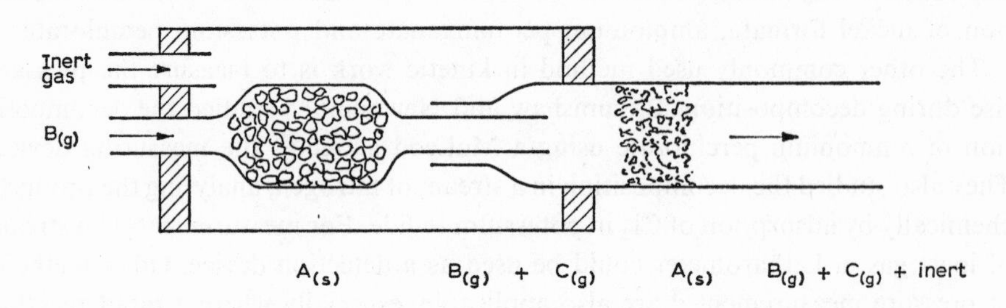

Fig. 38. Apparatus for the measurement of transport reactions using a pressure difference.

Gaseous diffusion, with subsequent decomposition on a hot wire, has been used for the purification of zirconium[304], uranium[305], chromium, niobium and tantalum[306]. The metal is transported as a volatile iodide, which then decomposes on the hot wire or other heated receiver. In addition to diffusion, convection is observed in large diameter tubes (> 2 cm) at several atmospheres pressure. The tube is placed in an inclined position with the hot end downward. Silicon in the form of its dihalides has been transported in this manner[306].

The observed kinetics must be interpreted with some care. For many reactions in closed tubes, gaseous diffusion is the rate determining step at pressures greater than 1 torr, whilst at higher pressures (> 3 at), thermal convection becomes appreciable.

2.3.4 *Isotope exchange*

The work on exchange of oxygen isotopes with metal oxides has already been discussed in Section 2.2.4b. Harrison *et al.*[307] have studied the exchange of ^{36}Cl in sodium chloride with gaseous chlorine. The apparatus is shown in Fig. 39. Chlorine is circulated through the apparatus by convection heating. It takes about 50 minutes to traverse the long path of the geiger counter C. This enables measurements on the initial fast exchange reaction to be made before mixing has occurred. An initial baking out of the system is required to prevent formation of HCl

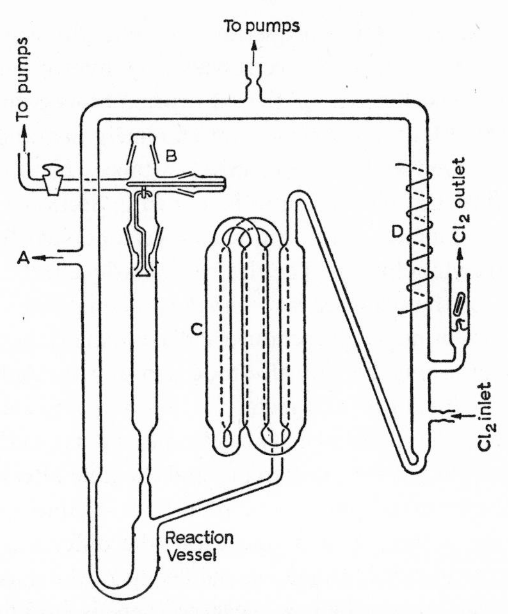

Fig. 39. Circulating system for studying exchange reactions. A, to quartz spiral manometer; B, device for loading sodium chloride into reaction vessel *in vacuo*; C, long path envelope of Geiger counter; D, convection heater. From ref. 307.

by reaction with desorbed water. Work on the HCl–KCl system has been reported by Clusius and Haimerl[308]. Harrison *et al.*[309] have also studied the oxidation of KI and NaCl with fluorine. The sodium chloride was doped with ^{36}Cl, and the rate of release of the isotope into the gas phase was observed. Electrical conductivity and visible–UV spectroscopy measurements were also found to correlate with the rate of reaction. Gregg and Leach[310] used ^{14}C to observe the oxidation of chromium by labelled CO and CO_2 at 600–800° C. The rate of disappearance of ^{14}C from the gas phase was measured and the total ^{14}C content checked by counting the activity of the chromium specimen at the end of the experiment.

3. The solid–liquid interface

3.1 ADSORPTION FROM SOLUTION

Adsorption from solution is discussed by Adamson[1], but with emphasis on the equilibrium aspects, rather than the kinetics. The subject can conveniently be divided into adsorption of non-electrolytes and adsorption of electrolytes. The former can be treated for dilute solutions, in a similar manner to adsorption of gases on solid surfaces. Multilayer adsorption has been observed however, so that

References pp. 270–278

in some instances the adsorption properties resemble physical adsorption rather than chemisorption. Adsorption of electrolytes may involve ion exchange or non-exchange processes. The kinetics of the adsorption process in solutions has not been widely studied. Many adsorptions are extremely fast and only equilibrium phenomena are observed and in many other instances mass transfer is the rate controlling step. Mass transfer in liquids is treated in Section 3.2.

Information on the kinetics of adsorption has been obtained from electrical measurements on electrolyte solutions. The electrocapillarity effect[1, 311] can be used to measure the charge on the adsorbed layer. As the potential at the surface is changed, the desorption of ions or molecules produces a maximum in the *differential capacity/voltage* curve. The dropping mercury electrode is most commonly used, but similar results have been obtained for solid metals[312]. With an impedance bridge as used by Grahame[313], the differential capacity can be measured directly. A slow uniform dropping rate is established for the mercury and the time interval from the falling of a drop to the attainment of a minimum in the audio frequency hum is measured. After estimating the surface area of the drop, the differential capacity can be calculated. As the frequency is increased, the height of the maxima on the differential capacity/voltage curve changes, because there is insufficient time for the adsorption equilibrium to be established at higher frequencies. Theoretical analyses of the changes have been developed[314, 315] and it is concluded that adsorption rates of 10^{-5} mole . cm^{-2} . sec^{-1} can be measured. Because the impedance of the double layer should be comparable with the resistance of the cell, frequencies of 0.1 to 1 megacycle are required.

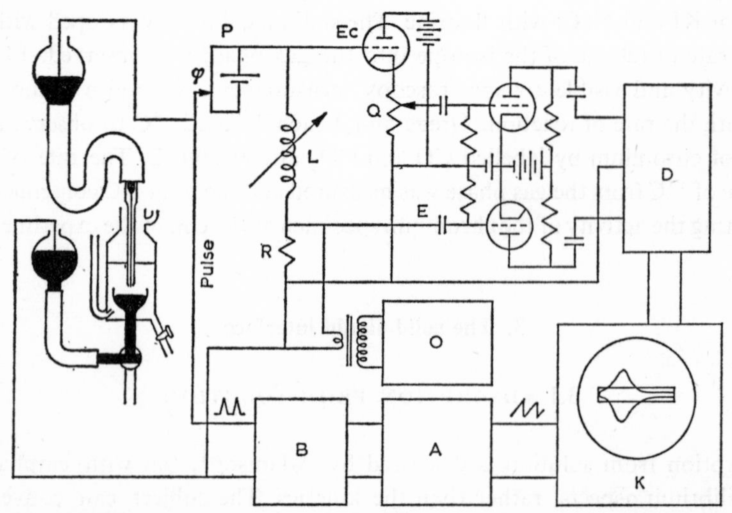

Fig. 40. Block diagram of measuring circuit. P, potentiometer; L, load inductance; R, small resistance; D, dual scoper; O, audio frequency oscillator; A, sawtooth wave generator; B, tuned amplifier; K, oscilloscope equipped with amplifiers; Q, gain control of cathode follower to read Q value of resonance circuit. From ref. 316.

A related method developed by Watanabe *et al.*[316] employs the dropping mercury electrode connected to a resonance circuit instead of an ac bridge. The total capacitance of the dropping mercury electrode increases with drop growth and the resonant ac voltage reaches a maximum at a certain instant in each dropping period. The following resonance condition is then satisfied

$$\frac{1}{WC} = \frac{WL}{2}\left[1+\left(1+\frac{4R_t^2}{WL^2}\right)^{\frac{1}{2}}\right]$$

$$Q_{\max} = \frac{E_{c,\max}}{E} = \frac{WL}{2R_t}\left[1+\left(1+\frac{4R_t^2}{WL^2}\right)^{\frac{1}{2}}\right]$$

where C is the double layer capacitance, R_t the solution resistance, L the load inductance and E the applied ac voltage. A block diagram of the circuit is shown in Fig. 40. E_c and E are both amplified and fed to the vertical axis of an oscilloscope. The gain ratio of the amplifier required to adjust both traces to the same height is the value $Q_{\max}$. Known resistance and capacity can then be substituted for the dropping mercury electrode, to give the same resonance point and Q value. The appropriate values of C and R give the double layer capacitance and solution resistance directly.

3.2 CATALYTIC REACTIONS

Catalytic reactions in the liquid phase have been widely studied, in general using similar methods, *i.e.* autoclaves, tubular reactors, to those employed for the study of gas phase reactions. The stirred tank reactor, in a batch or continuous form, has been widely used. Vigorous agitation is necessary to ensure complete mixing and to maintain a uniform distribution of catalyst throughout the solution. Catalyst is usually used in powder form to assist in its dispersion and to reduce pore diffusion limitations. The effect of mass transfer to the catalyst surface is usually estimated by varying the rate of agitation. At the point where the degree of agitation no longer affects the reaction rate the rate of mass transfer is considered to be unimportant. This does not affect the characteristics of pore diffusion, which are still influenced by the fact that mass transfer in liquids is slower than in gases.

A particular advantage of liquid phase reactions is the large heat transfer coefficients of liquids. This makes it possible to carry out highly exothermic reactions which would be difficult to control in the gas phase. Solid catalysts suspended in liquids (slurries) have been widely used for hydrogenation[317,318]. This can normally be treated as a liquid phase reaction because the liquid is essentially saturated with hydrogen.

Rates of mass transfer to the catalyst surface and pore diffusion can be calculated by the methods of Section 2.2.2 if the diffusion coefficients are known. However, the molecular theory of diffusion in liquids[319–321] is relatively undeveloped and it is not yet possible to treat diffusion in liquids with the same rigour as diffusion in gases. The complicating factors are that the diffusion coefficient varies with concentration and that the mass density is usually more constant than the molar density of the solution. An empirical equation, due to Wilke and Chang[322], which applies in dilute solution, gives

$$D(\text{cm}^2 \,.\, \text{sec}^{-1}) = 7.4 \times 10^{-10} \frac{T\sqrt{XM_2}}{\mu V_b^{0.6}}$$

where T is the temperature (°K), X is an empirical association parameter of the solvent, M_2 is the molecular weight of the solvent and μ is the solution viscosity (poises). V_b is the molar volume of the diffusing solute ($\text{cm}^3.\text{g}^{-1}.\text{mole}^{-1}$) obtained from Kopps Law[187]. For large molecules, the above equation does not hold and the Stokes–Einstein equation must be used. The diffusion coefficient in liquids at ordinary temperatures is about 10^{-4} times that for gaseous systems at atmospheric pressure. A recently published modification of the above equation[323] is claimed to be more widely applicable.

A mass transfer coefficient can be derived from the equation[324]

$$K_c d_p/D = 0.991(N_{Pe})^{\frac{1}{3}}$$

for $N_{Pe} > 500$, where N_{Pe} is the Peclet number and d_p is the particle diameter. At low fluid velocities $K_c d_p/D \rightarrow 2$.

3.3 NON CATALYTIC REACTIONS

The effects of diffusion in solid–liquid systems have been discussed in the previous section. Dissolution of sodium chloride is an example of a diffusion controlled reaction. In a dilute solution, the rate of dissolution R is given by

$$R = KS$$

where S is the total surface area. By Fick's law, the dissolution rate in $\text{mole}\,.\,\text{cm}^{-2}.\,\text{sec}^{-1}$ is $D(C_s - C)/\tau$, where D is the diffusion coefficient, C_s the concentration at the surface, C the concentration in the bulk of the solution, and τ the residence time. Then

$$R = DS(C_s - C)/\tau$$

τ is dependent on the rate of stirring, or, in a flow system, on the linear velocity of the fluid. It has also been found that the activation energy is about 4.5 kcal . mole^{-1} in diffusion controlled reactions. The dissolution of metals by iodine/potassium iodide solution[325,326] and of iron by hydrochloric acid[327] have been shown to be diffusion controlled.

Where chemical reaction is the rate determining step, the rate of dissolution will be proportional to the surface area and to the concentration of the second reagent

$$R = KS(C)$$

The rate of dissolution of marble ($CaCO_3$) in hydrochloric acid [328] is an example of a chemically controlled reaction. The rate can be measured from the evolution of carbon dioxide. Palmer and Clark[329] found that the rate of dissolution of silica in hydrofluoric acid was proportional to surface area and acid concentration. The rate of reaction was measured by noting the increase in conductivity of the solution

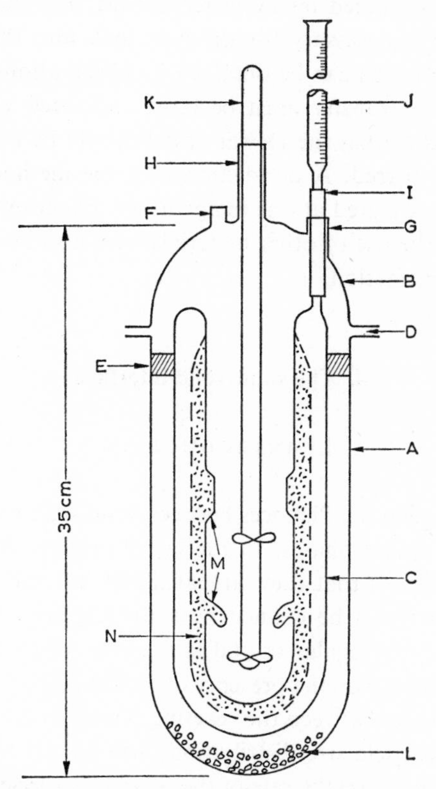

Fig. 41. Schematic diagram of ice calorimeter, A, outer jacket; B, cap; C, Dewar vessel; F, tube for addition of reagents; J, semi-microburette; M, protrusions to hold ice mantle; N, ice mantle. From ref. 332.

due to the formation of hydrofluorosilicic acid. Bradley[330] has studied the reaction of copper with a solution of sulphur in carbon tetrachloride. Reaction was followed by removing solution at intervals, separating off the copper sulphide and weighing the sulphur present after evaporation of the solution.

Exchange of silver with radioactive silver nitrate solution has been studied by Tingley *et al.*[331]. The ^{110}Ag isotope has a half life of 270 days. Various surface treatments were given to silver foils, which were then cemented into a wax block and suspended in the 0.1–0.005 *N* silver nitrate solution. After reaction, specimens were rinsed to remove solution and the activity determined with an end-window β–γ counter. Fluctuations of the counter efficiency were corrected by standardisation against a known ^{110}Ag source. The method of foil pretreatment was found to have a great effect on the kinetics of the exchange process.

Calorimetry has been used by Smith[332] to measure the rate of acetylation of cellulose. The apparatus is shown in Fig. 41. A coating of ice is formed inside the double wall of the Dewar vessel, the rest of the space and the burette containing water. As the reaction proceeds, the heat evolved causes some of the ice to melt and alter the volume occupied by the water and ice. The change is noted on the burette. A correction is necessary for the heat leak into the calorimeter in the absence of reaction, but this can be obtained by observations prior to the experiment. The rate of heat transfer must be large compared with the rate of heat evolution, which suggests that the Dewar vessel should be made of metal and the reaction mixture well stirred. A disadvantage of the method is that the rate of reaction can only be measured at one temperature. Experiments at other temperatures would require the use of other materials with a suitable melting point and volume change on liquefaction.

4. The solid–solid interface

4.1 INTRODUCTION

There are certain essential differences between solid state reactions and reactions involving gaseous or liquid phases. In the latter case, the kinetic motion of the reactant molecules ensures that they are available to one another for reaction under conditions which can be defined by statistical laws. Solid state reactions occur between apparently regular crystal lattices, in which the kinetic motion is very restricted and depends on the presence of lattice defects. Interaction can only occur at points of contact between the reacting phases and is therefore dependent on particle size and particle size distribution. The factors which govern the rate of a solid state reaction are (*i*) the rate of the boundary phase processes which lead to the consumption of the original lattices, and (*ii*) the rate of particle transfer through the product layer.

Any solid state reaction, however complex, must resolve itself into interactions between pairs of solid phases, the elementary processes occurring successively or simultaneously to give a variety of intermediate and final products. Because the entropy change is small, all solid state reactions are exothermic. This property forms the basis of the 'heating curve' method for detecting reactivity in solid mixtures.

Diffusion in solids is necessarily linked with reaction in solids; this topic has been reviewed by Barrer[333] and by Jost[334]. General reviews of solid state reactions have been given by Hüttig[335], Cohn[336] and by Welch[337].

4.2 X-RAY CRYSTALLOGRAPHY

The technique of determining the extent of a solid state reaction by X-ray powder photographs was pioneered by Jander[338]. Basically, the method consists of detecting the presence of a crystalline phase by its characteristic X-ray pattern. The amount of the phase present is determined from the intensity of the X-ray lines. Its disadvantage is that many products in solid state reactions are poorly crystallised and consequently difficult to detect. Furthermore, the presence of poorly crystallised material is likely to cause excessive scattering, thus preventing the detection of phases present in small concentration during the initial stages of reaction.

A more direct application of X-ray diffraction is the measurement of changes in X-ray patterns at elevated temperatures, whilst the reaction is occurring. A preliminary examination at a number of temperatures serves to pinpoint the temperature of reaction and to determine the change in the cell dimensions caused by thermal expansion. The rate of reaction can then be followed from the rate of change in intensity of the diffraction lines at the reaction temperature. Weiss and Rowland[339,340] have devised a technique for oscillating the goniometer across a selected X-ray diffraction maximum. This scanning technique permits repetitive automatic examination of the beam, and changes in the intensity due to lattice expansion can be detected. The limits of the goniometer travel are set as narrowly as possible, taking into account any change in the angle of the diffracted beam which occurs due to lattice expansion.

4.3 MICROSCOPY

The techniques of optical[341] and electron microscopy have been applied to solid state reactions, but it is difficult to obtain more than semi-quantitative results. Wagenblast *et al.*[342] have studied the formation and growth of carbon precipitation in iron by electron microscopy. The polymerisation of crystalline trioxane

References pp. 270–278

has been studied with an optical microscope[343]. After irradiation with 0.5 Mrad X-rays, the polymerisation proceeded at a measurable rate between 20–60° C. Single crystals were irradiated and placed on the microscope stage in a constant temperature bath of polybutene oil to follow the rate of polymerisation. The reaction was quite complex, but four or five distinct stages could be recognised as the temperature was raised.

Optical microscopy has been used to observe the rate of sintering of $NiO–Fe_2O_3$ and $MgO–Fe_2O_3$ systems[344]. Oxide spheres were placed in contact with a polycrystalline plate of the other component and given varied heat treatments. The specimens were then mounted in a lucite block and sectioned for microscopic examination. Temperature gradients are a problem, since the temperature sensing device is usually situated in the heating block and the sample is heated by conduction through the microscope slide. By selecting a relatively small area and using a slow heating rate, the area under observation will be nearly isothermal, but the error between the measured temperature and actual temperature of reaction still exists. A possible way round this objection is calibration of the temperature difference at various temperatures with substances whose transformation temperature is already known. Reese *et al.*[345] have developed a mechanical method, which permits detection of phase transformations by the recording of changes in light intensity. Polarized light is used to illuminate the specimen and this passes through the microscope and is directed by a beam splitter to a recording cell. The use of polarised light has the advantage that melting is distinguished by a large drop in the transmitted light intensity.

4.4 DIFFERENTIAL THERMAL ANALYSIS

The concept of "reaction temperature", when the heat produced in a solid state reaction causes an appreciable temperature rise, was used in early studies and is related to the kinetics of the reaction. This method has evolved into the technique of differential thermal analysis (DTA), which has developed rapidly in recent years. It is undoubtedly of great value as a qualitative tool for the study of solid state reactions. Several excellent comprehensive reviews of the subject have been made which cover every aspect of DTA techniques[346–349]. The review by Smothers and Chiang[349] contains an indexed list of over four thousand publications on DTA up to 1965.

The basic principle of DTA is the measurement of the temperature difference between a sample and a reference material, as they are simultaneously heated at a uniform rate. A solid state reaction will cause an evolution of heat, which will be shown as a temperature difference (ΔT) between the sample and the reference material. As this heat is dissipated to the surroundings, ΔT reduces to zero again. Measurement of temperature and ΔT over a suitable range will give a thermogram

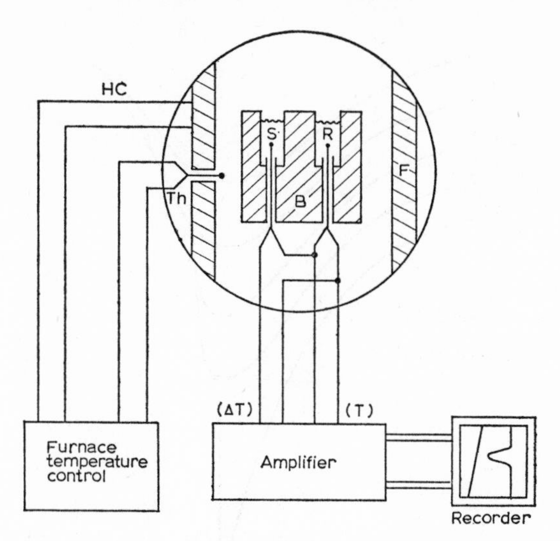

Fig. 42. Block diagram of an apparatus for differential thermal analysis. Area inside the circle is enlarged to show details of the thermocouples and sample block. B, sample block; F, furnace; R, reference material; S, sample; HC, furnace heating current; Th, furnace thermocouple.

characteristic of the reaction which has occurred. A block diagram to illustrate the essentials of a DTA apparatus is shown in Fig. 42.

Two thermocouples pass through the sample block and into cavities which contain the sample and the inert reference material respectively. These thermocouples are joined in opposition, so that the potential difference seen by the amplifier is a measure of the temperature difference (ΔT) between the sample and the reference material. One of the thermocouples is used to measure the sample temperature (T) directly. The output of both signals from the amplifier can be measured on a two pen recorder. It is usually necessary to control the furnace temperature automatically to ensure a uniform rate of heating or cooling. Many different types of commercial apparatus are available with various refinements and convenience features[334, 335].

The major problem with DTA stems from the difficulty of obtaining a uniform temperature throughout the sample. For this reason, there is rarely quantitative agreement between the results of different workers. The peak shape and area on the thermogram are affected by non-uniformity of furnace temperature, rate of heating, sample geometry, particle size, and dilution with inert material. In many instances, the atmosphere above the sample also influences the reaction.

All the factors mentioned above become more important in any attempt to make kinetic measurements for a solid state reaction. Kissinger[350] has attempted to use the temperature of maximum reaction rate to derive reaction order by using the following expression

$$dx/dt = A(1-x)^n \exp(-E/RT)$$

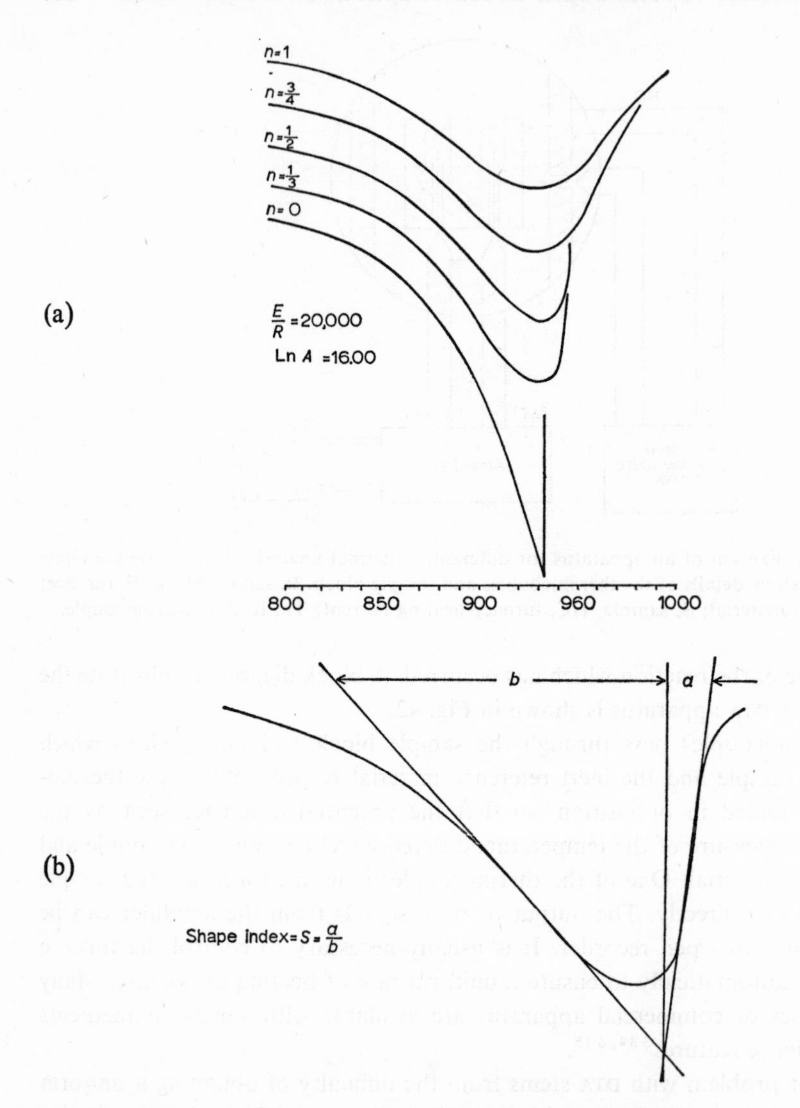

Fig. 43. (a) Effect of order of reaction on plots of reaction rate against temperature for constant heating rate, frequency factor and activation energy. $E/R = 20{,}000$, ln $A = 16.0$. From ref. 350. (b) Method of measuring reaction order from the shape index of DTA peak. Shape index $S = a/b$. From ref. 350.

differentiating the above expression and setting it equal to zero gives

$$\left[\frac{E\phi_2}{RT}\right]_{\mathrm{m}} = [An(1-x)^{n-1} \exp(-E/RT)]_{\mathrm{m}}$$

where subscript m refers to conditions at peak temperature. ϕ is the heating rate and the other symbols have their usual meaning. By various transformations, an expression is obtained which suggests that the reaction order, n, is related to the quantity of material not reacted at T_m. From the variation in the quantity reacted with T_m, Kissinger proposed the shape index shown in Fig. 43 to estimate the order of reaction. Changes in the activation energy and frequency factor only have a small effect on the shape of the curve. From the ratio of the slopes at the inflection points (see Fig. 43b) where the reaction rate is constant, the reaction order can be calculated from the expression

$$n = 1.26S^{\frac{1}{2}}$$

where the shape index $(S) = a/b$.

A basic assumption in Kissinger's analysis is that reaction occurs equally in all parts of the specimen. Disagreement between the results of various authors shows that this assumption is not generally true and the results are particularly influenced by particle size and sample geometry. Borchardt and Daniels[351] have derived an expression for the rate coefficient (k) derived from a DTA curve

$$k = \left[\frac{KAV}{N_0}\right]^{x-1} \cdot \frac{C_p \mathrm{d}\Delta T/\mathrm{d}t + K\Delta T}{[K(A-\alpha) - C_p \Delta T]^x}$$

where x is the order of reaction, K is the heat transfer coefficient of the sample cell, A is the total area under the curve, α is the area swept out in time t, V is the volume of reactant and N_0 the number of moles initially present. C_p is the specific heat of the sample at constant pressure and ΔT is the temperature difference between the sample and reference material. The quantities α, ΔT and $\mathrm{d}\Delta T/\mathrm{d}t$ are measured from the DTA curve as shown in Fig. 44.

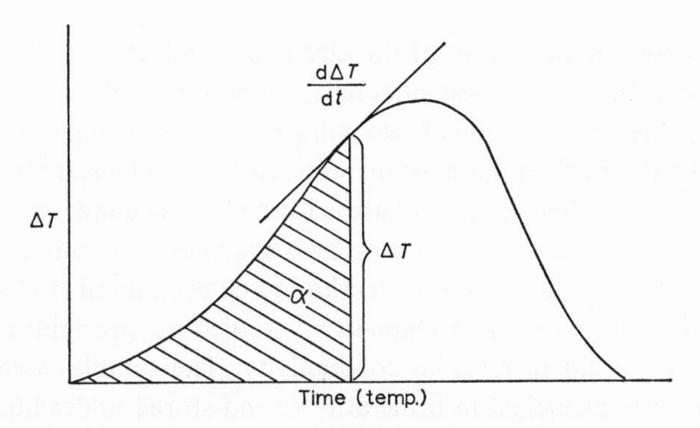

Fig. 44. Determination of rate coefficients from the DTA curve. At time t, α is the area swept out, ΔT is the temperature difference and $\mathrm{d}\Delta T/\mathrm{d}t$ is the slope of the curve. From ref. 351.

For a first order reaction the above equation simplifies to

$$k = \frac{C_p \mathrm{d}\Delta T/\mathrm{d}t + K\Delta T}{K(A-\alpha) - C_p \Delta T}$$

The original work[351] extends the discussion to more complex reactions and the determination of activation energies and heats of reaction. However, the equations were developed for homogeneous reactions in solution and required twelve assumptions some of which are very difficult to satisfy when applied to DTA studies of solid state reactions. These assumptions are (*i*) the heat transfer coefficients and heat capacities of reactants and products are equal and constant, and (*ii*) that the temperature is uniform throughout the sample and reference material. Freeman and Carroll[352] and Wendlandt[353] have suggested simplifications in Borchardt and Daniels'[351] procedure.

Borchardt[354] has also suggested a simplified procedure for determining initial reaction rates. If $(\mathrm{d}n/N_0)/\mathrm{d}t$ is the rate of reaction when the height of the DTA peak is ΔT, then

$$\frac{-\mathrm{d}n/N_0}{\mathrm{d}t} = \frac{\Delta T}{A}$$

It is assumed that the DTA peak can be approximated by a triangle of height ΔT_{max} and base b then

$$\frac{-\mathrm{d}n/N_0}{\mathrm{d}t} = \frac{2\Delta T}{b\Delta T_{max}}$$

4.5 ELECTRICAL CONDUCTIVITY MEASUREMENTS

The methods for determination of the electrical conductivity of solids are discussed in Section 2.1.7. For most materials, a change in electrical conductivity can be expected due to interaction of two solid phases. In a single phase transformation, the disorder in the solid must increase, causing an increase in conductivity which will be reversed when the new phase is formed. The change may be indicated by a change in temperature coefficient or by a sharp discontinuity.

Garn and Flaschen[355] have used the method to detect a number of polymorphic phase transformations. Fujita and Damask[356] studied the precipitation of carbon in iron which caused an increase in conductivity. The samples were carburised at high temperature, quenched in brine at 0° C and stored under liquid nitrogen. All resistance measurements were made at liquid nitrogen temperature, and compared with a well aged iron/carbon specimen and a standard resistance. Jaky *et al.*[357] have studied the formation of spinels from cadmium oxide/ferric oxide mixtures. A valve voltmeter was used to follow the conductivity changes and the results correlated with a number of other measurements.

4.6 MAGNETIC MEASUREMENTS

Experimental details for the techniques of magnetic susceptibility and magnetic resonance are described in Section 2.1.6. Experiments for solid state reactions are potentially simpler than adsorption studies if a controlled atmosphere is not required. Standard methods[134] for magnetic susceptibility determination can be used. Greatest sensitivity is achieved when a non ferromagnetic substance reacts to form a ferromagnetic substance or *vice versa.* If both are ferromagnetic the method becomes insensitive, but if the Curie points are sufficiently different it may be possible to choose a temperature above the Curie point of one of the reactants. Hofer *et al.*[358] have studied the decomposition of cobalt carbide into α cobalt and carbon; both are ferromagnetic but the specific magnetisation increases sharply during the decomposition. The disproportionation of ferrous oxide[359] can be followed by observing susceptibility changes. By quenching the stoichiometric composition, a homogeneous substance can be obtained. Disproportionation occurs above 300° C according to the equation

$$4FeO \rightarrow Fe_3O_4 + Fe$$

causing a rapid rise in the intensity of magnetisation. Many further references are to be found in the literature[360].

Turkevich *et al.*[361] have used EPR techniques to follow the interaction of manganous halides with zinc sulphide. Zinc sulphide shows no resonance, but on heating with manganous salts at 600° C the resonance lines appear. Measurements were made on a Varian spectrometer operating at 9.5 kMc . sec^{-1}, with the magnetic field modulated to 100 kc . sec^{-1}. A Varian multipurpose cavity was used with the sample extending beyond the confines of the cavity. The spin concentration was determined with an internal ruby standard.

4.7 DILATOMETRY

The measurement of dimensional or volume changes can be used in appropriate instances to observe the progress of a solid state reaction. Linear expansion can be measured by dial gauges, micrometers, interferometer, telescopes, linear differential transformers and from X-ray patterns. Except for the X-ray techniques, the reaction can be studied *in situ.* Non-isotropic materials probably require measurements in several orientations.

References pp. 270–278

A dilatometer constructed by Sauer[362] is shown in Fig. 45. The sensing element is a linear variable differential transformer, whose rectified output is recorded directly. A pivot arm connects the transformer (A) to the bearing arm (B) and a micrometer screw allows for calibration and adjustment of the transformer. By adjustment of the compensating screw, the pivot arm can be moved so that the contact point (B) is exactly one sample length from the fulcrum. Temperatures above and below ambient can be obtained by circulating gas round the specimen.

Volume expansion is most easily measured by fluid displacement and is particularly suitable for organic materials which are not generally hard or highly oriented. In an apparatus devised by Loasby[363], the sample is enclosed in a bellows and the intervening space filled with oil. Expansion is recorded as a linear motion of the bellows. The sample size is made as large as possible so that it is just out of contact with the bellows at the point of maximum expansion. Smaller pieces can be used instead of a single specimen. This allows a more rapid approach to thermal equilibrium, but at the expense of sensitivity.

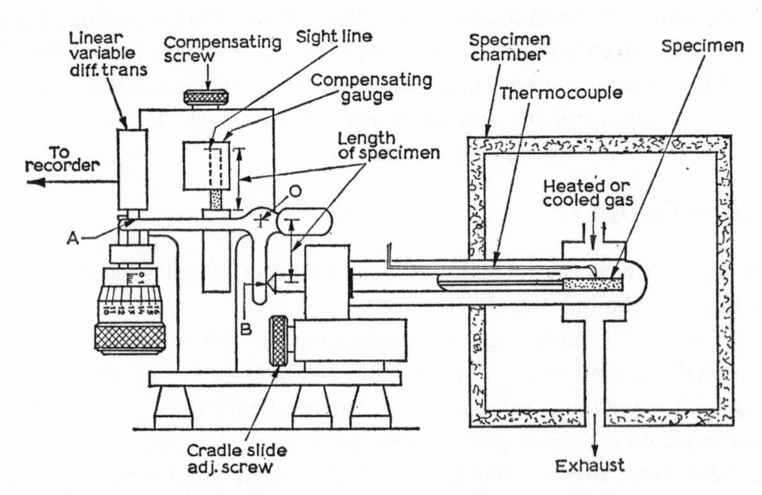

Fig. 45. Programmed dilatometer providing sub- and supra-ambient temperatures. H. A. Sauer, from ref. 348.

To avoid the need for calibration, Lieberman and Crandall[364] have devised a dilatometer which operates by measuring the change in light intensity after passing through two parallel diffraction gratings. As the sample expands, the gratings move parallel to one another and the light recorded by a photocell passes through a series of maxima and minima. Traverse through each maximum corresponds to a movement of one grating spacing. The accuracy of the measurement depends on the perfection of the grating and the accuracy of reading the maxima and minima.

Dannis[365] used a different optical method. The movement of a mirror, partially supported by the sample, was detected by light reflected onto a photoresistor bridge. The bridge was attached mechanically to a recorder pen and moved in response to the light, drawing a trace of the expansion.

4.8 OTHER METHODS

A number of other methods have been employed in specific systems, but the experimental details are limited and the references will only be mentioned briefly to indicate the possibilities.

Stone and Tilley[366] have shown that the spectral changes which occur on spinel formation can be related to the extent of reaction of the component oxides, although the kinetics were also followed by chemical analysis. Keyser *et al.*[367] have studied the reaction of zirconium silicate with calcium oxide by a radio tracer method. Wagenblast and Damask[368] have used internal friction measurements to study the rate of precipitation of carbon in iron. The rate of decrease of the Snoek peak can be related to the growth kinetics of the iron carbon precipitate[369,370].

5. The liquid–gas interface

5.1 INTRODUCTION

Reaction at liquid surfaces covers a wide field. However, special experimental methods have been developed mainly for the study of insoluble or partly soluble monolayers at the liquid–air interface. Adsorption and reaction of monolayers at the liquid–air interface and liquid–liquid interfaces can be studied by similar techniques. It is therefore convenient to treat these together. Comprehensive information about liquid interfaces is contained in publications by Adamson[1], Davies[371], Alexander[372], Davies and Rideal[373] and Gaines[374].

Kinetics of reaction at the interface are generally studied by the same methods as are used for the study of the monolayer itself. These are mainly the determination of surface area (A), film pressure (π), surface potential (ΔV), and surface viscosity. A number of relationships developed by Adamson[1] illustrate the type of kinetic expressions obtained for simple reactions.

(1) The rate of dissolution of a monolayer into an infinite medium may be determined by the rate of diffusion of material away from the surface region. Then

$$\ln (r/r_0) = 2K(Dt/\pi)^{\frac{1}{2}}$$

where r is the surface concentration per unit area and K is the proportionality constant.

References pp. 270–278

(2) For a chemical reaction occurring at a constant film pressure where the product is soluble

$$A = A_0 \exp(-kt)$$

where A is the surface area.

(3) At a constant film pressure where the products are insoluble and remain in monolayer form

$$(A - A_\infty)/(A_0 - A_\infty) = \exp(-kt)$$

It is assumed that the reactant molecule areas are additive and the substrate concentration remains essentially constant.

(4) A similar equation can be derived for the change in film pressure at total constant area

$$(\pi - \pi_\infty)/(\pi_0 - \pi_\infty) = \exp(-kt)$$

The above equation assumes that the film pressure is a linear function of composition.

(5) Measurement of surface potentials leads to similar equations. At constant area

$$(\Delta V - \Delta V_\infty)/(\Delta V_0 - \Delta V_\infty) = \exp(-kt)$$

If the total area varies during reaction

$$(A\Delta V - A_\infty \Delta V_\infty)/(A_0 \Delta V_0 - A_\infty \Delta V_\infty) = \exp(-kt)$$

The assumption is that the effective dipole moment, and hence the orientation, remain constant. This is most likely to be true at constant film pressure.

5.2 MEASUREMENT OF FILM PRESSURE–FILM AREA CHARACTERISTICS

(a) *Wilhelmy slide*

This is the simplest experimental method for the determination of film pressure, which is defined as the difference in surface tension between the pure solvent and the film covered surface. The equipment consists of a thin glass, platinum or mica slide suspended from a balance and dipping into the film covered surface. Any change in surface tension causes the slide to rise or fall until buoyancy compensation is reached.

The film pressure is given by

$$\pi = \Delta W / p$$

where ΔW is the difference between the total weight and the weight of the slide, p is the perimeter of the slide. ΔW can be determined by prior calibration of weight changes in terms of a scale deflection, or by changing the balancing weight to restore the balance to the null position. It is recommended that films should always be studied in compression[376]. Glass slides can be cleaned with chromic acid, mica slides by rinsing in alcohol after roughening with carborundum paper. The disadvantages of the Wilhelmy slide method are that the contact angle must remain constant during the experiment and that the reference surface tension, *i.e.* the clean surface value, cannot be measured during the experiment.

(*b*) *Film balance*

In most experiments, film pressures have been measured directly by means of a film balance or Langmuir–Adam trough. The principle of the method involves the direct measurement of the horizontal force on a float separating the film from

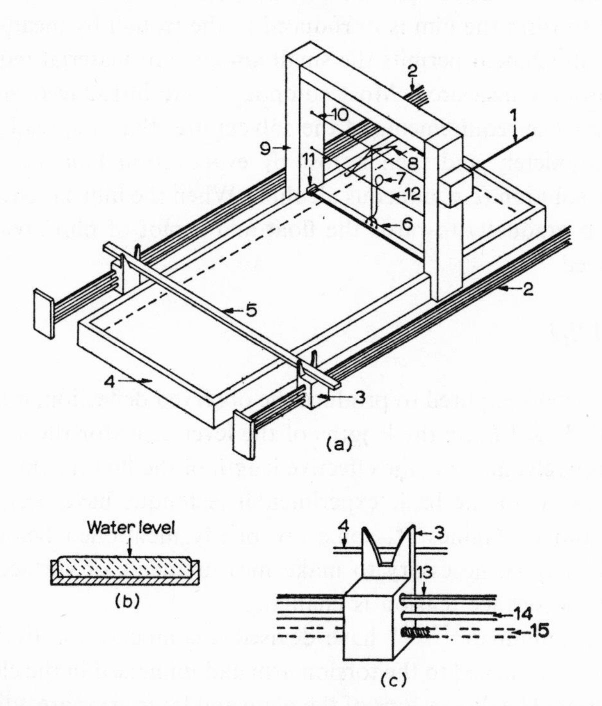

Fig. 46. A modern film balance. 1, torsion wire control; 2, sweep control; 3, sweep holder; 4, trough; 5, sweep; 6, float; 7, mirror; 8, calibration arm; 9, head; 10, main torsion wire; 11, gold foil; 12, wire for mirror; 13, elevation control; 14, guide; 15, traverse. From ref. 1.

clean solvent surface. A typical apparatus is illustrated in Fig. 46. The trough is a shallow, rectangular tray about 3 ft × 1 ft which can be filled with water or other aqueous substrate. A torsion wire is supported some distance from one end of the trough and attached to a float which rests on the surface of the water. The float is usually of mica or copper. It is connected to the sides of the trough by thin foils or threads, to prevent leakage of the film round the edges of the float. Any pressure on the float produces a twist in the torsion wire, which is observed optically. The float can be returned to the zero position by rotating the torsion head until the light image returns to its reference point. A calibration is obtained by hanging appropriate weights on the end of the calibration arm and noting the rotation of the torsion head necessary to return the light to the reference point.

The sweep device rests on the edges of the trough and is used to clean the surface before the experiment and to compress the film once it is formed. For accurate work, it is preferable that the sweep is attached to support rods (as shown in Fig. 46) which can be moved along the trough by means of a screw drive. It is usually necessary to place the film balance inside an air thermostat, with provision for operation from the outside.

The experimental procedure is to fill the trough so that the substrate is higher than the rim and clean the surface in front and behind the float with several sweep rods. Material to form the film is introduced to the trough by means of a syringe. A micrometer attachment permits the small amounts of material required (~ 0.1 ml) to be accurately measured. Most compounds are introduced in a solvent to assist spreading. The requirements of the solvent are that it spreads the material rapidly and completely and then is lost by evaporation (*e.g.* petroleum ether, benzene) or by solution (*e.g.* aqueous alcohol). When the film has been formed, the sweep is moved gradually towards the float and a plot of film area against film pressure obtained

$$\pi = wgl_c/l_f L$$

where w is the weight required to produce the observed deflection, g is the gravitational constant, l_c and l_f are the lengths of the lever arms for the calibrating arm and float respectively, and L is the effective length of the float in cm.

Many refinements of the basic experimental technique have been reviewed by Alexander[372] and by Gaines[374], some are briefly mentioned below. In certain experiments, it may be necessary to make measurements of surface pressure on a substrate whose surface tension is changing.

Matalon and Schulman[375,376] have devised a compensation device, in which a Wilhelmy slide is attached to the torsion arm and immersed in the clean substrate behind the barrier. The dimensions of the plate and lever arms are adjusted so that a change in the reference surface tension produces a force on the Wilhelmy plate which just compensates the force produced on the float. Many automatic recording balances have been constructed; one version described in detail by Gaines[374] incorporates a linear variable differential transformer driving the sweep device which restores the surface pressure to a previously set value. Alternatively, the sweep device may be driven at a constant speed and the output of the transformer recorded directly, as a surface pressure change.

The two major problems which arise in using a film balance are leakage past the ends of the barrier and contamination. There are several ways to prevent leakage; Langmuir's original method was to direct air jets on the gaps at the end of the float, but this method is unsatisfactory at high surface pressures. Thin ribbons of gold or platinum foil, waxed to prevent wetting, are more convenient, although the use of plastics such as Teflon is increasing. Because the ribbons can affect the sensitivity, other workers have used fine threads of nylon. Slight irregularities in the ribbons or threads can lead to incomplete contact with the surface. The resulting leaks can usually be overcome by stroking the water surface up to the ribbon, or sprinkling talc or sulphur powder in front of the ribbon. One way of detecting leaks is to sweep the rear part of the trough *towards* the float at the end of the experiment. Any deflection of the float is strong evidence for leaks or contamination.

Troughs are generally made of metal, coated with paraffin wax to make them hydrophobic. Stainless steel or chromium plated troughs coated in this manner are probably the most satisfactory. Glass or silica troughs give less contamination, but do not coat as easily with wax, are more fragile, and are difficult to produce to accurate dimensions. One disadvantage of waxed troughs is that the wax is soluble in most solvents used for spreading the film, thus providing a large possible source of contamination. A variety of plastic materials have been used to overcome this problem, but Teflon is superior to most of the others. It is sufficiently hydrophobic, but lacks dimensional stability and cannot be easily joined together.

5.3 SURFACE POTENTIAL MEASUREMENTS

The potential change ΔV which occurs when an insoluble monolayer is spread at a liquid–gas (or liquid–liquid) interface is called the surface potential of the film. It is frequently expressed in terms of a surface dipole moment μ, where

$$\mu = \Delta V/4\pi n$$

n being the number of molecules per cm^2 in the monolayer. An alternative form is

$$\mu = A\Delta V/12\pi$$

where A is expressed in Å^2 . molecule^{-1}, ΔV in millivolts and μ is obtained in millidebye units. The determination of surface potentials is often carried out simultaneously with measurements of area as the film is compressed. Film balances are often designed with provision for carrying out these measurements. Inhomogeneity of the film, or surface contamination, will cause a variation in ΔV over the film and surface potential measurements can be used to scan the film surface. Measurement of surface potentials can be accomplished by the ionising electrode or the vibrating condenser method. The former is simpler experimentally, but the latter can also be used for liquid–liquid interfaces.

(*a*) *The ionising electrode method*

This method has been extensively used by Schulman and Rideal[377] and is indicated diagrammatically in Fig. 47. A small amount of radioactive material, commonly polonium, is mounted on or near the air electrode. The α-particle radiation ionises the air between the electrode and the liquid, making it conducting. An electrometer can then be used to measure the potential difference across the gap. Normally, the potential difference is balanced against a standard cell with a potentiometer to obtain a null reading. The electrode E is usually silver–silver chloride, although others have been tried. The main requirement is freedom from appreciable drift during the duration of the experiment. There are references in the literature to the preparation of radioactive sources, but they are now readily available commercially.

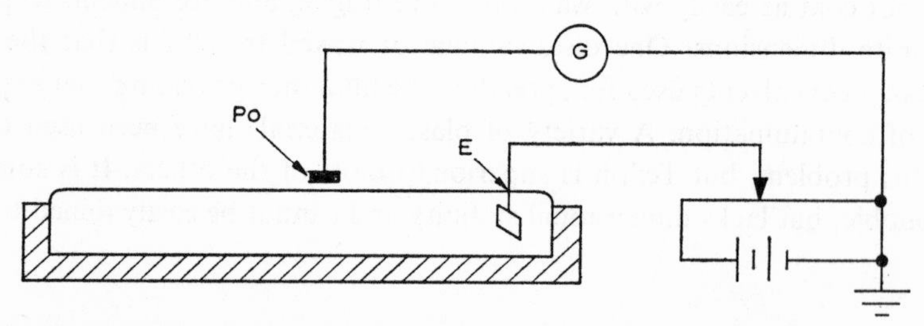

Fig. 47. Ionising electrode method for the determination of surface potentials. Po, polonium electrode; E, reference electrode; G, galvanometer. From ref. 1.

The range of polonium α-particles in air is 3–8 cm, so that effective ionisation can be achieved with the electrode up to 4 cm from the surface. If the radioactive material is covered with a protective coating, the effective range will be reduced. Polonium (^{210}Po) is convenient because it emits negligible β- and γ-radiation and special shielding is not required. Its disadvantage is its short half-life (140 days). Radium (^{226}Ra) has a half-life of 1590 years, but requires greater handling precautions because of its β- and γ-emission. Recently, plutonium (^{234}Pu, half-life 24000 years) and americium (^{241}Am, half-life 458 years) have been used because shielding problems are minimised. However, in handling any radioactive material the necessary precautions must be observed[246,249].

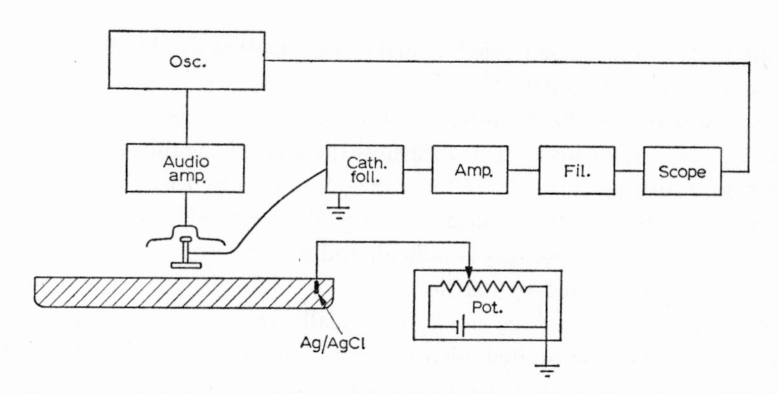

Fig. 48. Surface potential measurements by the vibrating plate method. The oscillator (Osc.) and amplifier (Audio amp.) are commercial units which drive the electrode. The oscilloscope (Scope) and high gain, low noise amplifier (Amp.) are also standard commercial instruments. From ref. 374.

(*b*) *The vibrating condenser method*

This method is similar to that used for measuring contact potential differences at the solid–gas interface (see Section 2.1.5). It was applied to monomolecular films by Yamins and Zisman[378]. An audiofrequency current is used to vibrate an electrode attached to the voice coil of a loudspeaker at between 200 and 1000 cps. The electrode is normally about 1 cm^2 in area and placed about 0.5 mm from the surface. The vibration of the electrode causes a variation in the capacity of the air gap, so that an alternating current is set up in the second circuit whose magnitude depends on the contact potential difference. This signal is amplified and detected by earphones. An opposing voltage is applied with a potentiometer until the hum disappears, giving the contact potential of the surface. A more refined electronic arrangement described by Gaines[374] is shown in Fig. 48. Here, the signal is amplified, filtered to improve the signal-to-noise ratio, and displayed on an oscilloscope.

Surface potential variations of about 0.1 mV can be detected, which is rather better than the ionising electrode method. However, the method is subject to more disturbances. Vibrations can be induced in the liquid surface which may disrupt the monolayer and give a noisy signal. Surface ripples on the liquid produce the same effect. The noise can be reduced by increasing the electrode–liquid spacing or by reducing the amplitude of vibration. Both of these alternatives reduce the sensitivity of the apparatus.

5.4 OTHER METHODS

There are a number of other methods which have been used for the study of monolayer properties, but not applied at present to the study of reaction kinetics.

These methods are discussed briefly; further information is obtainable from the references already mentioned[1,356–359].

Surface viscosity can be measured with a canal viscometer or with a torsion viscometer. The former instrument measures the flow of the film through a slit in a barrier on a film balance; the latter, the damping caused by the presence of a film on a body moving in the surface of the liquid. Interpretation of measurements on films with a constant viscosity is difficult and would be more so in the presence of viscosity changes caused by reaction.

Optical spectroscopy can be used, but usually requires a multiple reflection technique because of the limited adsorption by a monomolecular film. A discussion of the advantages and limitations of the multiple reflection technique has been given by Harrick[165]. Radioactive tracers can be used to study reaction in monomolecular films, the advantages and limitations are similar to those found in other types of heterogeneous reaction. In principle, it is possible to scan very small areas of film. For elements which emit weak β-particles, such as ^{14}C or ^{35}S, the rate of disappearance from the surface to the subphase can be studied because the range of radiation is only about 0.3 mm in water. However, back scattering occurs from the substrate which complicates the interpretation. The work by Mason *et al.*[379–381] on the exchange of iodide ion in solution with a surface monolayer of α-iodostearic acid illustrates the difficulties of the technique.

REFERENCES

1 A. W. Adamson, *Physical Chemistry of Surfaces*, Interscience, New York, 1960, p. 457.
2 D. O. Hayward and B. M. W. Trapnell, *Chemisorption*, Butterworths, London, 1964.
3 G. C. Bond, *Catalysis by Metals*, Academic Press, London, 1962, p. 49.
4 P. H. Emmett, *Catalysis*, Reinhold, New York, 1 (1954) 31.
5 C. Orr, Jr. and J. M. Dallevalle, *Fine Particle Measurement*, Macmillan, New York, 1962.
6 D. M. Young and A. D. Crowell, *Physical Adsorption of Gases*, Butterworths, London, 1962.
7 H. S. Taylor, *Discussions Faraday Soc.*, 8 (1950) 8.
8 W. G. Frankenburg, *J. Am. Chem. Soc.*, 66 (1944) 1827.
9 A. Couper and D. D. Eley, *Discussions Faraday Soc.*, 8 (1950) 172.
10 R. Rudham and F. S. Stone, *Trans. Faraday Soc.*, 54 (1958) 420.
11 J. A. Becker, E. J. Becker and R. G. Brandes, *J. Appl. Phys.*, 32 (1961) 411.
12 R. W. Roberts, *Brit. J. Appl. Phys.*, 14 (1963) 537.
13 I. Langmuir and D. S. Villars, *J. Am. Chem. Soc.*, 53 (1931) 486.
14 J. K. Roberts, *Proc. Roy. Soc. (London)*, *A*, 152 (1935) 445, 464.
15 H. D. Hagstrum, *Phys. Rev.*, 96 (1954) 336.
16 R. Gomer, *J. Chem. Phys.*, 21 (1953) 293.
17 R. H. Good and E. W. Muller, *Handbook of Physics*, Springer-Verlag, Berlin, 21 (1956) 176.
18 F. G. Allen, J. Eisinger, H. D. Hagstrum and J. T. Law, *J. Appl. Phys.*, 30 (1959) 1563.
19 H. D. Hagstrum, *J. Appl. Phys.*, 32 (1961) 1020.
20 J. T. Law, *J. Phys. Chem.*, 59 (1955) 543.
21 R. B. Anderson and P. H. Emmett, *J. Phys. Chem.*, 55 (1951) 337.
22 H. D. Hagstrum and C. D'Amico, *J. Appl. Phys.*, 31 (1960) 715.
23 H. E. Farnsworth and R. F. Woodcock, *Advances in Catalysis*, 9 (1957) 123.

24 D. SHOOTER AND H. E. FARNSWORTH, *J. Phys. Chem. Solids*, 21 (1961) 219.
25 F. G. ALLEN, T. M. BUCK AND J. T. LAW, *J. Appl. Phys.*, 31 (1960) 979.
26 J. T. LAW, *J. Phys. Chem. Solids*, 14 (1960) 9.
27 R. E. SCHLIER, *J. Appl. Phys.*, 29 (1958) 1162.
28 J. R. YOUNG, *J. Appl. Phys.*, 30 (1959) 1671.
29 E. D. HONDROS AND A. J. W. MOORE, *Acta Met.*, 8 (1960) 647.
30 R. E. SCHLIER AND H. E. FARNSWORTH, *J. Chem. Phys.*, 30 (1959) 917.
31 H. E. FARNSWORTH AND J. TUUL, *J. Phys. Chem. Solids*, 9 (1959) 48.
32 L. O. OLSEN, C. S. SMITH AND E. C. CRITTENDEN, JR., *J. Appl. Phys.*, 16 (1945) 425.
33 J. A. ALLEN, *Rev. Pure Appl. Chem.* (*Australia*), 4 (1954) 133.
34 W. ESPE, *Werkstoffkunde der Hochvakuumtechnik*, V.E.B. Deutscher Verlag der Wissenschaften, Berlin, 1959.
35 L. HOLLAND, *Vacuum Deposition of Thin Films*, Wiley, New York, 1962.
36 R. E. HONIG, *RCA Review*, 23 (1962) 567.
37 D. D. ELEY AND D. R. ROSSINGTON, *Chemisorption*, Ed. W. E. GARNER, Butterworths, London, 1957, p. 137.
38 D. D. ELEY AND D. SHOOTER, *J. Catalysis*, 2 (1963) 259.
39 E. GREENHALGH, D. O. HAYWARD AND B. M. W. TRAPNELL, *J. Phys. Chem.*, 61 (1957) 1254.
40 D. BRENNAN, D. O. HAYWARD AND B. M. W. TRAPNELL, *J. Phys. Chem. Solids*, 14 (1960) 117.
41 A. E. BELL, J. PRITCHARD AND K. W. SYKES, *Nature* (*London*), 191 (1961) 487.
42 R. W. ROBERTS, *Ann. New York Acad. Sci.*, 101 (1963) 766.
43 N. LAEGREID AND G. K. WEHNER, *Seventh Nat. Symposium on Vacuum Technology*, American Vacuum Society, Pergamon Press, Oxford, 1961, p. 286.
44 S. P. WOLSKY AND E. J. ZDANUK, *Seventh Nat. Symposium on Vacuum Technology*, American Vacuum Society, Pergamon Press, Oxford, 1961, p. 282.
45 D. SHOOTER AND H. E. FARNSWORTH, *J. Phys. Chem.*, 66 (1962) 222.
46 B. M. W. TRAPNELL, *Proc. Roy. Soc.* (*London*), *A*, 218 (1953) 566.
47 E. GREENHALGH AND B. M. W. TRAPNELL, *Advances in Catalysis*, 9 (1957) 238.
48 M. W. ROBERTS AND F. C. TOMPKINS, *Proc. Roy. Soc.* (*London*), *A*, 251 (1959) 369.
49 D. SHOOTER, *Ph. D. Thesis*, University of Nottingham, (1959).
50 O. BEECK, A. E. SMITH AND A. WHEELER, *Proc. Roy. Soc.* (*London*), *A*, 177 (1940) 62.
51 A. S. PORTER AND F. C. TOMPKINS, *Roy. Proc. Soc.* (*London*), *A*, 217 (1953) 544.
52 M. H. FRANCOMBE AND H. SATO (Eds.), *Single Crystal Thin Films*, Pergamon Press, Oxford, 1964.
53 D. W. PASHLEY, *Phil. Mag.*, 4 (1959) 316.
54 B. W. SLOOPE AND C. O. TILLER, *J. Appl. Phys.*, 33 (1962) 3458.
55 L. I. MAISSEL, *Physics of Thin Films*, Ed. G. HASS AND R. E. THUN, Academic Press, New York, 3 (1966) 61.
56 M. GREEN, J. A. KAFALAS AND P. H. ROBINSON, *Semiconductor Surface Physics*, Ed. R. H. KINGSTON, University of Pennsylvania Press, Philadelphia, 1957, p. 349.
57 T. A. VANDERSLICE AND N. R. WHETTEN, *J. Chem. Phys.*, 37 (1962) 535.
58 G. W. GOBELI AND F. G. ALLAN, *J. Phys. Chem. Solids*, 14 (1960) 23.
59 D. HANEMAN, *J. Phys. Chem. Solids*, 11 (1959) 205.
60 H. E. FARNSWORTH, *Ann. New York Acad. Sci.*, 101 (1963) 658.
61 P. J. BRYANT, *Ninth Nat. Symposium on Vacuum Technology*, American Vacuum Society, Macmillan, New York, 1962, p. 311.
62 R. E. SIMON, *Phys. Rev.*, 116 (1959) 613.
63 R. G. LYE, *Phys. Rev.*, 99 (1955) 1647A.
64 N. R. WHETTEN, *J. Appl. Phys.*, 34 (1963) 771.
65 R. H. KINDON AND I. LANGMUIR, *Phys. Rev.*, 22 (1923) 148.
66 C. W. OATLEY, *Proc. Phys. Soc.* (*London*), 51 (1939) 318.
67 A. E. J. EGGLETON AND F. C. TOMPKINS, *Trans. Faraday Soc.*, 48 (1952) 738.
68 G. K. WEHNER, *Advan. Electron. Electron Phys.*, 7 (1955) 239.
69 H. E. FARNSWORTH, R. E. SCHLIER, T. H. GEORGE AND R. M. BURGER, *J. Appl. Phys.*, 26 (1955) 252.
70 H. E. FARNSWORTH, *Advances in Catalysis*, 15 (1964) 31.

71 D. Haneman, *Phys. Rev.*, 119 (1960) 563.
72 G. S. Anderson, W. N. Mayer and G. K. Wehner, *J. Appl. Phys.*, 33 (1962) 2991.
73 W. B. Nottingham, *7th Ann. Conference Phys. Electronics*, M.I.T., Cambridge, Mass., March, 1947.
74 R. T. Bayard and D. Alpert, *Rev. Sci. Instr.*, 21 (1950) 571.
75 S. Dushman and J. M. Lafferty, *Scientific Foundations of Vacuum Technique*, Wiley, New York, 1962.
76 M. Pirani and J. Yarwood, *Principles of Vacuum Engineering*, Chapman and Hall, London, 1961.
77 R. W. Roberts and T. A. Vanderslice, *Ultra High Vacuum and its Applications*, Prentice Hall, New York, 1963.
78 J. H. Leck, *Pressure Measurement in Vacuum Systems*, Chapman and Hall, London, 1964.
79 J. M. Lafferty and T. A. Vanderslice, *Proc. Inst. Radio Engrs.*, 49 (1961) 1136.
80 A. S. Porter and F. C. Tompkins, *Proc. Roy. Soc. (London), A*, 217 (1953) 529.
81 W. E. Garner, F. S. Stone and P. F. Tiley, *Proc. Roy. Soc. (London), A*, 211 (1952) 472.
82 D. P. Marcus and A. Syverson, *Ind. Eng. Chem., Process Design Develop.*, 5 (1966) 397.
83 S. Yu. Elovich and G. M. Zhabrova, *Zh. Fiz. Khim.*, 13 (1939) 1761.
84 J. A. Becker and C. D. Hartman, *J. Phys. Chem.*, 57 (1953) 153.
85 J. Eisinger, *J. Chem. Phys.*, 30 (1959) 412.
86 T. W. Hickmott, *J. Appl. Phys.*, 31 (1960) 128.
87 G. Ehrlich, *J. Chem. Phys.*, 34 (1961) 29, 39.
88 G. Ehrlich, *Advances in Catalysis*, 14 (1963) 255.
89 H. D. Hagstrum, *Rev. Sci. Instr.*, 24 (1953) 1122.
90 T. W. Hickmott and G. Ehrlich, *J. Phys. Chem. Solids*, 5 (1958) 47.
91 R. W. Roberts, *Ann. New York Acad. Sci.*, 101 (1963) 766.
92 D. O. Hayward, N. Taylor and F. C. Tompkins, *Discussions Faraday Soc.*, 41 (1966) 75.
93 J. M. Lafferty, *J. Appl. Phys.*, 22 (1951) 299.
94 T. W. Hickmott, *Bull. Am. Phys. Soc.*, 3 (1958) 259.
95 T. W. Hickmott, *J. Chem. Phys.*, 32 (1960) 810.
96 T. N. Rhodin, *Advances in Catalysis*, 5 (1953) 39.
97 E. A. Gulbransen, *Advances in Catalysis*, 5 (1953) 119.
98 J. M. Thomas and B. R. Williams, *Quart. Rev. (London)*, 19 (1950) 231.
99 J. W. McBain and A. Baker, *J. Am. Chem. Soc.*, 48 (1926) 690.
100 S. L. Madorsky, *Vacuum Microbalance Techniques*, Plenum Press, New York, 2 (1962) 47.
101 P. J. Anderson and R. F. Horlock, *Trans. Faraday Soc.*, 58 (1962) 1993.
102 J. W. McBain and H. G. Tanner, *Proc. Roy. Soc. (London), A*, 125 (1929) 579.
103 S. J. Gregg, *J. Chem. Soc. (London)*, (1955) 1438.
104 A. W. Czanderna and J. M. Honig, *Anal. Chem.*, 29 (1957) 1206.
105 J. M. Honig, *Vacuum Microbalance Techniques*, Plenum Press, New York, 1 (1961) 55.
106 A. W. Czanderna, *Vacuum Microbalance Techniques*, Plenum Press, New York, 4 (1965) 175.
107 S. P. Wolsky and E. J. Zdanuk, *Vacuum Microbalance Techniques*, Plenum Press, New York, 2 (1962) 37.
108 J. M. Thomas and J. A. Poulis, *Vacuum Microbalance Techniques*, Plenum Press, New York, 3 (1963) 15.
109 C. N. Cochran, *Vacuum Microbalance Techniques*, Plenum Press, New York, 1 (1961) 23.
110 S. P. Wolsky and E. J. Zdanuck, *Vacuum Microbalance Techniques*, Plenum Press, New York, 1 (1961) 35.
111 D. Shooter, *Vacuum Microbalance Techniques*, Plenum Press, New York, 3 (1963) 117.
112 A. Langer, *Vacuum Microbalance Techniques*, Plenum Press, New York, 4 (1965) 231.
113 E. A. Gulbransen, K. F. Andrew and F. A. Bassart, *Vacuum Microbalance Techniques*, Plenum Press, New York, 3 (1963) 179.
114 R. F. Walker, *Vacuum Microbalance Techniques*, Plenum Press, New York, 1 (1961) 87.
115 N. J. Carrera, R. F. Walker, C. A. Steggerda and W. M. Nalley, *Vacuum Microbalance Techniques*, Plenum Press, New York, 3 (1963) 153.
116 R. V. Culver and F. C. Tompkins, *Advances in Catalysis*, 11 (1959) 67.

117 N. Cusack, *The Electrical and Mechanical Properties of Solids*, Longmans-Green & Co., London, 1958, pp. 36, 55, 79.
118 A. J. Dekker, *Solid State Physics*, Prentice Hall, Englewood Cliffs, N.J., 1959, p. 220 *et seq.*
119 A. L. Reimann, *Thermionic Emission*, Chapman and Hall, London, 1934, Chap. 3.
120 J. A. Becker, *Advances in Catalysis*, 7 (1955) 135.
121 R. H. Fowler, *Phys. Rev.*, 38 (1931) 45.
122 R. Suhrman, *Advances in Catalysis*, 8 (1956) 751.
123 E. W. Müller, *Physik. Z.*, 37 (1936) 838.
124 R. Gomer, *Advances in Catalysis*, 7 (1955) 93.
125 R. Gomer, *Field Emission and Field Ionisation*, Harvard University Press, Cambridge, Mass., 1961.
126 R. H. Fowler and L. W. Nordheim, *Proc. Roy. Soc.* (*London*), *A*, 119 (1928) 173.
127 J. R. Arthur, Jr. and R. S. Hansen, *Ann. New York Acad. Sci.*, 101 (1963) 756.
128 R. Culver, J. Pritchard and F. C. Tompkins, *2nd Intern. Cong. on Surface Activity*, Butterworths, London, 2 (1957) 243.
129 R. C. L. Bosworth and E. K. Rideal, *Proc. Roy. Soc.* (*London*), *A*, 162 (1937) 1.
130 T. H. Delchar, A. Eberhagen and F. C. Tompkins, *J. Sci. Instr.*, 40 (1963) 105.
131 J. C. P. Mignolet, *Discussions Faraday Soc.*, 8 (1950) 326.
132 L. A. Rudnitskii, N. V. Kulkova and M. I. Temkin, *Kinetics Catalysis* (*USSR*) (*Eng. Transl.*), 5 (1962) 154.
133 K. W. Bewig and W. A. Zisman, *Solid Surfaces and the Gas–Solid Interface*, *Advan. Chem. Ser.*, American Chemical Society, Washington, D.C., No. 33 (1961) 100.
134 P. W. Selwood, *Magnetochemistry*, Interscience, New York, 1956.
135 J. A. Sabatka and P. W. Selwood, *J. Am. Chem. Soc.*, 77 (1955) 5799.
136 L. E. Moore and P. W. Selwood, *J. Am. Chem. Soc.*, 78 (1956) 697.
137 P. Weiss, *J. Phys.*, 6 (1907) 661.
138 R. E. Dietz and P. W. Selwood, *J. Chem. Phys.*, 35 (1961) 270.
139 Ref. 134, p. 279.
140 D. E. O'Reilly, *Advances in Catalysis*, 12 (1960) 31.
141 F. Bloch, W. W. Hansen and M. Packard, *Phys. Rev.*, 70 (1946) 474.
142 R. V. Pound and W. D. Knight, *Rev. Sci. Instr.*, 21 (1950) 219.
143 N. Bloembergen, E. M. Purcell and R. V. Pound, *Phys. Rev.*, 73 (1948) 679.
144 H. S. Gutowsky, L. H. Meyer and R. E. McClure, *Rev. Sci. Instr.*, 24 (1953) 644.
145 W. M. H. Sachtler, *J. Chem. Phys.*, 25 (1956) 751.
146 W. C. Dunlap, Jr,, *Methods of Experimental Physics*, Eds. K. Lark-Horovitz and V. A. Johnson, Vol. 6B, Chap. 7.2, Academic Press, New York, 1959.
147 L. Hartshorn, *Radio Frequency Measurements by Bridge and Resonance Methods*, Wiley, New York, 1943.
148 C. G. Montgomery, *Techniques of Microwave Measurements*, McGraw Hill, New York, 1947.
149 N. B. Hannay, *Semiconductors*, Reinhold, New York, 1959.
150 T. J. Gray, *Chemistry of the Solid State*, Ed. W. E. Garner, Butterworths, London, 1955, p. 123.
151 P. Aigrain and C. Dugas, *Z. Elektrochem.*, 56 (1952) 363.
152 K. Hauffe and H. J. Engell, *Z. Elektrochem.*, 56 (1952) 366.
153 P. B. Weisz, *J. Chem. Phys.*, 20 (1952) 1483; 21 (1953) 1531.
154 J. T. Law, Ref. 149, p. 676.
155 J. S. Anderson, *Proc. Roy. Soc.* (*London*), *A*, 185 (1946) 69.
156 W. E. Garner, T. J. Gray and F. S. Stone, *Proc. Roy. Soc.* (*London*), *A*, 197 (1949) 314.
157 E. J. Schiebner, L. H. Germer and C. D. Hartman, *Rev. Sci. Instr.*, 31 (1960) 112.
158 L. H. Germer and C. D. Hartman, *Rev. Sci. Instr.*, 31 (1960) 784.
159 J. J. Lander and J. Morrison, *Ann. New York Acad. Sci.*, 101 (1962) 627.
160 R. P. Eischens and W. A. Pliskin, *Advances in Catalysis*, 10 (1958) 2.
161 V. Crawford, *Quart. Rev.* (*London*), 14 (1960) 378.
162 H. P. Leftin and M. C. Hobson, Jr., *Advances in Catalysis*, 14 (1963) 115.
163 L. H. Little, *J. Phys. Chem.*, 63 (1959) 1616.
164 H. L. Pickering and H. C. Eckstrom, *J. Phys. Chem.*, 63 (1959) 512.

165 N. J. Harrick, *Ann. New York Acad. Sci.*, 101 (1963) 928.
166 J. Fahrenfort and H. F. Hazebroek, *Z. Physik. Chem. (Frankfurt) N.S.*, 20 (1959) 105.
167 A. N. Webb, *Proc. 2nd Intern. Congr. Catalysis*, Editions Technip, *Paris*, 1 (1961) 1289.
168 H. P. Leftin and W. K. Hall, *J. Phys. Chem.*, 66 (1962) 1457.
169 J. B. Peri and R. B. Hannan, *J. Phys. Chem.*, 64 (1960) 1526.
170 H. A. Benesi and A. C. Jones, *J. Phys. Chem.*, 63 (1959) 179.
171 C. E. O'Neill and D. J. C. Yates, *J. Phys. Chem.*, 65 (1961) 901.
172 J. B. Peri, *Discussions Faraday Soc.*, 41 (1966) 121.
173 H. P. Leftin, *Rev. Sci. Instr.*, 32 (1961) 1418.
174 E. G. Boreskova, V. I. Lygin and K. V. Topchieva, *Kinetics Catalysis (USSR) (Eng. Transl.)*, 5 (1964) 991.
175 B. V. Liengme and W. K. Hall, *Trans. Faraday Soc.*, 62 (1966) 3230.
176 L. H. Little, N. Sheppard and D. J. C. Yates, *Proc. Roy. Soc. (London)*, *A*, 259 (1960) 242.
177 F. G. Ciapetta and C. J. Plank, *Catalysis*, Ed. P. H. Emmett, Reinhold, New York, 1 (1954) 315.
178 V. I. Komarewsky, C. H. Riesz and F. L. Morritz, *Techniques of Organic Chemistry*, Ed. A. Weissberger, Vol. 2, Interscience, New York, 1956, p. 5.
179 L. B. Ryland, M. W. Tamele and J. N. Wilson, *Catalysis*, Ed. P. H. Emmett, Reinhold, New York 7 (1960) 1.
180 *U.S. Patents*, 2882243, 2882244; *British Patents*, 777233, 909264, 909266.
181 M. Raney, *Ind. Eng. Chem.*, 32 (1940) 1199.
182 W. B. Innes, *Catalysis*, Ed. P. H. Emmett, Reinhold, New York, 1 (1954) 245.
183 A. Wheeler, *Advances in Catalysis*, 3 (1951) 250.
184 A. Wheeler, *Catalysis*, Ed. P. H. Emmett, Reinhold, New York, 2 (1955) 105.
185 O. A. Hougen and R. M. Watson, *Chemical Process Principles*, in *Kinetics and Catalysis*, Wiley, New York, 3 (1947).
186 O. A. Hougen, *Ind. Eng. Chem.*, 53 (1961) 509.
187 C. N. Satterfield and T. K. Sherwood, *The Role of Diffusion in Catalysis*, Addison-Wesley, Reading, Mass., 1963.
188 O. Levenspiel, *Chemical Reaction Engineering*, Wiley, New York, 1962.
189 D. Tholnese and H. Kramers, *Chem. Eng. Sci.*, 8 (1958) 271.
190 N. Wakao and J. M. Smith, *Chem. Eng. Sci.*, 17 (1962) 825.
191 J. L. Robertson and J. M. Smith, *Assoc. Inst. Chem. Eng. J.*, 9 (1963) 342.
192 M. Raja Rao and J. M. Smith, *Assoc. Inst. Chem. Eng. J.*, 9 (1963) 485.
193 P. B. Weisz and C. D. Prater, *Advances in Catalysis*, 6 (1954) 143.
194 E. W. Thiele, *Ind. Eng. Chem.*, 31 (1939) 916.
195 P. B. Weisz and J. S. Hicks, *Chem. Eng. Sci.*, 17 (1962) 265.
196 M. R. Jottrand, *Genie Chim.*, 95 (1966) 661.
197 G. W. Roberts and C. N. Satterfield, *Ind. Eng. Chem. Fundamentals*, 5 (1966) 317.
198 V. W. Weekman, Jr., *J. Catalysis*, 5 (1966) 44.
199 G. Wurzbacher, *J. Catalysis*, 5 (1966) 476.
200 Sir H. Melville and B. G. Gowenlock, *Experimental Methods in Gas Reactions*, MacMillan, London, 1964.
201 G. K. Boreskov, *Kinetics Catalysis (USSR) (Eng. Transl.)*, 3 (1962) 416.
202 M. J. Temkin, *Kinetics Catalysis (USSR) (Eng. Transl.)*, 3 (1962) 448.
203 G. P. Korneichuk, *Kinetics Catalysis (USSR) (Eng. Transl.)*, 3 (1962) 454.
204 I. P. Siderov, D. B. Kazarnovskaya and P. P. Andreichev, *Kinetics Catalysis (USSR) (Eng. Transl.)*, 3 (1962) 458.
205 G. I. Levi and V. E. Vasserberg, *Kinetics Catalysis (USSR) (Eng. Transl.)*, 3 (1962) 462.
206 J. J. Carberry, *Ind. Eng. Chem.*, 56 (1964) 36.
207 B. Gillespie and J. J. Carberry, *Ind. Eng. Chem. Fundamentals*, 5 (1966) 164.
208 D. V. Land, *J. Sci. Instr.*, 42 (1965) 444.
209 R. P. Chambers, N. A. Dougharty and M. Boudart, *J. Catalysis*, 4 (1965) 625.
210 C. E. Schwartz and J. M. Smith, *Ind. Eng. Chem.*, 45 (1953) 1209.
211 H. Kramers and K. R. Westertop, *Chemical Reactor Design and Operation*, Chapman and Hall, London, 1963, p. 225.

212 S. P. MUKHERJEE AND L. K. DORAISWAMY, *Brit. Chem. Eng.*, 10 (1965) 93.
213 F. E. FORD AND D. D. PERLMUTTER, *Chem. Eng. Sci.*, 19 (1964) 371.
214 D. J. TAJBL, J. B. SIMONS AND J. J. CARBERRY, *Ind. Eng. Chem. Fundamentals*, 5 (1966) 171.
215 R. J. KOKES, H. TOBIN, JR. AND P. H. EMMETT, *J. Am. Chem. Soc.*, 77 (1955) 5860.
216 P. H. EMMETT, *Advances in Catalysis*, 9 (1957) 645.
217 P. STEINGASZNER AND H. PINES, *J. Catalysis*, 5 (1966) 356.
218 G. M. SCHWAB AND A. M. WATSON, *J. Catalysis*, 4 (1965) 570.
219 J. W. HIGHTOWER, H. R. GERBERICH AND W. K. HALL, *J. Catalysis*, 7 (1966) 57.
220 W. K. HALL AND P. H. EMMETT, *J. Am. Chem. Soc.*, 79 (1957) 2091.
221 T. I. TAYLOR, *Catalysis*, Ed. P. H. EMMETT, Reinhold, New York, 5 (1957) 257.
222 P. H. EMMETT, P. SABATIER AND E. E. REID, *Catalysis Then and Now*, Franklin, Englewood, N. J., 1965, p. 69.
223 Ref. 200, p. 179.
224 Ref. 3, p. 149.
225 B. M. W. TRAPNELL, *Catalysis*, Ed. P. H. EMMETT, Reinhold, New York, 3 (1955) 1.
226 D. D. ELEY, *Proc. Roy. Soc. (London), A*, 178 (1941) 452.
227 C. KEMBALL, *Advances in Catalysis*, 11 (1959) 223.
228 Ref. 3, p. 183, 217.
229 Ref. 200, p. 184.
230 L. FARKAS AND H. SACHSSE, *Z. Physik. Chem.*, B23 (1933) 1.
231 A. FARKAS, *Ortho-hydrogen, Para-hydrogen and Heavy Hydrogen*, Cambridge University Press, Cambridge, England, 1935.
232 D. D. ELEY AND E. K. RIDEAL, *Proc. Roy. Soc. (London), A*, 178 (1941) 452.
233 D. D. ELEY AND P. R. NORTON, *Discussions Faraday Soc.*, 41 (1966) 135.
234 W. R. ALCORN AND J. K. SHERWOOD, *J. Catalysis*, 6 (1966) 288.
235 H. C. UREY AND G. K. TEAL, *Rev. Mod. Phys.*, 6 (1935) 34.
236 Ref. 200, p. 235.
237 D. H. WEITZEL AND L. E. WHITE, *Rev. Sci. Instr.*, 26 (1955) 290.
238 W. R. MOORE AND H. R. WARD, *J. Phys. Chem.*, 64 (1960) 832.
239 P. P. HUNT AND H. A. SMITH, *J. Phys. Chem.*, 65 (1961) 87.
240 M. VENUGOPLAN AND K. O. KUTSCHKE, *Can. J. Chem.*, 41 (1963) 548.
241 W. R. MOORE AND H. R. WARD, *J. Am. Chem. Soc.*, 80 (1958) 2909.
242 L. BACHMANN, E. BECHTOLD AND E. CREMER, *J. Catalysis*, 1 (1962) 113.
243 E. R. S. WINTER, *Advances in Catalysis*, 10 (1958) 196.
244 G. K. BORESKOV, *Advances in Catalysis*, 15 (1964) 285.
245 A. P. DZISJAK, G. K. BORESKOV, L. A. KASATKINA AND V. E. KOCHURIKHIN, *Kinetics Catalysis (USSR) (Eng. Transl.)*, 2 (1961) 355.
246 R. J. OVERMAN AND H. M. CLARK, *Radioisotope Techniques*, McGraw-Hill, New York, 1960.
247 P. E. YANKWICH, *Ann. Rev. Nucl. Sci.*, 3 (1953) 235.
248 C. L. COMAR, *Radioisotopes in Biology and Agriculture*, McGraw-Hill, New York, 1955.
249 B. J. WILSON (Ed.), *The Radiochemical Manual*, The Radiochemical Centre, Amersham, 1966.
250 E. J. WILSON, *Vacuum*, 4 (1954) 303.
251 R. I. REED (Ed.), *Mass Spectrometry*, Academic Press, London, 1965.
252 J. B. ROBERTSON, *Mass Spectrometry*, Methuen, London, 1954.
253 J. H. BEYNON, *Mass Spectrometry and its Applications to Organic Chemistry*, Elsevier, Amsterdam, 1960.
254 *Advances in Mass Spectrometry*, Pergamon Press, Oxford, from 1959.
255 *Ann. Symp. Am. Vacuum Society*, Pergamon Press, Oxford, from 1956.
256 W. D. DAVIS AND T. A. VANDERSLICE, *7th Nat. Symp. on Vacuum Technology*, American Vacuum Society, Pergamon Press, Oxford, 1961, p. 417.
257 H. SOMMER, H. A. THOMAS AND J. A. HIPPLE, *Phys. Rev.*, 82 (1951) 697.
258 D. ALPERT AND R. S. BURITZ, *J. Appl. Phys.*, 25 (1954) 28.
259 J. S. WAGENER AND P. T. MARTH, *J. Appl. Phys.*, 28 (1957) 1027.
260 D. CHARLES AND R. J. WARNECKE, JR., *6th Nat. Symp. on Vacuum Technology*, American Vacuum Society, Pergamon Press, Oxford, 1960, p. 34.
261 A. KLOPFER AND W. SCHMIDT, *Vacuum*, 10 (1960) 363.

262 E. J. Zdanuk, R. Bierig, L. G. Rubin and S. P. Wolsky, *Vacuum*, 10 (1960) 382.
263 G. R. Giedd and G. C. Roberts, *J. Sci. Instr.*, 38 (1961) 361.
264 Von G. Krause, *Vakuum Technik*, 11 (1962) 50.
265 D. Shooter and J. Welts, unpublished results.
266 W. A. Bennett, *J. Appl. Phys.*, 21 (1950) 143.
267 P. A. Redhead and C. R. Crowell, *J. Appl. Phys.*, 24 (1953) 331.
268 P. F. Váradi, L. G. Sebestyen and E. Rieger, *Vakuum Technik*, 7 (1958) 13.
269 P. F. Váradi, *7th Nat. Symp. on Vacuum Technology*, American Vacuum Society, Pergamon Press, Oxford, 1961, p. 149.
270 H. W. Ehlbeck, K. H. Loecherer, J. Ruf and H. J. Schütze, *7th Nat. Symp. on Vacuum Technology*, American Vacuum Society, Pergamon Press, Oxford, 1961, p. 407.
271 H. W. Ehlbeck, J. Ruf and H. J. Schütze, *8th Nat. Symp. on Vacuum Technology*, American Vacuum Society, Pergamon Press, Oxford, 1962, p. 567.
272 G. Reich, *7th Nat. Symp. on Vacuum Technology*, American Vacuum Society, Pergamon Press, Oxford, 1961, p. 396.
273 W. Bächler and G. Reich, *7th Nat. Symp. on Vacuum Technology*, American Vacuum Society, Pergamon Press, Oxford, 1961, p. 401.
274 W. Tretner, *Z. Angew. Phys.*, 11 (1959) 395.
275 W. Bleakney and J. A. Hipple, *Phys. Rev.*, 53 (1938) 521.
276 W. K. Huber and E. A. Trendelenburg, *8th Nat. Symp. on Vacuum Technology*, American Vacuum Society, Pergamon Press, Oxford, 1962, p. 592.
277 W. Paul and M. Raether, *Z. Physik*, 140 (1955) 262.
278 C. Wagner, *Z. Physik. Chem.*, B21 (1933) 25.
279 C. Wagner, *Z. Physik. Chem.*, B32 (1936) 447.
280 N. F. Mott, *J. Chim. Phys.*, 44 (1947) 172.
281 N. F. Mott, *Trans. Faraday Soc.*, 43 (1947) 429.
282 T. B. Grimley, *Chemistry of the Solid State*, Ed. W. E. Garner, Butterworths, London, 1955, p. 336.
283 O. Kubaschewski and B. E. Hopkins, *The Oxidation of Metals and Alloys*, Butterworths, London, 1962.
284 M. Wyn Roberts, *Quart. Rev. (London)*, 16 (1962) 71.
285 E. A. Gulbransen and K. F. Andrew, *Vacuum Microbalance Techniques*, Plenum Press, New York, 1 (1961) 1.
286 W. C. Tripp, R. W. Vest and N. M. Tallan, *Vacuum Microbalance Techniques*, Plenum Press, New York, 4 (1965) 141.
287 E. A. Gulbranson and K. F. Andrew, *Trans. Electrochem. Soc.*, 96 (1949) 364.
288 E. A. Gulbranson and K. F. Andrew, *J. Electrochem. Soc.*, 110 (1963) 476.
289 T. N. Rhodin, *J. Am. Chem. Soc.*, 72 (1950) 5102.
290 R. H. Condit and J. B. Holt, *Reactivity of Solids*, Ed. G. M. Schwab, Elsevier, Amsterdam, 1965, p. 334.
291 R. E. Carter and F. D. Richardson, *Trans. A.I.M.E.*, 203 (1955) 336.
292 M. T. Simnad and A. Spilners, *Trans. A.I.M.E.*, 203 (1955) 1011.
293 J. A. Davies, J. P. S. Pringle, R. L. Graham and F. Brown, *J. Electrochem. Soc.*, 109 (1962) 999.
294 J. A. Davies and B. Domeij, *J. Electrochem. Soc.*, 110 (1963) 849.
295 W. E. Garner, *Chemistry of the Solid State*, Ed. W. E. Garner, Butterworths, London, 1955, Chaps. 8, 9.
296 H. T. S. Britton, S. J. Gregg and G. W. Winsor, *Trans. Faraday Soc.*, 48 (1952) 63, 70.
297 L. L. Bircumshaw and J. Edwards, *J. Chem. Soc. (London)*, (1950) 1800.
298 L. L. Bircumshaw and F. M. Tayler, *J. Chem. Soc. (London)*, (1950) 3674.
299 L. L. Bircumshaw and T. R. Phillips, *J. Chem. Soc. (London)*, (1953) 703.
300 L. L. Bircumshaw and B. H. Newman, *Proc. Roy. Soc. (London)*, *A*, 227 (1954) 115.
301 H. Schafer, *Chemical Transport Reactions*, Academic Press, New York, 1964.
302 H. Schafer and H. J. Heitland, *Z. Anorg. Allgem. Chem.*, 304 (1960) 249.
303 H. Schafer, W. Gerhardt and A. Tebben, *Angew. Chem.*, 73 (1961) 27, 115.

304 Z. M. Shapiro, cited in B. Lustman and F. Kerse, *The Metallurgy of Zirconium*, McGraw-Hill, New York, 1955.
305 T. T. Magel, *U.S. Patent*, 2873108 (1947).
306 H. Schafer and B. Morcher, *Z. Anorg. Allgem. Chem.*, 290 (1957) 279.
307 L. G. Harrison, J. A. Morrison and G. S. Rose, *2nd Intern. Cong. on Surface Activity*, Butterworths, London, 2 (1957) 287.
308 C. Clusius and H. Haimerl, *Z. Physik. Chem.*, 51B (1942) 347.
309 L. G. Harrison, R. J. Adams, M. D. Baijal and D. J. Bird, *Reactivity of Solids*, Ed. G. M. Schwab, Elsevier, Amsterdam, 1965, p. 279.
310 S. J. Gregg and H. F. Leach, *Reactivity of Solids*, Ed. G. M. Schwab, Elsevier, Amsterdam, 1965, p. 303.
311 D. C. Grahame, *Chem. Rev.*, 41 (1947) 441.
312 T. I. Borisova, B. V. Ershler and A. N. Frumkin, *Zh. Fiz. Khim.*, 22 (1948) 925.
313 D. C. Grahame, *J. Am. Chem. Soc.*, 63 (1941) 1207.
314 A. Frumkin and V. I. Melik-Gaikasjan, *Dokl. Akad. Nauk SSSR*, 77 (1951) 855.
315 T. Berzins and P. Delahay, *J. Phys. Chem.*, 59 (1955) 906.
316 A. Watanabe, F. Tsuji and S. Ueda, *2nd Intern. Congress on Surface Activity*, Butterworths, London, 3 (1957) 94.
317 Ref. 187, p. 43.
318 D. V. Sokol'skii, *Hydrogenation in Solution*, Israel Program for Scientific Translation, Jerusalem, 1964.
319 R. M. Noyes, *Progress in Reaction Kinetics*, Ed. G. Porter, Pergamon Press, Oxford, 1 (1961) 129.
320 H. J. V. Tyrrell, *Diffusion and Heat Flow in Liquids*, Butterworths, London, 1961.
321 A. M. North, *Quart. Rev. (London)*, 20 (1966) 421.
322 C. R. Wilke and P. Chang, *Am. Inst. Chem. Eng. J.*, 1 (1955) 264.
323 K. A. Reddy and L. K. Doraiswamy, *Ind. Eng. Chem. Fundamentals*, 6 (1967) 77.
324 S. K. Friedlander, *Am. Inst. Chem. Eng. J.*, 7 (1961) 347.
325 Van Name and G. Edgar, *Am. J. Sci.*, 29 (1914) 237.
326 L. L. Bircumshaw and M. H. Everdell, *J. Chem. Soc. (London)*, (1942) 598.
327 F. K. Bell and W. A. Patrick, *J. Am. Chem. Soc.*, 43 (1921) 452.
328 C. V. King and C. L. Liu, *J. Am. Chem. Soc.*, 55 (1933) 1928.
329 W. G. Palmer and R. E. D. Clark, *Proc. Roy. Soc. (London)*, *A*, 149 (1935) 360.
330 R. S. Bradley, *Trans. Faraday Soc.*, 34 (1938) 278.
331 I. I. Tingley, I. H. S. Henderson and C. C. Coffin, *Can. J. Chem.*, 34 (1956) 14.
332 T. L. Smith, *J. Phys. Chem.*, 59 (1955) 385.
333 R. M. Barrer, *Diffusion in and through Solids*, Cambridge University Press, Cambridge, 1951.
334 W. Jost, *Diffusion in Solids, Liquids, Gases*, Academic Press, New York, 1952.
335 G. F. Hüttig, *Handbuch der Catalyse*, Ed. G. M. Schwab, Springer-Verlag, Wien, 6 (1943) 318.
336 G. Cohn, *Chem. Rev.*, 42 (1947) 527.
337 A. J. E. Welch, *Chemistry of the Solid State*, Ed. W. E. Garner, Butterworths, London, 1955, p. 297.
338 W. Jander and J. Petri, *Z. Elektrochem.*, 44 (1938) 747.
339 E. J. Weiss and R. A. Rowland, *Am. Minerologist*, 41 (1956) 117.
340 R. A. Rowland, E. V. Weiss and D. R. Lewis, *J. Am. Ceram. Soc.*, 42 (1959) 133.
341 W. C. McCrone, *Fusion Methods in Chemical Microscopy*, Wiley, New York, 1957.
342 H. Wagenblast, F. E. Fujita and A. C. Damask, *Acta Met.*, 12 (1964) 347.
343 H. B. van der Heyde, F. van de Craats, P. H. G. van Kasteren and A. de Wit, *Reactivity of Solids*, Ed. G. M. Schwab, Elsevier, Amsterdam, 1965, p. 164.
344 G. C. Kuczynski, *Reactivity of Solids*, Ed. G. M. Schwab, Elsevier, Amsterdam, 1965, p. 352.
345 D. R. Reese, P. N. Nordberg, S. P. Eriksen and J. V. Swintosky, *J. Pharm. Sci.*, 50 (1961) 177.
346 R. C. Mackenzie and B. D. Mitchell, *Analyst*, 87 (1962) 420.
347 C. B. Murphy, *Anal. Chem.*, 36 (1964) 347.
348 P. D. Garn, *Thermoanalytical Methods of Investigation*, Academic Press, New York, 1965.

349 W. J. SMOTHERS AND Y. CHIANG, *Handbook of Differential Thermal Analysis*, Chemical Publishing Co., New York, 1966.
350 H. E. KISSINGER, *Anal. Chem.*, 29 (1957) 1702.
351 H. J. BORCHARDT AND F. DANIELS, *J. Am. Chem. Soc.*, 79 (1957) 41.
352 E. S. FREEMAN AND B. CARROLL, *J. Phys. Chem.*, 62 (1958) 394.
353 W. W. WENDLANDT, *J. Chem. Educ.*, 38 (1961) 571.
354 H. J. BORCHARDT, *J. Inorg. Nucl. Chem.*, 12 (1960) 113.
355 P. D. GARN AND S. S. FLASCHEN, *Anal. Chem.*, 29 (1957) 268.
356 F. E. FUJITA AND A. C. DAMASK, *Acta Met.*, 12 (1964) 331.
357 K. JAKY, F. SOLYMOSI, I. BATTA AND Z. G. SZABO, *Reactivity of Solids*, Ed. G. M. SCHWAB, Elsevier, Amsterdam, 1965, p. 540.
358 L. J. E. HOFER, E. M. COHN AND W. C. PEEBLES, *J. Phys. Colloid Chem.*, 53 (1949) 661.
359 G. CHAUDRON AND J. BENARD, *Bull. Soc. Chim. France*, (1949) D117.
360 Ref. 134, p. 376 *et seq.*
361 J. TURKEVICH, S. LARACH AND P. N. YOCOM, *Reactivity of Solids*, Ed. G. M. SCHWAB, Elsevier, Amsterdam, 1965, p. 540.
362 H. A. SAUER, see ref. 348, p. 374.
363 R. G. LOASBY, *J. Sci. Instr.*, 38 (1961) 306.
364 A. LIEBERMAN AND W. B. CRANDALL, *J. Am. Ceram. Soc.*, 35 (1952) 304.
365 M. L. DANNIS, *J. Appl. Polymer Sci.*, 4 (1960) 249.
366 F. S. STONE AND R. J. D. TILLEY, *Reactivity of Solids*, Ed. G. M. SCHWAB, Elsevier, Amsterdam, 1965, p. 583.
367 W. L. DE KEYSER, R. WOLLAST, P. HANSEN AND G. NAESSENS, *Reactivity of Solids*, Ed. G. M. SCHWAB, Elsevier, Amsterdam, 1965, p. 658.
368 H. WAGENBLAST AND A. C. DAMASK, *J. Phys. Chem. Solids*, 23 (1962) 221.
369 C. A. WERT AND C. ZENER, *J. Appl. Phys.*, 21 (1950) 5.
370 C. A. WERT, *Modern Research Techniques in Metallurgy*, American Society of Metals, 1953, p. 225.
371 J. T. DAVIES, *Advances in Catalysis*, 6 (1954) 1.
372 A. E. ALEXANDER, *Techniques of Organic Chemistry*, Ed. A. WEISSBERGER, Vol. 1, Interscience, New York, 1959, p. 727.
373 J. T. DAVIES AND E. K. RIDEAL, *Interfacial Phenomena*, Academic Press, New York, 1963.
374 G. L. GAINES, JR., *Insoluble Monolayers at Liquid–Gas Interfaces*, Interscience, New York, 1966.
375 R. MATALON AND J. H. SCHULMAN, *J. Colloid Sci.*, 4 (1949) 89.
376 R. MATALON, *J. Colloid Sci.*, 8 (1953) 53.
377 J. H. SCHULMAN AND E. K. RIDEAL, *Proc. Roy. Soc.* (*London*), *A*, 130 (1931) 259.
378 H. G. YAMINS AND W. A. ZISMAN, *J. Chem. Phys.*, 1 (1933) 656.
379 R. F. ROBERTSON, C. A. WINKLER AND S. G. MASON, *Can. J. Chem.*, 34 (1956) 716.
380 W. RABINOVICH, R. F. ROBERTSON AND S. G. MASON, *J. Colloid Sci.*, 13 (1959) 600.
381 S. G. MASON AND W. RABINOVICH, *Proc. Roy. Soc.* (*London*), *A*, 249 (1959) 90.

Chapter 4

The Detection and Estimation of Intermediates

R. P. WAYNE

1. Introduction

Although it may be possible to represent any chemical process by an equation showing the reactants on the one side, and the final products on the other, yet it is clear that in very many cases the reaction proceeds through more than one discrete stage. In such processes, the reactants in the intermediate steps are species different from the original reactants, and the present chapter is concerned with the way in which the nature of the intermediate species, and its concentration, may be determined. Information about the intermediate may be of considerable importance in the elucidation of the reaction mechanism, and, in particular, a knowledge of the concentration of the intermediate may be required to establish the kinetic details of the reaction.

The intermediates of greatest importance, and with which we are concerned here, are (*i*) more or less stable molecular species, (*ii*) free radicals and atoms, (*iii*) excited molecules, free radicals or atoms and (*iv*) ions. If, in fact, the intermediate is a stable substance, then it is amenable to analysis by any of the techniques applicable to chemical investigations, and the problem may be one of isolation of the intermediate (*i.e.* preventing its participation in the processes which normally follow its formation) rather than one of analysis. For this reason, we shall discuss in Section 1.1 some methods by which stable intermediates have been isolated from chemical reactions. If isolation is not possible, then the methods available for the study of stable intermediates are identical with those used to study unstable species. Unstable intermediates will be discussed briefly in Sections 1.2–1.4, with reference to the systems in which they are formed. Subsequent sections of this chapter will then explore a number of particular experimental techniques, and describe the ways in which the technique has been used to investigate the labile intermediates of the types discussed in Sections 1.2–1.4. Each method is not necessarily applicable to all three kinds of labile species, or, indeed, to all members of any one class. Again, the methods may be restricted further with respect to the phase in which the reaction is conducted: clearly mass spectrometry is not suited to the investigation of short-lived species in solution. Examples have been chosen, therefore, with regard to the way in which they illustrate the applications of a method: the author hopes he has chosen well!

References pp. 336–342

Cases arise where the method of investigation uses techniques of more than one category. This is particularly true of the "chemical" methods described in Section 5, in which the effect of the active intermediate on foreign reactants is studied, for example, by mass spectrometry or optical absorption or emission spectroscopy. The principle process involved in the estimation of the intermediate is, however, a chemical one, and this determines the heading under which the method is described.

Certain general observations may, perhaps, be made at this stage. If a reaction is in a steady state, then it is usually valid to assume that the concentration of reaction intermediates is also stationary. The high reactivity of many intermediate species results in low stationary concentrations unless the rate of intermediate formation is unusually high (*e.g.* as it is in flames). In order, therefore, to study reactive intermediates, a number of special experimental techniques have been developed. Flash photolysis may be used isothermally to produce high intermediate concentrations photochemically, or adiabatically to produce the intermediates thermally, and the rise and decay of intermediate species may be followed by one or other of the techniques available (in this case spectroscopy is normally used). Again, shock waves may be used to produce large concentrations of reactive species by thermal decomposition of reactants. Such studies necessitate analytical methods for the intermediates which have a response comparable with the lifetime of the intermediate. The non-stationary techniques have been complemented by steady-state methods, such as flow techniques, designed to follow fast reactions. The active species —no longer truly "intermediates"—may be produced in high concentration within the apparatus, and used as the reactants in suitable chemical processes. Isolation of individual steps has proved a powerful aid to the understanding of complex chemical reactions, but the problem remains of the determination of the active species involved.

It is often desirable to investigate the intermediates without appreciably disturbing the reaction system. Methods such as optical spectroscopy, or, to a lesser degree, mass spectrometry, which affect only a small fraction of the total number of species present, satisfy this requirement. On the other hand, the chemical methods of Section 5 rely directly on the participation of intermediates in processes not concerned with the original reaction. In general, then, allowance must be made for the perturbation caused by the foreign substance, or the experiment designed so that the results are applicable to the system free from additive.

1.1 STABLE INTERMEDIATES

A number of reactions proceed *via* intermediates which would be recognised as stable species outside the reaction system. Typical of these are the aldehydes and peroxides formed in the combustion of hydrocarbons. If during the course of the

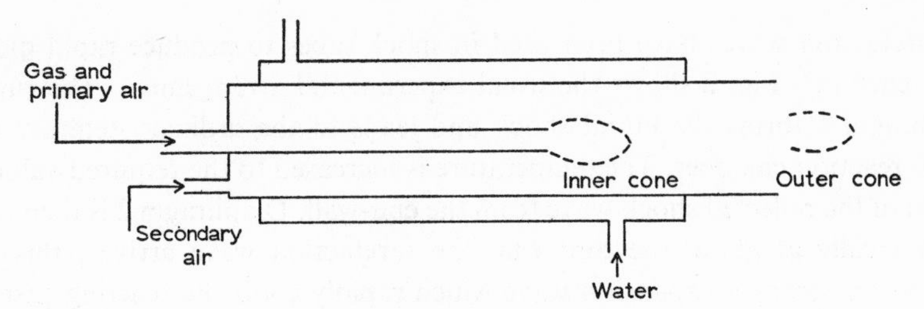

Fig. 1. Separated flame burner.

reaction the rate of reaction of intermediate can be reduced drastically by physical means—reduction of the pressure or temperature, or both—or by chemical quenching (addition of an inhibitor), it may prove possible to extract quantities of intermediate sufficient for characterization. Smithells and Ingle[1] suggested in 1892 that if the inner and outer cones of a Bunsen flame were separated out, then it would be possible to determine the intermediate substances formed in the inner cone by analysis of the gases in the outer cone. The separated flame technique has been perfected by Gaydon[2,3], who produced "struck-back" hydrocarbon flames in a burner consisting of two concentric water-cooled steel tubes (Fig. 1). The rapid chilling quenches the interconal gases, and Gaydon was able to show by chemical analysis of the collected products that they contain large amounts of both formaldehyde and peroxidic substances, together with some acid.

Direct "on-line" analytical techniques may also be used. Typical of these studies is that made by Bradley *et al.*[4] of stabilised cool flames in the oxidation of acetaldehyde and propionaldehyde. A fine quartz probe was attached to an A.E.I. MS 10 mass spectrometer, and could be moved through the cool flame. In this way composition profiles were obtained for reactant aldehyde and oxygen and also for carbon dioxide, carbon monoxide, formaldehyde, methane and methanol.

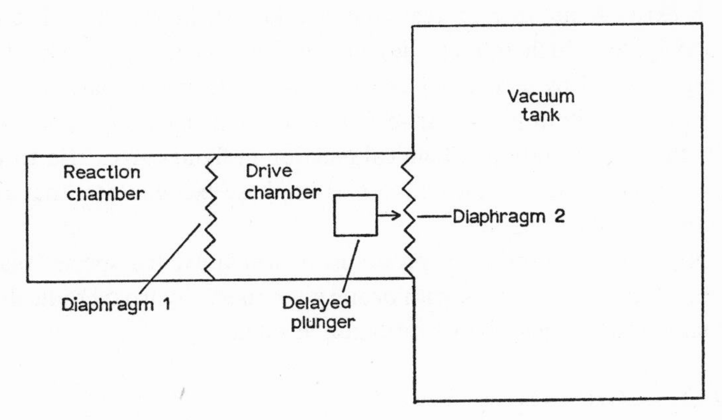

Fig. 2. Chemical shock tube.

References pp. 336–342

Rarefaction waves have been used in shock tubes to produce rapid quenching of a reaction[5]. Fig. 2 shows the usual experimental arrangement. The rupture of diaphragm 1 forms the initial shock and leads to the ordinary temperature rise in the reaction chamber. The temperature is increased to the required value by the return of the reflected shock wave from the end-wall. Diaphragm 2 is then ruptured mechanically at about the time that the rarefaction wave arrives; this rupture produces a stronger expansion wave which rapidly cools the reacting gases.

1.2 FREE RADICALS AND ATOMS

Of those intermediates susceptible to quantitative determination, perhaps the most important class is made up of free radicals and atoms. Although most members of the class have one unpaired electron and are therefore in doublet states, such species as atomic oxygen (^3P in the ground state, ^1D in the first electronically excited state) and methylene (also a triplet in the ground state) belong, generically, in the same category.

Free radicals and atoms appear as intermediates in a wide range of processes. Almost all photochemical reactions initiated by radiation in the visible or near ultraviolet regions of the spectrum are propagated by radicals or atoms (but see Section 1.4 for the photochemical formation of ions). Similarly, combustion processes are believed to occur predominantly by a radical mechanism, as are a large number of other thermal reactions.

1.3 EXCITED SPECIES

Molecules, free radicals, atoms, and, indeed, ions, may possess excess energy by virtue of excitation in electronic, vibrational, rotational or translational modes where such exist. A number of reactions are known in which such energy-rich species participate. Much interest lies in the fundamental problem of how the excess energy carried by a reactant can overcome the energy barrier to reaction.

The energy of excitation may arise from chemical reaction, by absorption of light, or by thermal excitation. Chemical reaction in flames gives rise to electronically excited species, and it is the emission from these excited states that give rise to the characteristic "flame bands"[3].

Ozone enters into a number of processes in which excited species play a part. The reaction of atomic hydrogen with ozone gives rise to hydroxyl radicals of which an appreciable fraction may be vibrationally excited[6]

$$H+O_3 \rightarrow OH^{\dagger}+O_2 \qquad (1)$$

It is possible that the vibrationally excited hydroxyl radicals may decompose ozone[7], and vibrationally excited nitrogen may also decompose ozone[8,9]

$$N_2^\dagger + O_3 \rightarrow N_2 + O_2 + O \tag{2}$$

although in both these cases there is some danger of misinterpretation of the results[10].

A process of great importance in the atmosphere is the ultraviolet photolysis of ozone, and the reaction involves[11] a number of excited species, $O(^1D)$, $O_2(^1\Delta_g)$ and O_2^*

$$O_3 + h\nu \rightarrow O(^1D) + O_2(^1\Delta_g) \tag{3}$$
$$O(^1D) + O_3 \rightarrow O_2^* + O_2 \tag{4}$$
$$O_2^* + O_3 \rightarrow 2O_2 + O(^1D) \tag{5}$$
$$O_2(^1\Delta_g) + O_3 \rightarrow 2O_2 + O(^3P) \tag{6}$$
$$O(^3P) + O_3 \rightarrow 2O_2 \tag{7}$$

all of which react with ozone. O_2^* may possibly be a vibrationally excited molecule of oxygen.

Translationally "hot" species may also have enhanced reaction rates. Reuben and Linnett[12] have proposed that such "hot" oxygen atoms are formed in the thermal decomposition of nitrous oxide, and that these react at an accelerated rate with further nitrous oxide. Williams and Ogg[13] have shown that hydrogen atoms with 42 kcal.mole^{-1} of translational energy react at an enhanced rate with hydrogen iodide, and have demonstrated a similar effect in the reactivity of "hot" methyl radicals.

Spectroscopic methods of identification and estimation of excited species are clearly of great importance, although several other techniques have been employed (mass spectrometric, chemical, calorimetric, *etc.*, and, for radicals, ESR).

1.4 IONS

Many inorganic and organic reactions in solution occur *via* ionic or semi-ionic intermediates. However, the stationary concentration of the intermediates is usually very small, and methods of producing large instantaneous concentrations are not of general applicability. In special cases suitable techniques are available—for example, ESR spectroscopy may be used to detect unstable oxidation states of transition metals, and optical spectroscopy may occasionally be used for the investigation of ionic organic species—although the method normally has to be devised for the particular problem. It is intended, therefore, specifically to exclude ionic intermediates in solution in the present discussion. The omission is probably

References pp. 336–342

justified since, in any case, elucidation of liquid-phase ionic mechanisms is normally achieved by inference from the nature of the reactants and products and from the overall kinetics of the process rather than by an explicit investigation of the nature and rate of reaction of the intermediate.

Ionic intermediates in the gas phase are also of considerable importance, and may be studied explicitly by the same sort of methods used for uncharged intermediates, as well as by special techniques relying on electrical properties. Many flames are known to be conductors of electricity, and it is established that ions may be formed in flames as a result of chemical reaction (chemi-ionization). Direct thermal ionization is not expected to be significant in flames except for metallic elements of low ionization potentials (*e.g.* the alkali metals) and nitric oxide (which has the low ionization potential of 9.4 eV). In shock tubes, however, where much higher temperatures may be produced, considerable thermal ionization can occur in other materials. Photo-ionization may be produced by short wavelength radiation, and if the energy of the radiation is greater than that required for ionization, then excited states of ions may be produced. Ionic intermediates are also produced by electron or heavy particle bombardment of neutral reactants, and again the ions may be in ground or excited states. The processes described lead to the production of "primary" ions, which may themselves react with neutral molecules to form "secondary" ions. Such ion–molecule reactions have recently elicited much interest in so far as they are important in upper atmosphere photochemistry, problems of rocket exhausts and re-entry into atmospheres, radiation chemistry, plasma magnetohydrodynamics and processes within mass spectrometers.

2. Classical spectroscopy

The use of the term spectroscopy in this section is taken to imply, perhaps somewhat arbitrarily, the study of electronic, vibrational and rotational transitions in reaction intermediates. In principle, spectroscopy is the ideal tool with which to study reaction intermediates, since not only unequivocal identification, but also determination of the absolute concentration and energy state of the intermediate should be possible. Both emission and absorption studies may be made, and these will be discussed separately. The small intensities absorbed or emitted also leave the system virtually unaffected by the investigation. However, a number of practical problems do arise which to some extent offset the theoretical advantages of spectroscopic study. Perhaps the most severe difficulty is that of obtaining a sufficiently strong absorption or emission from the reaction intermediate. The difficulty may result from the operation of a number of factors. The oscillator strength may be small for the transition, or, if a strong transition does exist, it may be in an inaccessible spectral region. Thus the presence of relatively high concentrations of intermediate may be needed before detection can be undertaken by spectroscopic methods. A second related problem is concerned with the resolving power of the instrument employed. What may appear in absorption under low resolution as one weak, fairly broad, band, may turn out at high resolution to be a series of strong, widely spaced bands. Any quantitative measurements made at low resolution, and assuming a weak transition, will be in error. Accurate quantitative estimation will require a comparison of line shapes and intensities using instruments capable of resolving fully the structure of the bands, and will take into account the effects on the line shape of differing pressures and temperatures. Such considerations are of particular importance if the energy distribution in the intermediate is not the equilibrium distribution. Determination of absolute concentrations of intermediates requires also a knowledge of the absolute transition probability, and this information may be difficult of access. Much of the discussion which follows in this chapter is therefore concerned with experimental techniques devised to overcome certain of the problems associated with spectroscopic investigations (see also Section 8).

Kinetic spectroscopy is applicable to the study of non-stationary concentrations of reaction intermediates. The greatest success of kinetic spectroscopy has been in the investigation of reaction intermediates produced by flash photolysis. Flash photolysis was developed by Norrish and Porter[14,15] in order to produce instantaneous high concentrations of reactive species. In its most usual form, the apparatus is arranged as shown diagrammatically in Fig. 3. The quartz cell is filled with the reaction mixture, and exposed to the intense flash from the "photolytic" flash tube.

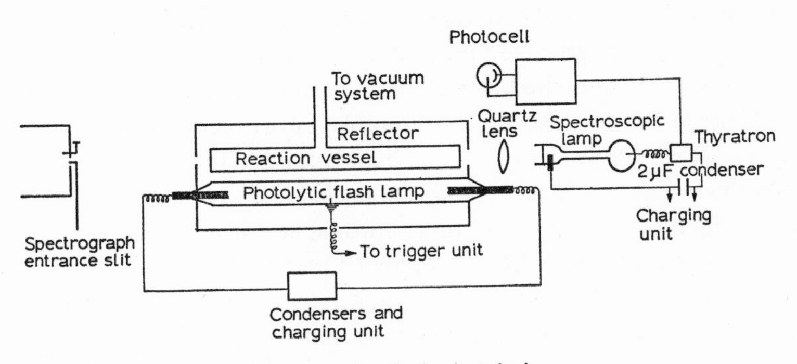

Fig. 3. Apparatus for flash photolysis.

After a definite delay, the duration of which is determined by the setting of the electronic delay unit, the much weaker "spectroscopic" flash tube is fired. The almost continuous radiation from this source passes through the reaction mixture, the absorption spectrum of which may be recorded photographically in the spectrograph. Experiments with different delay times, but identical reaction mixtures, give information about the variation with time of the contents of the reaction cell (reactants, products and intermediates). One obvious modification to the technique is to replace the photographic plate by one or more photomultipliers, equipped with collimating slits and placed in the focal plane of the spectrograph[16]. The spectroscopic flash tube may then be replaced by a xenon arc lamp, and the output from each photomultiplier displayed as a trace on an oscilloscope screen. The modification is clearly useful for emission studies, since all that need be done to go from absorption to emission is to turn off the xenon arc. A good review of the application of spectroscopy to flash photolysis is given by Norrish and Thrush[17].

A similar application has been made of kinetic spectroscopy to shock-tube studies. "Spectroscopic" flash tubes have been used to obtain a complete spectrogram of species present at any particular point behind the shock front[18,19].

2.1 EMISSION SPECTROSCOPY

The usefulness of emission spectroscopy in the quantitative investigation of reaction intermediates has been somewhat underestimated, and, indeed, Herzberg[20] says "emission spectra cannot in general serve as a quantitative measure of the concentration of free radicals". The obstacle to the employment of emission spectroscopy is that the excitation of the reaction intermediate is necessary. It is true that in many reactions excited species do not appear, and that emission spectroscopy is inapplicable in these cases. In a great many other reactions, the intermediates may be excited, and the excitation may have a profound effect on the chemical reactivity of the species. The real danger to misinterpretation of the emission spectra of reactive intermediates is that the results shall be taken to be representative of unexcited as well as excited species. In the "cool flame" of the acetaldehyde–oxygen reaction, for example, ultraviolet bands of formaldehyde appear. However, only one quantum of light is emitted for every million acetaldehyde molecules consumed, and it is probable that the chemiluminescent process is a minor side reaction[21]. Perhaps, then, it would be fair to say that emission spectroscopy gives qualitative information about intermediates appearing in a reaction. If measurement of the absolute emission intensity is possible, and the transition probability is known, then the concentration of excited species can be calculated and their contribution to the reaction processes estimated.

The production of excited species in flames has already been mentioned in Section 1.3. A hydrogen–oxygen flame exhibits a well defined band system in the near ultraviolet, which has been shown to originate from electronically excited hydroxyl radicals[22,23]. A very wide variety of emitting species has been identified in flames: we are not concerned here with the chemical implications of the results obtained, and the reader must be referred to more specific articles which have appeared elsewhere[24,25,58].

Transition probabilities are known for some of the emitting species in flames (*e.g.* OH[26], CN[27], CH[28] and C_2[29]). A method has been developed recently by

Dalby and Bennett[63–67] which has given accurate probabilities for a series of transitions. The technique is described briefly on pp. 291–2. Accurate determination of concentrations may still be hindered by self-absorption of the radiation, particularly in the case of the hydroxyl radical. Penner and co-workers have overcome the difficulty by the use of a double path technique, and are able to determine the rotational temperature and concentration of hydroxyl radicals in both flame[30] and shock-tube[31] studies. The single and double path emissivities are compared simultaneously, the double path beam being chopped to give modulation at about 5 μsec intervals[31]. The method of correction for line widths and Doppler broadening is discussed[30].

A further requirement for measurement of absolute concentrations of excited species in flames is that the volume from which emission is collected be known. The simplest experimental arrangement for flames at atmospheric pressures is to focus the radiation from the flame onto the entrance slit of a spectrograph. Reasonable assumptions can be made about the thickness of the emitting layer, and Ausloos and van Tiggelen[32, 33] have used the arrangement successfully in semi-quantitative determinations of excited OH, NH, NO and NH_2 in flames emitting the bands of these species.

Studies of low-pressure flames offer several advantages. In particular, the flame can be maintained flat, and the light from different parts of the reaction zone studied separately; the reaction volume from which light is collected is determined with much greater accuracy for such flames. At low pressures, chemiluminescent processes are more important than thermal excitation, collisional quenching of excited species is reduced, and self-absorption is diminished. A typical investigation of the low pressure flame is that of Gaydon and Wolfhard[34]: quantitative measurements of the C_2 emission were made.

So far, we have been concerned mainly with emission of radiation from electronically excited states. Emission may also arise from vibrational transitions in various reaction systems. The species HO_2 has long been postulated as an important chain carrier in combustion reactions, although emission from electronically excited HO_2 has yet to be demonstrated unequivocally. However, Tagirov[35] has observed radiation in flames at a frequency of 1305 cm^{-1} which he ascribes to transitions from vibrationally excited HO_2. Investigations of vibrational quenching processes are of great interest, and if the vibrationally excited species emit infrared radiation, then emission spectrometry may be the most satisfactory way of following the reaction. Davidson *et al.*[36, 37] describe a shock-tube study of the relaxation of vibrationally excited carbon monoxide, in which the overtone $2 \rightarrow 0$ emission at 2.335 μ was followed. Hooker[38] presents a similar study of the 3287 cm^{-1} parallel band of acetylene.

No discussion of emission spectroscopy would be complete without mention of the "atomic flame" studies of Michael Polanyi. In these now-classical investigations, Polanyi and his co-workers (ref. 39 and the references cited therein) observed the chemiluminescent reaction between alkali metals and halogens. The processes leading to excitation were suggested to be

$$Na+Cl_2 \rightarrow NaCl+Cl \quad (8)$$
$$Cl+Na_2 \rightarrow NaCl^{\dagger} + Na \quad (9)$$
$$NaCl^{\dagger} + Na \rightarrow NaCl+Na^* \quad (10)$$

(for the sodium–chlorine reaction)

$NaCl^{\dagger}$ represents a molecule of vibrationally excited sodium chloride, while Na^* is an electronically excited, ^2P, sodium atom which emits the resonance radiation. It will be remembered that these studies made very significant contributions to fundamental theories of reaction kinetics, especially with regard to the understanding of the potential energy surface describing reactions, activated complex and products.

Emission spectroscopy has been used recently in an elegant attempt to elucidate the mechanism of the energy transfer process (10). Moulton and Herschbach[40] have examined the emission from a triple molecular beam experiment. Molecular beams of bromine and atomic potassium cross each other, and vibrationally excited KBr is formed, which is then collimated into a further beam.

$$K+Br_2 \rightarrow KBr^{\dagger}+Br \quad (11)$$

The beam of KBr now meets a beam of atomic sodium, and the radiation emitted turns out to be of the *potassium* resonance lines, so that the energy transfer process must be written

$$KBr^{\dagger}+Na \rightarrow K^*+NaBr \quad (12)$$

John Polanyi and his collaborators[41–43] have used the infrared emission from vibrationally excited HCl to give information about the potential energy surface for the reaction

$$H+Cl_2 \rightarrow HCl^{\dagger}+Cl \quad (13)$$

The observations of the various vibrational bands may be used to determine the initial energy distribution of the excited hydrogen chloride. This knowledge of the energy distribution may then be used to test empirical contour maps for the reaction surface.

The participation in reaction of electronically excited oxygen molecules (in the $^1\Sigma_g^+$ and $^1\Delta_g$ states) has of late drawn much interest, and emission spectroscopy has played an important part in the identification and estimation of the singlet molecules. Hornbeck and Herman[44] observed the emission from the $O_2(^1\Sigma_g^+-^3\Sigma_g^-)$

'atmospheric band' system in hydrocarbon flames, and the same radiation has been detected in carbon monoxide–oxygen explosion flames[45]. The singlet oxygen molecules may also be produced in discharge-flow systems operated at low pressures[10], and measurement of the intensity of radiation has given quantitative data about the rates of reaction of the metastable molecules. Transition probabilities are known for both the $^1\Delta_g$–$^3\Sigma_g^-$ (ref. 46) and $^1\Sigma_g^+$–$^3\Sigma_g^-$ (ref. 47) systems (whose (0, 0) bands lie at 12700 A and 7619 A respectively), so that concentrations may be calculated from absolute intensity measurements[48]. Measurement of absolute intensities in low pressure flow systems has been simplified by the determination of the absolute intensity–wavelength distribution curve for the "air afterglow" reaction[49]

$$O+NO \rightarrow NO_2+h\nu \tag{14}$$

A photomultiplier whose spectral sensitivity is known may be calibrated against the air afterglow produced from known pressures of atomic oxygen and nitric oxide. If the geometry of the apparatus is unaltered when the intensity of the "unknown" emission is measured, then the absolute intensity may be calculated without explicit measurement of the emitting volume, photomultiplier acceptance angle and so on. In many cases, the desired kinetic information can be obtained from relative concentration measurements under differing experimental conditions. Such studies, using the emission of radiation from $O_2(^1\Delta_g)$ and $O_2(^1\Sigma_g^+)$, have been described by several workers[50–53].

Absolute intensity measurements have been made for the chemiluminescent reaction between nitric oxide and ozone[54]

$$NO+O_3 \rightarrow NO_2^*+O_2 \tag{15}$$

$$NO_2^* \rightarrow NO_2+h\nu \tag{16}$$

An estimate for the relative rates of radiation and quenching may be made and the rate constant of (15) calculated. It was shown that the pre-exponential factors for reaction (15) and for the non-radiative process

$$NO+O_3 \rightarrow NO_2+O_2 \tag{17}$$

were similar and that the differences in rate of (15) and (17) could be ascribed to the different, measured, activation energies.

In the example just described, excited nitrogen dioxide is certainly a reaction intermediate, although perhaps not in the usual sense. However, if we are to admit to the scope of this discussion excited species which do not "react" further, then we should include also all fluorescence phenomena. Fluorescence studies do indeed yield valuable information about the species excited, their energy states, and the rates at which they may radiate or be quenched. Nevertheless, it seems inappropriate in the present context to do more than point out the importance of such investigations. There is, however, one special use of fluorescence which is pertinent to our thesis, and it is well illustrated by experiments of von Hartel and Polanyi[55]. These workers, investigating the reactions of atomic sodium, measured the atom concentrations by study of the resonance fluorescence from the reaction mixture excited by the D lines of a sodium discharge lamp.

The discussion of emission spectroscopy will be concluded by a description of a rather unusual application. Bay and Steiner[56] have measured atomic hydrogen concentrations in the presence of molecular hydrogen by microwave excitation of the atomic hydrogen line spectrum. With low power fed into the gas (*ca.* 5 watts), there is not enough energy available for the dissociation of molecular hydrogen and subsequent excitation. Thus the measured intensities of the atomic hydrogen lines correspond to the concentrations of atoms already present in the reaction mixture. The method is curiously similar to that adopted to detect atoms and free radicals by mass spectrometry (see Section 3).

2.2 ABSORPTION SPECTROSCOPY

The difficulties associated with the use of absorption spectroscopy as a method for investigating reaction intermediates have already been touched upon. Even where the concentration of intermediate is high, and the spectroscope resolution adequate, the experimental conditions may be unfavourable to absorption techniques. Thus in flames, at atmospheric pressures, the concentration of intermediates may be relatively high, but the reaction zone is very thin and absorption will be weak. The light beam used must also be sufficiently narrow to keep it within the reaction zone, and a complication arises in that the gradient in refractive indices of the high-temperature reaction gases tends to deflect the beam. Further, it may be difficult to obtain a background source of higher brightness than the flame.

Notwithstanding the obstacles, however, some absorption studies of combustion processes have been made. Molecular intermediates, such as aldehydes and acids, have been identified in the slow combustion of propane[57]. Hydroxyl radicals can be observed in the absorption spectra of several flames[58]. The greatest success in the application of absorption spectroscopy to flame studies has been in investigations of diffusion flames. Wolfhard and Parker[59,60] studied the diffusion flames in oxygen of hydrogen, ammonia, hydrocarbons and carbon monoxide. In every case they were able to observe absorption by hydroxyl radicals, and they observed also the absorption of NH in the ammonia flame (NH_2 appeared in emission only). Molecular oxygen, and in suitable cases the reactants, could be detected by their absorption spectra, so that a clear picture of the structure of the diffusion flame emerged. Garton and Broida[61] have made vacuum ultraviolet studies of flames burning in a nitrogen atmosphere, but without definite result.

One of the first radicals to be investigated by absorption spectroscopy was the hydroxyl radical. Bonhoeffer and Reichardt[62] detected the absorption of the $A^2\Sigma$–$X^2\Pi$ transition of OH in water heated to 1600° C. Oldenburg and his collaborators made an exhaustive study of the absorption by OH resulting from electrical or thermal dissociation of water vapour and hydrogen peroxide. An absolute value for the oscillator strength of the transition is derived in the paper by Dwyer and Oldenburg[26], and the paper gives references to Oldenburg's earlier work.

Direct calculation of extinction coefficients (and hence oscillator strengths) requires a knowledge of both intensity of absorption and concentration of absorbing species. For the transition described above, concentrations of OH were calculated from the known fractional dissociation of water at elevated temperatures. In a few other instances, it may be possible to estimate the concentration of the intermediate from chemical considerations. Thus, Lipscomb *et al.*[76] were able to calculate the extinction coefficient at $\lambda = 2577$ A of the radical ClO from flash photolysis studies of chlorine dioxide. It was shown that the disappearance of ClO obeyed a second order rate law, so that

$$\frac{1}{D} = \frac{1}{D_0} + \frac{kt}{\varepsilon}$$

where D is the optical density ($= \varepsilon[\text{ClO}]$) at time t and D_0 the optical density at $t = 0$. The optical density of the reaction mixture was determined by plate photometry at a number of delay times, and the value of D_0 obtained by extrapolation of the results. The assumption was made that, apart from some ClO_3 (for which corrections were applied), there was material balance between ClO and ClO_2, and that $[\text{ClO}]_0$ could be calculated from the amount of ClO_2 destroyed. The values of D_0 and $[\text{ClO}]_0$ lead immediately to an estimate of ε.

Extinction coefficients obviously may be determined for suitable absorptions of species whose concentration is already known from non-spectroscopic investigations. The results obtained in this way are limited by the accuracy and availability of other methods for the estimation of the species in question. For many reaction intermediates, direct calculation of concentration is not possible, and until recently only theoretical estimates were available for the oscillator strengths of absorptions involving such species.

Dalby and Bennett have now calculated oscillator strengths from spontaneous radiative lifetimes of several excited species. Excited unstable species are produced by electron bombardment of suitable parent molecules. After the electron beam is turned off the intensity of radiation decays, and from observations of the decay rate the absolute lifetime for the radiative transition may be calculated. The experimental method and theory are given in the first paper[63]. Among the oscillator strengths determined have been those for the $A\,^2\Delta \leftarrow X\,^2\Sigma$ and $B\,^2\Sigma \leftarrow X\,^2\Sigma$ systems of CH[64], the $A\,^2\Pi \leftarrow X\,^3\Sigma$ system of NH[64], the $B\,^2\Sigma \leftarrow X\,^2\Sigma$ system of CN[65] and the $A\,^2\Sigma \leftarrow X\,^2\Pi$ system of OH[66]. A modified form of the experiment using flash photolysis has been employed[67] to study transitions in CF_2.

Hydrogen and deuterium arcs, and high pressure xenon arcs are convenient, near-continuous, light sources for absorption studies. Kaufman and Del Greco[68,69] have, however, used a rather special light source in experiments on hydroxyl radicals prepared in a low pressure flow system by the reaction

$$H+NO_2 \rightarrow OH+NO \tag{18}$$

They excite a discharge in a flowing argon–water vapour mixture with a low power microwave generator, and use this as a source of the hydroxyl emission. The light is reflected to give three traversals of the flow tube (giving an optical path of 9 cm) within a width of 1.3 cm in the direction of flow, and is then incident upon the slit of a photoelectric grating monochromator. Individual rotational states may be isolated, but the Doppler widths of both emission and absorption lines (*ca.* 0.01 A) are less than the resolution of the monochromator (*ca.* 0.2 A). Hence a correction is applied for the differing Doppler widths of emitting and absorbing molecules. The rotational temperatures are measured by comparison of the relative absorption of different rotational lines. The temperatures were found to be *ca.* 100° C for the emitting OH excited by the microwave discharge, and *ca.* 40° C for the OH produced in the flow tube by reaction (18). Absolute oscillator strengths were obtained by calculating OH concentrations from the amount of NO_2 consumed.

Hydroxyl emission sources have been applied also in shock tube studies of hydroxyl radical reactions[70–72]. In this case the required time resolution (see refs. 18, 19) was obtained by using a flash discharge in water vapour. Similar methods have been described in which the C_2 Swan bands are excited by a discharge in butane[73].

Microwave absorption spectroscopy has been used to investigate ^{16}OH, ^{18}OH and ^{16}OD produced by electric discharges in the appropriate water vapour[74,75]. The absorptions lie in the region 7.7 to 37 kMc.sec^{-1} and arise from transitions between the Λ-doublets, $^2\Pi_{\frac{1}{2}}$ and $^2\Pi_{\frac{3}{2}}$ (the pure rotation spectrum for OH, a light radical, is in the far infrared). Zeeman modulation was used with a conventional microwave spectrometer.

The most spectacular applications of absorption spectroscopy have been, of course, to flash photolysis used in either adiabatic (*i.e.* "explosion") or isothermal (*i.e.* "photochemical") studies. A brief description of the technique is to be found in the introduction to this section, and reference is made to a review article[17]. Relative intensities of absorption may be determined by plate photometry of photographic records, or by photoelectric measurements using a monochromator. The qualitative information about the species present as intermediates may go far in elucidating a reaction mechanism, while the measures of the relative concentrations of such species may yield useful kinetic data. Known transition probabilities, determined in the ways indicated previously, may be used to calculate absolute concentrations of intermediate in suitable cases.

Energy-rich species also may be observed by absorption spectroscopy in flash-photolytic experiments. A brief survey of this work will be given since previously it has received less adequate notice than the more familiar studies of ground state radicals. The flash photolysis of chlorine dioxide[76], nitrogen dioxide[76] and ozone[77,78] leads in each case to the appearance of molecular oxygen in high vibrational levels of the ground electronic state (up to 21 vibrational quanta in the case of ozone photolysis[78]), which may be detected as an extension of the Schumann–Runge absorption spectrum to longer wavelengths (2000–4000 A). The excited oxygen seems to be produced in each case by the reaction of atomic oxygen, formed in the primary photolysis, with a reactant molecule. Examination of the rotational fine structure of the vibrational bands shows that the molecules are rotationally cold and that the reactions are isothermal. The importance of absolute measurements is emphasised by the recent determination by Fitzsimmons and Bair[79] of the transition probabilities for high vibrational levels of the Schumann–Runge system. Their probabilities, coupled with the observations of McGrath and Norrish[77], suggest that the concentration of highly excited oxygen is small. Unfortunately this does not resolve the question as to whether the concentration is small because the process leading to the production of excited molecules is a minor one, or whether, as McGrath and Norrish themselves propose, excited oxygen is consumed by reaction with ozone. Other secondary reactions in photochemical experiments which give rise to vibrationally excited products are those of atomic halogens with ozone. $ClO^{\dagger}$ with 5 quanta of excitation, and $BrO^{\dagger}$ with 4, have been observed in the flash photolysis of chlorine or bromine in the presence of ozone[80].

Primary photochemical processes also may yield vibrationally excited products. Flash photolysis[81] of cyanogen, cyanogen bromide and cyanogen iodide in each case yields a product whose absorption spectrum identifies it as a vibrationally hot CN radical with up to 6 quanta of excitation. Vibrationally excited nitric oxide in the ground electronic state (detected by the $A\,^2\Sigma^+ – X\,^2\Pi$ absorption system) is produced in the flash photolysis of nitric oxide itself[82] and of the nitrosyl halides[83]. Relaxation studies have been particularly fruitful in these systems, since it has been possible to make absolute concentration measurements. The 0–1 absorption band appears in the ordinary absorption spectrum, and under equilibrium conditions

$$[NO^{\dagger}, v = 1] = [NO, v = o]e^{-h\nu/kT}$$

where ν is the vibration frequency. Thus a calibration can be obtained for any chosen rotational line in the band and applied to the flash photolysis experiments. A similar calibration has been made for vibrationally excited carbon monoxide in order to make quantitative studies of the relaxation of nitric oxide by carbon monoxide[84]

$$[NO^{\dagger}, v = 1]+[CO, v = 0] \rightarrow [NO, v = 0]+[CO^{\dagger}, v = 1] \qquad (19)$$

Absorption spectra of electronically excited states may be observed in flash photolysis studies. Porter[85] has established the existence of the triplet state in a wide range of organic compounds in the liquid and gaseous phases. For example, the first triplet state of anthracene is populated by radiationless conversion from a photochemically excited singlet molecule, and may be observed by the absorption to the second triplet level. Absolute measurements of the triplet concentration may be made by determinations, from the absorption spectra, of the depletion of the singlet state. Similar results have been obtained with a variety of hydrocarbons, ketones, quinones and dyestuffs.

Callear and Norrish[86] have used the technique to study the rise and decay of metastable Hg ($6\ ^3P_0$) atoms populated from Hg ($6\ ^3P_1$) atoms excited in mercury vapour by the flash. A number of transitions in absorption from Hg ($6\ ^3P_0$) were identified, and the $6\ ^3D_1 \leftarrow 6\ ^3P_0$ line at 2967 A was chosen for plate photometry. Herzberg and Shoosmith[87] have flash photolysed diazomethane in the presence of a large excess of inert gas, and observed the absorption spectrum of excited, singlet, methylene (the state expected from photolysis of diazomethane). Absorption from the triplet state is also observed, and it is favoured by long time delays or high dilutions, observations which suggest that the triplet is indeed the lower energy state. In the absence of excess inert gas[88], the singlet CH_2—which is highly excited vibrationally—decomposes to electronically excited $CH(^2\Sigma^-)$, identified by its absorption in the $^2\Pi \leftarrow {}^2\Sigma^-$ system at about 3000 A. After flash photolysis of hydrazoic acid[89], strong absorption of the lowest singlet state of NH, $^1\Delta$, is observed, although in the photolysis of hydrazine only the $^3\Sigma^-$, ground state, of NH may be detected.

3. Mass spectrometry

3.1 UNCHARGED SPECIES

The use of mass spectrometers suggests itself as a convenient method for the investigation of unstable reaction intermediates. The mass spectrometer can not only measure the concentration of an intermediate but also establish its chemical identity. Eltenton[90,91] first applied the technique to the study of free radicals, and from the earliest experiments useful results have been obtained. Thus Eltenton in his first paper[90] showed that the pyrolysis of methane on a carbon filament leads to the formation of methyl radicals, and not methylene as believed formerly, and he established that his apparatus was sensitive to methylene by demonstrating the formation of the latter during the pyrolysis of dilute mixtures of diazomethane in helium.

One of the problems in the mass spectrometric identification of unknown species is that of near-coincidence in mass between two different species. Thus CH_4, NH_2 and O all have "identical" mass-numbers, and the number of overlapping species is increased if any are isotopically substituted. Unequivocal identification requires the use of high resolution spectrometers capable of resolving to as much as one part in 10,000. On the other hand, if the reaction system is such that certain species cannot be present, then lower resolution instruments may be suitable. In many cases a compromise must be reached. For example, mass spectrometers carried by rockets for investigations in the upper atmosphere must of necessity be light. Radio-frequency spectrometers of the kind described by Bennett[92] satisfy the weight restriction, although the ultimate mass resolution is probably less than one in 100 with this kind of instrument.

Sensitivity may also affect the choice of experiment. The concentration of intermediate is frequently too small for detection in static thermal or photochemical reaction systems. Investigations are therefore restricted either to those reactions in which exceptionally high intermediate concentrations are found—for example, in flames—or in systems designed to produce high concentrations. The latter group includes flow systems as well as the non-stationary methods such as flash-photolysis and shock-tube studies. Use of non-stationary methods may itself impose restrictions on the minimum time-resolution of the instrument employed.

Time-of-flight mass spectrometers, developed by Wiley and McLaurin[93], seem particularly well adapted to this kind of study. Commercial instruments are now capable of making around 10^4 complete scans per second, and the limitation on cycling time becomes one of obtaining a statistically significant sample. Few or no ions of a particular species may be present during a very small time interval, so that information obtained from a single scan is not meaningful. Several scans must be made, thus reducing the effective time resolution. To some extent a time-of-flight instrument has the advantage over a conventional magnet spectrometer that every ion produced is "seen", although magnet spectrographs recording the entire spectrum photographically, or special spectrometers with detectors for each mass number of interest, may also be used. From the point of view of the detection of unstable intermediates one very real advantage of the time-of-flight spectrometer is the open construction of the ion source, and this feature will be considered later.

If the intermediate species is uncharged, then ions must be formed from it within the spectrometer. Although some experiments have been performed using charge-exchange ionization (for example, by Tal'roze[94,95]), most of the work to be described here has used a low-energy electron impact ionization procedure. A beam of electrons, produced from a heated filament, is allowed to collide with the species under investigation, the two beams usually being at right angles. Of course,

References pp. 336–342

if a parent of the intermediate is also present in the reaction mixture, then electron impact may produce ions of the intermediate from this source as well as from the intermediate itself. Thus suppose the species of interest is a radical R, and its precursor in a reaction is RX, then we may write the processes

$$R+e \rightarrow R^{+}+2e \tag{20}$$

$$RX+e \rightarrow R^{+}+X+2e \tag{21}$$

together with similar reactions for the fragment X. Fortunately, if the electron beam is produced with sufficiently closely defined energy, then it is possible to distinguish between R^+ produced in (20) and (21). The appearance potential of R^+ in (21) will be greater, by an amount depending on the dissociation energy of R–X, than the appearance potential of R^+ in (20). Fig. 4 represents the variation

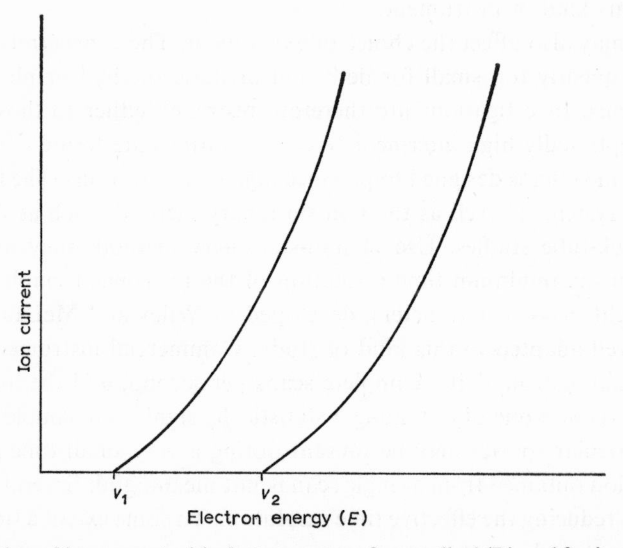

Fig. 4. Variation of ion current with electron energy for a radical (R) and for its precursor (RX).

of ion current with electron energy (E) for the two processes (20) and (21). V_1 is the ionization potential of R in (20) while V_2 is the appearance potential of R^+ from RX in (21). If E lies between V_1 and V_2, then any ion current observed will be a result of the radical R entering the spectrometer, while if E is greater than V_2 both ionization processes (20) and (21) contribute to the ion current. The sensitivity of the apparatus to radicals depends on the shape and on the separation of the two curves in Fig. 4, and the most favourable electron energy is not immediately obvious. Robertson[96] has shown that the minimum concentration of R(C_R) that can be detected in a concentration C_{RX} of RX is given by the expression

$$\frac{C_{RX}}{C_R} = \frac{2k^3T^3+(E-V_1)k^2T^2}{2k^3T^3+(V_2-E)k^2T^2} \exp\left\{\frac{V_2-E}{kT}\right\}$$

where k, T are the Boltzmann constant and absolute temperature of the electron beam respectively.

It will be realised that investigations of the kind described may be used in the calculation of dissociation energies and ionization potentials; a discussion of such work is, however, outside the scope of this chapter.

Several factors conspire to make the interpretation of the results obtained by mass spectrometry less simple than the preceding treatment might suggest. The radical-ions, R^+, may be produced with a lower appearance potential than V_2 by a number of processes other than (20), and the occurrence of these will depend not only upon the reaction being investigated but also on the design of the spectrometer. Amongst these processes may be mentioned:

(*i*) formation of R by the pyrolysis of reactants or products on the filament used as the electron emitter;
(*ii*) ionization of products giving rise to R^+ with a low appearance potential;
(*iii*) formation of R^+ by ion-pair processes;
(*iv*) formation of radicals in the ionization chamber by electron impact, and then ionization in reaction (20).

The presence of surfaces in the ionization chambers may lead to a decrease in concentration of radicals, although in special circumstances wall decomposition of RX may occur, thus enhancing the radical concentration. LeGoff[97] has given the expression

$$R = R_0/(1+\gamma S)$$

to relate the observed radical concentration R to the initial concentration R_0 in a system with a ratio of wall area to openings in the ionization chamber of S and a wall recombination efficiency of γ. The value of an instrument with a low value of S—such as the time-of-flight spectrometer—is clearly understood.

Excess energy in RX may lead to a reduction in the ionization potential, so that in experiments where "hot" species may be present due allowance must be made for this reduction. This feature may, of course, be turned to advantage in the investigation of excited species. Foner and Hudson[98] investigated the appearance potential for O_2^+ ions in oxygen which had passed through a Wood's discharge tube at a pressure of about 4 mm Hg. Their appearance potential curves for the discharge on and off are reproduced in Fig. 5, and show clearly that some O_2 species is formed with a lower ionization potential than ground state molecular oxygen (the decrease in ion intensity at higher energies when the discharge is on is accounted for in terms of O_2 decomposition to O atoms). Analysis of the curves shows that the energy separation of the two oxygen states is 0.93 ± 0.1 eV, and

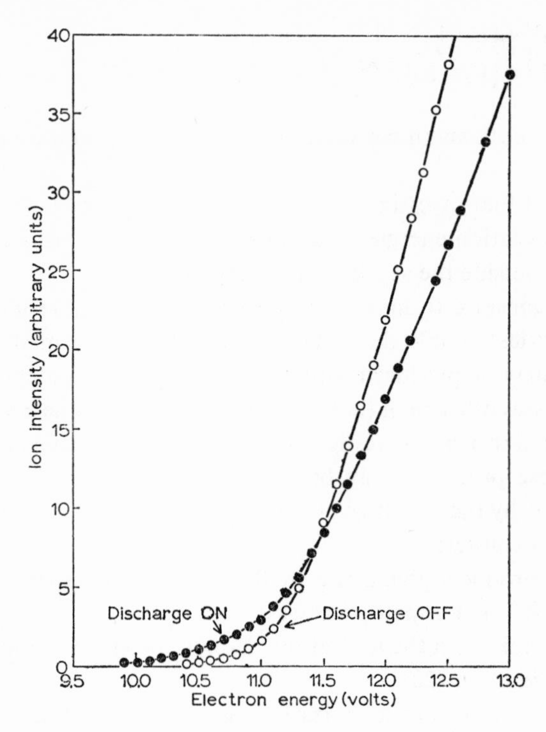

Fig. 5. Appearance potential curves for O_2^+ ions from oxygen with the electric discharge on and off.

comparison of this figure with the spectroscopic value of 0.98 eV for the separation of $O_2(^1\Delta_g)$ and ground state oxygen, $O_2(^3\Sigma_g^-)$ suggests that the excited state involved is $O_2(^1\Delta_g)$. A similar, and rather more complete, study of electronically excited molecular oxygen has been made subsequently by Herron and Schiff[99].

So far, no mention has been made of the difficulties of sampling from reaction systems. Mass spectrometers work at pressures of about 10^{-5} torr, above which pressure ion–molecule and other interfering process make operation impossible. Most gas-phase chemical reactions are carried out at pressures considerably in excess of this value—say in the range 1 torr to several atmospheres—and some provision must be made for introducing the reaction mixture into the spectrometer at the reduced pressure. The simplest form of sampling device is a pinhole through which the mixture expands. However, unless abnormally fast pumping systems are available, the required pressure differential can be maintained only with relatively small pinholes (say 10–100 μ in diameter). Thus a large fraction of species entering the ionization chamber of the spectrometer have made several collisions with the walls of the orifice. Probably they have entered the orifice from a boundary layer near the walls of the material dividing reaction system from spectrometer, and may, therefore, be unrepresentative of the species within the bulk of the reaction mixture.

Suitable choice of wall material may reduce, but not eliminate, the effect. A molecular beam technique, involving several orifices, can be used to isolate only those particles which have not made collisions during sampling, although the problem of the boundary layer at the first orifice remains. Several studies have been made of molecular beam formation in mass spectrometer applications[100-103], and Greene *et al.*[102-103] have made an investigation of sampling from 1 atmosphere flames by the molecular beam technique. They find that a simple picture of beam formation, involving isentropic expansion of gas through the first orifice into a supersonic molecular beam, is justified. They also observe mass separation, an effect expected in molecular beams, and have verified the relation

$$I_{x^+} = k_x P_x M_x$$

between ion intensity (I_{x^+}), partial pressure of the species (P_x) and molecular weight (M_x).

Special procedures must be adopted in the determination of the absolute sensitivity of the mass spectrometer to the unstable species.

Some measure of the concentrations of unstable species can be obtained from the shape of appearance potential curves, although usually it is not possible to distinguish clearly the contributions from precursors and intermediates. It is also necessary to make assumptions about ionization cross-sections. However, in suitable cases some information may be obtained. Foner and Hudson[98], for example, made the reasonable assumption that the ionization cross-sections of $O_2(^1\Delta_g)$ and $O_2(^3\Sigma_g^-)$ were the same, and were able to place limits on the concentration of $O_2(^1\Delta_g)$ in their oxygen flow of between 10 % and 20 %. Again Jackson and Schiff[106] were able to obtain approximate concentrations of N atoms in discharge-flow experiments by this procedure.

A typical determination is that of Lossing and Tickner[104] for the methyl radical. Methyl radicals were produced by the pyrolysis of mercury dimethyl diluted by helium, and the mass spectrum showed that only CH_3, mercury, ethane and a trace of methane were formed. Sensitivity calibrations were obtained in the usual way for the stable substances, and then the "net" peak at mass 15, after subtraction of the contributions from mercury dimethyl, ethane and methane, was determined. At high temperatures of pyrolysis, where the methyl radicals were most abundant, the sensitivity for the mass 15 peak of the methyl radical could then be calculated on the basis of 100 % carbon balance. As discussed earlier, wall reactions may lead to appreciable disappearance of the radical under observation, and such effects must be taken into account when calculating the sensitivity of the apparatus. Corrections of these kinds were applied to the experiments described above when Ingold and Lossing[105] discovered that part of the methane observed was produced by reaction in the ionization chamber. The smallest relative concentration of radicals which can be determined accurately (*i.e.* where several species give rise to the same mass peak) is of the order of a few per cent, a value which emphasises the restriction of mass spectrometric studies to systems containing relatively high concentrations of intermediates.

Approaches similar to those of Lossing have been made by several other workers. Kistiakowsky *et al.*[107,108] have measured nitrogen atom concentrations in discharge-flow systems by determining the changes in the mass 14 and mass 28 peaks when the discharge is turned on or off, and, by using relatively low energy (*ca.* 22.5 eV) electrons, they were able to eliminate as far as possible any contribution to the mass 14 peak from molecular nitrogen. A correction was applied for N atom recombination during sampling. Phillips and Schiff[7] extended the method to the measurement of atomic hydrogen concentrations in discharge-flow experiments. The high relative concentration of atomic to molecular hydrogen (1 : 1) enables accurate determination of instrument sensitivity to atomic hydrogen from changes in mass 1 and mass 2 peaks with the discharge on and off.

The chemical methods for the estimation of unstable species (Section 5) may be used in suitable cases to determine the concentration of the species in a reaction system analysed also by mass spectrometry, and thus the mass spectrometer sensitivity can be calculated. The mass spectrometer may now be used to investigate the reaction system actually under study. Several investigations of atom concentrations have been made in this way. Phillips and Schiff[8,109] used the reaction of atomic nitrogen with nitric oxide to measure atomic nitrogen concentrations. Herron and Schiff[99] calibrated their mass spectrometer against known O atom concentrations determined by titration with nitrogen dioxide, and were able to show that only one atom in 21 passing through the sampling orifice is ionized, the rest recombining on the wall. (Details of these experiments will be found in Section 5).

Various research groups have tried to overcome some of the difficulties outlined earlier and have developed methods for the investigation of particular types of reaction systems. The remainder of this section is devoted to typical arrangements used by some of the several workers in the field.

A considerable number of thermal reactions has been studied mass spectrometrically by Lossing and his co-workers at the N.R.C., Ottawa, using an apparatus whose earliest form[104] is shown in Fig. 6. The arrangement consists of a heated quartz reactor through which the reaction mixture flows at low pressure (*e.g.* in the first experiments of Lossing and Tickner the mixture was 8–14 μ Hg of mercury dimethyl in a helium carrier at 6–20 torr). The sampling orifice is punctured, by means of a Tesla discharge, in the top of a 15 μ thick quartz cone projecting into the reaction area: orifices made in this way appeared to be perfectly circular and to have the edges fire polished. The orifice diameter was usually in the range 30–50 μ. Quartz was chosen to reduce the probability of radical reaction on the surface. The residence time of reactants in the heated part of the tube was only about 0.001 sec, and the contact time could not be calculated with any degree of

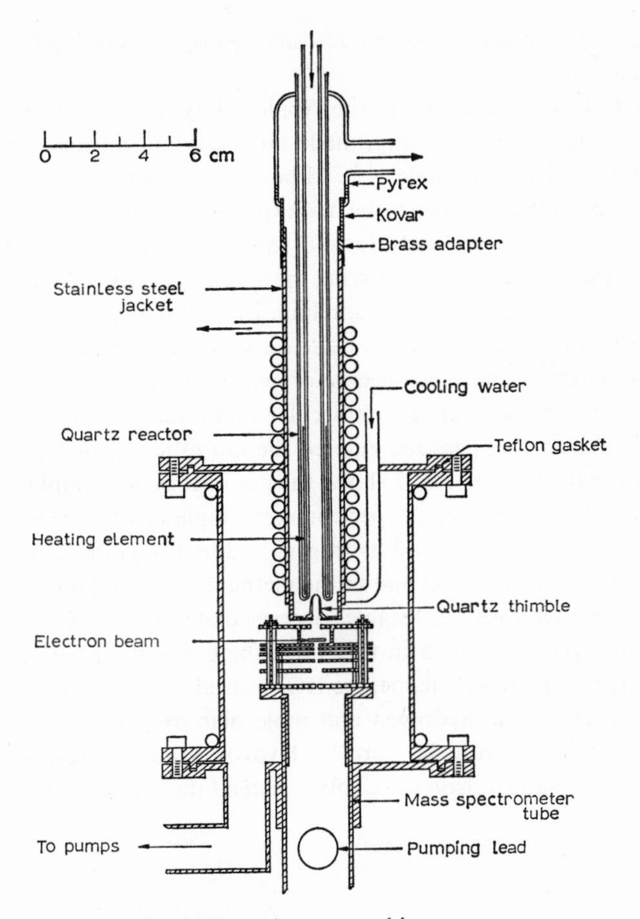

Fig. 6. Thermal reactor and ion source.

accuracy. In order to make possible quantitative measurements of rates of radical reactions, a modification to the apparatus was introduced by Ingold and Lossing[105] who arranged for the furnace to be moveable with respect to the sampling orifice. In this way they were able to ascribe changes in molecule and radical concentrations to the change in reaction time when the furnace was moved. A wide variety of radicals has been produced thermally and studied by Lossing's group since 1952: Lossing has given a list[110] of the radicals investigated up to 1963.

A similar flow and mass spectrometer inlet system has been developed by Lossing for the investigation of photochemical systems. Mercury photosensitization was used so that the concentration of radicals should be within the limits of sensitivity of the mass spectrometer. With the light intensities used in experiments with acetone[111], 80 % of the reactant could be decomposed with an exposure time of 0.002 sec. Later modifications[112,113] permitted variation of the exposure time by altering the flow velocity of reactants, or, more conveniently, by varying the length of the illuminated zone.

Foner and Hudson, working in the Applied Physics Laboratory of the Johns Hopkins University, Silver Spring, made an important advance in the study of unstable intermediates in flames[114]. Although Eltenton[91] had successfully detected some free radicals in flames, he had experienced difficulty in the detection of oxygenated radicals. Foner and Hudson employed a molecular beam sampling device to ensure that the species entering the spectrometer would have been as free as possible from collisions during sampling. In addition, they employed a beam-chopping technique together with phase sensitive detection to discriminate between true reaction intermediates and species formed on the ionizing filament or on the walls of the ionizing chamber. The molecular beam was chopped at 200 cps with a magnetically driven chopper, and entered the ionization chamber co-axially with the ionizing electron beam. Fig. 7 shows the gas inlet system employed, and Fig. 8 is a block diagram of the mass spectrometer–amplifier arrangement. Using this apparatus[114], atomic oxygen and hydrogen, and hydroxyl radicals, were detected in the hydrogen–oxygen flame; and in the methane–oxygen flame, methyl radicals were found together with the "stable" intermediates C_2H_2, CO, CH_4O, CH_2O or C_2H_2 and C_4H_2. Although in this work there was no positive evidence for HO_2 formation in the H_2–O_2 flame, the HO_2 radical was subsequently found in the reaction between atomic hydrogen and molecular oxygen, both at pressures of around 30 torr[115] and around 1 torr[116]. Hydroxyl radicals were also detected in both cases. Foner and Hudson were able to extend the use of their apparatus in an

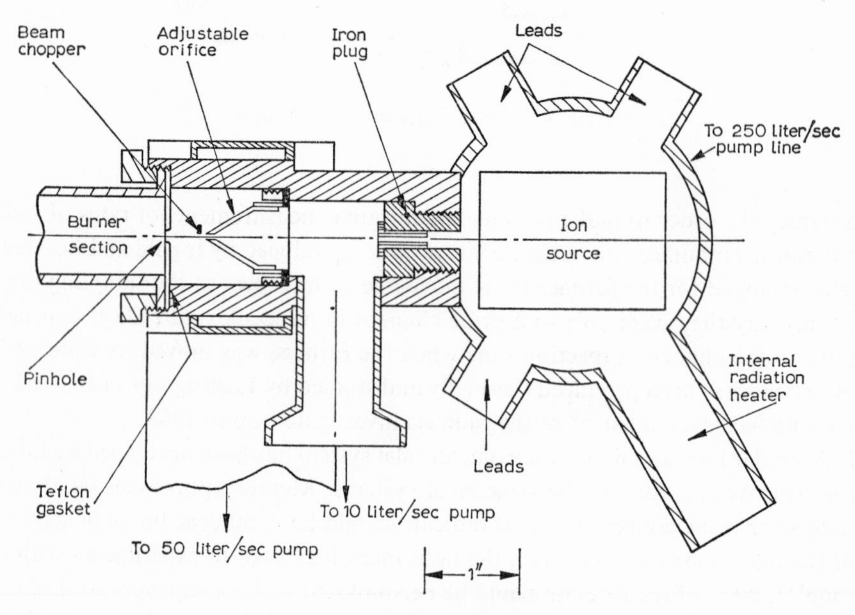

Fig 7. Chopped beam inlet system.

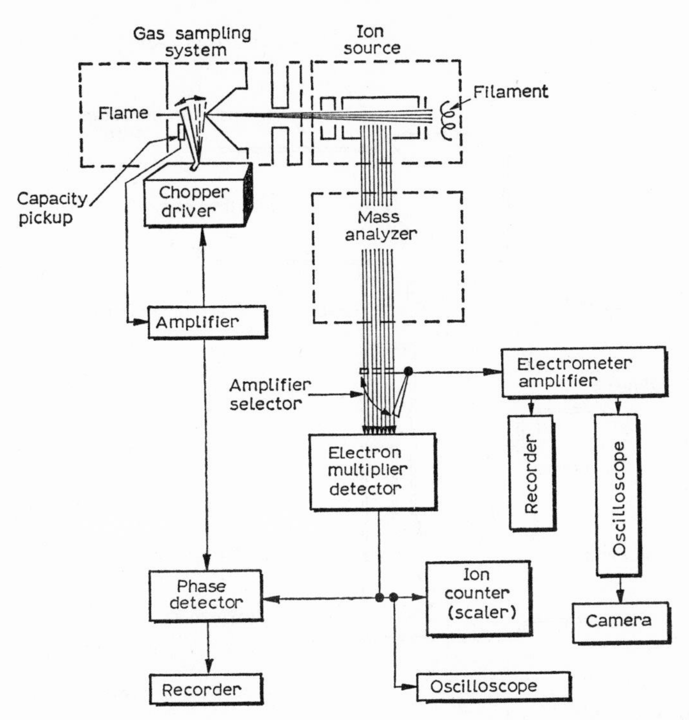

Fig. 8. Block diagram of mass spectrometer employing a chopped beam technique.

investigation of metastable, electronically excited oxygen molecules, $O_2(^1\Delta_g)$, as described earlier[98].

Molecular beam formation from flames at 1 atmosphere has been discussed. Greene *et al.*[102,103] used several slits together with differential pumping between the stages. Typical dimensions of the apertures and pressures used by these workers are given in Table 1: the apparatus to which they refer is shown in Fig. 9.

Schiff and his colleagues have used, *inter alia*, mass spectrometric methods to study atom reactions in discharge flow systems: reactions of atomic oxygen[99,117], nitrogen[8,106,109] and hydrogen[7] have been investigated so far by Schiff's group. The basic apparatus used is described in detail in ref. 109. A sampling leak, consisting of a 20 × 20 μ cylindrical hole in a pyrex thimble, was fixed in a position relatively far downstream in the flow tube. One reactant, usually atoms diluted by their molecular precursor, flowed down this tube, and the second reactant could be admitted to the flow tube through an inlet jet whose position in the main tube was adjustable. Since distance in a flow experiment corresponds to time of reaction, the mass spectrometer leak samples the reaction mixture at various reaction times, and rates of atom reactions can be calculated from the experimental data. The apparatus has been used to study the reactions $N+O_3$, $N+NO$, $O+O_3$, $O+NO_2$[109], $H+NO_2$, $H+O_3$[7], $O+O_2+O_2$[117], together with investigations of the

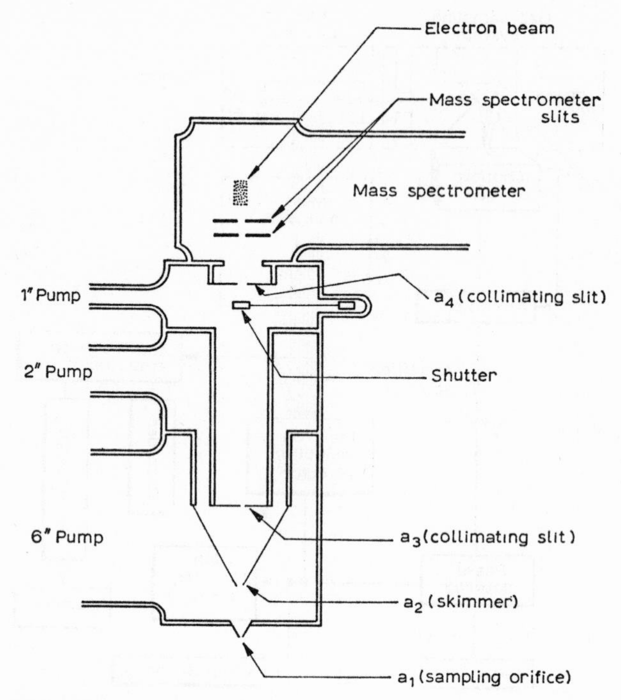

Fig. 9. Sampling device for flames burning at one atmosphere.

TABLE 1

OPERATING CONDITIONS OF APPARATUS SHOWN IN FIG. 9

(a) Dimensions

Figure reference	*Physical size (mm)*	*Distance from electron beam (mm)*
a_1	0.25 (diameter)	400
a_2	0.25 (diameter)	394
a_3	0.50×9	344
a_4	0.75×12	113

(b) Pressures

Chamber	*Pressure (torr)*
Source	7×10^2
Stage 1	10^{-2}
Stage 2	3×10^{-5}
Stage 3	3×10^{-6}
Ion source	10^{-6}

reactivity of excited molecular nitrogen[8] and molecular oxygen[117] with ozone.

The systems so far considered have involved taking samples for mass spectrometric analysis from a reaction mixture containing a high stationary concentration of unstable intermediates. We now consider work in which the intermediates are present only for short periods of time. Some early experiments were carried out by Léger and Ouellet[118] using a conventional magnet spectrometer capable of scanning a 40 peak spectrum in 5 milliseconds. They studied the cool flame produced as an ether–oxygen mixture is expanded into a heated reaction vessel: in this kind of reaction the radical concentration must be low, and although "stable" molecular intermediates were observed, no radicals could be detected.

The greater time resolution of the time-of-flight mass spectrometer (subject to the restrictions discussed earlier for this instrument) suggests its use for following rapid reactions. Kistiakowsky first used this type of spectrometer to investigate reactions in flash photolysis[119] and in shock waves[120]. Kistiakowsky and Kydd[119] used a Bendix time-of-flight mass spectrometer in a study of the flash photolysis of ketene and of nitrogen dioxide. The ions in a time-of-flight spectrometer are formed close to a grounded electrode, which in these experiments was made the wall of the reaction cell containing the effusion pinhole. Beyond the pinhole there is little resistance to gas flow because of the absence of focussing slits, and it was calculated that more than 90 % of the ions formed within the spectrometer come from species in direct flight from the pinhole. The flash lamp was operated at an energy of 500 joules, and the flash had a duration of about 10 μsec. Several consecutive mass spectra were recorded on a rotating drum camera. The opening of the mechanical shutter of the camera triggered the flash lamp, and the exact instant of the flash was recorded on the film by a little light piped from the flash. The repetition frequency employed was 10 $kc.sec^{-1}$, and the problem of statistical fluctuations of ion intensity was overcome by averaging the results of several almost identical runs. Some trouble was experienced initially with interference to the spectrometer during the flash, but, although the exact nature of the disturbance is not clear, electrical screening of the lamp reduced the effect sufficiently for observations to be made during the flash. The absence of detectable CH_2 after flash photolysis of ketene is discussed by Kistiakowsky and Kydd in terms of the high reactivity of CH_2 towards ketene, and they are able to give an estimate of the reaction probability on collision. Some evidence was obtained for the formation of 'hot' cyclopropanone, some of which could be stabilised to a relatively long-lived radical C_3H_4O. A discussion is given also for the results of nitrogen dioxide photolysis.

A similar type of experimental arrangement is described by Bradley and Kistiakowsky[120] in a study of the intermediates formed in shock tubes. The thermal decomposition of nitrous oxide was investigated, in the temperature range 1780–2000° K, and the concentrations of reactants, products and atomic oxygen were measured every 100 μsec.

3.2 IONIC SPECIES

It is natural that mass spectrometric techniques should have been applied to species that are already charged, and a number of studies have been made of both negative and positive ion formation in chemical reactions. One of the earliest investigations by mass spectrometry of ions produced outside the spectrometer was that of Brasefield[121] who produced ions in an electric discharge. Although such systems are of great importance in the study of ion–molecule reactions, they do not really fall within the scope of this chapter, and most of the results to be discussed concern chemi- or photo-ionization.

The problems of sampling, already discussed for uncharged species, are augmented in the case of ions. Boundary potentials are often set up at the wall containing the sampling orifice, and an ion sheath may be formed at the wall (the sheath is a positive space charge, since the species of highest mobility are electrons). Except under particularly unfavourable conditions, the boundary potential will not result in appreciable acceleration of positive ions (or retardation of negative ions). However, some electrostatic lens effect may operate at the orifice, thus falsifying the sample. Again, the loss of positive ions to the surface may be enhanced during passage through the orifice. Pahl[122] has considered in detail the possible misfortunes which may befall ions during passage through a sampling orifice.

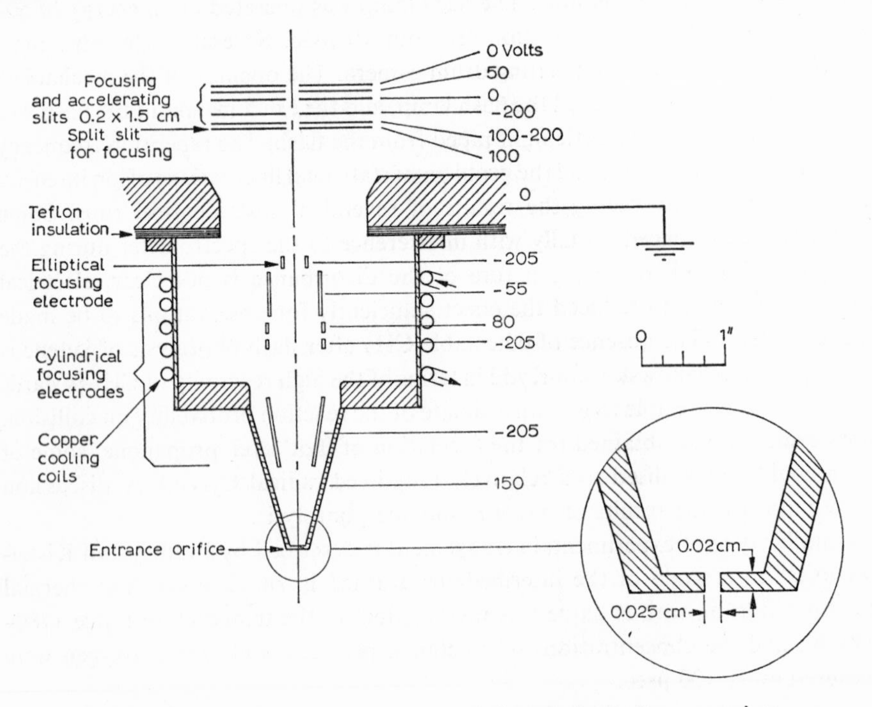

Fig. 10. Device for sampling ions. Typical operating voltages are given.

Bascombe, Green and Sugden[123] have shown the necessity of using large sampling apertures in order to get a mass spectrum representative of ions in their atmospheric-pressure flame, rather than of a species in a boundary layer at the walls of the system. Spokes and Evans[124] have also examined ion sampling processes in some detail, and carried out an extensive analysis of the data. Their results suggest strongly that boundary effects will be of importance.

The particular experimental technique employed to investigate ions in flames has depended upon the pressure region to be studied. Calcote and Reuter[125] have described a mass spectrometric study of ion profiles in flames in the pressure region 1–6 torr. They employ a stainless steel or copper probe with a sampling orifice of dimensions less than the mean free path of the molecules in the flames, to minimise disturbance from collisions with the metal during sampling. An intermediate pumping chamber attains a pressure of 10^{-4} to 10^{-5} torr, and contains an ion lens system to focus the ions onto the entrance slit of the mass spectrometer. A Bennett type RF mass spectrometer is employed, partly on account of the lightness of construction, but also because of the absence of magnets which might affect the ion plasma in the flame. Fig. 10 shows the entrance section of the mass spectrometer (with approximate operating voltages), and the inset gives an expanded section of the orifice. The authors give a large list of positive ions detected. Other accounts of this work have appeared elsewhere[126,127].

A number of positive ions have been identified in the reaction of atomic oxygen with C_2H_2, C_2D_2 and C_2D_4[128,129] using apparatus essentially similar to that of Calcote.

In the pressure range 10–100 torr, van Tiggelen *et al.*[130–134] have successfully carried out the mass analysis of flames. They have attempted to reconcile the requirements of a large sampling orifice and a low mass spectrometer pressure by using an interrupted sampling technique. A large sampling channel (2 mm) is used with a rotating disc pierced in one place only: gas only passes through for a brief period in each revolution. Two low pressure chambers are used, the first at 10^{-2} torr, and the second (the mass spectrometer itself) at 10^{-5} torr. The types and relative abundances of ions in several different flame mixtures have been discovered.

Sugden and his research group in Cambridge have investigated flames burning at 1 atmosphere[123,135–138]. A flame burns horizontally against a thin platinum foil clamped in a water-cooled block. A sampling hole is pierced in the platinum, and a range of hole sizes and foil thickness may be employed. A system of ion lenses increases the transmission of the apparatus by about two orders of magnitude, but also enhances the difficulties of maintaining constant sensitivity. The spectrometer employed was a 60° sector type of 12 in. radius. Much data has been obtained from these studies, not only about the ions present in flames at atmospheric pressures, but also about the ways in which samples may be falsified during introduction to the spectrometer.

Negative ions, as well as positive ions, appear in flames, and each main group of workers has identified negative ions by mass spectrometry. Thus Calcote *et al.*[139] have obtained profiles for a whole series of negative ions in acetylene–oxygen flames at low pressures. Feugier and van Tiggelen[140] identify an even greater variety of negative ions in the intermediate pressure range, and Knewstubb and Sugden[141] report a similar variety at 1 atmosphere, but only well downstream of seeded flames. Knewstubb[142] has shown, using the very high sampling rates which appear to be critical for this kind of experiment, that negative ions do appear in unseeded hydrocarbon flames. However, the negative ions are observed only in the preheating and very early reaction zones. Fontijn and Miller[143] observe negative ions also in their atomic-diffusion flame apparatus (*cf.* refs. 128, 129).

Photoionization phenomena are of considerable interest, not only because they may occur in the upper atmosphere, but also because accurate appearance potentials may possibly be determined from the photochemical threshold. Lossing and Tanaka[144] in some preliminary experiments found that they could obtain only very low ion currents with the restricted band widths of radiation needed to obtain accurate appearance potentials. Hurzeler *et al.*[145,146] working with low source pressures and using an electron multiplier appear to have overcome this difficulty. However, Steiner *et al.*[147] conclude that exact measurement of the ionization potential by the photolytic method is not possible. RF mass spectrometric techniques have been applied by Herzog and Marmo[148] to photoionization of several gases: investigations were performed with energies up to 11.4 eV, the cut-off of the lithium fluoride windows.

Upper atmosphere chemical processes have been mentioned above, and rocket-borne mass spectrometers have been used to investigate the various species—molecules, atoms and ions—present in the upper atmosphere. Bennett type RF mass spectrometers have been especially useful in view of their small weight compared with conventional magnet spectrometers. Johnson[149] describes a typical mass spectrometer–rocket probe of the atmosphere.

Heavy particle bombardment of uncharged species may lead to ionization, and it is believed that many radiochemical reactions may involve ionic intermediates. Melton and Rudolph[150–152] have made a mass spectrometric study of ions produced in gases by particle bombardment, and Melton[153] has investigated the secondary reactions of the ions in relation to radiochemical processes. Considerable information may also be obtained about ion–molecule reactions in these systems, and the reader is referred to a review by Melton[154] for detailed treatment.

4. ESR Spectrometry

Species possessing one or more unpaired electron spins may display the phenomenon of paramagnetism, and measurement of paramagnetic susceptibility affords a method for the estimation of such species present as reaction intermediates. Muller *et al.*[155,156] first used the method in a study of the dissociation of hexaphenylethane in benzene solution at different temperatures; the technique may be extended to the investigation of triplet states produced in fluorescence processes[157]. Unfortunately, direct determination of paramagnetic susceptibility has some important drawbacks. Correction must be made for the inherent diamagnetism of the species studied, and traces of ferromagnetic impurity will increase greatly the apparent paramagnetic susceptibility. The measurements do not in themselves yield any information about the nature of the paramagnetic species. Electron spin resonance (ESR) techniques overcome each of these difficulties: the diamagnetic component of the susceptibility appears only as a perturbation, ferromagnetic resonance is readily resolved from paramagnetic resonance, and the hyperfine structure and position of the resonance spectrum can lead to the identification of the paramagnetic species. Absolute concentrations may be measured by calibration of the spectrometer with a stable free radical such as diphenyl–picryl–hydrazyl (DPPH) or molecular oxygen.

In ESR spectrometry, the sample under investigation is placed in a magnetic field. The energy levels for a free electron become split, corresponding to the two orientations of the electron spin with respect to the field. The situation is more complex in a real free radical, since coupling may exist between electrons and neighbouring nuclei of non-zero spin: such coupling leads to the observed hyperfine structure, but to a first approximation, the energy levels are similar to those for a free electron. Transitions between the energy levels (subject to the selection rule $\Delta M_s = \pm 1$, $\Delta M_I = 0$) may occur, and can be observed by the absorption of radiation. The energy separation, ΔE, between states is given by

$$\Delta E = g_e \beta_e H$$

where H is the magnetic field, β_e is the Bohr magneton for the electron ($eh/4\pi mc$) and g_e is a molecular constant analogous to the Landé splitting factor for the electron in an atom (= 2.0023 for an electron uncoupled to any orbital angular momentum). Since $\Delta E = h\nu$, the frequencies at which absorption is to be expected may be calculated at any field strength. Most spectrometers have operated with magnetic fields of around 3000 gauss, which correspond to resonant frequencies of 9,000 Mc.sec^{-1} (*i.e.* a wavelength of 3 cm—the so-called *X*-band), although low-field instruments are sometimes used[158] in cases where the sample would absorb microwave radiation too strongly. The frequency at which resonance is made to occur affects also the sensitivity of the apparatus. If the two levels are in thermal equilibrium, then the relative populations of the two states are given by

$$\frac{n_{\text{upper}}}{n_{\text{lower}}} = \exp(-\Delta E/kT)$$

Hence the highest sensitivity is obtained with large ΔE (equivalent to large H) and low temperature. At the energy corresponding to X-band resonance, and at room temperature, the ratio is about 0.998. It may be shown that the ultimate sensitivity of a spectrometer working in the X-band at room temperature is approximately $10^{12}\ \Delta H.\tau^{-\frac{1}{2}}$ spins where ΔH is the width of the absorption line, and τ is the time taken to sweep through the line. The actual sensitivity is at least ten times less than this theoretical value; ΔH may be taken to be about 10 gauss and τ about 10 seconds, so that the minimum concentration of free radicals which may be detected is about 10^{-9} molar. Thus ESR spectrometers are normally not sufficiently sensitive for the estimation of intermediates present at their steady state concentrations in photochemical and thermal systems, although in specially favourable circumstances radicals may be detected. For example, Harle and Thomas[159] followed the concentration of radicals formed during the oxidation of octadecene by freezing samples withdrawn from the reaction mixture at different time intervals. The inhibiting action on the reactions of phenyl α-naphthylamine could also be studied.

One of the most successful techniques which has been used to produce concentrations of radicals measurable by ESR methods is that of 'trapping' (*cf.* Section 8). In condensed phases, the rate of radical recombination may be limited by diffusion, and at sufficiently high viscosities, the rate of production of radicals may exceed the rate of destruction, and the concentration of radicals will then build up. X-ray and γ-ray irradiation of glasses may lead to production of fragments with enough kinetic energy to escape from the 'cage' and thus avoid primary recombination. Schneider *et al.*[160] performed the first investigation by ESR spectrometry using X-ray irradiation. They irradiated polymethylmethacrylate, and obtained a resonance spectrum similar to that observed in polymerization studies of methylmethacrylate. Defects in crystals, produced by radiation damage, may also be studied by the resonance techniques[161]; work has been done on F-centres[162], V-centres[163,164], U-centres[164] and interstitial atoms[165], and such defect centres may be of importance in determining the course of chemical reaction in irradiated solids.

Fragmentation of molecules by ultraviolet radiation is perhaps of greater interest than fragmentation by X-rays or γ-rays, since the energy of the radiation may be controlled more closely, and the number of alternative reaction paths is much reduced. As a direct result of the smaller energy available, the kinetic energy of the photolytic fragments is less, and primary recombination may limit the concentration of free radicals if the medium is too viscous. Low temperature hydrocarbon glasses have proved useful for trapping studies of photolytically formed radicals, since the viscosity of the glass may be changed by alteration of the constitution of the hydrocarbon mixture or by variation of the temperature. An ether–isopentane–ethanol (EPA) mixture is well suited to these experiments. The earliest ESR spectra of radicals produced photochemically were made by Ingram *et al.*[166]. Ultraviolet photolysis at $\lambda = 2537$ A of ethyl iodide, benzylamine and benzyl chloride in a hydrocarbon glass produced radicals (although irradiation at $\lambda = 3650$ A did not).

Some measure of the mobility of the radicals within the glass may be obtained, since elevation of the temperature from 77° K to 90° K led to the recombination of the radicals. The variation in viscosity of substances which are glass-like at room temperature is less easily controlled, although some experiments have been described[167] in which radicals are trapped in plastic films.

Secondary radicals may be formed by reaction between primary radicals and solvent. Thus hydroxyl radicals are produced by the photolysis of hydrogen peroxide in an isopropanol glass[168]. At 110° K a seven-line spectrum, with intensities of 1 : 6 : 15 : 20 : 15 : 6 : 1, was observed. Such a structure is to be expected from an unpaired electron which interacts equally with six protons, and since it is known that there is no interaction with the proton of the –OH group in alcohol radicals, the spectrum suggests that the radical is $(CH_3)_2\dot{C}OH$.

Electron resonance spectroscopy has been used also in the investigation of vinyl polymerization in the solid phase. Thus Bamford *et al.*[169,170] have studied the radicals produced during the photochemically initiated solid phase polymerization of methacrylic acid, and the work demonstrates the scope of ESR methods. The way in which the radical concentration varies with the temperature at which the reaction is carried out yields valuable information about the reaction mechanism. The spectrum normally consists of nine lines, and is believed to originate from the methacrylic acid radical. If, however, the polymerization is carried out at temperatures below −5° C, then the spectrum is replaced by a more complex one containing thirteen lines. The new spectrum is ascribed[169] to an abnormal conformation of the methacrylic acid radical.

Polymerization studies have, in fact, been an important application of ESR spectroscopy. Although the stationary radical concentrations in radical polymerizations are only just within the limit of sensitivity of ESR spectrometers, two types of polymerization process are self-trapping, so that the radicals may be detected. A number of cross-linked vinyl polymers form a 'gel' during the course of polymerization, and the active radicals present at this stage of the reaction may be immobilised within the gel. Fraenkel *et al.*[171] made the initial observations on systems of this kind, using a number of monomers and initiators. Bamford *et al.*[172] showed that similar steric trapping occurred when the polymer formed was insoluble in the monomer. Thus polyacrylonitrile is insoluble in acrylonitrile, and is precipitated out during the course of reaction. Quite high concentrations of radicals—up to 10^{17} spins per gram—become trapped in the dead polymer cage. Kinetic studies, such as those of Atherton *et al.*[173], may be made by determination of the trapped radical concentrations after differing reaction times.

Intermediate species in the gas phase may also be studied by electron resonance methods, although the spectra obtained are usually more complex than those observed with condensed phases. In small gaseous radicals and atoms with degenerate orbital states, the orbital angular momentum will not be quenched, and may make a contribution to the paramagnetism of the species. Even in orbitally non-degener-

References pp. 336–342

ate molecules an interaction between the magnetic moment and the spin moment may occur. The exact nature of the contribution depends on the coupling case involved. If the coupling is described by the Hund (a) case, then microwave absorption in the presence of a magnetic field can arise from transitions between the different M_J values (subject to the selection rule $\Delta M_J = \pm 1$). However, some decoupling in the magnetic field may occur: that is to say, there may be an incipient Paschen–Back effect. With magnetic fields large compared with the internal field of the species, the electronic spin and orbital angular momentum will be quantised separately along the direction of the field. Such a decoupling effect is observed in the resonance spectrum of nitrogen dioxide[174], but not at all with nitric oxide[175]: molecular oxygen shows appreciable but not complete decoupling[176]. Hyperfine structure of the resonance spectrum of gases may result from interactions both with nuclear spin and with molecular rotation. A number of resonances are observed from each main electronic level, depending upon the multiplicity of the species. Fig. 11 shows the transitions expected in the three sublevels $^3P_0$, $^3P_1$ and $^3P_2$ of ground state atomic oxygen. It will be appreciated that while the gas-phase resonance spectra may be more complex than those of condensed-phase systems, the information carried by the spectra is also more complete. In suitable cases, both the chemical identity and the electronic state of the resonant species can be established. Absolute concentrations may be obtained by calculation from absolute absorption intensities, or, preferably, by calibration of the spectrometer with a suitable gas-phase standard (see below). Considerable effort has gone into relating line shapes and absorption intensities with perturbing factors such as field modulation (see ref. 177).

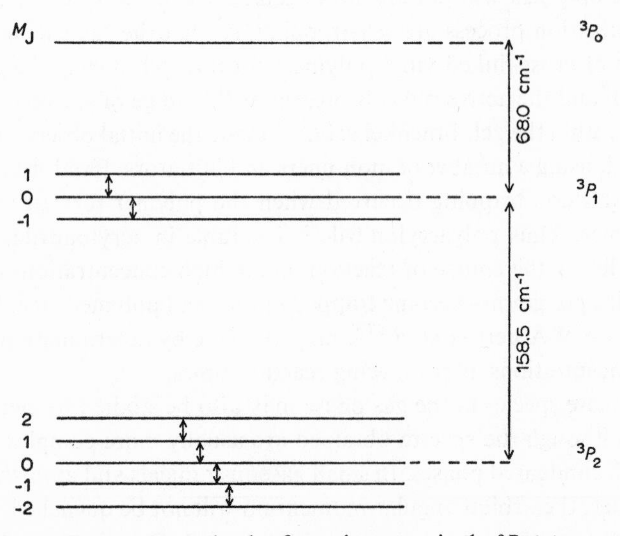

Fig. 11. Zeeman energy levels of atomic oxygen in the ^3P states.

The initial gas-phase studies of unstable species were of atomic hydrogen[178], oxygen[179] and nitrogen[180] produced in Wood's tube electric discharges. F, Cl, Br and I atoms have been observed by EPR (electron *paramagnetic* resonance, a term used to include electronic paramagnetism from sources which may not involve electron spin). (A list of references is given in a paper by Westenberg and de Haas[181]). The first radical, as distinct from atomic, species to be observed was hydroxyl[182]. The spectrum consists of a number of Λ-doublets, but arises from an electric dipole rather than a magnetic dipole transition. Thus a microwave cavity must be used in which the external magnetic field may be applied at right angles to the direction necessary for magnetic dipole studies. A cylindrical cavity, used in what is described as the TE_{011} mode, has proved popular for gas-phase radical experiments, since it can be employed for simultaneous studies of magnetic and electric dipole interactions.

EPR spectra have been obtained subsequently of SH[183], SeH[184], TeH[184], SO[183,185] and NF_2[186]. Carrington and Levy[187] have recently observed EPR spectra from the short-lived radicals ClO and BrO, produced in the reaction of atomic oxygen with the halogen. Excited NS($^2\Pi_{\frac{3}{2}}$, $J = \frac{3}{2}$), formed in the reaction of atomic nitrogen with hydrogen sulphide, has been identified[187].

Gas-phase EPR studies have proved useful in the qualitative way described above, and they have been used also in quantitative kinetic studies. Low pressure discharge flow methods are eminently suitable, and the microwave cavity can be incorporated in the flow tube. Krongelb and Strandberg[188] used the method to measure the rate of recombination of atomic oxygen. The spectrometer was calibrated using molecular oxygen: that this procedure is valid was shown later by Westenberg and de Haas[181], who checked the calibration for both O and N by titration with NO_2 and NO (see Section 5).

Spectrometer calibration for species such as hydroxyl has been carried out with nitric oxide, which also couples with the electric rather than the magnetic field. Westenberg[189] has calculated the theoretical basis of such calibrations, and used the method in flow-tube studies of hydroxyl reactions[190]. Dixon-Lewis *et al.*[191] have measured absolute OH concentrations in a similar way, and used the results to elucidate the kinetics of the reactions OH+OH, OH+H_2 and OH+CO.

EPR spectrometry has been applied also to flame studies. Westenberg and Fristrom[192] measured H and O atom profiles in a number of hydrogen and hydrocarbon flames. The sampling device was a quartz probe inserted into the low-pressure flame. The probe acted as a sonic orifice, and the decompression time was of the order of a few microseconds (see Section 3 for a discussion of sampling problems). Gases in the microwave cavity were effectively at room temperature. Some decay of atoms took place down the sampling tube: a correction was made by extrapolation of results obtained with the cavity at various distances from the probe.

The relatively high sensitivity of EPR spectrometry, compared with other methods for atom and radical detection, suggests its use in experiments designed to follow quantitatively the decay of active species under non-stationary conditions. The approach would be similar to that used in optical spectroscopy combined with flash photolysis, but might possess added advantages in sensitivity and absolute calibration; and, further, species not showing optical absorption in convenient spectral regions might be studied. Research using flash photolytic radical production coupled with EPR spectrometry has been started recently[193], and the technique is likely to be a valuable new tool in the field of reaction kinetics.

5. 'Chemical' methods

This section deals with methods for the detection or estimation of intermediates which depend upon the addition of foreign substances to the reaction system. By definition, such techniques cannot but perturb the system under examination, and the results obtained must be modified where necessary if they are to apply to the unperturbed case.

Experimental confirmation of free radical participation in chemical reactions was first obtained by Paneth and Hofeditz[194] by the 'chemical' method of mirror removal. This most familiar technique has been used subsequently in many studies of free radicals. The apparatus consisted of a flow tube operated at pressures of about 1–2 torr. In the original experiments of Paneth and Hofeditz[194] the reactant was lead tetramethyl, and the carrier gas was specially purified hydrogen. A lead mirror was deposited in the downstream part of the tube by local application of heat. If now the tube was heated upstream from the mirror, the first mirror gradually disappeared, suggesting that methyl radicals, produced pyrolytically from lead tetramethyl, removed the first mirror by the reaction

$$4CH_3+Pb \rightarrow Pb(CH_3)_4 \tag{22}$$

With sufficiently high flow rates (16 metres sec^{-1}) the radicals may be made to survive for up to 30 cm in the tube. It was shown[195] that the lifetime of methyl (or ethyl) radicals is more or less independent of the nature of the carrier gas, and that the rate of decay of the radicals is not affected by the presence of iron or platinum (*i.e.* the mirror removal cannot be ascribed to atomic hydrogen rather than the alkyl radicals).

There are two main ways in which the mirror technique has been applied. In the first, the concentration of radicals is determined by following the rate of mirror removal by one of a number of methods. These include determination of mirror opacity, change in radioactivity from a radioactive mirror, measurement of electrical conductivity, and so on. In the other method, the products formed by the removal of the mirrors are characterized, thus giving an indication of both the nature and concentration of the radicals concerned. Thus Paneth and Hofeditz[194] demonstrated the presence of methyl radicals by characterization of zinc dimethyl formed from zinc mirrors, and ethyl radicals were detected in a similar manner[196]. The formation of the dark red polymeric telluroformaldehyde, $(CH_2Te)_x$, has been taken as a characteristic product of the reaction between methylene and a tellurium mirror. Rice *et al.*[197,198] used the test diagnostically to show that methylene is formed by heating diazomethane, while the products of pyrolysis of methane at 1200° C yield only ditellurium dimethyl on reaction with a tellurium mirror. In fairness, it must be said that the results are not unequivocal, and a certain amount of conflict exists between this and other data.

There appears to be some specificity of attack on the different mirrors. Rice and Glasebrook find[198,199] that although Zn, Cd, Bi, Tl and Pb mirrors are attacked by alkyl radicals, they are not affected by methylene; on the other hand, methylene does remove Te, Sb, Se and As mirrors. Once again, there is some lack of agreement between different groups of workers. For example, Norrish and Porter[200] find that Bi *is* removed by methylene, while both methylene and alkyl radicals attack Pb and Zn but slowly. Atomic hydrogen[201] removes mirrors of As, Sb, Te, Ge or Sn, but not mirrors of Pb or Bi. Burton *et al.*[202–205] have utilised the differing reactivity of Pb towards alkyl radicals and atomic hydrogen to determine concentrations of the latter: they used two mirrors, the first a heavy one of lead to remove all alkyl radicals, and a second light one of antimony to look for atomic hydrogen. To detect methyl radicals in the presence of large amounts of atomic hydrogen, bismuth mirrors have been employed[206,207].

The apparent conflict between the results of different workers pinpoints the essential difficulties of the mirror technique. Not only is it difficult to deposit mirrors of a uniform and reproducible thickness, but also the activity of the mirror may depend upon the method of deposition, the age of the glass tube and a number of other factors. Poisoning of the metal surface seems to play a part in decreasing the reproducibility of the mirrors, and, in particular, oxygen inactivates most mirror deposits. Tellurium is not as sensitive to traces of oxygen as most metals[208], and is therefore a more satisfactory mirror material; and "guard" mirror techniques may be used to ensure complete reaction of the radicals. Nevertheless, the mirror technique is at best semi-quantitative, and is now only occasionally used.

A few other reactions between intermediate species in the gas phase and a solid reactant have been used for the estimation of the intermediate (*e.g.* Melville and Robb[209,210], have used, quantitatively, the "blueing" of molybdenum oxide by atomic hydrogen). However, the great majority of chemical reactions used for the determination of intermediates are carried out homogeneously, and most are gas-phase reactions.

The basic requirement of any chemical reaction designed to investigate unstable intermediates is that it should yield, in a reproducible and known manner, a product susceptible of characterization. A few typical investigations may be described.

References pp. 336–342

Polanyi *et al.*[211,212] have identified ethyl, phenyl and methyl radicals by allowing them to react with chlorine or iodine, and then identifying the halide formed. The reaction of radicals (and hydrogen atoms) with iodine has been used in a number of studies, including photochemical ones[213,214]. A similar method has been employed by Szwarc[215], using toluene as a radical trap. Benzyl radicals are readily formed from toluene, and disappear from the system as dibenzyl without undergoing any side-reactions. The amount of dibenzyl formed may then be determined, and the extent of the radical reaction with toluene inferred.

Complete inhibition of chain reactions by reactants such as nitric oxide or oxygen in the gas phase, or hydroquinone in solution, is good evidence for supposing the reaction to be propagated by radicals or atoms. Considerable information about the chain length may be obtained from a comparison of the rates of reaction in the presence and absence of inhibitor; the studies by Hinshelwood (see ref. 216) of nitric oxide inhibition (*e.g.* of ethane and diethyl ether decomposition) are particularly noteworthy in this regard. The rate of radical formation may be determined either from the time required to consume all the inhibitor, or by measuring the rate of formation of any stable products of the inhibition reaction. The latter method depends directly on the stability of products, and is therefore of rather limited application, particularly in thermal reactions, and, further, gives information about the rate of radical formation only in the absence of the chain processes.

In one of the most elegant applications of gas-phase inhibition by nitric oxide, Birss, Danby and Hinshelwood[217] have studied the thermal dissociation of *t*-butyl peroxide. The low temperatures required for pyrolysis permitted mass spectrometric determination of *t*-butyl nitrite, and a fairly complete kinetic analysis of the system was possible. The rate of decomposition of peroxide was related to the consumption of nitric oxide and to the appearance of butyl nitrite during the inhibition period, and curves were obtained which showed the acetone and ethane concentrations as a function of time during and after inhibition.

A reaction which does not necessarily perturb the stationary concentration of hydrogen atoms is the conversion of *para*- to *ortho*-hydrogen

$$H+H_2(para) \rightarrow H_2(ortho)+H \tag{23}$$

The kinetics of the reaction have been investigated[218,219], and it is therefore possible to determine atomic hydrogen concentrations from the rate at which *ortho*-H_2 appears in a system initially containing only *para*-H_2. *Ortho–para* hydrogen ratios may be determined by vapour phase chromatography using charcoal columns[219], and this technique probably replaces the earlier thermal conductivity measurements (for a description of the latter see ref. 220). The rate coefficients of reaction (23) cannot be fitted to a simple Arrhenius expression; Schulz and LeRoy[219] find that their results can be expressed by

$$\log_{10} k_{23} = 12.45-3.49\times10^3/T+3.83\times10^5/T^2$$

where k_{23} is in l. mole^{-1}. sec^{-1} units.

The technique has been used by Melville and his colleagues in studies of a number of reactions: the reaction of atomic hydrogen with hydrazine[220], and olefins[221] and the photolysis of ammonia[222]. Atomic hydrogen was formed by mercury-sensitised photolysis of molecular hydrogen. Farkas and Sachsse had also applied the *o–p*-H_2 conversion previously to determine the atomic hydrogen concentration in mercury-sensitised photochemical systems[223], and Patat[224] and Sachsse[225] used the conversion to detect H atoms formed in the pyrolysis of ethane. An assumption is made in the latter work that catalysis of the *para*-hydrogen conversion by methyl radicals is negligible compared with the effect due to atomic hydrogen. It is not clear how far such an assumption is justified since the conversion is, in fact, catalysed by all free radicals, including "stable" ones such as nitric oxide and even molecular oxygen. Indeed, the method has been used to determine radical concentrations, as in the photochemical experiments of West[226]. The method is obviously most valuable where the nature of the reaction intermediates is already established, and a measure of their concentration is required. A second restriction exists in that if reaction (23) is not to affect the concentration of intermediate (a situation only possible for atomic hydrogen) then it must proceed more rapidly than the fastest step of the chemical reaction under investigation.

Isotope exchange reactions can be used in the same way as the *ortho–para* hydrogen interconversion, and the results may be interpreted more readily. The obvious reaction to compare with the *ortho–para* exchange is

$$H+D_2 \rightarrow HD+D \tag{24}$$

and in systems containing molecular hydrogen, the process

$$D+H_2 \rightarrow HD+H \tag{25}$$

also leads to the formation of HD. The rate coefficients for these processes are known[227,228], so that measurement of the rate at which HD appears (mass spectrometrically, for example) can be used to calculate the atomic hydrogen concentration. If, however, the experimental conditions are such that H_2, D_2 and HD rapidly reach their equilibrium concentrations, then the method becomes unsuitable. Fenimore and Jones[229] experienced this difficulty in flame studies: at all points within the flame the HD concentrations were at their equilibrium values. These workers found that the reaction

$$H+D_2O \rightarrow HD+OD \tag{26}$$

could be used instead of (24) to measure atomic hydrogen concentrations. At flame temperatures (26) proceeds much more slowly than (24): Fenimore and Jones use a value of $k_{26} = 10^{12} \exp(-25{,}000/RT)$ l.mole^{-1}.sec^{-1} for reasons which they discuss[229].

The use of other isotopically labelled compounds may be of great value in the study of unstable intermediates; in all cases, however, due allowance should be made for the differing chemical reactivities of the isotopically substituted compounds.

Flame studies may also make use of a number of straightforward chemical reactions which would proceed too slowly at lower temperatures. Thus Fenimore and Jones[230] were able to measure atomic oxygen concentrations in hydrogen and hydrogen–carbon monoxide flames from the rate of the reaction

$$O+N_2O \rightarrow 2NO \tag{27}$$

Nitric oxide concentrations at given points within the flame were measured mass spectrometrically with known small concentrations of nitrous oxide injected into the reaction mixture. Hence the atomic oxygen concentration could be determined using the rate coefficient for (27), $k_{27} = 2\times10^{11}$ exp $(-32{,}000/RT)$ l.mole^{-1}.sec^{-1}. It was shown that nitric oxide itself would not decompose appreciably under the conditions of the experiment.

Sugden and his co-workers[231–236] have developed a number of special techniques for the estimation of intermediates in flames, with particular reference to atomic hydrogen and hydroxyl radicals. In each case the technique involves the addition to the reaction mixture of traces of metal salts, which lead to the emission of radiation in the flame. The basis of the first method[232] is a comparison of the relative intensities of the lithium and sodium resonance lines emitted when salts of these metals are added in equal concentrations to the flame. Lithium hydroxide is stable at the flame temperatures, and since water is one of the combustion products the lithium concentration is modified by the equilibrium reaction

$$Li+H_2O \rightleftharpoons LiOH+H \tag{28}$$

The corresponding reaction for sodium is negligible, so that measurement of the intensities of the resonance lines for the two metals may be combined with an estimated equilibrium constant for (28) to give the atomic hydrogen concentration. The method is satisfactory in that only trace amounts of the alkali metal salt need be added, and no appreciable perturbation of the reaction system occurs. Values for the atomic hydrogen concentrations obtained by the method described were compared with those calculated when chlorine was added to a flame already containing traces of sodium[232]. An equilibrium is again established as a result of the reactions

$$Na+HCl \rightleftharpoons NaCl+H \tag{29}$$

and the intensity of the atomic sodium emission is diminished. The concentration of HCl may be estimated from the total amount of chlorine added, and the value combined with the intensity measurements to calculate the atomic hydrogen concentration. Although in this case the reaction system is probably affected by the relatively large amounts of chlorine added, the agreement between atomic hydrogen concentrations measured by the two methods is remarkably good.

James and Sugden[233] have shown that the radiation emitted from flames containing traces of alkali metals consists of a continuum as well as the atomic emission lines. The continuum extends from the red to near ultraviolet; its intensity depends little upon the temperature, but may be correlated with the concentration of the hydroxyl radical in the flame. It is concluded that the origin of the continuum is the radiative process

$$A + OH \rightarrow AOH + h\nu \tag{30}$$

where A is an alkali metal. The intensity of the continuum can therefore be used as a measure of the relative hydroxyl concentration in a flame.

Resonance emission from alkali metals itself seems to arise from two sources. Padley and Sugden[234] have shown that in addition to thermal excitation of the sodium D lines, the processes

$$H + H + Na \rightarrow H_2 + Na^* \tag{31}$$

$$H + OH + Na \rightarrow H_2O + Na^* \tag{32}$$

lead to emission of radiation. At "low" temperatures in flames (*e.g.* $\sim 1500°$ K) the bulk of the emission does, in fact, arise from reactions (31) and (32). Padley and Sugden have determined the rate coefficients for the reactions, and with a typical low temperature flame find $k_{31} \sim 7 \times 10^9$ $l^2.mole^{-2}.sec^{-1}$, and $k_{32} \sim 2 \times 10^{10}$ $l^2.mole^{-2}.sec^{-1}$. This quantitative data now makes possible calculation of atomic hydrogen and hydroxyl radical concentrations in the flame.

Addition of traces of copper salts to a flame produces radiation from a number of emitters. Bulewicz and Sugden have shown that both CuH[235] and CuOH[236] bands appear, and that the balanced processes

$$Cu + H + X \rightleftharpoons CuH + X \tag{33}$$

$$Cu + OH + X \rightleftharpoons CuOH + X \tag{34}$$

control the formation of the hydride and hydroxide. Thus determination of the intensities of the CuH and CuOH bands gives yet another measure of relative H or OH concentrations.

Three very fast reactions provide methods for the measurement of atomic oxygen, hydrogen and nitrogen concentrations; the techniques are particularly applicable to low pressure flow systems, and have been used with great success over the past decade. The reactions are

$$O + NO_2 \rightarrow O_2 + NO \tag{35}$$
$$H + NO_2 \rightarrow OH + NO \tag{36}$$
$$N + NO \rightarrow N_2 + O \tag{37}$$

The rate coefficients at room temperature are

$k_{35} = 1.5 \times 10^{9}$ l . mole^{-1}.sec^{-1} (ref. 109)
$k_{36} = 2.9 \times 10^{10}$ l . mole^{-1}.sec^{-1} (ref. 7)
$k_{37} = 1.3 \times 10^{10}$ l . mole^{-1}.sec^{-1} (ref. 109)

The reactions are so rapid under normal conditions of concentration that, with the flow velocities used in these experiments, they may be regarded as proceeding to completion. Thus estimation of one of the products or reactants leads to a value for the concentration of the atomic species. Phillips and Schiff[109] employed a mass spectrometric technique to this end (see Section 3 for the details of the mass spectrometer–flow tube apparatus). To measure atomic nitrogen concentrations, an excess of nitric oxide over that required for complete reaction with the atomic nitrogen was added. The height of the mass 30 peak was then determined with the discharge on and off: the difference in peak heights represents the amount of nitric oxide consumed in (37), and thus the atomic nitrogen concentration. Atomic oxygen concentrations were determined by similar observations on the mass 46 peak when nitrogen dioxide was added to the reaction system.

Visual or photoelectric observations of chemiluminescent emission from reactions involving atoms have, in fact, been much more widely applied than mass spectrometry as methods for the study of the titration reactions (35), (36) and (37). Strutt (Lord Rayleigh)[237] observed in 1910 a chemiluminescent emission (the "air afterglow") in air subjected to an electric discharge. It has been shown that the emission results from the reactions

$$O + NO + M \rightarrow NO_2^* + M \tag{38}$$
$$NO_2^* \rightarrow NO_2 + h\nu \tag{39}$$

and the intensity of emission is proportional to the product of the nitric oxide and atomic oxygen concentrations. Measurement of intensities of the emission when a trace quantity of nitric oxide is added therefore gives relative values for atomic oxygen concentrations. Nitric oxide is not consumed in the overall reaction since every molecule of NO consumed in the radiative process (38), (39), or in the equivalent non-radiative reaction, is regenerated in the fast step (35). Indeed, the absolute quantum yield for the chemiluminescent reaction has been determined[238], so that it is possible to use the intensity measurements to calculate absolute atom concentrations. In practice, however, the difficulties and uncertainties in the measurement of absolute light intensities make this method unsatisfactory. The use of nitrogen dioxide as a reactant furnishes a more suitable way of determining absolute concentrations. Nitrogen dioxide reacts stoichiometrically with atomic oxygen in (35), to produce nitric oxide, which may itself participate in the chemiluminescent processes (38), (39). If, however, sufficient nitrogen dioxide is added to remove all the atomic oxygen, then no chemiluminescence can appear. Thus the intensity of emission increases with increasing concentrations of NO_2 up to a maximum, reached when $[NO_2] = \frac{1}{2}[O]$. Further increase in $[NO_2]$ reduces the intensity of the chemiluminescence, which is extinguished when $[NO_2] = [O]$. The maximum intensity, or more particularly the "end-point" of the titration, may be determined photoelectrically with considerable accuracy. One disadvantage attached to the use of nitrogen dioxide as a reactant is the error liable to be introduced into determinations of reactant pressure or flow rate by the presence of dinitrogen tetroxide. The equilibrium constant for the NO_2/N_2O_4 reaction is, of course, known, and corrections to the measured pressures and flows can be made. Some special devices can be used for making the measurements without direct reference to the equilibrium. For example, the flow rate of nitrogen dioxide may be adjusted to reach the end-point of the titration; the NO_2 comes from a trap containing liquid N_2O_4, and the flow rate is calculated from the rate of loss of weight. A particularly interesting modification is given by Reeves *et al.*[239], who use both nitric oxide and nitrogen dioxide. The latter gas is added in the amount required to produce the *maximum* intensity of chemiluminescence, and then shut off. Nitric oxide is now added instead, the intensity matched photoelectrically with the maximum for NO_2, and the flow rate of NO measured. This flow rate may be seen to be one quarter that of the atomic oxygen.

Hydrogen atoms may be estimated[240,241] in a similar way by using the titration reaction (36). Nitric oxide formed in this reaction can participate in the chemiluminescent process[241]

$$H + NO + M \rightarrow HNO^* + M \qquad (40)$$

$$HNO^* \rightarrow HNO + h\nu \qquad (41)$$

Emission from HNO* lies mainly in the near infrared, and may be detected with suitable photomultipliers. Titration by nitrogen dioxide to the end-point (*i.e.* extinction of chemiluminescence) then gives the atomic hydrogen concentration. Clyne and Thrush[240] point out that the stoichiometry implied in equation (36) is only correct in the absence of molecular hydrogen, and that up to three molecules of NO_2 may be consumed for every hydrogen atom if much H_2 is present; they show, however, that at room temperature and with $[H]/[H_2] > \frac{1}{4}$ equation (36)

References pp. 336–342

represents the situation fairly faithfully. Similar observations on the overall stoichiometry of the reaction have been presented by Phillips and Schiff[7].

The recombination of atomic nitrogen gives rise to a yellow chemiluminescence[107], which is the main component of the glow of 'active' nitrogen

$$N+N+M \rightarrow N_2^*+M \quad (42)$$

$$N_2^* \rightarrow N_2+h\nu \quad (43)$$

The disappearance of the yellow emission can be used, therefore, to mark the end-point of reaction (37) when nitric oxide is used as titrant. It may be noted that since the titration reaction produces one atom of oxygen for every molecule of nitric oxide consumed, it also provides a means of producing known atomic oxygen concentrations.

The detection of excited species by chemical means will now be considered. If a species is energetically rich, it may take part in a chemical reaction more rapidly than does the normal species. In particular, if a reaction is exothermic for an excited species, but endothermic for the ground state reactant, then it may proceed isothermally only with the excited species. Typical of such reactions is that of atomic oxygen with water vapour; the reaction

$$O(^3P)+H_2O \rightarrow 2OH \quad (44)$$

is 17 kcal . mole^{-1} endothermic, while the reaction

$$O(^1D)+H_2O \rightarrow 2OH \quad (45)$$

is 29 kcal . mole^{-1} exothermic. A water vapour–ozone mixture shows absorption due to hydroxyl radicals on flash photolysis[80], suggesting that excited, ^1D, oxygen atoms are produced in the photolysis of ozone by ultraviolet radiation. Direct measurement of the quantum yield for ozone decomposition gives a similar result: the presence of water vapour greatly increases the quantum yield for photolysis by ultraviolet radiation, and it is believed that hydroxyl radicals produced in (45) are the chain carriers[11]. Water does not affect the quantum yield for photolysis by red light[242], at which wavelength there is only sufficient energy for the formation of $O(^3P)$.

Foote and Wexler[243] have investigated the reaction of singlet molecular oxygen (in the $^1\Delta_g$ or $^1\Sigma_g^+$ states) with a number of olefins. Sodium hypochlorite and hydrogen peroxide were added to solutions of the olefin: these compounds are known to react with the formation of singlet oxygen[244]. The products observed are identical with those seen in the photoinitiated autoxidation of the olefins, thus suggesting that singlet molecular oxygen is important in the autoxidations. Two examples of the olefins employed are 2,5-dimethylfuran in methanolic solution and 2,3-dimethylbutene-2. The reactions are shown below

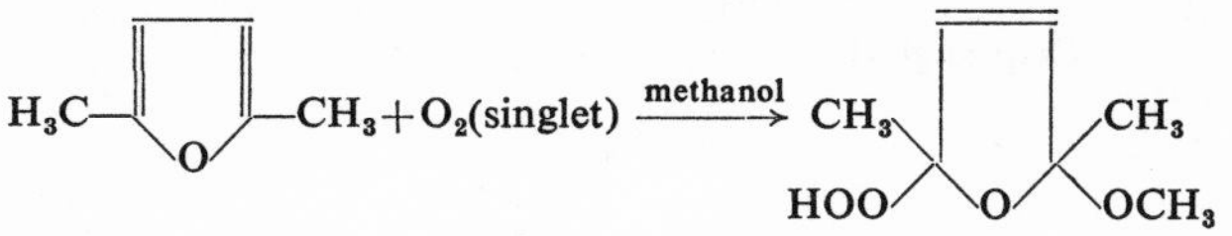

$$H_3C\text{-}(C_4H_2O)\text{-}CH_3 + O_2(\text{singlet}) \xrightarrow{\text{methanol}} \text{CH}_3,\ \text{HOO} > (\text{dihydrofuran ring, O}) < \text{CH}_3,\ \text{OCH}_3 \quad (46)$$

and

$$(H_3C)_2C{=}C(CH_3)_2 + O_2(\text{singlet}) \rightarrow (H_3C)(H_2C{=})C\text{-}C(CH_3)_2\text{-OOH} \quad (47)$$

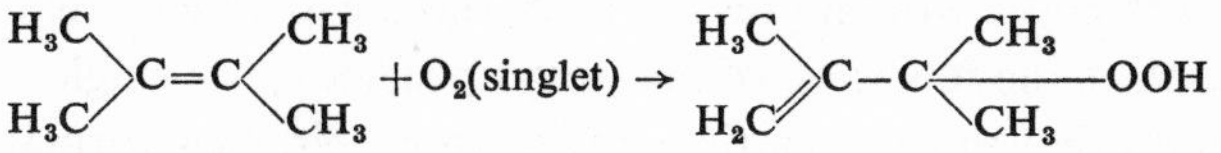

The reaction of 2,3-dimethylbutene-2 with $O_2(^1\Delta_g)$ has been used by Winer and Bayes[245] to study the decay of $O_2(^1\Delta_g)$ in the gas phase. Concentrations of the peroxidic product were measured by vapour-phase chromatography.

Ozone is potentially a useful reactant in this sort of study. The reaction

$$O_3 \rightarrow O_2 + O(^3P) \quad (48$$

is 23 kcal . mole^{-1} endothermic, and it seems that a number of excited species with more than around this amount of excess energy can pass on the energy to ozone. For example, $O_2(^1\Delta_g)$ and $O_2(^1\Sigma_g^+)$ both decompose ozone[10,246], and a similar reaction is postulated for the reaction of molecular oxygen in high vibrational levels of the electronic ground state[77] [yielding, in this case, $O(^1D)$ in (5)]. Phillips and Schiff[8] have used the reaction with vibrationally excited nitrogen

$$N_2^\dagger + O_3 \rightarrow N_2 + O_2 + O \quad (2)$$

to measure concentrations of the vibrationally excited species in a discharge-flow system. It must be said that the results could be misleading, since under the experimental conditions used, traces of atomic hydrogen could lead to the amount of decomposition of ozone ascribed to reaction (2)[10].

A chemical test of a quite different nature is used for the identification of excited methylene, CH_2. Methylene in the ground state is a triplet, but the photolysis of diazomethane or ketene (the standard methods for CH_2 production) may lead to the formation of appreciable amounts of the excited singlet methylene. The fraction of singlet methylene has been determined[247–249] by a study of the stereospecific addition of CH_2 to double bonds (*e.g.* the addition of methylene to *trans*-butene-2 to yield dimethylcyclopropane[249]). It is assumed that spin will be conserved in the addition; thus triplet methylene will form initially a triplet adduct (*i.e.* a biradical), and rotation may occur around the 2 : 3 C–C bond in the radical before ring closure. On the other hand, singlet methylene may form the cyclopropane directly, so that stereospecific addition mav be expected from the singlet and non-specific addition from the triplet.

$$\begin{matrix} H_3C \\ H \end{matrix} \!\!>\! C{=}C \!<\!\! \begin{matrix} H \\ CH_3 \end{matrix} + CH_2(\text{singlet}) \rightarrow \text{cis-1,2-dimethylcyclopropane } (CH_3, CH_3) \tag{49}$$

It seems that the interpretation of the results from these experiments is open to question. In the first place the retention of the stereospecificity, for singlet methylene only, requires both that the assumptions made about biradical formation for triplet and direct addition for singlet are accurate, and that rotation about the C–C bond is rapid compared with ring closure for the biradical. Secondly, addition of singlet CH_2 in reaction (49) leads initially to an excited singlet adduct which may itself isomerise unless it is quenched, and the experiment must be performed under conditions favourable to the quenching process. However, these same conditions favour the quenching of the singlet methylene itself, so that the proportion of singlet methylene detected may differ very appreciably from that produced.

6. Miscellaneous physical techniques

6.1 PRESSURE MEASUREMENT

Recombination of atoms in a constant volume clearly leads to a decrease in pressure, and Smallwood[250] was able to follow the recombination of hydrogen atoms manometrically. A mixture of atomic and molecular hydrogen, produced in an electric discharge, was compressed suddenly, and the pressure changes measured with a recording diaphragm gauge. Elias[292] has extended the technique to the measurement of H, N and O atom concentrations in low pressure flow experiments. A section of the flow system may be isolated rapidly from the main flow, and the pressure changes within the isolated region followed with a sensitive transducer. It is claimed that [N] and [O] may be measured within 10 % and [H] within 15 %. The value of the technique is that it is absolute, and it has been used to confirm the validity of atom concentration determinations which employ the chemical titration reactions (Section 5).

6.2 THERMAL CONDUCTIVITY MEASUREMENT

Thermal conductivity measurements offer a further possible method for detecting unstable reaction intermediates. Such studies have been made by Senftleben *et al.* to detect atomic hydrogen in the mercury sensitised photolysis of molecular hydrogen[251], and atomic chlorine in the photochemical decomposition of molecular chlorine[252].

6.3 CALORIMETRY

Bonhoeffer[253] first realised the possibility of detecting hydrogen atoms by measurement of the heat released when they recombine on a catalytic surface. The atoms are destroyed, so the applications of the technique are to systems where the species are generated continuously. All the studies to be described refer to discharge-flow experiments; even in these cases diffusion effects caused by the recombination process can lead to errors unless the flow velocity is sufficiently great[258]. The heat of recombination is known accurately, so that if the heat losses may be calculated or measured, the atom concentration can be determined. A few flow experiments have been described in which atomic hydrogen is measured by a determination of the temperature increase of a greater or smaller mass; thermocouples are normally employed in this kind of study to measure the temperature change. However, for atomic hydrogen, the preferred method of study is by the use of an isothermal hot wire calorimeter. The method, which was developed by Tollefson and LeRoy[254] for quantitative work, uses what is effectively a resistance thermometer in a current-passing arm of a balanced bridge circuit. The recombination of atoms on the hot wire would lead to an increase in temperature, so that to keep the bridge balanced, the electrical power dissipated in the catalytic probe must be reduced. It is apparent that the rate of flow, Φ, of atomic hydrogen is given by

$$\Phi = \frac{R}{J\Delta H}(i_0^2 - i^2)$$

where R is the resistance of the probe at the constant temperature, i_0 and i the currents required to maintain the temperature without and with the atoms present, ΔH the heat of recombination per mole of the atoms and J the mechanical equivalent of heat. It is assumed in using the above relation that all the atoms recombine on the probe. At room temperature the recombination efficiency of atomic hydrogen on materials such as platinum is low, but increases with temperature[255], and careful design of the apparatus is necessary. Larkin and Thrush[256,257] describe an experimental arrangement designed to eliminate uncertainties in the fraction of atoms recombining on the probe, and their apparatus is shown in Fig. 12. Two calorimeters were used, each consisting of 50 cm of 34 swg rhodium–platinum wire. The calorimeters were mounted in tandem about 3 cm apart, and the second, downstream, calorimeter was used to determine the efficiency of the first (which was always > 80 %). Kinetic measurements were effected by making the pair of calorimeters moveable up- or down-stream in the flow tube.

References pp. 336–342

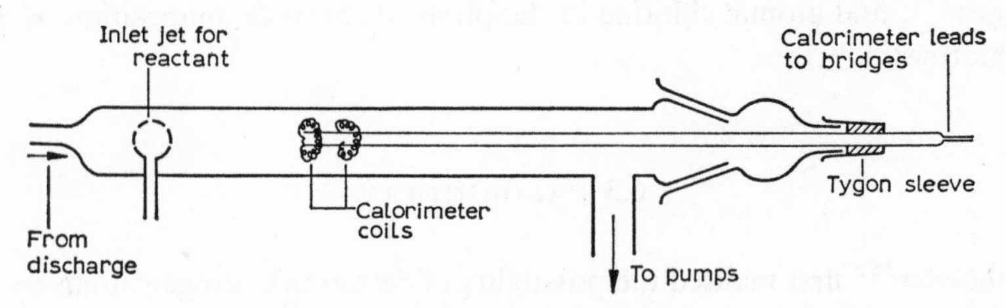

Fig. 12. Double isothermal wire calorimeter probe for low pressure flow system.

Atomic oxygen concentrations may be determined by the calorimetric method if the surface of the catalytic probe is chosen suitably. Linnett *et al.*[259,260] have studied the recombination of oxygen atoms under various conditions using a thermocouple coated with silver. In the presence of atomic oxygen, the surface of the silver is oxidised, and the oxide layer appears to be a particularly efficient catalyst for the O-atom recombination. The calorimetric determinations were shown[261] to be related directly to atom concentrations measured with the Wrede–Harteck gauge to be described in Section 6.4.

Absolute atomic oxygen concentrations are best measured with the isothermal resistance probe. Elias *et al.*[262] describe investigations of atomic oxygen reactions using a silver coated platinum wire coil.

Atomic oxygen and hydrogen have been the species most widely studied by the calorimetric technique, although relative concentrations of a few other atomic species have also been measured. For example, Schwab used the heating of a thermocouple to measure relative chlorine[263] and bromine[264] atom concentrations. However, one of the most interesting developments of the method has been discovery that excited, energy-rich, species may be estimated calorimetrically. Elias *et al.*[262] reported that the apparent concentration of atomic oxygen measured with the catalytic probe was about 10 % higher than that measured with a Wrede–Harteck gauge (Section 6.4) and 25 % higher than the value obtained by NO_2 titration (Section 5). The effect was ascribed to appreciable concentrations of metastable $O_2(^1\Delta_g)$ present in the products leaving the electric discharge, and, indeed, some heat is released even after the *atomic* oxygen has all been removed by reaction with ethylene. So far as measurements of atom concentrations are concerned, it is desirable to find a surface which is an efficient catalyst for the O-atom recombination, but which does not deactivate the metastable molecules. A fresh silver oxide surface has this property, although in use it becomes progressively more sensitive to excited molecular oxygen. On the other hand, where it is the metastable species which are of direct interest, an oxidised cobalt or nickel surface is found to be especially suitable. Atomic oxygen may be removed from the products of the electric discharge by passing the effluent gases over a mercury mirror. Bader and Ogryzlo[265] studied calorimetrically the reactions of both $^1\Delta_g$ and $^1\Sigma_g^+$ metastable molecular oxygen. Arnold *et al.*[266] describe a similar kinetic study using a nickel coated calorimetric detector.

Vibrationally excited species also may give up their energy to a suitable catalytic probe. Morgan *et al.*[9] report calorimetric studies of vibrationally excited molecular nitrogen, and compare the results obtained with those from mass spectrometric investigations.

6.4 WREDE–HARTECK GAUGES

Wrede[267] and Harteck[268] first applied the property of effusion to the measurement of atom concentrations. A catalyst, effective for the atom recombination reaction, is placed within an enclosed volume which is connected to the reaction system by one or more small orifices. Molecules and atoms effuse into the enclosed volume, while only molecules effuse out. Thus an equilibrium is set up, and mass balance requires a pressure difference, Δp, between the two sides of the orifice, given by

$$\Delta p = \alpha p \left(1 - \frac{\sqrt{2}}{2}\right)$$

where p is the total gas pressure and α the fraction of atomic species. Hence, if the chemical identity of the atomic species is known, measurement of Δp and p gives its concentration. Neither foreign reactant gases nor excited species affect the value of Δp, so that the method commends itself for use in circumstances where such species may be present. There are, however, certain limitations to the employment of the technique. First, the technique cannot distinguish between different atomic species. Secondly, the accuracy and sensitivity limit depends on the sensitivity with which Δp is measured. Groth and Warneck[269] have described recently the use of a diaphragm manometer capable of measuring a pressure difference of 10^{-4} torr; at $p = 1$ torr, this corresponds to an atom concentration of 0.03 %. Greaves and Linnett[261] use a special Pirani gauge which enables them to measure atom concentrations to ± 0.1 % at $p = 0.1$ torr. A third factor which is of importance is the rate at which equilibration occurs between the enclosed volume and reaction system. In order to minimise mass transfer through the orifice (which would lead to erroneous, low, values of Δp), the size of the orifice must be kept small compared with the mean free path. Thus the equilibration rate may be slow, and it is essential to use as small an enclosed volume as possible. It has been shown[261] that with a volume as small as 3 cm^3, the time required for 99 % equilibration with a single orifice of diameter equal to one-tenth of the mean free path is 4 sec in oxygen at $p = 0.1$ torr, but more than 7 min at $p = 1$ torr. The pressure range within which concentration measurements may be made is therefore restricted, and the problem is the more acute the heavier the atom. Nevertheless, useful results have been obtained for oxygen (*e.g.* ref. 261), nitrogen[270] and even chlorine[270a] atoms.

A number of workers have attempted to use multiple orifices to reduce equilibration time, and Beckey and Warneck[270b] use fritted discs, although it is possible that errors will result from the multiple surface reflections of atoms within the long tubes which comprise the disc. To overcome such objections, Groth and Warneck[269] have developed orifices in the form of long, narrow slots, and gauges using this system appear to be relatively fast and accurate. Greaves and Linnett[261] give a useful account of the possible sources of error in effusion gauge determinations of atom concentrations.

7. Electrical methods for charged species

The detection and estimation of ions in the gas phase is carried out very conveniently by the mass spectrometric methods described in Section 3. However, electrically charged species may be detected also by methods depending on the electrical properties peculiar to these species. This section gives a very brief outline of some of the techniques which have been employed in the study of charged species in chemical reactions. Flames offer the most readily investigated chemi-ionization system, and most of the work reported here derives from flame studies.

The conductivity of a gas containing charged species depends upon the number of charge-carrying species and their mobility. Explicit mobility studies of the charged species in flames have, in fact, been made[271], although electrical conductivity is more usually the property measured. The information to be obtained from conductivity measurements has been described by Wilson[272]. Langmuir probe methods are well suited to flames since they can be designed to have quite good spatial resolution. The technique is usually applied in a "single-probe" manner: that is, a probe is inserted into a flame and a large grid placed some distance above the probe. Electron currents to a positive probe or positive ion currents to a negative probe may then be measured. Calcote[273-275] has recently put the determination of charged species by such methods on a quantitative footing, and he discusses[274] the interpretation of the probe current curves. The use of small probes is desirable for a number of reasons: (*i*) it is difficult to obtain saturation electron currents on a large probe without saturating the grid electrode, (*ii*) large probes imply relatively large currents which may affect the plasma under study, (*iii*) better spatial resolution will be obtained with small probes. However, most electrical insulators become either semiconductors or thermal emitters at the elevated temperatures of flames, so that there is some difficulty in building small probes. Calcote[274] has overcome the difficulty by using water cooled insulators for his probes, and the use of boron nitride insulators, which have good high-temperature insulation properties, has been reported[276]. Langmuir probe studies have been used also in indirect determinations of negative ion concentrations[275]. The electron concentration, measured with a probe, in low-pressure acetylene–oxygen flames, is only a few per cent of the positive ion concentration, and the negative ion concentration may be obtained by difference. The very high apparent concentrations of negative ions calculated in this manner do, however, appear to be at variance with mass spectrometric values. Travers and Williams[276] believe that single probe methods are not particularly suitable for the measurement of electron concentrations, and describe a double probe technique which they have developed. Unfortunately, it seems that although the double probe technique can be used to measure positive ion concentrations (with less spatial resolution than the single probe technique) yet it is unsuitable for the measurement of electron concentrations. In reply to the criticisms of the use of single probes, Calcote has said[277] "Therefore, for determining local electron concentrations in a flame, one is left with two choices: the use of single probes which is fraught with experimental difficulties, or to make no measurements at all". It seems, then, that at the present time single probe experiments are at least as useful as double probe studies. The use of Langmuir probes has been extended to the study of ionization in shock tubes[278].

Another way in which the conductivity of a flame, and hence the electron concentration, may be determined is by the use of microwaves. Studies of this kind have the great advantage that there is no interference with the plasma, but at the same time the spatial resolution is rather poor (since it depends on the size of the wave guides used). In a series of papers Sugden *et al.* have examined microwave methods for the determination of electron concentrations in flames containing traces of alkali metals. The earliest crude results[279] were extended to studies of coal-gas flames[280]. In the latter experiments the flame was placed in the path of the microwave radiation and the attenuation determined. Sugden and Thrush[281] improved the sensitivity of the microwave technique by introducing the conducting column of flame gases into a resonant cavity. The "Q" of the cavity is reduced, and the conductivity of the gases may be calculated from the "Q" values. Sufficient sensitivity was available in this apparatus to make possible determinations of the electron concentration in "pure" acetylene–air flames (*i.e.* free from alkali metal salts). Refinements in microwave technology have subsequently allowed much greater sensitivity in the instruments. Thus Shuler and Weber[282] used a bridge circuit, slightly off balance, with the flame between two electromagnetic horns in one arm of the bridge. Concentrations as low as 5×10^{-8} electrons cm^{-3} could be detected with this apparatus, and hence the electron density could be measured in "pure" hydrogen–oxygen and acetylene–oxygen flames. Sugden[283] has given a review of this work, together with the simplified—but sufficiently accurate—theory required for the interpretation of the results.

The microwave absorption of an electron cloud may have a resonance character if a magnetic field be applied. The condition for this "cyclotron" resonance is that the angular microwave frequency, ω, should be equal to the cyclotron frequency, ω_c, given by $\omega_c = eH/mc$. The collision frequency, ν, of electrons (with atoms or molecules) is required for an evaluation of the electron density from microwave attenuation measurements. However, the integral absorption in a cyclotron resonance experiment is a function of the electron concentration alone, so that the cyclotron method is, in this respect, more satisfactory than the microwave attenuation method. The relative half-width, ΔH, of the cyclotron resonance curve does, in fact, give a measure of the collision frequency, and it may be shown that $\Delta H/H = 2\nu/\omega$. Cyclotron resonance in flames was first observed by Schneider and Hofmann[284] with low pressure acetylene flames. The resonance curve becomes broader and less intense as the pressure (and, therefore, ν) increases, and with an acetylene–air flame, operating in the pressure range 30–200 torr, resonance could be detected only when potassium chloride was added to the flame. Stable flames could be established in the pressure range 6–25 torr with acetylene–oxygen, and at the lowest pressures resonance could be observed without added alkali metal. Bulewicz and Padley have used the cyclotron resonance technique to investigate flames of acetylene[285,286], cyanogen[285] and a variety of other fuels[287]. Useful information has emerged about the collision cross-sections of electrons with other species; in the study of acetylene flames[286] the cross-sections were obtained for collision with nitrogen, helium, argon, neon, water and carbon dioxide as well as with the reactant gases.

Radio frequencies (say in the range 10–100 Mc.sec^{-1}) are of the same order as the collision frequencies of the more massive ions in flames, and experiments using these frequencies can yield information about both electrons and heavier ions. Smith and Sugden[288] first applied the technique to flames, and gave an analysis of the effects expected on the resonant circuit characteristics when electrons or ions, or both, are present in the flame. The resonant frequency and the selectivity of the tuned circuit may be affected by the presence of charged species in the flame. At a frequency of about 100 Mc.sec^{-1} the resonant frequency shift and selectivity change induced by electrons in the flame show a parallel effect, whereas if massive ions are present in sufficient concentration they will markedly affect the resonant frequency but have little effect on the selectivity. From observations made on a hydrogen–air flame containing alkali metal salts it was possible to infer the presence of ions. The results indicated that the ions were twenty times as numerous as the electrons, and by dividing the ions equally into alkali metal and hydroxyl ions (to retain a charge balance) it was possible to estimate collision cross-sections for these species. More recently, workers[289,290] have tended to use the rf resonance method for its relatively high sensitivity to electrons rather than to exploit its potential use in the study of heavy ions. One reason for this bias is that changes in the ohmic resistance of the flame (which results mainly from the electrons) will swamp the effect of the heavier ions on the resonant frequency. Sugden[291] has, however, suggested that an rf experiment be performed with a flame containing an alkali and a few percent of a halogen. Such a flame should contain a large preponderance of heavy ions over electrons and both concentrations and collision frequencies should be measurable.

8. Trapped radicals

The great chemical reactivity of most free radicals results in relatively low steady concentrations of the species, even under conditions favouring their formation. Methods available for the estimation and detection of radicals may not be sensitive enough to detect the small concentrations encountered, and techniques must be sought which enable radicals to be accumulated without reaction until a concentration has been built up sufficient for detection by the instruments available. A brief discussion of work on "trapped" radicals therefore concludes the present chapter. The application of the technique of trapping has, indeed, been mentioned already in the section describing electron spin resonance (Section 4), and it is intended now to amplify the remarks made in that section with particular reference to optical spectroscopy of trapped species. The customary caveat will be issued at this stage: except for systems which are self-trapping, the results of trapped-radical studies refer to a system which differs from the normal reaction system, and care must be exercised in the extrapolation of results to the non-trapped case.

A reduction in temperature will decrease the rate of any radical reaction which has an appreciable activation energy, and it has long been understood that radical reactions become increasingly simple at lower temperatures. However, radicals combine with little or no activation energy, and except at temperatures very close to the absolute zero, recombination of radicals will be a significant process. By trapping radicals within a solid matrix, their diffusion towards one another may be impeded and recombination made unimportant. Such effects are observed in polymeric systems, where radicals may become trapped within a high polymer matrix even at room temperature (a survey is given in Chapter 12 of reference 293). An estimate of the maximum concentration of radicals that may be obtained by trapping may be arrived at by assuming that each reactive species must be isolated from its neighbours by a matrix cage. Jackson and Montroll[294] and Golden[295] have concluded that the radical concentration could not exceed 10 to 14 % of the total, depending upon the packing. In the event, the highest radical concentrations observed have been generally less than one-tenth of this theoretical estimate. One factor which may place a limit on the maximum concentration attainable is the effect of the heat released on those occasions when two radicals *do* recombine: local diffusion may be initiated to an extent sufficient to set off a "chain" of recombination events. Fontana[296] has shown that atomic nitrogen is stable in a molecular nitrogen matrix up to concentrations of 0.04 mole per cent, although at higher concentrations the sample becomes unstable.

Consideration will be given now to the requirements of a matrix that radical stabilisation may occur and to the ways in which radicals may be introduced into the matrix. In order that stabilisation of active species may occur, it is desirable that the matrix material

(*i*) is inert with respect to the active species;

(*ii*) is sufficiently rigid to prevent diffusion of the active species; and

(*iii*) has sites or holes suitable for the accommodation of the active species.

The vapour pressure of the matrix must be low at the operating temperature ($< 10^{-3}$ torr), although it may be convenient if the vapour pressure is high enough at room temperature to permit handling as a gas. Further restrictions on the matrix material are imposed by the method of detection which is to be employed. In particular, for spectroscopic investigation the matrix material must not absorb appreciably, and scattering must be reduced to a minimum. Some degree of compromise may have to be accepted in the properties required. Thus glasses of ether–isopentane–alcohol (EPA) are clear, and may be used for near-ultraviolet studies of trapped aromatic radicals[297], although the glass may absorb too strongly for use in infrared studies, and it may be too reactive to act as a good matrix for aliphatic radicals. On the other hand, the inert gases and nitrogen, which are chemically ideal, frequently form highly scattering microcrystalline deposits when frozen. The temperature at which the matrix is formed may affect its scattering properties: Becker and Pimentel[298] find, for example, that xenon forms a fairly transparent matrix if deposited at 60° K, but a highly scattering one if formed below 50° K. Scattering may be less of a problem in infrared studies than it is in visible or ultraviolet experiments, although against the reduced scattering must be set the much smaller oscillator strength of vibrational, rather than electronic, transitions.

Radicals may be trapped in inert gas matrices in one of two main ways. First, the radicals may be prepared by a gas-phase reaction (*i.e.* by photochemical, thermal, electrical or chemical processes) and allowed to condense in the presence of the matrix material on a cold surface. Secondly, a matrix containing a precursor of the radical may be deposited, and the radical produced in the matrix by a solid-phase reaction. Photolysis of the precursor is the most controlled way of forming the radical, although the precursors are then restricted to photosensitive materials. Again, the excess energy available in a photochemical reaction may be sufficient for the radical(s) to escape from the matrix cage and primary recombination may occur. Gamma ray or X-ray radiolysis, or electron bombardment, of the solid is not restricted to photosensitive precursors, nor is the excess energy likely to be insufficient to allow the radicals to escape from the cage. In general, however, radiolysis and electron bombardment are too indiscriminating in their attack, and photolysis remains the method of choice.

Radicals may be produced also by secondary reactions with a reactive matrix of the radicals first produced. For example, Milligan and Jacox[299] have prepared HCO (DCO) by the photolysis of HI (DI) or H_2S (D_2S) in a matrix of solid carbon monoxide. Formation and trapping of radicals from the gas phase and in the solid are reviewed by Thrush and Pimentel respectively in Chapters 2 and 4 of ref. 293.

Radicals trapped in solid matrices may be detected by those of the general methods suited to use with condensed phases. Electron spin resonance experiments are sensitive and may yield information about the nature and absolute concentrations of the radical species: typical ESR studies of trapped radicals have been presented already in Section 4. Calorimetric determinations may be used for the accurate absolute measurement of concentrations of radicals whose nature is known. Broida and Lutes[300] describe an experiment in which a large and rapid temperature rise is measured when solid nitrogen containing nitrogen atoms is allowed to warm up. The sudden temperature rise is ascribed to the recombination of atomic nitrogen when the sample has warmed up sufficiently for diffusion to take place. A knowledge of the heat capacity of the glass thus allows calculation of the initial atomic nitrogen concentration. Minkoff *et al.*[301] have devised a more sophisticated calorimeter, and have explored in detail the acceptability of calorimetric concentration determinations.

The study of trapped radicals by optical spectroscopy has been particularly fruitful. Unstable intermediates may be built up in concentrations sufficient to allow not only detection by ultraviolet absorption spectroscopy, but also investigation by the much less sensitive infrared techniques. In recent years, infrared matrix isolation spectroscopy (in the hands of Pimentel, Milligan, Jacox, and others) has proved a powerful tool in the determination of the structures, chemical reactivities and the various molecular parameters of radical species. However, before going on to discuss typical examples of this work, a brief description must be given of some earlier experiments which led to the development of matrix isolation spectroscopy. That solids may emit radiation has been known for a considerable time. Studies of phosphorescent solids have been carried out since the late nineteenth century: phosphorescence generally involves electronically excited molecules, and its study may yield data about the excited species. Vegard[302] was amongst the first to investigate the spectra of trapped radicals: his experiments involved the bombardment of solid nitrogen with electrons and positive ions, which resulted in an intense luminescence. The processes leading to the luminescence are complex, and after an exhaustive study, Peyron and Broida[303] came to the conclusion that a simple explanation of the observed emission was not possible. The recent interest in absorption by trapped radicals appears to have been triggered by the observation in 1951 of Rice and Freamo[304] that a blue solid is obtained if the pyrolysis products of hydrazoic acid are condensed rapidly at liquid nitrogen temperature. From that time on, absorption studies of an increasing number of radical, and even atomic, species has been reported. Ramsay gives a survey of the more important work published up to 1960 (Chapter 6, ref. 293).

References pp. 336–342

Solid state spectra are, in general, broader than the corresponding gas-phase spectra, and there may be some frequency shift of the various spectroscopic features. Thus some care must be exercised before absorptions in the solid phase are ascribed to particular species on the basis of gas-phase identifications. Further complications introduced by the matrix include splittings and changes in oscillator strength of optical transitions, but these effects may not themselves interfere with correct identification of the absorbing species. The nature of the matrix shifts is not understood, although some empirical data has been collected. McCarty and Robinson[305] examined the matrix shifts in different matrices of the $^3P_1$–$^1S_0$ resonance line of atomic mercury, and this work has been extended by Brewer *et al.*[306]. Meyer[307] finds that the absorption features of atomic sodium or potassium in argon, krypton or xenon matrices are shifted to the blue, and that there are signs of a triplet structure. If the matrix sample be warmed up, the absorptions broaden and shift to the red. Barger and Broida[308,309] have made similar studies of radical species in the region 3000–4500 A. A Knudsen cell was used to produce a molecular beam of known density of C, C_2 and C_3. In the case[308] of C_3 trapped in Ar, Kr, Xe, O_2, N_2, CO_2 and SF_6 matrices it was found that the line widths of the $^1\Pi_f \leftarrow {}^1\Sigma_g^+$ transition varied from 40 to 500 cm^{-1}, while the line shifts were in the range 410 to 1065 cm^{-1} (all positive, except with a carbon dioxide matrix, in which a red shift was observed). There was no definite effect of temperature, nor did there appear to be any great effect of the matrix on intensities. Experiments[309] on the species C_2 were less successful, and identifications were possible only for the species trapped in carbon dioxide. With this matrix, 5 bands of the $A^3\Pi_g \leftarrow X^3\Pi_u$ (Swan) system, and 4 bands of the forbidden $A^3\Pi_g \leftarrow X^1\Sigma_g^+$ system were observed.

The blue solid prepared by trapping the pyrolysis (or electric discharge) products of HN_3 has been mentioned already, and the solid was one of the first subjects of infrared matrix isolation study. Dows *et al.*[310] were able to ascribe most of the infrared absorption features of the solid to HN_3, NH_4N_3 and NH_3, and from the behaviour of the spectrum on warming the solid up from 90° K, they were able to show that further absorption bands were associated with the species $(NH)_x$ ($x > 2$ and may be 4). During the warming up process, di-imide, N_2H_2, appears, and the presence of imidogen, NH, at 90° K was inferred. Milligan and Jacox[311] obtained the infrared absorption of NH and DH by the photolysis of HN_3 and DN_3 in Ar, N_2 and CO matrices at 4° K, 14° K and 20° K. It was shown that an appreciable concentration of NH was present in the ground, $^3\Sigma^-$, state, and from the reactivity observed in the CO matrix it was inferred that a side reaction is the appearance of H atoms by rupture of the H–N bond in HN_3. Identification of NH and N_2H_2 in the photolysis of HN_3–N_2 matrices at 20° K has been made unequivocal by studying the effect of substitution in HN_3 of H by D and ^{14}N by ^{15}N in the different positions[312]. Both *cis-* and *trans-*forms of N_2H_2 were observed. NH appears as well as NH_2 in the vacuum ultraviolet photolysis[313] of NH_3 in Ar or N_2 matrices (14° K). In a CO matrix, the photolysis of NH_3 leads to the infrared absorption of HCO which further demonstrates the production of atomic hydrogen in the photolytic process.

Matrix isolation spectroscopy has been used also in numerous attempts to study the species CH_2. The first investigations [314] of the photolysis of diazomethane in Ar and N_2 matrices (at 20° K) revealed new absorption features, which were ascribed to methylene and a CH_2N_2 tautomer, and a later study [315] using ultraviolet and infrared absorption spectroscopy appeared to confirm these identifications. However, after publication[87,88] of Herzberg's data on the gas-phase spectrum of CH_2, it was appreciated that the solid-phase features could not result from CH_2 absorption. A recent high resolution infrared investigation[316] suggests that the unidentified absorption is due to methylene-imine ($CH_2{=}NH$), formed together with HCN by the photolysis of CH_2N_2 dimers. This result indicates the importance of orientation in matrix photolysis, since $CH_2{=}NH$ is not formed in the gas phase, nor is it formed by the matrix photolysis of diazirine. The absence of absorption attributed to CH_2 is not to be taken as evidence that CH_2 is not produced photochemically from CH_2N_2, but rather as evidence of the extremely high reactivity of CH_2. Indeed, Moore and Pimentel[317] find that CH_2 reacts with N_2 in the matrix: the photolysis of $CH_2{}^{15}N^{14}N$ in solid $^{14}N_2$ leads to the appearance of $CH_2{}^{14}N^{14}N$, and the matrix photolysis of diazirine yields some diazomethane. On the other hand, CF_2, formed in high concentration by the matrix photolysis of CF_2CO does not react[318] with CO or CO_2. The species $CH_2{=}NH$ ($CD_2{=}ND$) is formed also by the photolysis of methyl azide (methyl d_3-azide) in Ar (4.2° K) or CO_2 (50° K) matrices[319]. Photolysis[320] in similar matrices of the azides FN_3, ClN_3 and BrN_3 leads to compounds whose infrared spectra show them to be NF, NCl and NBr: in the case of NBr there is considerable matrix deactivation to the ground $^3\Sigma^-$ state. Cyanogen azide (NCN_3) may be photolysed at wavelengths greater than 2800 A to form NCN in Ar, N_2, CO or CO_2 matrices[321]. The radical was identified by its infrared absorption spectrum taken together with ultraviolet absorption data. Isotope studies confirmed the identification, and it was possible to estimate the force constants and thermodynamic properties of the species. If the photolysis is carried out in a CO matrix at wavelengths shorter than 2800 A the infrared and ultraviolet absorption of CCO may be observed[322].

Finally mention should be made of an interesting and useful method, involving a "rotating cryostat" developed by Thomas *et al.*[323], for the preparation of trapped hydrocarbon radicals. Beams of sodium atoms and halohydrocarbon molecules are condensed in a vacuum of $< 10^{-5}$ torr on opposite sides of the exterior surface of a rapidly rotating stainless steel drum containing liquid nitrogen. Radicals formed by the reaction

$$Na + RX \rightarrow NaX + R$$

are trapped in a matrix of excess halohydrocarbon and their spectra can be obtained. By a slight modification, reactions of R with substances such as oxygen can be investigated.

The examples given show the way in which studies of trapped radicals can complement other investigations of intermediate species. The techniques now at the disposal of the chemist are such that any intermediate participating in a chemical reaction should be susceptible to detection, identification and quantitative estimation, although in many cases a great deal of persistence and good luck will be needed to achieve these objects.

REFERENCES

1 A. Smithells and H. Ingle, *Trans. Chem. Soc.*, 61 (1892) 216.
2 A. G. Gaydon, *Proc. Roy. Soc. (London), A*, 179 (1942) 439.
3 A. G. Gaydon, *Spectroscopy and Combustion Theory*, Chapman and Hall, London, 1948, pp. 38–58.
4 J. N. Bradley, G. A. Jones, G. Skirrow and C. F. H. Tipper, *Symp. Combustion, 10th* (1965) 139.
5 H. S. Glick, J. J. Klein and W. Squire, *J. Chem. Phys.*, 27 (1957) 850.
6 D. Garvin and M. Boudart, *J. Chem. Phys.*, 23 (1955) 784.
7 L. F. Phillips and H. I. Schiff, *J. Chem. Phys.*, 37 (1962) 1233.
8 L. F. Phillips and H. I. Schiff, *J. Chem. Phys.*, 36 (1962) 3283.
9 J. E. Morgan, L. F. Phillips and H. I. Schiff, *Discussions Faraday Soc.*, 33 (1962) 118.
10 M. A. A. Clyne, B. A. Thrush and R. P. Wayne, *Nature*, 199 (1963) 1057.
11 R. G. W. Norrish and R. P. Wayne, *Proc. Roy. Soc. (London), A*, 288 (1965) 200.
12 B. G. Reuben and J. W. Linnett, *Trans. Faraday Soc.*, 55 (1959) 1543.
13 R. R. Williams and R. A. Ogg, *J. Chem. Phys.*, 15 (1947) 691.
14 R. G. W. Norrish and G. Porter, *Nature*, 164 (1949) 658.
15 G. Porter, *Proc. Roy. Soc. (London), A*, 200 (1950) 284.
16 R. G. W. Norrish, G. Porter and B. A. Thrush, *Symp. Combustion, 5th* (1955) 651.
17 R. G. W. Norrish and B. A. Thrush, *Quart. Rev. (London)*, 10 (1956) 149.
18 G. Schott and N. Davidson, *J. Am. Chem. Soc.*, 80 (1958) 1841.
19 C. E. Campbell and I. Johnson, *J. Chem. Phys.*, 27 (1957) 316.
20 G. Herzberg, *Spectra of Diatomic Molecules*, D. Van Nostrand, Princeton, N. J., 1950, p. 472.
21 J. E. C. Topps and D. T. A. Townend, *Trans. Faraday Soc.*, 42 (1946) 345.
22 W. W. Watson, *Astrophys. J.*, 60 (1924) 145.
23 D. Jack, *Proc. Roy. Soc. (London), A*, 115 (1927) 373.
24 A. G. Gaydon, *Spectroscopy of Flames*, Chapman and Hall, London, 1957.
25 A. G. Gaydon, *Quart. Rev. (London)*, 4 (1950) 1.
26 R. J. Dwyer and O. Oldenburg, *J. Chem. Phys.*, 12 (1944) 351.
27 J. W. White, *J. Chem. Phys.*, 6 (1938) 294.
28 G. Stephenson, *Proc. Phys. Soc. (London), A*, 64 (1951) 666.
29 L. Brewer, P. W. Gilles and F. A. Jenkins, *J. Chem. Phys.*, 16 (1948) 797.
30 S. S. Penner and E. K. Björnerud, *J. Chem. Phys.*, 23 (1955) 143.
31 W. Hooker, M. Lapp, D. Weber and S. S. Penner, *J. Chem. Phys.*, 25 (1956) 1087.
32 P. Ausloos and A. van Tiggelen, *Bull. Soc. Chim. Belges*, 61 (1952) 569.
33 P. Ausloos and A. van Tiggelen, *Bull. Soc. Chim. Belges*, 62 (1953) 223.
34 A. G. Gaydon and H. G. Wolfhard, *Proc. Roy. Soc. (London), A*, 201 (1950) 570.
35 R. B. Tagirov, *Zh. Fiz. Khim. (Moscow)*, 30 (1956) 949.
36 M. W. Windsor, N. Davidson and R. Taylor, *J. Chem. Phys.*, 27 (1957) 315.
37 M. W. Windsor, N. Davidson and R. Taylor, *Symp. Combustion, 7th* (1959) 80.

38 W. Hooker, *Symp. Combustion, 7th* (1959) 949.
39 M. G. Evans and M. Polanyi, *Trans. Faraday Soc.*, 35 (1939) 178.
40 M. C. Moulton and D. R. Herschbach, *J. Chem. Phys.*, 44 (1966) 3010.
41 J. K. Cashion and J. C. Polanyi, *J. Chem. Phys.*, 30 (1959) 1097.
42 J. K. Cashion and J. C. Polanyi, *Proc. Roy. Soc.* (*London*), *A*, 258 (1960) 529, 564, 570.
43 P. E. Charters and J. C. Polanyi, *Discussions Faraday Soc.*, 33 (1962) 107.
44 G. A. Hornbeck and R. C. Herman, *Ind. Eng. Chem.*, 43 (1951) 2739.
45 R. C. Herman, H. S. Hopfield, G. A. Hornbeck and S. Silverman, *J. Chem. Phys.*, 17 (1949) 220.
46 R. M. Badger, A. C. Wright and R. F. Whitlock, *J. Chem. Phys.*, 43 (1965) 4345.
47 R. W. Nicholls, *Ann. Geophys.*, 20 (1964) 144.
48 M. A. A. Clyne, B. A. Thrush and R. P. Wayne, *Photochem. and Photobiol.*, 4 (1965) 957.
49 A. Fontijn, C. B. Meyer and H. I. Schiff, *J. Chem. Phys.*, 40 (1964) 64.
50 R. A. Young and G. Black, *Symp. on Chemiluminescence, Durham, N.C.*, 1965, preprints, p. 1.
51 J. S. Arnold, R. J. Browne and E. A. Ogryzlo, *Photochem. and Photobiol.*, 4 (1965) 963.
52 R. E. March, S. G. Furnival and H. I. Schiff, *Photochem. and Photobiol.*, 4 (1965) 971.
53 L. W. Bader and E. A. Ogryzlo, *Discussions Faraday Soc.*, 37 (1964) 46.
54 M. A. A. Clyne, B. A. Thrush and R. P. Wayne, *Trans. Faraday Soc.*, 60 (1964) 359.
55 H. von Hartel and M. Polanyi, *Z. Physik. Chem.*, B11 (1930) 97.
56 Z. Bay and W. Z. Steiner, *Z. Physik. Chem.*, B3 (1929) 149.
57 D. M. Newitt and L. M. Baxt, *J. Chem. Soc.*, (1939) 1711.
58 A. G. Gaydon and H. G. Wolfhard, *Flames*, Chapman and Hall, London, 1960.
59 H. G. Wolfhard and W. G. Parker, *Proc. Phys. Soc.* (*London*), *A*, 62 (1949) 722.
60 H. G. Wolfhard and W. G. Parker, *Proc. Phys. Soc.* (*London*), *A*, 65 (1952) 2.
61 W. R. S. Garton and H. P. Broida, *Fuel*, 32 (1953) 519.
62 K. F. Bonhoeffer and H. Reichardt, *Z. Physik. Chem.*, 139 (1928) 75.
63 R. G. Bennett and F. W. Dalby, *J. Chem. Phys.*, 31 (1959) 434.
64 R. G. Bennett and F. W. Dalby, *J. Chem. Phys.*, 32 (1960) 1716.
65 R. G. Bennett and F. W. Dalby, *J. Chem. Phys.*, 36 (1962) 399.
66 R. G. Bennett and F. W. Dalby, *J. Chem. Phys.*, 40 (1964) 1414.
67 F. W. Dalby, *J. Chem. Phys.*, 41 (1964) 2297.
68 F. Kaufman and F. P. Del Greco, *J. Chem. Phys.*, 35 (1961) 1895.
69 F. P. Del Greco and F. Kaufman, *Discussions Faraday Soc.*, 33 (1962) 128.
70 S. H. Bauer, G. L. Schott and R. E. Duff, *J. Chem. Phys.*, 28 (1958) 1089.
71 G. L. Schott and J. L. Kinsey, *J. Chem. Phys.*, 29 (1958) 1177.
72 G. L. Schott, *J. Chem. Phys.*, 32 (1960) 710.
73 S. H. Bauer, F. Waelbroek and W. Tsang, *Abstr. of 135th meeting of Am. Chem. Soc.*, (1959) 36R.
74 G. C. Dousmanis, T. M. Sanders and C. H. Townes, *Phys. Rev.*, 100 (1955) 1735.
75 T. M. Sanders, A. L. Schawlow, G. C. Dousmanis and C. H. Townes, *J. Chem. Phys.*, 22 (1954) 245.
76 F. J. Lipscomb, R. G. W. Norrish and B. A. Thrush, *Proc. Roy. Soc.* (*London*), *A*, 233 (1956) 455.
77 W. D. McGrath and R. G. W. Norrish, *Proc. Roy. Soc.* (*London*), *A*, 242 (1957) 265.
78 N. Basco and R. G. W. Norrish, *Can. J. Chem.*, 38 (1960) 1769.
79 R. V. Fitzsimmons and E. J. Bair, *J. Chem. Phys.*, 40 (1964) 451.
80 W. D. McGrath and R. G. W. Norrish, *Proc. Roy. Soc.* (*London*), *A*, 254 (1960) 317.
81 N. Basco, J. E. Nicholas, R. G. W. Norrish and W. H. J. Vickers, *Proc. Roy. Soc.* (*London*), *A*, 272 (1963) 147.
82 N. Basco, A. B. Callear and R. G. W. Norrish, *Proc. Roy. Soc.* (*London*), *A*, 260 (1960) 459.
83 N. Basco and R. G. W. Norrish, *Proc. Roy. Soc.* (*London*), *A*, 268 (1962) 291.
84 N. Basco, A. B. Callear and R. G. W. Norrish, *Proc. Roy. Soc.* (*London*), *A*, 269 (1962) 180.
85 G. Porter, *Proc. Chem. Soc.*, (1959) 291.
86 A. B. Callear and R. G. W. Norrish, *Proc. Roy. Soc.* (*London*), *A*, 266 (1962) 299.

87 G. Herzberg and J. Shoosmith, *Nature*, 183 (1959) 1801.
88 G. Herzberg, *Proc. Roy. Soc. (London), A*, 262 (1961) 291.
89 G. Herzberg, *Report of XII Conseil de Chimie d'Institut Solvay, Bruxelles*, Interscience, New York, 1962, p. 70.
90 G. C. Eltenton, *J. Chem. Phys.*, 10 (1942) 403.
91 G. C. Eltenton, *J. Chem. Phys.*, 15 (1947) 455.
92 W. H. Bennett, *J. Appl. Phys.*, 21 (1950) 143.
93 W. C. Wiley and I. H. McLaurin, *Rev. Sci. Instr.*, 26 (1955) 1150.
94 V. L. Tal'roze, *Pribory i Tekhn. Eksperim.*, 5 (1957) 116.
95 V. L. Tal'roze, *Pribory i Tekhn. Eksperim.*, 6 (1960) 78.
96 A. J. B. Robertson, *Mass Spectrometry*, Methuen, London, 1954.
97 P. LeGoff, *Applied Mass Spectrometry*, The Institute of Petroleum, London, 1954, pp. 120–126.
98 S. N. Foner and R. L. Hudson, *J. Chem. Phys.*, 25 (1956) 601.
99 J. T. Herron and H. I. Schiff, *Can. J. Chem.*, 36 (1958) 1159.
100 C. W. Nutt, J. S. M. Botterill, G. Thorpe and G. W. Penmore, *Trans. Faraday Soc.*, 55 (1959) 1500.
101 C. W. Nutt, G. W. Penmore and A. J. Biddlestone, *Trans. Faraday Soc.*, 55 (1959) 1516.
102 F. T. Greene, J. Brewer and T. A. Milne, *J. Chem. Phys.*, 40 (1964) 1488.
103 T. A. Milne and F. T. Greene, *Symp. Combustion, 10th* (1965) 153.
104 F. P. Lossing and A. W. Tickner, *J. Chem. Phys.*, 20 (1952) 907.
105 K. U. Ingold and F. P. Lossing, *J. Chem. Phys.*, 21 (1953) 1135.
106 D. S. Jackson and H. I. Schiff, *J. Chem. Phys.*, 23 (1955) 2333.
107 J. Berkowitz, W. A. Chupka and G. B. Kistiakowsky, *J. Chem. Phys.*, 25 (1956) 457.
108 G. B. Kistiakowsky and G. G. Volpi, *J. Chem. Phys.*, 27 (1957) 1141.
109 L. F. Phillips and H. I. Schiff, *J. Chem. Phys.*, 36 (1962) 1509.
110 F. P. Lossing, *Mass Spectrometry*, Ed. C. A. McDowell, McGraw-Hill, New York, 1963, p. 457.
111 J. B. Farmer, F. P. Lossing, D. G. H. Marsden and E. W. R. Steacie, *J. Chem. Phys.*, 23 (1955) 1169.
112 F. P. Lossing, D. G. H. Marsden and J. B. Farmer, *Can. J. Chem.*, 34 (1956) 701.
113 P. Kebarle and F. P. Lossing, *Can. J. Chem.*, 37 (1959) 389.
114 S. N. Foner and R. L. Hudson, *J. Chem. Phys.*, 21 (1953) 1374.
115 S. N. Foner and R. L. Hudson, *J. Chem. Phys.*, 21 (1953) 1608.
116 S. N. Foner and R. L. Hudson, *J. Chem. Phys.*, 23 (1955) 1974.
117 H. I. Schiff and A. Mathias, *Discussions Faraday Soc.*, 37 (1964) 38.
118 E. G. Léger and C. Ouellet, *J. Chem. Phys.*, 21 (1953) 1310.
119 G. B. Kistiakowsky and P. H. Kydd, *J. Am. Chem. Soc.*, 79 (1957) 4825.
120 J. N. Bradley and G. B. Kistiakowsky, *J. Chem. Phys.*, 35 (1961) 256.
121 C. S. Brasefield, *Phys. Rev.*, 31 (1928) 52.
122 M. Pahl, *Z. Naturforsch.*, 12a (1957) 632.
123 K. N. Bascombe, J. A. Green and T. M. Sugden, *Advances in Mass Spectrometry*, Ed. R. M. Elliott, Pergamon, Oxford, 1963, p. 60.
124 G. N. Spokes and B. E. Evans, *Symp. Combustion, 10th* (1965) 639.
125 H. F. Calcote and J. L. Reuter, *J. Chem. Phys.*, 38 (1963) 310.
126 H. F. Calcote, *Symp. Combustion, 8th* (1962) 184.
127 H. F. Calcote, *Combustion and Flame*, 1 (1957) 385.
128 A. Fontijn and G. L. Baughman, *J. Chem. Phys.*, 38 (1963) 1784.
129 A. Fontijn, W. J. Miller and J. M. Hogan, *Symp. Combustion, 10th* (1965) 545.
130 J. Deckers and A. van Tiggelen, *Combustion and Flame*, 1 (1957) 281.
131 J. Deckers and A. van Tiggelen, *Nature*, 181 (1958) 1460.
132 J. Deckers and A. van Tiggelen, *Nature*, 182 (1958) 863.
133 J. Deckers and A. van Tiggelen, *Symp. Combustion, 7th* (1959) 254.
134 S. de Jaegere, J. Deckers and A. van Tiggelen, *Symp. Combustion, 8th* (1962) 155.
135 P. F. Knewstubb and T. M. Sugden, *Nature*, 181 (1958) 474.
136 P. F. Knewstubb and T. M. Sugden, *Nature*, 181 (1958) 1261.

137 P. F. Knewstubb and T. M. Sugden, *Symp. Combustion, 7th* (1959) 247.
138 P. F. Knewstubb and T. M. Sugden, *Proc. Roy. Soc. (London), A*, 255 (1960) 520.
139 H. F. Calcote, S. C. Kurzius and W. J. Miller, *Symp. Combustion, 10th* (1965) 605.
140 A. Feugier and A. van Tiggelen, *Symp. Combustion, 10th* (1965) 621.
141 P. F. Knewstubb and T. M. Sugden, *Nature*, 196 (1962) 1312.
142 P. F. Knewstubb, *Symp. Combustion, 10th* (1965) 623.
143 A. Fontijn and W. J. Miller, *Symp. Combustion, 10th* (1965) 623.
144 F. P. Lossing and I. Tanaka, *J. Chem. Phys.*, 25 (1956) 1031.
145 H. Hurzeler, M. G. Inghram and H. E. Stanton, *J. Chem. Phys.*, 27 (1957) 313.
146 H. Hurzeler, M. G. Inghram and J. D. Morrison, *J. Chem. Phys.*, 28 (1958) 76.
147 B. Steiner, C. F. Geise and M. G. Inghram, *J. Chem. Phys.*, 34 (1961) 189.
148 R. F. Herzog and F. F. Marmo, *J. Chem. Phys.*, 27 (1957) 1202.
149 C. Y. Johnson, *Ann. Geophys.*, 17 (1961) 100.
150 P. S. Rudolph and C. E. Melton, *J. Phys. Chem.*, 63 (1959) 916.
151 P. S. Rudolph and C. E. Melton, *J. Chem. Phys.*, 30 (1959) 847.
152 C. E. Melton and P. S. Rudolph, *J. Chem. Phys.*, 32 (1960) 1128.
153 C. E. Melton, *J. Chem. Phys.*, 33 (1960) 647.
154 C. E. Melton, *Mass Spectrometry of Organic Ions*, Ed. F. W. McLafferty, Academic Press, New York, 1963, pp. 65–112.
155 E. Muller, I. Muller-Rodloff and W. Bunge, *Ann.*, 520 (1935) 235.
156 E. Muller, I. Muller-Rodloff and W. Bunge, *Ann.*, 521 (1936) 89.
157 D. F. Evans, *Nature*, 176 (1955) 777.
158 M. A. Garstens, L. S. Singer and A. H. Ryan, *Phys. Rev.*, 96 (1954) 53.
159 O. L. Harle and J. R. Thomas, *J. Am. Chem. Soc.*, 79 (1957) 2973.
160 E. E. Schneider, M. J. Day and G. Stein, *Nature*, 168 (1951) 645.
161 P. W. Atkins and M. C. R. Symons, *The Structure of Inorganic Radicals*, Elsevier, Amsterdam, 1967.
162 G. Fehrer, *Phys. Rev.*, 105 (1957) 1122.
163 W. Kanzig, *Phys. Rev.*, 99 (1955) 1890.
164 C. J. Delbecq, B. Smaller and P. H. Yuster, *Phys. Rev.*, 104 (1956) 599.
165 J. H. E. Griffiths, J. Owen and I. M. Ward, *Defects in Crystalline Solids*, Phys. Soc., London, 1955, p. 81.
166 D. J. E. Ingram, W. G. Hodgson, C. A. Parker and W. T. Rees, *Nature*, 176 (1955) 1227.
167 D. Bijl and A. C. Rose-Innes, *Nature*, 175 (1955) 82.
168 J. F. Gibson, D. J. E. Ingram, M. C. R. Symons and M. G. Townsend, *Trans. Faraday Soc.*, 53 (1957) 914.
169 C. H. Bamford, A. D. Jenkins and J. C. Ward, *J. Polymer Sci.*, 48 (1960) 37.
170 C. H. Bamford, G. C. Eastmond and Y. Sakai, *Nature*, 200 (1963) 1284.
171 G. K. Fraenkel, J. M. Hirshon and C. Walling, *J. Am. Chem. Soc.*, 76 (1954) 3606.
172 C. H. Bamford, A. D. Jenkins, D. J. E. Ingram and M. C. R. Symons, *Nature*, 175 (1955) 894.
173 N. M. Atherton, H. W. Melville and D. H. Whiffen, *Trans. Faraday Soc.*, 54 (1958) 1300.
174 J. G. Castle and R. Beringer, *Phys. Rev.*, 80 (1950) 114.
175 R. Beringer and J. G. Castle, *Phys. Rev.*, 78 (1950) 581.
176 R. Beringer and J. G. Castle, *Phys. Rev.*, 81 (1951) 82.
177 C. A. Barth, A. F. Hildebrandt and M. Patapoff, *Discussions Faraday Soc.*, 33 (1962) 162.
178 R. Beringer and E. B. Rawson, *Phys. Rev.*, 87 (1952) 228.
179 E. B. Rawson and R. Beringer, *Phys. Rev.*, 88 (1952) 677.
180 M. A. Heald and R. Beringer, *Phys. Rev.*, 96 (1954) 645.
181 A. A. Westenberg and N. de Haas, *J. Chem. Phys.*, 40 (1964) 3087.
182 H. E. Radford, *Phys. Rev.*, 122 (1961) 114.
183 C. C. McDonald, *J. Chem. Phys.*, 39 (1963) 2587.
184 H. E. Radford, *J. Chem. Phys.*, 40 (1964) 2732.
185 J. M. Daniels and P. B. Dorain, *J. Chem. Phys.*, 40 (1964) 1160.
186 L. H. Piette, F. A. Johnson, K. A. Bouman and C. B. Colburn, *J. Chem. Phys.*, 35 (1961) 1481.

187 A. Carrington and D. H. Levy, *J. Chem. Phys.*, 44 (1966) 1298.
188 S. Krongelb and M. W. P. Strandberg, *J. Chem. Phys.*, 31 (1959) 1196.
189 A. A. Westenberg, *J. Chem. Phys.*, 43 (1965) 1544.
190 A. A. Westenberg and N. de Haas, *J. Chem. Phys.*, 43 (1965) 1550.
191 G. Dixon-Lewis, W. E. Wilson and A. A. Westenberg, *J. Chem. Phys.*, 44 (1966) 2287.
192 A. A. Westenberg and R. M. Fristrom, *Symp. Combustion, 10th* (1965) 473.
193 P. W. Atkins and K. A. McLauchlan, Private communication.
194 F. A. Paneth and W. Hofeditz, *Ber.*, 62B (1929) 1335.
195 F. A. Paneth and W. Lautsch, *Ber.*, 64B (1931) 2708.
196 F. A. Paneth and W. Lautsch, *Ber.*, 64B (1931) 2702.
197 F. O. Rice and M. D. Dooley, *J. Am. Chem. Soc.*, 56 (1934) 2747.
198 F. O. Rice and A. L. Glasebrook, *J. Am. Chem. Soc.*, 56 (1934) 2381.
199 F. O. Rice and A. L. Glasebrook, *J. Am. Chem. Soc.*, 55 (1933) 4329.
200 R. G. W. Norrish and G. Porter, *Discussions Faraday Soc.*, 2 (1947) 97.
201 T. G. Pearson, P. L. Robinson and E. M. Stoddart, *Proc. Roy. Soc.* (*London*), *A*, 142 (1933) 275.
202 M. Burton, *J. Am. Chem. Soc.*, 58 (1936) 692.
203 M. Burton, *J. Am. Chem. Soc.*, 58 (1936) 1645.
204 H. Henkin and M. Burton, *J. Am. Chem. Soc.*, 60 (1938) 831.
205 L. May, H. A. Taylor and M. Burton, *J. Am. Chem. Soc.*, 63 (1941) 249.
206 G. M. Harris and A. W. Tickner, *J. Chem. Phys.*, 15 (1947) 686.
207 G. M. Harris and A. W. Tickner, *Can. J. Res.*, 26B (1948) 343.
208 F. O. Rice, *Chem. Rev.*, 17 (1935) 53.
209 H. W. Melville and J. C. Robb, *Proc. Roy. Soc.* (*London*), *A*, 196 (1949) 479, 494.
210 H. W. Melville and J. C. Robb, *Proc. Roy. Soc.* (*London*), *A*, 202 (1950) 181.
211 E. Horn and M. Polanyi, *Z. Physik. Chem.*, B25 (1934) 151.
212 E. Horn, M. Polanyi and D. W. G. Style, *Trans. Faraday Soc.*, 30 (1934) 189.
213 E. Gorin, *Acta Physicochim. URSS*, 9 (1938) 681.
214 E. Gorin, *J. Chem. Phys.*, 7 (1939) 256.
215 M. Szwarc, *J. Chem. Phys.*, 17 (1949) 431.
216 L. A. K. Staveley and C. N. Hinshelwood, *Trans. Faraday Soc.*, 35 (1939) 845.
217 F. W. Birss, C. J. Danby and C. N. Hinshelwood, *Proc. Roy. Soc.* (*London*), *A*, 239 (1957) 154.
218 A. Farkas, *Z. Physik. Chem.*, B10 (1930) 419.
219 W. R. Schulz and D. J. LeRoy, *J. Chem. Phys.*, 42 (1955) 3869.
220 E. A. B. Birse and H. W. Melville, *Proc. Roy. Soc.* (*London*), *A*, 175 (1940) 164.
221 H. W. Melville and J. C. Robb, *Proc. Roy. Soc.* (*London*), *A*, 196 (1949) 445.
222 E. A. B. Birse and H. W. Melville, *Proc. Roy. Soc.* (*London*), *A*, 175 (1940) 187.
223 L. Farkas and H. Sachsse, *Z. Physik. Chem.*, B27 (1935) 111.
224 F. Patat, *Z. Physik. Chem.*, B32 (1936) 274, 294.
225 H. Sachsse, *Z. Physik. Chem.*, B31 (1935) 79, 87.
226 W. West, *J. Am. Chem. Soc.*, 57 (1935) 1931.
227 W. R. Schulz and D. J. LeRoy, *Can. J. Chem.*, 42 (1964) 2480.
228 B. A. Ridley, W. R. Schulz and D. J. LeRoy, *J. Chem. Phys.*, 44 (1966) 3344.
229 C. P. Fenimore and G. W. Jones, *J. Chem. Phys.*, 62 (1958) 693.
230 C. P. Fenimore and G. W. Jones, *J. Chem. Phys.*, 62 (1958) 178.
231 C. G. James and T. M. Sugden, *Proc. Roy. Soc.* (*London*), *A*, 227 (1955) 312.
232 E. M. Bulewicz, C. G. James and T. M. Sugden, *Proc. Roy. Soc.* (*London*), *A*, 235 (1956) 89.
233 C. G. James and T. M. Sugden, *Proc. Roy. Soc.* (*London*), *A*, 248 (1958) 238.
234 P. J. Padley and T. M. Sugden, *Proc. Roy. Soc.* (*London*), *A*, 248 (1958) 248.
235 E. M. Bulewicz and T. M. Sugden, *Trans. Faraday Soc.*, 52 (1956) 1475.
236 E. M. Bulewicz and T. M. Sugden, *Trans. Faraday Soc.*, 52 (1956) 1481.
237 R. J. Strutt, *Proc. Phys. Soc.* (*London*), 23 (1910) 66, 147.
238 A. Fontijn, C. B. Meyer and H. I. Schiff, *J. Chem. Phys.*, 40 (1964) 64.
239 R. R. Reeves, G. Manella and P. Harteck, *J. Chem. Phys.*, 32 (1960) 639.
240 M. A. A. Clyne and B. A. Thrush, *Trans. Faraday Soc.*, 57 (1961) 1305.

241 M. A. A. Clyne and B. A. Thrush, *Discussions Faraday Soc.*, 33 (1962) 139.
242 G. B. Kistiakowsky, *Z. Physik. Chem.*, 117 (1925) 337.
243 C. S. Foote and S. Wexler, *J. Am. Chem. Soc.*, 86 (1964) 3879.
244 A. U. Khan and M. Kasha, *J. Chem. Phys.*, 39 (1963) 2105.
245 A. M. Winer and K. D. Bayes, *J. Phys. Chem.*, 70 (1966) 302.
246 A. Mathias and H. I. Schiff, *Discussions Faraday Soc.*, 37 (1964) 38.
247 P. S. Skell and R. C. Woodworth, *J. Am. Chem. Soc.*, 78 (1956) 4496.
248 D. W. Setser and B. S. Rabinovitch, *Can. J. Chem.*, 40 (1962) 1425.
249 R. W. Carr and G. B. Kistiakowsky, *J. Phys. Chem.*, 70 (1966) 118.
250 H. M. Smallwood, *J. Am. Chem. Soc.*, 56 (1934) 1542.
251 H. Senftleben and O. Riechemeir, *Phys. Z.*, 30 (1929) 745.
252 H. Senftleben and E. Germer, *Ann. Physik*, 2 (1939) 847.
253 K. F. Bonhoeffer, *Z. Physik. Chem.*, 113 (1924) 199, 492.
254 E. L. Tollefson and D. J. LeRoy, *J. Chem. Phys.*, 16 (1948) 1057.
255 B. J. Wood and H. Wise, *J. Phys. Chem.*, 65 (1961) 1976.
256 F. S. Larkin and B. A. Thrush, *Discussions Faraday Soc.*, 37 (1964) 112.
257 F. S. Larkin and B. A. Thrush, *Symp. Combustion, 10th* (1965) 398.
258 W. R. Schulz and D. J. LeRoy, *Can. J. Chem.*, 40 (1962) 2413.
259 J. W. Linnett and D. G. H. Marsden, *Proc. Roy. Soc. (London), A*, 234 (1956) 489, 504.
260 J. C. Greaves and J. W. Linnett, *Trans. Faraday Soc.*, 54 (1958) 1323.
261 J. C. Greaves and J. W. Linnett, *Trans. Faraday Soc.*, 55 (1959) 1338.
262 L. Elias, E. A. Ogryzlo and H. I. Schiff, *Can. J. Chem.*, 37 (1959) 1680.
263 G. M. Schwab and H. Friess, *Z. Elektrochem.*, 39 (1933) 586.
264 G. M. Schwab, *Z. Physik. Chem.*, B27 (1935) 452.
265 L. W. Bader and E. A. Ogryzlo, *Discussions Faraday Soc.*, 37 (1964) 46.
266 J. S. Arnold, R. J. Browne and E. A. Ogryzlo, *Photochem. and Photobiol.*, 4 (1965) 963.
267 E. Wrede, *Z. Instrumentenkunde*, 48 (1928) 201.
268 P. Harteck, *Z. Physik. Chem.*, 139 (1928) 98.
269 W. Groth and P. Warneck, *Z. Physik. Chem.*, 10 (1957) 323.
270 H. Blades and C. A. Winkler, *Can. J. Chem.*, 29 (1951) 1022.
270a W. H. Rodebush and W. C. Klingelhoefer, *J. Am. Chem. Soc.*, 55 (1933) 130.
270b H. D. Beckey and P. Warneck, *Z. Naturforsch.*, 10a (1955) 62.
271 T. Kinbara and H. Ikegami, *Combustion and Flame*, 1 (1957) 199.
272 H. A. Wilson, *Modern Physics*, Blackie, London, 1944, p. 321.
273 H. F. Calcote, *Symp. Combustion, 8th* (1962) 184.
274 H. F. Calcote, *Symp. Combustion, 9th* (1963) 622.
275 H. F. Calcote, S. C. Kurzius and W. J. Miller, *Symp. Combustion, 10th* (1965) 605.
276 B. E. L. Travers and H. Williams, *Symp. Combustion, 10th* (1965) 657.
277 H. F. Calcote, *Symp. Combustion, 10th* (1965) 688.
278 D. L. Turcotte and W. Friedman, *Symp. Combustion, 10th* (1965) 673.
279 E. Andrew, D. W. E. Axford and T. M. Sugden, *Trans. Faraday Soc.*, 44 (1948) 427.
280 H. Belcher and T. M. Sugden, *Proc. Roy. Soc. (London), A*, 201 (1950) 480; *A*, 202 (1950) 17.
281 T. M. Sugden and B. A. Thrush, *Nature*, 168 (1951) 703.
282 K. E. Shuler and J. Weber, *J. Chem. Phys.*, 22 (1954) 491.
283 T. M. Sugden, *Discussions Faraday Soc.*, 19 (1955) 68.
284 J. Schneider and F. W. Hofmann, *Phys. Rev.*, 116 (1959) 244.
285 E. M. Bulewicz and P. J. Padley, *J. Chem. Phys.*, 35 (1961) 1590.
286 E. M. Bulewicz, *J. Chem. Phys.*, 36 (1962) 385.
287 E. M. Bulewicz and P. J. Padley, *J. Chem. Phys.*, 36 (1962) 2231.
288 H. Smith and T. M. Sugden, *Proc. Roy. Soc. (London), A*, 211 (1952) 31.
289 P. F. Knewstubb and T. M. Sugden, *Trans. Faraday Soc.*, 54 (1958) 372.
290 A. J. Borgers, *Symp. Combustion, 10th* (1965) 627.
291 T. M. Sugden, *Symp. Combustion, 10th* (1965) 636.
292 L. Elias, *J. Chem. Phys.*, 44 (1966) 3810.
293 A. M. Bass and H. P. Broida (Eds.), *Formation and Trapping of Free Radicals*, Academic Press, New York, 1960.

294 J. L. Jackson and E. W. Montroll, *J. Chem. Phys.*, 28 (1958) 1101.
295 S. Golden, *J. Chem. Phys.*, 29 (1958) 61.
296 B. J. Fontana, *J. Chem. Phys.*, 31 (1959) 148.
297 I. Norman and G. Porter, *Proc. Roy. Soc. (London), A*, 230 (1955) 399.
298 E. D. Becker and G. C. Pimentel, *J. Chem. Phys.*, 25 (1956) 224.
299 D. E. Milligan and M. E. Jacox, *J. Chem. Phys.*, 41 (1964) 3032.
300 H. P. Broida and O. S. Lutes, *J. Chem. Phys.*, 24 (1956) 484.
301 G. J. Minkoff, F. I. Scherber and J. S. Gallagher, *J. Chem. Phys.*, 30 (1959) 753.
302 L. Vegard, *Nature*, 114 (1924) 357.
303 M. Peyron and H. P. Broida, *J. Chem. Phys.*, 30 (1959) 139.
304 F. O. Rice and M. Freamo, *J. Am. Chem. Soc.*, 73 (1951) 5529.
305 M. McCarty and G. W. Robinson, *J. Chim. Phys.*, 56 (1959) 723.
306 L. Brewer, B. Meyer and G. D. Brabson, *J. Chem. Phys.*, 43 (1965) 3973.
307 B. Meyer, *J. Chem. Phys.*, 43 (1965) 2986.
308 R. L. Barger and H. P. Broida, *J. Chem. Phys.*, 43 (1965) 2364.
309 R. L. Barger and H. P. Broida, *J. Chem. Phys.*, 43 (1965) 2371.
310 D. A. Dows, G. C. Pimentel and E. Whittle, *J. Chem. Phys.*, 23 (1955) 1606.
311 D. E. Milligan and M. E. Jacox, *J. Chem. Phys.*, 41 (1964) 2838.
312 K. Rosengren and G. C. Pimentel, *J. Chem. Phys.*, 43 (1965) 507.
313 D. E. Milligan and M. E. Jacox, *J. Chem. Phys.*, 43 (1965) 4487.
314 D. E. Milligan and G. C. Pimentel, *J. Chem. Phys.*, 29 (1958) 1405.
315 T. D. Goldfarb and G. C. Pimentel, *J. Am. Chem. Soc.*, 82 (1960) 1865.
316 C. B. Moore, G. C. Pimentel and T. G. Goldfarb, *J. Chem. Phys.*, 43 (1965) 63.
317 C. B. Moore and G. C. Pimentel, *J. Chem. Phys.*, 41 (1964) 3504.
318 D. E. Milligan, D. E. Mann, M. E. Jacox and R. A. Mitsch, *J. Chem. Phys.*, 41 (1964) 1199.
319 D. E. Milligan, *J. Chem. Phys.*, 35 (1961) 1491.
320 D. E. Milligan and M. E. Jacox, *J. Chem. Phys.*, 40 (1964) 2461.
321 D. E. Milligan, M. E. Jacox and A. M. Bass, *J. Chem. Phys.*, 43 (1965) 3149.
322 M. E. Jacox, D. E. Milligan, N. G. Moll and W. E. Thompson, *J. Chem. Phys.*, 43 (1965) 3734.
323 See A. Thomas, *Oxidation and Combustion Reviews*, Ed. C. F. H. Tipper, Vol. 2, Elsevier, Amsterdam, 1967, p. 257 for further details, references and a discussion of trapped radicals and their relevance to combustion chemistry.

Chapter 5

The Treatment of Experimental Data

D. MARGERISON

1. Introduction

The majority of kinetic investigations are carried out with the object of gaining insight into the mechanism by which chemical change occurs. Very often, the investigation may be solely concerned to establish the series of steps by which the reactants are converted into the products—the reaction mechanism, as it is called. In other cases where the reaction mechanism is known, studies of the kinetics of the reaction may be employed to throw light on the details of the interaction of the reactant molecules. These objectives demand that the experimental data be summarized in equations of the same form as those which appear in theoretical treatments of reaction kinetics. It is the purpose of this article to explain the procedures employed but we shall not consider the whole range of reaction types. In fact, we shall confine our attention to homogeneous reactions occurring in the gaseous and liquid phases though it should be noted that some of the methods discussed can be applied to reactions in which a surface is involved.

The key parameter throughout all discussions of reaction kinetics is the instantaneous value of the reaction rate. This may be defined in as many ways as there are reactants or products; for example, in the case of a reaction between two reactants A and B to give two products P and Q, there are four alternative definitions of the reaction rate

$$\frac{-\mathrm{d}[\mathrm{A}]}{\mathrm{d}t}, \frac{-\mathrm{d}[\mathrm{B}]}{\mathrm{d}t}, \frac{\mathrm{d}[\mathrm{P}]}{\mathrm{d}t} \text{ or } \frac{\mathrm{d}[\mathrm{Q}]}{\mathrm{d}t}$$

where [A], [B], [P] and [Q] represent the instantaneous values of the molar concentrations of the various species at time t. It is clear that each rate can be found by constructing the tangent to the appropriate concentration–time curve at the particular time under consideration. It should be noted that the alternative definitions of the reaction rate are not usually equal in value though all are positive. Consequently, in any discussion involving the term reaction rate, it is necessary to define precisely which of the various alternatives is used. However, in the author's opinion, there is a strong case for confining discussion to the rate of change of one or other of the *reactant* concentrations whenever this is possible since, from a

References p. 407

fundamental point of view, it is the interaction of the reactants which determines any particular reaction rate. We shall follow this practice throughout this chapter and hence we shall refer extensively to the reaction rate as the rate of disappearance of a particular reactant, *viz.*

$$\frac{-\mathrm{d}[\text{reactant}]}{\mathrm{d}t}$$

In most of the discussion, we shall consider that our reaction involves only two reactants A and B giving products P and Q, in which case we shall identify the reaction rate with the rate of disappearance of the reactant A, *viz.*

$$\frac{-\mathrm{d}[\mathrm{A}]}{\mathrm{d}t}$$

Now, it is known from the results of many kinetic investigations that a large number of factors may influence the rate of a reaction. Some of these factors are:
(1) the concentrations of the reactants;
(2) the concentrations of the products;
(3) the concentrations of any catalysts added (*i.e.* the concentrations of any substances which do not appear in the reaction stoichiometry and which are therefore recoverable unchanged in amount at the end of the reaction);
(4) the temperature;
(5) the total pressure on the system;
(6) the viscosity of the reaction mixture;
(7) the dielectric constant of the reaction mixture;
(8) the total ionic strength of the reaction system;
(9) the intensity of radiation absorbed by the reaction system.
It is the basic objective of every experimental kinetic investigation to discover the quantitative relationship between the instantaneous reaction rate and the corresponding values of those factors on which it depends. In simple cases, the instantaneous rate of disappearance of reactant A can be expressed as the product of two functions, one dependent on the concentrations of the reactants only and the second dependent on the other reaction parameters such as temperature, [catalyst], pressure, etc. *i.e.*

$$\frac{-\mathrm{d}[\mathrm{A}]}{\mathrm{d}t} = k(T, [\text{catalyst}], P \text{ etc.}) \times \mathrm{f}([\text{reactants}]) \quad (1)$$

In these simple cases, the explicit form of the second function for a reaction involving two reactants A and B is

$$\mathrm{f}([\mathrm{A}], [\mathrm{B}]) = [\mathrm{A}]^a[\mathrm{B}]^b$$

so that eqn. (1) reduces to

$$\frac{-\mathrm{d}[\mathrm{A}]}{\mathrm{d}t} = k[\mathrm{A}]^a[\mathrm{B}]^b$$

The quantities a and b are termed the *orders with respect to* A *and* B while the function k (T, [catalyst], P etc.), usually contracted to k, is termed the *rate coefficient*. There are, however, many reactions in which the expression for the instantaneous rate of disappearance of A is more complicated, *e.g.*

$$\frac{-\mathrm{d}[\mathrm{A}]}{\mathrm{d}t} = \sum_{\mathrm{i}} k_i \, \mathrm{f}_i([\text{reactants}], [\text{products}]) \tag{2}$$

or

$$\frac{-\mathrm{d}[\mathrm{A}]}{\mathrm{d}t} = \sum_{\mathrm{i}} k_i \, \mathrm{f}_i([\text{reactants}], [\text{products}]) - \sum_{\mathrm{i}} k_i' \mathrm{f}_i'([\text{reactants}], [\text{products}]) \tag{3}$$

or

$$\frac{-\mathrm{d}[\mathrm{A}]}{\mathrm{d}t} = \frac{k\mathrm{f}([\text{reactants}], [\text{products}])}{1 \pm k'\mathrm{f}'([\text{reactants}], [\text{products}])} \tag{4}$$

These four expressions for the reaction rate have one feature in common, namely that the influence of the concentrations of reactants and products on the reaction rate is separated from the influence of the other reaction parameters. The reason for proceeding in this way is that these rate expressions suggest possible mechanisms for the reaction in question, a feature which will be dealt with later on, in Volume 2.

Although the rate of the reaction is the parameter in kinetic studies which provides the link between the experimental investigation and the theoretical interpretation, it is seldom measured directly. In the usual *closed or static* experimental system, the standard procedure is to follow the change with time of the concentrations of reactants and products† in two distinct series of experiments. In the first series, the initial concentrations of the reactants and products are varied with the other reaction variables held constant, the object being to discover the exact relationship between rate and concentration. In the second series, the experiments are repeated at different values of the other reaction variables so that the dependence of the various rate coefficients on temperature, pressure, ionic strength etc., can be found. It is with the methods of examining concentration—time data obtained in closed systems in order to deduce these relationships that we shall be concerned in this chapter. However, before embarking on a description of these

† Throughout this chapter, the term "product" refers to any substance which is not a reactant; thus, any reference to products must be taken to include intermediates.

procedures, we should make it clear that kinetic studies are not confined to closed systems but can be carried out on *open* systems either of the *tube-flow* type or *capacity-flow* type. We shall not consider these methods here since they are described elsewhere.

As far as the first objective of kinetic studies is concerned, namely the discovery of the relationship between rate and concentration, much time and effort can be saved by a preliminary examination of the data.

Suppose that our data consist of the concentrations of all reactants and products at a series of times under a wide variety of initial conditions together with the corresponding volumes of the reaction system. From these data, we can establish whether or not the disappearance of a given number of moles of reactant A is always accompanied by the disappearance of a constant number of moles of reactant B and the appearance of a constant number of moles of the products P and Q irrespective of the extent to which the reaction has taken place and under all conditions of temperature, pressure etc. For this purpose, we calculate for each set of concentration–time determinations a series of mean stoichiometric ratios at different extents of reaction, thus

$$\frac{V_0[\mathrm{B}]_0 - V[\mathrm{B}]}{V_0[\mathrm{A}]_0 - V[\mathrm{A}]}$$

$$\frac{V[\mathrm{P}] - V_0[\mathrm{P}]_0}{V_0[\mathrm{A}]_0 - V[\mathrm{A}]}$$

$$\frac{V[\mathrm{Q}] - V_0[\mathrm{Q}]_0}{V_0[\mathrm{A}]_0 - V[\mathrm{A}]}$$

where V_0 is the initial volume of the system, V is the volume at time t and $[\mathrm{A}]_0$, $[\mathrm{B}]_0$, $[\mathrm{P}]_0$ and $[\mathrm{Q}]_0$ are the initial concentrations. If these ratios vary as the reaction proceeds or alter when, for example, the temperature is changed, then the rate expression is certainly more complicated than the simple form (1). On the other hand, if these ratios remain constant throughout the whole series of experiments so that we can write the stoichiometric equation as

$$\nu_\mathrm{A}\mathrm{A} + \nu_\mathrm{B}\mathrm{B} = \nu_\mathrm{P}\mathrm{P} + \nu_\mathrm{Q}\mathrm{Q},$$

where ν_A, ν_B, ν_P and ν_Q are the stoichiometric coefficients, we cannot conclude that the rate expression is simple. Before proceeding to test the data for consistency with eqn. (1) in the case of a reaction of invariant stoichiometry, it is worth while examining the initial slope of the curve showing the concentration of A as a function of time, *i.e.*, the initial rate. If variations of the initial product concentration between, say, zero and that produced by the complete disappearance of the reactant A result in alterations in the initial rate, then clearly the expression for the rate of the reaction must include terms in [P] and/or [Q] and hence eqn. (1) cannot possibly represent the data. One other property of the simple rate expressions generalized by eqn. (1) is that the maximum rate occurs when the concentration of the reactants is at a maximum, that is at zero time. Consequently, if the rate of the reaction increases to a maximum value over a period of time, it is certain that an expression such as (2), (3) or (4) will be required.

Thus if a preliminary examination of the concentration–time data shows any one of the following three features:

(1) a variable stoichiometry;

(2) an initial rate which is affected by the presence of product;

or (3) an initial rate which is less than the rate observed at a later stage in the reaction,

then the methods employed to process the data are those appropriate to expressions of the type (2), (3) or (4). Alternatively if we observe:

(1) an invariant stoichiometry;

(2) an initial rate unaffected by the presence of product;

and (3) the maximum rate at the beginning of the reaction,

then we attempt to process the data on the basis of eqn. (1).

However, it is often the case that not all the concentrations which are required for the calculation of the reaction stoichiometry are available. It may be that experimentally it is only practicable to measure the concentration of, say, the reactant A at a series of times subsequent to the reaction being started. In this situation, our only help comes from the initial rate. As before, if the initial rate of disappearance of A is affected by the presence of product and/or is less than the rate observed somewhat later in the reaction, we know immediately that the more complicated expressions are required. The converse observations, however, do not necessarily imply that the rate can be represented by the simple equation, (1). In those cases where the preliminary examination of the data indicates that the maximum rate occurs at zero time and is unaffected by the presence of product, the data are examined on the basis of eqn. (1) first; if inconsistent rate coefficients are found together with non-simple orders, the data are then re-examined on the basis of the more complicated expressions. This same procedure is adopted when the data consist of a series of concentrations and times but where, for experimental reasons, the early concentration–time values are either unobtainable or sufficiently unreliable as to preclude any reasonable estimates of the initial rate being made; the data are examined on the basis of the simple expression first and in the event of inconsistencies re-examined on the basis of the more complicated expressions. It should be clear that improvements in the experimental technique designed to reduce the uncertainties in the initial rates can more than repay the effort involved.

The same general point can be made but with greater force on the type of experiment in which a single physical property characteristic of the reaction system as a whole is measured at a series of times (*e.g.* pressure–time measurements in

References p. 407

gaseous reactions) and in which the stoichiometry of the reaction is *assumed* in order to deduce the concentration of a particular reactant or product from the measured property. With such limited information on a reaction, it is difficult to reach valid conclusions as to the form taken by the rate expression.

2. Testing the data for consistency with equations of the type $-\mathrm{d}[\mathrm{A}]/\mathrm{d}t = k[\mathrm{A}]^a [\mathrm{B}]^b$

Equations of this type are the simplest to deal with. The quantities a and b are termed the *orders of the reaction with respect to* A *and* B as mentioned previously while their sum, $(a+b)$, is termed the *total order of the reaction.* The equation can be generalized without difficulty to cover the case in which the concentrations of more than two reactants enter the rate equation. It must be emphasized that the various orders cannot be equated to or deduced from the stoichiometric coefficients but must be obtained from the kinetic data by the methods to be explained.

In attempting to fit our concentration–time data to equations of this type, we have to find *rate coefficients* and *orders* which remain constant over the time scale of any particular experiment *and* which do not alter throughout a series of experiments in which the initial concentrations of the reactants are varied under otherwise constant conditions. Furthermore, if we take the view that the main purpose of processing the data is to cast the results into a form capable of interpretation, the values which a or b can take are limited to 0, 0.5, 1, 1.5 and 2 with minor exceptions to include such simple fractions as $\frac{1}{4}$ etc.; it is probably true to say that very often these special cases can be anticipated from the prior knowledge available to the investigator and so little difficulty is experienced in such situations. If values for the orders a and b other than small integers or simple fractions are found, it can be concluded that a rate expression more complicated than

$$\frac{-\mathrm{d}[\mathrm{A}]}{\mathrm{d}t} = k[\mathrm{A}]^a[\mathrm{B}]^b$$

is necessary to represent the data adequately. For example, if it were found that the expression

$$\frac{-\mathrm{d}[\mathrm{A}]}{\mathrm{d}t} = k[\mathrm{A}]^{1.39}$$

represented a certain set of concentration–time data, it is likely that the data would equally well fit the expression

$$\frac{-\mathrm{d}[\mathrm{A}]}{\mathrm{d}t} = k[\mathrm{A}]+k'[\mathrm{A}]^2$$

if the values of k and k' were chosen correctly. For the purpose of elucidating the reaction mechanism, it is preferable to express the data by an equation of the second type rather than to constrain the data to the equation

$$\frac{-\mathrm{d}[\mathrm{A}]}{\mathrm{d}t} = k[\mathrm{A}]^a$$

2.1 REACTIONS WITH KNOWN INVARIANT STOICHIOMETRY

The concentration–time data from a reaction whose stoichiometry has been studied independently of its rate and found to be invariant can be analysed much more rigorously than the data from a reaction whose stoichiometry is unknown. Every effort should be made to avoid this latter situation since the advantages of a knowledge of the stoichiometry more than compensate for the time spent in obtaining the additional experimental observations which are required. Nowhere is this more obvious than when the stoichiometry is *known and invariant* so that the reaction can be written

$$\nu_{\mathrm{A}}\mathrm{A}+\nu_{\mathrm{B}}\mathrm{B} = \nu_{\mathrm{P}}\mathrm{P}+\nu_{\mathrm{Q}}\mathrm{Q}$$

It follows from the conditions of conservation that

$$[\mathrm{B}] = \frac{V_0[\mathrm{B}]_0-(\nu_{\mathrm{B}}/\nu_{\mathrm{A}})(V_0[\mathrm{A}]_0-V[\mathrm{A}])}{V} \tag{5}$$

and

$$[\mathrm{P}] = \frac{V_0[\mathrm{P}]_0+(\nu_{\mathrm{P}}/\nu_{\mathrm{A}})(V_0[\mathrm{A}]_0-V[\mathrm{A}])}{V} \tag{6}$$

etc.

and, therefore, if the concentration of A is known, the concentrations of the other reactants and the products can be calculated from the volume of the system and their initial concentrations. The above equations simplify to

$$[\mathrm{B}] = [\mathrm{B}]_0-(\nu_{\mathrm{B}}/\nu_{\mathrm{A}})([\mathrm{A}]_0-[\mathrm{A}]) \tag{7}$$

$$[\mathrm{P}] = [\mathrm{P}]_0+(\nu_{\mathrm{P}}/\nu_{\mathrm{A}})([\mathrm{A}]_0-[\mathrm{A}]) \tag{8}$$

etc.

for a reaction in which there is no change in volume as the reaction proceeds. In this case, the four alternative definitions of reaction rate are simply related to one another as can be seen by differentiating the above equations with respect to time.

References p. 407

The results are

$$-\frac{d[B]}{dt} = -\frac{\nu_B}{\nu_A}\frac{d[A]}{dt} \tag{9}$$

$$\frac{d[P]}{dt} = -\frac{\nu_P}{\nu_A}\frac{d[A]}{dt} \tag{10}$$

etc.

Thus, for a reaction with constant stoichiometry

$$A = 2P$$

proceeding under constant volume conditions with a rate of disappearance of A given by

$$-\frac{d[A]}{dt} = k[A]^a$$

the rate at which product appears is

$$\frac{d[P]}{dt} = 2k[A]^a$$

Since most reactions are studied under conditions where the volume remains constant or almost so, we shall confine the remainder of the discussion of this chapter to these cases.

In order to rigorously test the data for consistency with the rate expression

$$\frac{-d[A]}{dt} = k[A]^a[B]^b \tag{11}$$

with a particular pair of a and b values, we require the corresponding integrated expression. The problem is that in a particular case we have no prior knowledge of the orders a and b and so we are unable to select the appropriate integrated equation. Since the value of $(a+b)$ seldom exceeds 2 and since a and b cannct usually take on values other than 0, 0.5, 1.0, 1.5, or 2.0, there are a limited number of possible integrated forms of eqn. (11) for a given ν_B/ν_A. In principle, therefore, we could proceed by inserting the concentration–time data into each of the possible integrated equations until we obtained a constant rate coefficient. We shall discuss this method more critically later on but for the moment it is obvious that, if some way can be found of discovering the limits between which a

and b are likely to lie, much laborious calculation can be saved. This is the topic we shall discuss first.

2.1.1 *Estimation of the orders a and b*

(a) Estimation of total order, $a+b$

We first recast eqn. (11) in terms of the dimensionless quantities

$$\alpha = \frac{[\mathrm{A}]}{[\mathrm{A}]_0} \tag{12}$$

$$I = \frac{[\mathrm{B}]_0}{[\mathrm{A}]_0} \tag{13}$$

α is thus the fraction of the reactant A remaining at time t and I is the initial concentration ratio. Substituting in eqn. (11) using eqns. (7), (12) and (13), we obtain

$$-\frac{\mathrm{d}\alpha}{\mathrm{d}t} = k[\mathrm{A}]_0^{a+b-1}\alpha^a(I-r+r\alpha)^b \tag{14}$$

where r has been written for the stoichiometric ratio $\nu_\mathrm{B}/\nu_\mathrm{A}$. Suppose we choose the initial concentrations $[\mathrm{A}]_0$ and $[\mathrm{B}]_0$ to be in the ratio of the corresponding stoichiometric coefficients, thus

$$\frac{[\mathrm{B}]_0}{[\mathrm{A}]_0} = \frac{\nu_\mathrm{B}}{\nu_\mathrm{A}}$$

or $I = r$

Then eqn. (14) becomes

$$-\frac{\mathrm{d}\alpha}{\mathrm{d}t} = k[\mathrm{A}]_0^{a+b-1}r^b\alpha^{a+b} \tag{15}$$

which is a rather simple equation. If now we define the quantity $\mathrm{d}\theta$ by the equation

$$\mathrm{d}\theta = k[\mathrm{A}]_0^{a+b-1}r^b\,\mathrm{d}t$$

eqn. (15) reduces to

$$-\frac{\mathrm{d}\alpha}{\mathrm{d}\theta} = \alpha^{a+b} \tag{16}$$

References p. 407

We shall call the dimensionless quantity θ the *reduced time*. It is related to actual time through the equation

$$\begin{aligned}\theta &= \int_0^\theta \mathrm{d}\theta \\ &= \int_{t_0}^t k[\mathrm{A}]_0^{a+b-1} r^b \mathrm{d}t \\ &= k[\mathrm{A}]_0^{a+b-1} r^b (t-t_0) \end{aligned} \tag{17}$$

where t_0 stands for the time at which the reaction commenced. Integration of eqn. (16), thus

$$-\int_1^\alpha \frac{\mathrm{d}\alpha}{\alpha^{a+b}} = \int_0^\theta \mathrm{d}\theta$$

gives

$$\frac{1}{1-(a+b)}\{1-\alpha^{1-(a+b)}\} = \theta \qquad a+b \neq 1 \tag{18}$$

or

$$-\ln \alpha = \theta \qquad a+b = 1 \tag{19}$$

It is perhaps worth mentioning that eqn. (19) is only a special case of eqn. (18) so that it is not necessary to separate the general discussion of reactions with any value of the total order from those with a total order of unity; in future, therefore, we shall use the general form (18).

(*i*) *Determination of order from the plots of* α *against* $\log_{10} \theta$

From any of the above equations, we see that α is a unique function of θ for a given value of $a+b$. Fig. 1 shows the characteristic curves of α against $\log_{10} \theta$ for $(a+b)$ values: 0, 0.5, 1.0, 1.5, 2.0; the numerical data for these curves are given in Appendix 1.

From these curves, the total order of any reaction following the rate expression (11) can be found. First, we make the initial concentration ratio identical with the ratio of the corresponding stoichiometric coefficients. Second, we convert the observed values of the concentrations of A to the dimensionless quantity α by simply dividing by $[\mathrm{A}]_0$. Third, we plot α against $\log_{10} (t-t_0)$ on transparent graph paper. Finally we superimpose this graph on one of the curves of α against $\log_{10} \theta$ (clearly the same scale intervals must be used for both graphs); the curve of α against $\log_{10} \theta$ which best fits the experimental data gives the value of $(a+b)$.

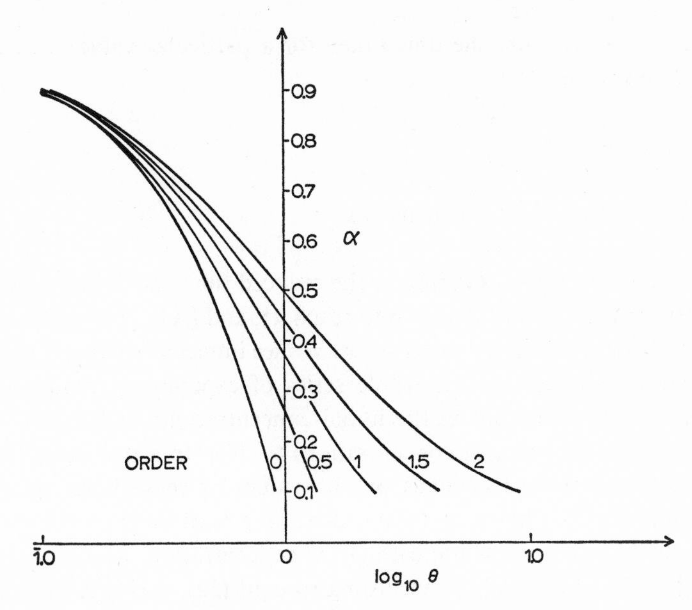

Fig. 1. Variation of α with $\log_{10}\theta$ for simple orders.

It should be noted that the displacement of the $\log_{10}(t-t_0)$ values from the $\log_{10}\theta$ values is related to the rate coefficient k, the initial concentration $[A]_0$, and the stoichiometric ratio r by the expression

$$\log_{10}\theta - \log_{10}(t-t_0) = \log_{10}k + (a+b-1)\log_{10}[A]_0 + b\log_{10}r \tag{20}$$

as can be seen from eqn. (17). Examination of the five curves of Fig. 1 shows that it is not possible to discriminate between the five possible values of $(a+b)$ if only small extents of reactions are studied; for this reason, it is essential that the reaction be followed until at least 60 or 70% of the reactants have been converted into products. Although it is unlikely that there are any complications in the case of reactions with an invariant stoichiometry, it is desirable to confirm that the same curve of α against $\log_{10}\theta$ fits the data obtained in a series of experiments at different values of $[A]_0$ and $[B]_0$ subject to the condition that their ratio remains constant and equal to r. As a further check on the consistency of the data, a straight line of slope $(a+b-1)$ should be obtained when the scale displacements are plotted against the logarithms of the initial concentrations of A.

(ii) Alternative methods based on eqns. (15), (16), (17), (18) and (19)

Fractional life method (I). It follows from either eqn. (18) or (19) that θ is fixed for a given value of α and $(a+b)$. Therefore, if, in a series of experiments at different values of the initial concentrations of A and B throughout which the ratio

References p. 407

$[B]_0/[A]_0 = r$, we measure the time taken for a particular value of α, α^*, to be reached, we can write

$$\theta_i(\alpha^*) = \text{constant} \tag{21}$$

or

$$k[A]_{0,i}^{a+b-1} r^b (t_i(\alpha^*) - t_0) = \text{constant} \tag{22}$$

where $\theta_i(\alpha^*)$ and $(t_i(\alpha^*) - t_0)$ represent the reduced time and actual time required in the ith experiment for the value of α to reach α^*, and $[A]_{0,i}$ represents the initial concentration of A used in that experiment. We see immediately that if $(a+b) = 1$, the time interval throughout the whole series of experiments remains constant independent of any variations in the initial concentrations. In general, however, the time taken for the reactant concentrations to drop to a particular fraction of their initial concentrations depends on the values of these latter quantities as shown in eqn. (22). To obtain the total order of the reaction from the variation of this time interval or fractional life with initial concentration, we simply plot $\log_{10}(t_i(\alpha^*) - t_0)$ against $\log_{10}[A]_{0,i}$; according to eqn. (22), this is a straight line of slope $1-(a+b)$ and hence $(a+b)$ can be obtained immediately. The choice of the value of α^* is determined primarily by the need to avoid serious errors in the estimation of the time interval $t_i(\alpha^*) - t_0$. It follows from eqns. (18) and (22) that, other things being equal, the closer α^* is to unity, the smaller is the time interval which must be measured; this is merely a statement of the obvious fact that, the less the fractional decomposition of the reactants, the less the time required. Clearly the relative error in the time interval increases rapidly as the time interval decreases with the result that the uncertainty in the value of the slope of the logarithmic plot of time interval against initial concentration becomes too large to enable a useful estimate of $(a+b)$ to be made. From this point of view, it is desirable to choose a fairly low value of α^* so that the relative error in the smallest time interval measured is small. On the other hand if α^* is made very small, say 0.1 or 0.2, the reaction has to be followed to so large an extent that the first method of obtaining $(a+b)$ is far easier in general than the fractional life method under discussion. For these reasons, therefore, values of α^* outside the range 0.5 to 0.8 are seldom used. In fact, apart from special situations where the time interval $t_i(\alpha^*) - t_0$ is particularly easy to measure experimentally, the fractional life method has but one advantage over the method based on the superimposition of $\alpha/\log_{10}(t-t_0)$ and $\alpha/\log_{10}\theta$ curves, *viz.*, that each reaction need not be followed to conversions much exceeding 50 %.

Fractional life method (II). If, for some reason, it is not possible to cover a wide range of initial concentration conditions as is clearly necessary for the application of the above method (this could happen when one of the reactants has a limited solubility), essentially the same method can be applied to the data of a single experiment. We select a series of values of α $(\alpha_0, \alpha_1, \alpha_2, \ldots \alpha_j \ldots)$ such that

$$\alpha_1 = f\alpha_0$$
$$\alpha_2 = f\alpha_1$$
$$\vdots$$
$$\alpha_j = f\alpha_{j-1}$$
$$\vdots$$

where f is a constant less than 1 and α_0 has been written in place of the initial value of α of unity to preserve a consistent symbolism. Note that we use the running subscript j to specify the general value of α or t in a *particular experiment*; we shall reserve the subscript i to distinguish one experiment from another. From the graph of α against time, we read the corresponding times $t_0, t_1, t_2, \ldots t_j \ldots$ It follows from eqns. (17) and (18) that

$$\frac{1}{1-(a+b)}\{\alpha_{j-1}^{1-(a+b)}-\alpha_j^{1-(a+b)}\} = k[\mathrm{A}]_0^{a+b-1}r^b\{t_j-t_{j-1}\}$$

$$\frac{\alpha_{j-1}^{1-(a+b)}}{1-(a+b)}\{1-f^{1-(a+b)}\} = k[\mathrm{A}]_0^{a+b-1}r^b\{t_j-t_{j-1}\} \tag{23}$$

If we write τ_j for the jth time interval t_j-t_{j-1} in which α decreases by a factor f of its value at the beginning of the interval, eqn. (23) can be recast in the form

$$\log_{10}\tau_j = \text{constant}+\{1-(a+b)\}\log_{10}\alpha_{j-1} \tag{24}$$

Note that the first interval τ_1 is the time taken for α to drop from α_0 to α_1, that is from 1 to f. The procedure to estimate $(a+b)$ is thus:

(1) plot the experimental values of α against $(t-t_0)$ and draw a smoothed curve through the points;
(2) select a value of f equal to 0.8 or thereabouts and read off from the above curve the times at which α reaches the value f (*i.e.* α_1), f^2 (*i.e.* α_2), f^3 (*i.e.* α_3) etc.;
(3) hence calculate the time intervals for α to change from α_0 (*i.e.* 1) to α_1, α_1 to α_2, α_2 to α_3 etc., and plot the logarithms of these time intervals against the logarithms of the values of α at the *start* of these intervals;
(4) measure the slope of the resulting straight line and hence, as shown in eqn. (24), calculate the value of $(a+b)$.

Instead of utilizing the parameters α_j, the actual concentrations of the reactant A, $[\mathrm{A}]_j$, can be used; since $[\mathrm{A}]_j = \alpha_j[\mathrm{A}]_0$, eqn. (24) becomes

$$\log_{10}\tau_j = \text{constant}+\{1-(a+b)\}\log_{10}[\mathrm{A}]_{j-1} \tag{25}$$

References p. 407

The two constants of eqns. (24) and (25) differ by $\{(a+b)-1\}\log_{10}[A]_0$. Eqns. (23), (24), or (25) show that, as α or [A] decreases throughout the particular experiment, the time interval for a given fractional change increases if $(a+b) > 1$, remains constant if $(a+b) = 1$ and decreases if $(a+b) < 1$. The choice of f is determined in part by the need to avoid large uncertainties in the interpolated values of τ_j; if f is too close to unity, the points on the $\log_{10}\tau_j$, $\log_{10}\alpha_{j-1}$ (or $\log_{10}[A]_{j-1}$) graph become widely scattered about the true straight line thus giving rise to a large degree of uncertainty in the value of the total order. At the same time, unlike the former fractional life treatment of several experiments, too low a value of f cannot be used; if a low value of f is chosen, the reaction has to be followed to very high degrees of conversion to obtain a reasonable number of points to define the slope of the straight line of $\log_{10}\tau_j$ against $\log_{10}\alpha_{j-1}$; even with a value $f = 0.8$, the reaction has to be followed to nearly 70 % conversion for five points to be obtained on this graph. Clearly this method of treating the data of a single experiment has few advantages over the superimposition method.

(iii) Method of tangents

The slope of the tangent to the α, t curve at any time is related to the value of α at that time by eqn. (15)

$$-\frac{d\alpha}{dt} = k[A]_0^{a+b-1}r^b\alpha^{a+b}$$

provided that the initial concentration ratio $[B]_0/[A]_0$ has been chosen equal to the stoichiometric ratio ν_B/ν_A. Consequently if a series of tangents are drawn to the α, t curve at a series of values of α and their slopes measured, the total order $(a+b)$ can be obtained from a logarithmic plot of $(-d\alpha/dt)$ against α since

$$\log_{10}(-d\alpha/dt) = \text{constant}+(a+b)\log_{10}\alpha \tag{26}$$

From a practical point of view, there is little virtue in converting all the experimental values of [A] to α when utilizing this method. In terms of the directly measured quantities, eqn. (26) becomes

$$\log_{10}(-d[A]/dt) = \text{constant}+(a+b)\log_{10}[A] \tag{27}$$

The two constants of eqns. (26) and (27) differ by

$$\{(a+b)-1\}\log_{10}[A]_0$$

The construction of tangents to concentration–time curves is greatly facilitated by the use of a front-surfaced plane mirror to find the normal. It is still, however, difficult and, therefore, the points on the graphs corresponding to either eqns. (26) or (27) are likely to show considerable scatter about a straight line. This scatter arising from incorrect construction of the tangents is liable to introduce considerable uncertainty into the estimated value of $(a+b)$. For this reason, the application of this method to the problem of estimating the total reaction order in the case of a comparatively simple reaction with constant, known stoichiometry (the case under discussion) has no advantages over either the superimposition or fractional life methods. With complicated reactions, however, the construction of tangents is sometimes the only way of obtaining essential clues as to the type of kinetic expression required and, therefore, this procedure is not without its merits.

Before dismissing the tangent method, we should mention a slight variant which becomes possible when small concentrations of product are experimentally detectable (as, for example, in a reaction yielding a coloured product with a very high extinction coefficient). In these circumstances, the appearance of the product can be followed under conditions where the concentrations of the reactants remain effectively constant at their initial values and so we can write

$$\left\{\frac{\mathrm{d}[\mathrm{P}]}{\mathrm{d}t}\right\}_{\text{initial}} = -\frac{\nu_{\mathrm{P}}}{\nu_{\mathrm{A}}}\left\{\frac{\mathrm{d}[\mathrm{A}]}{\mathrm{d}t}\right\}_{\text{initial}}$$

$$= \frac{\nu_{\mathrm{P}}}{\nu_{\mathrm{A}}} k[\mathrm{A}]_0^{a+b} r^b \qquad (28)$$

This equation defines the initial rate of formation of product for a reaction proceeding at constant volume in which the initial concentration ratio I is identical with the stoichiometric ratio r. It follows from eqn. (15) by writing $\alpha = 1$ and $\mathrm{d}\alpha = [\mathrm{A}]_0^{-1}\,\mathrm{d}[\mathrm{A}]$, or more directly from eqn. (11) and the definition of r. Clearly the product concentration–time curve is almost linear in the very early stages of the reaction (say from $\alpha = 1.0$ to 0.95) and, therefore, provided it can be accurately defined by making a series of accurate determinations of [P] and t in this region, the measurement of initial rate presents no problem. A series of such measurements at differing values of $[\mathrm{A}]_0$ (and $[\mathrm{B}]_0$ to keep $I = r$) enables the value of $(a+b)$ to be found on the basis of the analogous equation to (27), *viz.*

$$\log_{10}(\mathrm{d}[\mathrm{P}]/\mathrm{d}t)_{\text{initial}} = \text{constant} + (a+b)\log_{10}[\mathrm{A}]_0$$

This method is very useful in that the total order of the reaction can be estimated without having to follow the reaction to large extents; it must be remembered, however, that it requires accurate measurements of small concentrations of product and a procedure for mixing the reactants in a much shorter time than that required for 1% of either to disappear—from this latter point of view, the method can only be applied to fairly slow reactions.

References p. 407

(b) Estimation of the order, a

The methods by which the order with respect to A, a, may be estimated are merely modifications of those described in the previous section. Instead of choosing $[B]_0$ such that $[B]_0/[A]_0 = r$ as required for the estimation of total order, we now arrange that $[B]_0 \gg [A]_0$ so that the disappearance of A is followed in the presence of an effectively constant concentration of B. More precisely, we arrange that

$$\frac{[B]_0}{[A]_0} \gg \frac{\nu_B}{\nu_A}$$

or

$$I \gg r.$$

Eqn. (14),

$$-\frac{d\alpha}{dt} = k[A]_0^{a+b-1}\alpha^a(I-r+r\alpha)^b$$

now becomes

$$-\frac{d\alpha}{dt} \doteqdot k[A]_0^{a+b-1}\alpha^a I^b \tag{29}$$

since the quantity $(I-r+r\alpha)$ is practically identical with I. The value taken by $(I-r+r\alpha)$ ranges from I at $t = t_0$ to $(I-r)$ at $t = \infty$ and so, provided that the initial concentration ratio is at least an order of magnitude greater than the stoichiometric ratio, is practically time invariant. If we redefine the reduced time as

$$\theta = k[A]_0^{a+b-1}I^b(t-t_0) \tag{30}$$

eqn. (29) becomes

$$-\frac{d\alpha}{d\theta} \doteqdot \alpha^a \tag{31}$$

and, therefore, as before α is a unique function of θ for any given value of a. Since eqn. (31) is identical in mathematical properties to eqn. (16), all procedures based on the latter equation for the determination of $(a+b)$ apply equally well to the estimation of a from the former. It should be remembered, however, that eqn. (31) describes the dependence of α on θ for a series of experiments in which $I \gg r$ but throughout which I may either vary from one experiment to another or remain constant. Thus, one method of estimating a would be to determine α as a function of t in a series of experiments at *different* values of $[A]_0$, $[A]_{0,1}$,

$[\mathrm{A}]_{0,2} \ldots [\mathrm{A}]_{0,i}$, but at the *same* value of $[\mathrm{B}]_0$, the relative values being such that in every case $[\mathrm{B}]_0/[\mathrm{A}]_{0,i}$ was always greater than $\nu_\mathrm{B}/\nu_\mathrm{A}$. If the fundamental kinetic equation, eqn. (11), correctly describes the disappearance of the reactants, each of the resulting curves of α against $\log_{10}(t-t_0)$ should superimpose on the same curve of α against $\log_{10}\theta$ shown in Fig. 1 with a scale displacement of

$$\log_{10} k[\mathrm{A}]_{0,i}^{a-1}[\mathrm{B}]_0^b$$

On the other hand, there is a good case for performing the series of experiments at different values of $[\mathrm{A}]_0$ *and* $[\mathrm{B}]_0$ chosen such that the ratio, $[\mathrm{B}]_{0,i}/[\mathrm{A}]_{0,i}$, is the same throughout and considerably greater than $\nu_\mathrm{B}/\nu_\mathrm{A}$. Again each of the curves of α against $\log_{10}(t-t_0)$ should superimpose on the same curve of α against $\log_{10}\theta$ of Fig. 1 if eqn. (11) applies, but now each is displaced by

$$\log_{10} k[\mathrm{A}]_{0,i}^{a+b-1}I^b$$

Since both k and I are constants and a is known from the superimposition, an estimate of b can be obtained from the slope of the linear plot of scale displacement against $\log_{10}[\mathrm{A}]_{0,i}$.

The fractional life methods can also be used to determine a. As already indicated, the fractional life method of treating the data of a single experiment has few advantages over the superimposition method and consequently need not be considered further. On the other hand, the measurement of a particular fractional life of the reactant A in a series of experiments in which its initial concentration is varied provides a means by which a can be determined which sometimes offers experimental advantages. The only point to remember is that throughout the series of experiments $[\mathrm{B}]_0$ must be kept constant at such a value that $[\mathrm{B}]_0/[\mathrm{A}]_{0,i} \gg \nu_\mathrm{B}/\nu_\mathrm{A}$. From the condition that

$$\theta(\alpha^*) = \text{constant}, \alpha^* = \text{constant}$$

and eqn. (30), we have the expression

$$k[\mathrm{A}]_{0,i}^{a-1}[\mathrm{B}]_0^b(t_i(\alpha^*)-t_0) = \text{constant}$$

which clearly enables a to be found. Note that if $[\mathrm{B}]_0$ were varied with $[\mathrm{A}]_0$ so that I remained constant, these experiments would yield $(a+b)$ not a. Suitable values of α^* lie in the range 0.5 to 0.8.

Finally, it should be obvious that the method of tangents can be equally well applied to the determination of a utilizing eqn. (29) in precisely the same way as eqn. (15).

References p. 407

(c) *Estimation of the order, b*

The simplest procedure is to deduce the value of b from the values of $(a+b)$ and a determined in the manner previously described.

Alternatively, it may be obtained independently by essentially the same procedure as outlined for the determination of a. It is clear that we can write for every equation in α an exactly analogous equation in β where β represents the dimensionless ratio $[B]/[B]_0$ and so all previous arguments concerning α are applicable with minor modifications to β. Thus the order with respect to B can be determined by following the disappearance of B in the presence of a large excess of A using similar arguments to those previously described.

2.1.2 *Calculation of the rate coefficient*

Although the rate coefficient can be obtained as a by-product of the methods described for the estimation of the orders a and b, it is best obtained from the integrated form of the general rate equation

$$-\frac{d[A]}{dt} = k[A]^a[B]^b$$

In terms of the parameters α, I, r and t, we may write

$$-\int_1^{\alpha} \frac{d\alpha}{\alpha^a(I-r+r\alpha)^b} = k[A]_0^{a+b-1}\int_{t_0}^{t} dt$$

which becomes

$$\int_{\alpha}^{1} \frac{d\alpha}{\alpha^a(\alpha+c)^b} = k[A]_0^{a+b-1}r^b(t-t_0) \tag{32}$$

where

$$c = \frac{I}{r} - 1 \tag{33}$$

Table 1 lists the forms of the integrated expressions for common values of a and b; in certain cases, the integral differs according to whether c is positive or negative and therefore both forms have been given in the table; in other cases, the general expression becomes indeterminate when c equals zero and a special form of the integral must be used—where appropriate these special cases are also shown in the table. It should be appreciated that when c is less than zero, that is, when A is in excess of B, the reaction stops when the value of α reaches $-c$; at this point, the concentration of reactant B is zero.

TABLE 1

$-\frac{d[A]}{dt}$		*Integrated form of rate equation*	
k		$1-\alpha = k[A]_0^{-1}(t-t_0)$	(34)
$k[A]^{\frac{1}{2}}$		$1-\alpha^{\frac{1}{2}} = \frac{k}{2}[A]_0^{-\frac{1}{2}}(t-t_0)$	(35)
$k[A]$		$\ln\frac{1}{\alpha} = k(t-t_0)$	(36)
$k[A]^{\frac{3}{2}}$		$\frac{1}{\alpha^{\frac{1}{2}}}-1 = \frac{k}{2}[A]_0^{\frac{1}{2}}(t-t_0)$	(37)
$k[A]^2$		$\frac{1}{\alpha}-1 = k[A]_0(t-t_0)$	(38)
$k[B]^{\frac{1}{2}}$		$(1+c)^{\frac{1}{2}}-(\alpha+c)^{\frac{1}{2}} = \frac{k}{2}[A]_0^{-\frac{1}{2}}r^{\frac{1}{2}}(t-t_0)$	(39)
$k[A]^{\frac{1}{2}}[B]^{\frac{1}{2}}$		$\ln\left\{\frac{(1+c)^{\frac{1}{2}}+1}{(\alpha+c)^{\frac{1}{2}}+\alpha^{\frac{1}{2}}}\right\} = \frac{k}{2}r^{\frac{1}{2}}(t-t_0)$	(40)
$k[A][B]^{\frac{1}{2}}$	$c>0$	$\ln\left\{\frac{(\alpha+c)^{\frac{1}{2}}+c^{\frac{1}{2}}}{\alpha^{\frac{1}{2}}[(1+c)^{\frac{1}{2}}+c^{\frac{1}{2}}]}\right\} = \frac{k}{2}c^{\frac{1}{2}}[A]_0^{\frac{1}{2}}r^{\frac{1}{2}}(t-t_0)$	(41)*
	$c=0$	$\frac{1}{\alpha^{\frac{1}{2}}}-1 = \frac{k}{2}[A]_0^{\frac{1}{2}}r^{\frac{1}{2}}(t-t_0)$	(42)
	$c<0$	$\tan^{-1}\left\{\frac{1+c}{-c}\right\}^{\frac{1}{2}} - \tan^{-1}\left\{\frac{\alpha+c}{-c}\right\}^{\frac{1}{2}} = \frac{k}{2}(-c)^{\frac{1}{2}}[A]_0^{\frac{1}{2}}r^{\frac{1}{2}}(t-t_0)$	(43)*
$k[A]^{\frac{3}{2}}[B]^{\frac{1}{2}}$	$c>0$ $c<0$	$\left\{\frac{\alpha+c}{\alpha}\right\}^{\frac{1}{2}} - \left\{1+c\right\}^{\frac{1}{2}} = \frac{k}{2}c[A]_0 r^{\frac{1}{2}}(t-t_0)$	(44)*
	$c=0$	$\frac{1}{\alpha}-1 = k[A]_0 r^{\frac{1}{2}}(t-t_0)$	(45)
$k[B]$		$\ln\left\{\frac{1+c}{\alpha+c}\right\} = kr(t-t_0)$	(46)
$k[A]^{\frac{1}{2}}[B]$	$c>0$	$\tan^{-1}\left(\frac{1}{c}\right)^{\frac{1}{2}} - \tan^{-1}\left(\frac{\alpha}{c}\right)^{\frac{1}{2}} = \frac{k}{2}c^{\frac{1}{2}}[A]_0^{\frac{1}{2}}r(t-t_0)$	(47)*
	$c=0$	$\frac{1}{\alpha^{\frac{1}{2}}}-1 = \frac{k}{2}[A]_0^{\frac{1}{2}}r(t-t_0)$	(48)
	$c<0$	$\ln\left\{\frac{(1+c)^{\frac{1}{2}}(\alpha^{\frac{1}{2}}+(-c)^{\frac{1}{2}})}{(\alpha+c)^{\frac{1}{2}}(1+(-c)^{\frac{1}{2}})}\right\} = \frac{k}{2}(-c)^{\frac{1}{2}}[A]_0^{\frac{1}{2}}r(t-t_0)$	(49)*
$k[A][B]$	$c>0$ $c<0$	$\ln\left\{\frac{\alpha+c}{\alpha(1+c)}\right\} = kc[A]_0 r(t-t_0)$	(50)*
	$c=0$	$\frac{1}{\alpha}-1 = k[A]_0 r(t-t_0)$	(51)
$k[B]^{\frac{3}{2}}$		$\frac{1}{(\alpha+c)^{\frac{1}{2}}} - \frac{1}{(1+c)^{\frac{1}{2}}} = \frac{k}{2}[A]_0^{\frac{1}{2}}r^{\frac{3}{2}}(t-t_0)$	(52)
$k[A]^{\frac{1}{2}}[B]^{\frac{3}{2}}$	$c>0$ $c<0$	$\frac{1}{(1+c)^{\frac{1}{2}}} - \frac{\alpha^{\frac{1}{2}}}{(\alpha+c)^{\frac{1}{2}}} = \frac{k}{2}c[A]_0 r^{\frac{3}{2}}(t-t_0)$	(53)*
	$c=0$	$\frac{1}{\alpha}-1 = k[A]_0 r^{\frac{3}{2}}(t-t_0)$	(54)
$k[B]^2$		$\frac{1}{\alpha+c} - \frac{1}{1+c} = k[A]_0 r^2(t-t_0)$	(55)

* See p. 386 *et seq.*

In principle, the procedure for the calculation of the rate coefficient is simple. For each run, the values of α and t are substituted in that integrated form of eqn. (11) which corresponds to the orders a and b suggested by the preliminary examination of the data. Knowing the values of I, r, c, $[A]_0$ and t_0, a series of values of k can then be calculated which, provided that they show no systematic increase or decrease with decreasing α, can be averaged to give a mean value. Alternatively, the appropriate function of α shown in Table 1 can be plotted against t and, provided that the experimental points show no systematic deviation from a straight line, an average value of k can be obtained from the gradient. If the values of k obtained in either of these ways do not vary from run to run, then not only do we obtain a reliable value for the rate coefficient but also we confirm the orders a and b deduced from the preliminary examination of the data. On the other hand, if a systematic drift of the calculated rate coefficient is observed within any one particular run or from run to run, it can be concluded that either the incorrect integrated form of eqn. (11) has been chosen or that one of the more complicated rate expressions such as eqns. (2), (3) or (4) is required; which of these two interpretations is correct can be decided by attempting to fit the data to one of the other integrated forms of eqn. (11). If one cannot be found which gives a constant value for the rate coefficient, then another type of rate expression must be investigated.

Within this general scheme, there are several different methods of performing the calculation of the rate coefficient. As we shall show, each method weights the experimental determinations rather differently and so each method yields a different result for the average value of the rate coefficient. The question as to which of these methods yields the best value of the rate coefficient will be deferred until the methods themselves are discussed. In the first place, we shall confine our attention to the treatment of the data from a single run. In other words, we shall consider first the possible methods of calculating a value for the rate coefficient given one set of values of t and α, *viz.* (t_0, α_0), (t_1, α_1), (t_2, α_2), ... (t_j, α_j), ... (t_n, α_n). Note that the first pair of values has been written (t_0, α_0) instead of $(t_0, 1)$ to preserve consistency in the symbolism.

(*a*) *Treatment of the data from a single run*

Method 1. Calculation of the rate coefficient from pairs of values of α and t.

(*i*) *Pairing each value of t and α with the initial values t_0 and α_0*

Each of the integrated forms of the rate equation may be written

$$f(\alpha_0)-f(\alpha) = k'(t-t_0) \tag{56}$$

where

$$k' = mk$$

The quantity m is a constant whose value can be found from Table 1 for a given set of values of a, b and c. If we take the n pairs

$$\begin{matrix} t_0, \alpha_0 & t_0, \alpha_0 & \ldots & t_0, \alpha_0 & \ldots & t_0, \alpha_0 \\ t_1, \alpha_1 & t_2, \alpha_2 & & t_j, \alpha_j & & t_n, \alpha_n \end{matrix},$$

we can calculate n values of k', *viz.* $k'_1, k'_2 \ldots k'_j, \ldots k'_n$, using the expressions

$$\mathrm{f}(\alpha_0)-\mathrm{f}(\alpha_1) = k'_1(t_1-t_0)$$

$$\mathrm{f}(\alpha_0)-\mathrm{f}(\alpha_2) = k'_2(t_2-t_0)$$

$$\vdots$$

$$\mathrm{f}(\alpha_0)-\mathrm{f}(\alpha_j) = k'_j(t_j-t_0)$$

$$\vdots$$

$$\mathrm{f}(\alpha_0)-\mathrm{f}(\alpha_n) = k'_n(t_n-t_0)$$

The average value of the rate coefficient is clearly

$$k_{\mathrm{av}} = \frac{\sum\limits_{j=1}^{j=n} k_j}{n} = \frac{\sum\limits_{j=1}^{j=n} k'_j}{nm}$$

From the above definitions of $k'_1, k'_2, \ldots k'_j \ldots k'_n$, it follows that

$$k_{\mathrm{av}} = \frac{1}{nm}\left\{\left(\frac{1}{t_1-t_0}+\frac{1}{t_2-t_0}+\ldots\frac{1}{t_n-t_0}\right)\mathrm{f}(\alpha_0) - \frac{1}{t_1-t_0}\mathrm{f}(\alpha_1) - \frac{1}{t_2-t_0}\mathrm{f}(\alpha_2)\ldots - \frac{1}{t_n-t_0}\mathrm{f}(\alpha_n)\right\}$$

$$= \frac{1}{nm}\left\{\left(\sum_{j=1}^{j=n}\frac{1}{t_j-t_0}\right)\mathrm{f}(\alpha_0) - \sum_{j=1}^{j=n}\frac{1}{t_j-t_0}\mathrm{f}(\alpha_j)\right\}$$

It can be seen that the various values of $\mathrm{f}(\alpha)$ do not appear in the expression on the same footing—each value of $\mathrm{f}(\alpha)$ is multiplied by a factor which varies with the value of α. These factors are termed *weights* since they determine the contribution made by each term to the average value. Clearly, the quantity $\mathrm{f}(\alpha_0)$ is heavily overweighted relative to the subsequent quantities $\mathrm{f}(\alpha_j)$. Furthermore, as the time interval separating a particular value of α from the initial value increases, so the corresponding value of the function $\mathrm{f}(\alpha)$ contributes less and less to the average

References p. 407

value of k. In fact, the mean value of k is determined essentially by the early concentration observations and the heavily overweighted initial value if this method of averaging is employed. The main objection to the use of this method arises from the fact that large errors in the values of $t_j - t_0$ are probable at the early stages of the reaction owing to the experimental difficulty of determining the exact value of t_0. Consequently the weighting factors in the most important contributions to k_{av} are subject to fairly considerable errors particularly if the early time intervals are very small. This, in turn, leads to considerable uncertainties in k_{av} and, therefore, this method of treating the data cannot be recommended.

(ii) Using successive pairs of t and α

To avoid undue emphasis on the initial values of t and α, an average value of k can be calculated by taking the pairs

$$\begin{matrix} t_0, \alpha_0 & t_1, \alpha_1 & \ldots & t_{j-1}, \alpha_{j-1} & \ldots & t_{n-1}, \alpha_{n-1} \\ t_1, \alpha_1 & t_2, \alpha_2 & & t_j, \alpha_j & & t_n, \alpha_n \end{matrix}$$

Writing, as before, k_j for the rate constant derived from the jth pair, we have

$$k_{av} = \frac{\sum_{j=1}^{j=n} k_j}{n}$$

$$= \frac{\sum_{j=1}^{j=n} k'_j}{nm}$$

where k'_j is now given by

$$f(\alpha_{j-1}) - f(\alpha_j) = k'_j(t_j - t_{j-1})$$

Performing the summation, we find

$$k_{av} = \frac{1}{mn}\left\{\frac{1}{t_1 - t_0} f(\alpha_0) + \sum_{j=1}^{j=n-1} \frac{2t_j - t_{j-1} - t_{j+1}}{(t_j - t_{j-1})(t_{j+1} - t_j)} f(\alpha_j) - \frac{1}{t_n - t_{n-1}} f(\alpha_n)\right\}$$

It is clear that each value of $f(\alpha)$ is weighted by the factor

$$\frac{2t_j - t_{j-1} - t_{j+1}}{(t_j - t_{j-1})(t_{j+1} - t_j)}$$

with the exception of the first and last values of the set, $f(\alpha_0)$ and $f(\alpha_n)$. Apart from the two times t_0 and t_n, any particular time t_j is not usually very different from the average of the two times t_{j-1} and t_{j+1}; *i.e.*, $t_j \sim \frac{1}{2}(t_{j-1}+t_{j+1})$, and so the contributions of all values of $\mathrm{f}(\alpha)$ intermediate between the first and last are seriously underweighted. Indeed, in the special case when the time intervals between successive concentration determinations are equal so that $t_j = \frac{1}{2}(t_{j-1}+t_{j+1})$ the weighting factor becomes zero for all values of j from 1 to $(n-1)$ inclusive. In this situation, none of the intermediate concentration terms contribute to the mean value of k and we obtain

$$k_{\mathrm{av}} = \frac{1}{mn}\left\{\frac{1}{t_1-t_0}\,\mathrm{f}(\alpha_0) - \frac{1}{t_n-t_{n-1}}\,\mathrm{f}(\alpha_n)\right\}$$

$$= \frac{1}{m(t_n-t_0)}\,\{\mathrm{f}(\alpha_0)-\mathrm{f}(\alpha_n)\}$$

As far as this method of obtaining an average value of k is concerned, the intermediate concentration determinations at equally spaced time intervals might as well not have been made!

The general conclusion to be drawn from this analysis is that there is little point in performing the laborious operation of calculating individual k values from successive pairs of observations in order to obtain a mean value since this procedure places undue weight on the initial and final observations which in practice are often the least accurate.

Method 2. Calculation of the rate coefficient using a graphical method.

As we have already pointed out, each of the integrated forms of the rate equation may be written

$$\mathrm{f}(\alpha_0)-\mathrm{f}(\alpha) = k'(t-t_0)$$

where k' is proportional to the rate coefficient k. Clearly a plot of $\mathrm{f}(\alpha)$ against $(t-t_0)$ should be linear with an intercept equal to $\mathrm{f}(\alpha_0)$ at $(t-t_0) = 0$ and a slope, $-k'$. From the slope and the value of m, the rate coefficient follows at once. We shall not consider at this point the arguments for and against constraining the line to pass through $\mathrm{f}(\alpha_0)$ at $(t-t_0) = 0$. For the moment, we shall adopt this procedure since the object of this section is merely to present *some* of the problems which arise in the treatment of data; the recommended method is always to make use of the data of a number of experiments as described in the next section.

In view of our previous discussion, it is clearly of importance to evaluate the contribution of each experimental observation to the average value of k so obtained. This depends on the criterion according to which the straight line through the points is constructed. If we assume that the line is drawn so that the algebraic sum of the deviations, δ_j, of the experimental points from the line is close to zero, we can deduce the contribution of each point to the average value of k as follows.

References p. 407

We define the deviation of the jth point as

$$\delta_j = f(\alpha_j) - \{f(\alpha_0) - k'_{av}(t_j - t_0)\}$$

where k'_{av} is the slope of the line drawn through the experimental points according to the condition that

$$\sum_{j=1}^{j=n} \delta_j = 0$$

(In practice, it would not be possible to make $\sum_{j=1}^{j=n} \delta_j$ exactly equal to zero; however, this condition approximates closely to that used intuitively in constructing by eye the 'best' straight line through a set of experimental points). From these two equations, we see that

$$\sum_{j=1}^{j=n} f(\alpha_j) = \sum_{j=1}^{j=n} f(\alpha_0) - k'_{av} \sum_{j=1}^{j=n} (t_j - t_0)$$

or

$$k'_{av} = \frac{nf(\alpha_0) - \sum_{j=1}^{j=n} f(\alpha_j)}{\sum_{j=1}^{j=n} (t_j - t_0)}$$

Therefore

$$k_{av} = \frac{nf(\alpha_0) - \sum_{j=1}^{j=n} f(\alpha_j)}{m \sum_{j=1}^{j=n} (t_j - t_0)}$$

Thus with the exception of the initial value, every experimental observation of α is equally weighted. In the special case where the observations of α are made at equally spaced time intervals, the above expression is particularly easy to comprehend. If we write τ for the first time interval $t_1 - t_0$, 2τ for $t_2 - t_0$ and so on, then

$$\begin{aligned}\sum_{j=1}^{j=n} (t_j - t_0) &= \sum_{j=1}^{j=n} j\tau \\ &= \frac{n(n+1)\tau}{2}\end{aligned}$$

Therefore

$$k_{\mathrm{av}} = \frac{2}{m(n+1)\tau}\left\{\mathrm{f}(\alpha_0) - \frac{\sum_{j=1}^{j=n} \mathrm{f}(\alpha_j)}{n}\right\}$$

$$= \frac{1}{m(t_{(n+1)/2} - t_0)}\left\{\mathrm{f}(\alpha_0) - \frac{\sum_{j=1}^{j=n} \mathrm{f}(\alpha_j)}{n}\right\}$$

Thus, the average value of k obtained graphically using a set of observations made at equally spaced time intervals is determined by the ratio of the difference between the value of $\mathrm{f}(\alpha_0)$ and the unweighted average of the $\mathrm{f}(\alpha_j)$ to the time elapsed to exactly midway in the experiment.

The graphical method of determining the average value of k from the data of a single experiment is thus clearly superior to the numerical methods discussed earlier.

(b) Treatment of the data from a series of runs

The use of the data of a single experiment to obtain an average value of the rate coefficient is inadvisable. In the first place, any estimate of the precision of the rate coefficient is necessarily of limited significance; it is for this reason that we have not devoted any attention to this quantity in our previous discussion. In the second place, although unlikely in the case under discussion, complications in the kinetic behaviour may well be missed if the α, t data are only available for one pair of $[\mathrm{A}]_0$ and $[\mathrm{B}]_0$ values.

In order to discuss the treatment of the data obtained from a number of experiments, we must be quite clear as to what our aims are. We require:

(1) an objective method of computing the rate coefficient from the α, t data;
and (2) an estimate of its standard error.

We shall suppose that our data consist of r sets of concentration–time determinations, each set constituting a separate experiment. We shall label the separate experiments: run 1, run 2, . . . run i . . . run r; although it is of no immediate importance, we shall suppose that this set of r experiments can be divided into a number of sub-sets such that the runs within each sub-set are replicates—that is, the runs constituting a given sub-set correspond to ostensibly the same initial concentrations, $[\mathrm{A}]_0$ and $[\mathrm{B}]_0$, and the same conditions of temperature, pressure etc. Within each experiment, we shall use the subscript j to designate each pair of observations of t and α, j running from zero for the initial values to n_i, the total number of pairs of observations (excluding the initial pair) in the ith run. We thus have a set of observations of the form given in Table 2.

As we have seen, a variety of values for the rate coefficient can be obtained from the data of a single run according to the calculation method used. In this section, we shall adopt the criterion that for a given run the best value of the rate

References p. 407

TABLE 2

Run 1	...	*Run i*	...	*Run r*
t_{10} α_0		t_{i0} α_0		t_{r0} α_0
t_{11} α_{11}		t_{i1} α_{i1}		t_{r1} α_{r1}
t_{12} α_{12}		t_{i2} α_{i2}		t_{r2} α_{r2}
$\vdots$		$\vdots$		$\vdots$
t_{1j} α_{1j}		t_{ij} α_{ij}		t_{rj} α_{rj}
$\vdots$		$\vdots$		$\vdots$
t_{1n_1} α_{1n_1}		t_{in_i} α_{in_i}		t_{rn_r} α_{rn_r}

coefficient is that value which minimizes the weighted sum of the squares of the deviations of the experimental points from the equation

$$f(\alpha) = l - mk(t - t_0)$$

the weights being calculated from the experimental errors in each value of α. In this equation, $f(\alpha)$ is the function suggested by the preliminary examination of the data, l is a constant, k is the rate coefficient and m is another constant whose value can be found from Table 1. In the computation work which follows, the equations of Table 1 must be rearranged so as to correspond to the form of the above equation to avoid confusion over signs.

For example, eqn. (36)

$$\ln 1/\alpha = k(t - t_0)$$

is clearly of the form

$$\ln \alpha = \text{constant} - k(t - t_0)$$

and so $f(\alpha)$ in this particular case is $\ln(\alpha)$. On the other hand, eqn. (38)

$$(1/\alpha) - 1 = k[\text{A}]_0(t - t_0)$$

must be written

$$-1/\alpha = \text{constant} - k[\text{A}]_0(t - t_0)$$

so that $f(\alpha)$ in this case is $-1/\alpha$.

The important point to notice is that the graph of $f(\alpha)$ against $(t - t_0)$ is *not* constrained to pass through the value of $f(\alpha_0)$ at $t - t_0 = 0$. The reason for this is that there are usually *systematic errors* in the reaction time resulting from the fact that the reaction rate does not reach instantaneously the value corresponding to the rate coefficient characteristic of the run as a whole. The simplest example of such an error is that which arises when the two reactants are mixed together at a temperature different from that at which the reaction is to be studied. Obviously there is a time lag between mixing and attainment of thermostat temperature so that the observed value of the time taken for α to drop from unity to a particular value α' is not the same as the time which would have been required had the reaction mixture reached thermostat temperature instantaneously. Provided that determinations of α are not commenced until thermal equilibrium has been attained, such an effect can be represented as a systematic error in the reaction time as shown in the schematic diagram of Fig. 2. This shows the type of plot found when the reactants are mixed together at a lower temperature than that at which the reaction is studied kinetically. In the period of time indicated by Δt_0, the temperature of the reaction mixture increases and the rate of disappearance of A cannot be characterized by a single value of the rate coefficient. Once the reaction has reached the thermostat temperature, the plot of $f(\alpha)$ against $t-t_0$ becomes linear but its intercept at $t-t_0 = 0$ is clearly not $f(\alpha_0)$; its slope, however, is equal to mk. Therefore, to find the rate coefficient, we use the equation

$$f(\alpha) = l - mk(t-t_0)$$

rather than eqn. (56)

$$f(\alpha) = f(\alpha_0) - mk(t-t_0)$$

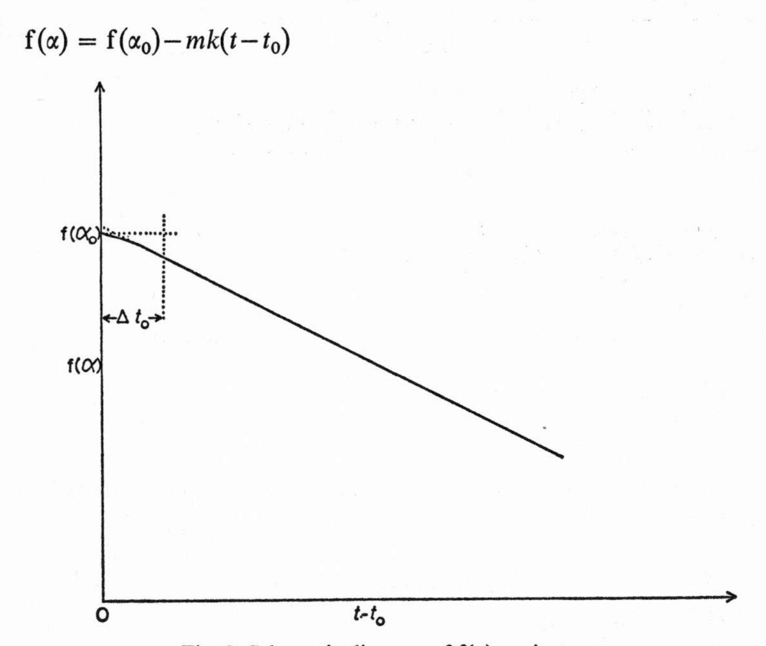

Fig. 2. Schematic diagram of $f(\alpha)$ against $t-t_0$.

References p. 407

care being taken to avoid including measurements of α obtained in the very early stages of the reaction where the rate is not characterized by a single value of the rate coefficient. It is not suggested that the time taken for temperature equilibration is the sole factor introducing systematic errors in the reaction time; indeed, it is sometimes possible to design the experimental procedure so that such errors are practically eliminated. The point is that the calculated value of the rate coefficient is unaffected by systematic errors in the reaction time provided that the data are not constrained to eqn. (56).

For the ith run, therefore, we determine the parameters of the weighted least squares line through the points $\{(t_{ij}-t_{i0}), \mathrm{f}(\alpha_{ij})\}$, *viz.* l_i and k'_i in the equation

$$\mathrm{f}(\alpha_{ij}) = l_i - k'_i(t_{ij}-t_{i0}) \tag{57}$$

where $k'_i = m_i k_i$. Writing

$$y_{ij} = \mathrm{f}(\alpha_{ij})$$

and

$$x_{ij} = t_{ij}-t_{i0}$$

and assigning a weight w_{ij} to the jth point, these quantities are obtained from the following equations

$$k'_i = \frac{(\sum_{j=1}^{j=n_i} w_{ij}x_{ij})(\sum_{j=1}^{j=n_i} w_{ij}y_{ij}) - (\sum_{j=1}^{j=n_i} w_{ij})(\sum_{j=1}^{j=n_i} w_{ij}x_{ij}y_{ij})}{D_i} \tag{58}$$

$$l_i = \frac{(\sum_{j=1}^{j=n_i} w_{ij}y_{ij})(\sum_{j=1}^{j=n_i} w_{ij}x_{ij}^2) - (\sum_{j=1}^{j=n_i} w_{ij}x_{ij})(\sum_{j=1}^{j=n_i} w_{ij}x_{ij}y_{ij})}{D_i} \tag{59}$$

where

$$D_i = (\sum_{j=1}^{j=n_i} w_{ij})(\sum_{j=1}^{j=n_i} w_{ij}x_{ij}^2) - (\sum_{j=1}^{j=n_i} w_{ij}x_{ij})^2 \tag{60}$$

In addition, two other quantities should be calculated. These are

$$\text{(a)} \quad E_i = \sum_{j=1}^{j=n_i} w_{ij}\delta_{ij}^2 \tag{61}$$

where δ_{ij}, the deviation of the experimental value of $\mathrm{f}(\alpha_{ij})$ from the value calculated from eqn. (57), is given by

$$\begin{aligned} \delta_{ij} &= \text{observed } \mathrm{f}(\alpha_{ij}) - \text{calculated } \mathrm{f}(\alpha_{ij}) \\ &= \mathrm{f}(\alpha_{ij}) - \{l_i - k'_i(t_{ij}-t_{i0})\} \\ &= y_{ij} - \{l_i - k'_i x_{ij}\} \end{aligned} \tag{62}$$

δ_{ij} is termed a residual. Eqns. (58) and (59) give the values of k'_i and l_i which make E_i a minimum in accordance with the criterion given earlier.

(b) $$s(k'_i) = \left\{\frac{E_i \sum_{j=1}^{j=n_i} w_{ij}}{(n_i-2)D_i}\right\}^{\frac{1}{2}} \tag{63}$$

$s(k'_i)$ being an estimate of the *standard error* in k'_i.

In writing these expressions, it is implicitly assumed that the observations α_{ij} are subject to random errors and that the reaction times $(t_{ij}-t_{i0})$ are free from random errors. The assumption that the random errors occur solely in α is reasonable since, in the majority of kinetic studies, time can be measured with much greater precision than concentration. When the random errors in the determination of time are commensurate with the random errors in the determination of concentration, the above equations are no longer valid; we shall not discuss this situation in view of the difficulties which arise but would suggest that, if possible, every effort should be made to avoid it by redesigning the experimental procedure.

(*i*) *The weights* w_{ij}

The real problem in utilizing expressions (58) to (63) is the assignment of the weights, w_{ij}. To proceed further, it is necessary to digress into a discussion of some basic statistical ideas.

Let us return to our basic parameter, α. Suppose that it were possible to make a very large number of determinations of α at some particular value of the reaction time, $t-t_0$, by performing a large number of identical experiments. Suppose further that all systematic errors had been eliminated and that the observed values of α were distributed normally about a mean value of $\bar{A}$; *i.e.*, the frequency with which a value of α lying between α and $\alpha+\mathrm{d}\alpha$ is found is given by

$$\frac{1}{\sigma(\alpha)\sqrt{2\pi}} \exp\left\{\frac{(\alpha-\bar{A})^2}{2\sigma^2(\alpha)}\right\} \mathrm{d}\alpha$$

The quantity $\sigma^2(\alpha)$ is termed the *variance of the population of* α or more simply the *variance of* α and is the average of the squares of the deviations of the α values from the *population* mean, $\bar{A}$; *i.e.*

$$\sigma^2(\alpha) = \frac{1}{R}\sum_{i=1}^{i=R}(\alpha_i-\bar{A})^2 \dagger$$

† This equation must not be confused with the equation giving an *estimate* of the variance of the population, $s^2(\alpha)$, from a *limited* sample of r α-values, *viz.*

$$s^2(\alpha) = \frac{1}{r-1}\sum_{i=1}^{i=r}(\alpha_i-\bar{\alpha})^2$$

where $\bar{\alpha}$ is the *sample* mean.

References p. 407

where R, the number of α_i values making up the population, is extremely large. The variance of the population is a measure of the scatter of the observed values of α at the time $(t-t_0)$ about the population mean. An alternative measure of the width of the distribution is the square root of the variance, $\{\sigma^2(\alpha)\}^{\frac{1}{2}}$, or more simply $\sigma(\alpha)$; this quantity is termed the *standard deviation* of the distribution and has the advantage that it characterizes the width of the distribution on the same scale as α.

Now, if, as a result of performing a large number of experiments, we observed a random scatter of the α_{ij} values about some mean $\bar{\alpha}_j$, it follows that the values of $f(\alpha_{ij})$ would also be scattered about some mean value. For purely random errors, it can be shown that the weight which should be assigned to each value of $f(\alpha_{ij})$ in our previous formulae is inversely proportional to the variance of the corresponding population of $f(\alpha_{ij})$, $\sigma^2\{f(\alpha_{ij})\}$; *i.e.*

$$w_{ij} = \frac{\sigma^2}{\sigma^2\{f(\alpha_{ij})\}} \tag{64}$$

The constant of proportionality, σ^2, may be regarded as the variance of a set of observations of unit weight. If eqn. (58) giving the best value of the rate coefficient is examined carefully, it will be seen that the presence of a constant multiplier in the weighting factors makes no difference at all to the calculated value of k_i'; the same statement can be made about eqns. (59) and (63) giving the values of l_i and $s(k_i')$. Therefore, for purely computational purposes, we omit the constant of proportionality, σ^2, from eqn. (64) and write

$$w_{ij} = \frac{1}{\sigma^2\{f(\alpha_{ij})\}} \tag{65}$$

Obviously the greater the value of the variance of $f(\alpha_{ij})$, the less is the weight given to that point. Provided that the random error in $f(\alpha_{ij})$ originates from the random errors of observation of α_{ij} and provided that these are small so that $\sigma^2(\alpha_{ij})$ is small, we can write

$$\sigma^2\{f(\alpha_{ij})\} = \left\{\frac{\partial f(\alpha)}{\partial \alpha}\right\}^2_{\alpha=\alpha_{ij}} \sigma^2(\alpha_{ij}) \tag{66}$$

The values of $\partial f(\alpha)/\partial\alpha$ follow at once from the rate equation under test; for orders a and b

$$\frac{\partial f(\alpha)}{\partial \alpha} = \frac{1}{\alpha^a(\alpha+c)^b} \tag{67}$$

Combining eqns. (65), (66) and (67), we obtain

$$w_{ij} = \frac{\alpha_{ij}^{2a}(\alpha_{ij}+c)^{2b}}{\sigma^2(\alpha_{ij})} \tag{68}$$

which, irrespective of the values of a and b, shows that

$$w_{ij} \to 0 \quad \text{as} \quad \alpha_{ij} \to 0 \qquad (c+\text{ve})$$

or

$$w_{ij} \to 0 \quad \text{as} \quad \alpha_{ij} \to -c \qquad (c-\text{ve})$$

In other words, the experimental determinations made towards the end of a reaction contribute very little to the least squares estimate of the rate coefficient. Eqn. (68) is of little use in computation, however, unless we can make some statement about $\sigma(\alpha_{ij})$. There are two clearcut situations which we can discuss. The first of these situations is that the value of $\sigma(\alpha_{ij})$ does not depend on j and so within any one run, $\sigma(\alpha_{ij})$ is a constant. In other words, were it possible to determine a large number of values of each α_{ij}, the width of the normal distribution would be the same in each case. Since it is only relative weights which are important in computation, we can omit such constant factors from the weights [as was done in eqn. (65)] and write

$$w_{ij} = \alpha_{ij}^{2a}(\alpha_{ij}+c)^{2b}$$

when $\sigma(\alpha_{ij})$ is independent of j. The alternative situation which permits of discussion is that which arises when $\sigma(\alpha_{ij})$ is proportional to α_{ij} so that $\sigma(\alpha_{ij})/\alpha_{ij}$ remains constant throughout the run; this situation is described by saying that the *relative* error in α_{ij} is constant. In this case

$$w_{ij} = \alpha_{ij}^{2(a-1)}(\alpha_{ij}+c)^{2b}$$

The choice between these two alternatives can only be made by considering the experimental technique employed to determine the concentration of A.

A simple example of common occurrence will illustrate the difference between the two weighting procedures. Consider a simple first order disappearance of A whose rate equation is

$$-\frac{\mathrm{d}\alpha}{\mathrm{d}t} = k\alpha \text{ in differential form,}$$

or

$$-\ln \alpha = k(t-t_0) \text{ in integral form}$$

so that $\mathrm{f}(\alpha) = \ln \alpha$.

References p. 407

Then, if each value of α can be written as $\alpha \pm \sigma(\alpha)$ where $\sigma(\alpha)$, the standard deviation of α, is independent of the value of α, the weight of each value of $f(\alpha)$ decreases rapidly as α decreases in accordance with the equation

$$w = \alpha^2$$

On the other hand, if the relative error in α remains constant so that the value of $\sigma(\alpha)$ in the expression $\alpha \pm \sigma(\alpha)$ is proportional to α, the weight of each value of $f(\alpha)$ is the same irrespective of the value of α; in this case

$$w = 1 \quad \text{for all} \quad \alpha$$

It is clear that the two methods of weighting differ greatly and in consequence the rate coefficient calculated from a given set of data on the basis of the first method is not exactly the same as that calculated using the second method.

This simple example illustrates the very real problems which arise when it is decided to weight individual observations in computing some type of average quantity; it is for this reason that the problem is often avoided by allocating equal weights to all observations irrespective of the improbability of this being justified in the case of most non-linear functions of α. Be that as it may, we should mention that the above criterion for deciding the weights, *viz.*

$$w_{ij} = \frac{1}{\sigma^2\{f(\alpha_{ij})\}}$$

is not the only one used in practice. Often, weighting is carried out on the basis

$$w_{ij} = \frac{1}{\sigma\{f(\alpha_{ij})\}} \tag{69}$$

which is rather less severe than that based on the reciprocal of $\sigma^2\{f(\alpha_{ij})\}$. These two methods of weighting are sometimes described as weighting according to (a) the reciprocals of the squares of the error bars or (b) the reciprocals of the error bars themselves. Error bars are used in the graphical representation of data to indicate the precision with which each point is known when only one of the variables is subject to error. As far as the author is aware, there is no single convention employed by all scientific workers for the construction of error bars. Assuming that the error is in the quantity plotted on the ordinate, the most common method used is to construct a line parallel to the ordinate of length twice the standard deviation† such that its mid-point is coincident with the value in question. In this

---

† If the point in question is the mean of several determinations, the length of the error bar should be equal to twice the *standard deviation of the mean* or *the standard error* as it is more often called.

case, weighting in accordance with the reciprocals of either the squares of the error bars or the error bars themselves is precisely equivalent to the two procedures given.

To summarize this matter of weighting, we can say that there are three different systems of assigning values to w_{ij}. These are:

(1) $w_{ij} = 1$

(2) $w_{ij} = \dfrac{1}{\sigma^2\{\mathrm{f}(\alpha_{ij})\}}$

(3) $w_{ij} = \dfrac{1}{\sigma\{\mathrm{f}(\alpha_{ij})\}}$

In most cases where non-linear functions of α are being considered, the first method is unsatisfactory and so the choice lies between methods (2) and (3). Of these two procedures, method (2) is the more soundly based. It must be said, however, that weighting is, to a considerable extent, a matter of personal judgement and so the use of method (2) is not automatic in every case. The one certain rule that can be given is that every worker should state exactly which method of weighting has been used and briefly indicate his reasons for his choice. Without this information, it is very difficult to assess the reliability of the quoted rate coefficient.

(ii) The quantities k'_i and $s(k'_i)$

Assuming that the previous calculations have been carried out, we now have a set of r values of k'_i and $s(k'_i)$. Since some of the runs correspond to different initial concentration conditions, the value of k'_i [and $s(k'_i)$] are not comparable directly. We therefore calculate the rate coefficient, k_i, and an estimate of its standard error, $s(k_i)$, for each run using the simple expressions

$$\left.\begin{aligned} k_i &= \frac{k'_i}{m_i} \\ s(k_i) &= \frac{s(k'_i)}{m_i} \end{aligned}\right\} \qquad (70)$$

Naturally, it is highly unlikely that all the calculated values of k_i will be identical and so the question arises as to whether the differences between them are those which result from the random error of measurement or whether they indicate that the rate equation under investigation does not represent the data adequately.

Before discussing this important question, let us consider briefly the significance of the basic quantities k'_i and $s(k'_i)$; it is then not difficult to understand the significance of the related quantities k_i and $s(k_i)$.

References p. 407

First, consider k_i'. It can be shown that the line fitted through the points (x_{ij}, y_{ij}) passes through the weighted means of the values of x_{ij} and y_{ij}, $\bar{x}_i$ and $\bar{y}_i$, and that

$$k_i' = \frac{\sum_{j=1}^{j=n_i} w_{ij}(x_{ij}-\bar{x}_i)y_{ij}}{\sum_{j=1}^{j=n_i} w_{ij}(x_{ij}-\bar{x}_i)^2}$$

where

$$\bar{x}_i = \frac{\sum_{j=1}^{j=n_i} w_{ij}x_{ij}}{\sum_{j=1}^{j=n_i} w_{ij}}$$

and

$$\bar{y}_i = \frac{\sum_{j=1}^{j=n_i} w_{ij}y_{ij}}{\sum_{j=1}^{j=n_i} w_{ij}}$$

Thus every value of y_{ij}, that is, $f(\alpha_{ij})$, contributes to the average value of k_i' in accordance with (1) a weighting factor dependent on the random errors of measurement of α and (2) a second weighting factor dependent on its distance from the weighted mean of the x_{ij} values. Furthermore, the value of $f(\alpha_0)$ does not appear in the equation for k_i' since we have deliberately avoided constraining the line through the point $\{t_0, f(\alpha_0)\}$.

Second, consider the quantity $s(k_i')$. This is an estimate of the *standard error* of k_i' based on (n_i-2) *degrees of freedom.* We use the quantity standard error to describe the width of the distribution of k_i' values which we would observe if we were able to make a vast number of separate determinations; we use the term standard error rather than standard deviation to remind us that the distribution under consideration is a distribution of mean values, each of which is itself derived from a population. The term (n_i-2) degrees of freedom signifies that, out of our n_i experimental pairs of observations, only (n_i-2) are available to give us an estimate of the precision of the measurement, the other two having been 'lost' in fixing the two parameters, l_i and k_i', of our fitted line. Now, clearly we can never obtain a vast number of determinations of a statistic such as k_i' in order to obtain a value for its standard error; necessarily, the number of observations which we can make is limited and so we cannot do any better than make an estimate of what the width of the distribution would be for a vast population. This estimate is extremely useful for it enables us to calculate from the sample mean, *i.e.* k_i', the limits between which the population mean is likely to lie. Suppose we designate the population mean by the symbol, K_i'. We now introduce the statistic, t,

$$t = \frac{k'_i - K'_i}{s(k'_i)}$$

The point is that, corresponding to the number of degrees of freedom used in estimating $s(k'_i)$, we can look up in statistical tables the absolute value of t which is exceeded with a given frequency. In this way, we can establish for a given confidence level the limits between which K'_i must lie. A numerical example taken from Appendix 2 will make this clearer.

Example. Taking run 1 of the set of experiments, we find that

$$k'_1 = 7.097 \times 10^{-4}\ \text{sec}^{-1}\,\dagger$$
$$s(k'_1) = 0.023 \times 10^{-4}\ \text{sec}^{-1}$$
$$n_1 = 22$$

Our object in this example is to establish the 90% confidence limits for K'_1; that is, the limits between which K'_1 is likely to lie with 90% probability. From tables, 90% of all values of t based on 20 degrees of freedom lie between ± 1.72 and so the 90% confidence limits for the difference between k'_1 and the population average are

$$\pm 1.72 \times s(k'_1)$$
$$= \pm 0.040 \times 10^{-4}\ \text{sec}^{-1}$$

In other words, on the basis of the data of run 1, we expect with 90% confidence that the 'true' value of the rate coefficient (the population average) will lie between

$$(7.057 \quad \text{and} \quad 7.137) \times 10^{-4}\ \text{sec}^{-1}$$

Returning to the general case, it should be obvious that, if we can establish confidence limits for each value of k'_i, we can calculate confidence limits for each value of the rate coefficient.

(iii) Homogeneity of the rate coefficients

To this point, our discussion has been confined to describing the methods of processing the data of a single run and to offering some explanation of the significance of the derived quantities. However, we have repeatedly stressed that valid conclusions can only be reached if the data of several runs are compared. It is the purpose of this section to demonstrate how this comparison is best carried out.

We have r values of the rate coefficient together with a corresponding number of estimates of standard error. This set of values can be divided into u groups

---

† These data are characterized by $m_i = 1$ and so in this particular case k'_1 can be identified with the rate coefficient.

References p. 407

corresponding to determinations at different initial concentrations, there being r_1 values in the first group, r_2 values in the second group, ... r_v values in the vth group, ... and r_u values in the uth and final group. Clearly

$$r = \sum_{v=1}^{v=u} r_v$$

For example, the six first-order rate coefficients listed in Appendix 2 can be divided naturally into 3 groups according to the initial concentration of isoprene.

TABLE 3

Run	*[Isoprene]*$_0$ mole.litre^{-1}	$10^4\ k$ sec^{-1}	$10^4\ s(k)$ sec^{-1}
1		7.097	0.023
2	1.690	6.944	0.015
3		7.275	0.010
4	3.220	7.155	0.036
5		7.050	0.033
6	0.815	7.091	0.012

In other words, the runs within a group are replicate experiments. The problem which we have to solve is whether or not there is any significant difference between the rate coefficients of the various groups. If there is no significant difference between the rate coefficients so that the values are not dependent on the initial concentrations of the reactants, we confirm that the rate expression under investigation represents the data adequately. In such a case where the values from a series of sub-sets do not differ significantly from one another, we can pool the whole set to obtain a grand average—a set of this type is called an *homogeneous* set. We can also obtain an estimate of the standard error of the grand average as will be shown shortly.

We should note in passing that we do not really need a statistical analysis of the data presented in Table 3 to convince us that the six rate coefficients constitute an homogeneous set. The sub-set of runs 1, 2 and 3, which were carried out, as far as was possible, under identical conditions, shows a scatter of values overlapping those obtained under different initial concentrations (runs 4, 5 and 6).

We approach the general problem in the following way. Firstly, we relabel our rate coefficients, k_{vi}, where v specifies the group and i specifies the particular k within the group; thus, k_{13} specifies the third value of the rate coefficient in the first group. Secondly, we evaluate the average of the rate coefficients,

(a) for each group, $\bar{k}_v$

and

(b) for the whole set, $\bar{k}$

In calculating these averages, each value of the rate coefficient should be weighted by the reciprocal of the square of its standard error; if we write $w(k_{vi})$ for the weight assigned to k_{vi}, we have

$$w(k_{vi}) = \frac{1}{\sigma^2(k_{vi})}$$

In practice, we do not know the value of $\sigma^2\,(k_{vi})$ and so we use the estimate $s^2(k_{vi})$ calculated from eqns. (63) and (70); *i.e.*

$$w(k_{vi}) = \frac{1}{s^2(k_{vi})}$$

The average of the rate coefficients of the vth group is given by

$$\bar{k}_v = \frac{\sum_{i=1}^{i=r_v} w(k_{vi})k_{vi}}{\sum_{i=1}^{i=r_v} w(k_{vi})} \tag{71}$$

and, since there are u groups in all, we obtain u values, $\bar{k}_1, \bar{k}_2, \ldots \bar{k}_v, \ldots \bar{k}_u$. The average of the rate coefficients of the whole set is clearly

$$\bar{k} = \frac{\sum_{v=1}^{v=u}\sum_{i=1}^{i=r_v} w(k_{vi})k_{vi}}{\sum_{v=1}^{v=u}\sum_{i=1}^{i=r_v} w(k_{vi})} \tag{72}$$

Now, without going deeply into the underlying arguments, we can see that we should be able to relate the scatter of the group means about the set mean to the distribution of the individual rate coefficients within groups, if all groups are part of the same population. If there is no significant difference between groups, it can be shown that

$$\frac{\sum_{v=1}^{v=u}\sum_{i=1}^{i=r_v} w(k_{vi})(\bar{k}_v-\bar{k})^2}{u-1}$$

and

$$\frac{\sum_{v=1}^{v=u}\sum_{i=1}^{i=r_v} w(k_{vi})(k_{vi}-\bar{k}_v)^2}{r-u}$$

References p. 407

each provide an estimate of the same quantity, namely, the variance of a rate coefficient of unit weight. Thus, their ratio, F, defined as

$$F = \frac{\sum_{v=1}^{v=u} \sum_{i=1}^{i=r_v} w(k_{vi})(\bar{k}_v - \bar{k})^2/(u-1)}{\sum_{v=1}^{v=u} \sum_{i=1}^{i=r_v} w(k_{vi})(k_{vi} - \bar{k}_v)^2/(r-u)} \tag{73}$$

should be unity. In practice, this ratio may be less or greater than unity. If it is less than unity, the scatter of the group means about the set mean is less than would be anticipated from the scatter of the individual rate coefficients and so the whole set is homogeneous. (This is exemplified by the data given in Table 3 and Appendix 2). If F is greater than unity, the set may still be homogeneous but there are limits to the value F can take. These limits are determined by the confidence level chosen and by the values $(u-1)$ and $(r-u)$ appearing in the expression for F—the 'degrees of freedom' associated with the numerator and denominator. Thus, returning to our example, the maximum value of F which is likely to be obtained in, say, 95† cases out of 100 for 2 degrees of freedom in the numerator and 3 degrees of freedom in the denominator, is 9.55†† if there is no difference between the three groups.

If the calculated value of F exceeds the critical value, it is probable that the data cannot be represented by the expression under examination. However, before turning to other rate expressions, the data of the replicate experiments should be examined to make sure that no hidden or incompletely controlled factors systematically influence the values of k_{vi} obtained. The procedure is described in the next section.

From this discussion, one very important point with practical consequences emerges. This concerns the use of replicate experiments. If no replicate experiments are performed, it is not possible to define the basic error of the measurements, *i.e.* the distribution of the individual rate coefficients of replicate experiments about their means. In terms of our previous expression for F, the lack of replicate experiments means that there are as many 'groups' as there are experiments in the set (*i.e.* $u = r$) and so any value of F is consistent with the whole set being homogeneous; the scatter of the rate coefficients about the set mean merely defines the variance of the distribution of a rate coefficient of unit weight. It is clear that replication of initial conditions is highly desirable.

The above calculations are often set out in the form of an *Analysis of Variance* Table as shown below

† Other confidence levels may be chosen depending on the judgement of the investigator.
†† A table of F is given in the *Handbook of Chemistry and Physics.*

Distribution	Sum of squares	Degrees of freedom	Mean square
of group means from set mean ("between" groups)	$\sum_{v=1}^{v=u} \sum_{i=1}^{i=r_v} w(k_{vi})(\bar{k}_v - \bar{k})^2$ (s.s.1)	$u-1$	$\frac{\text{s.s.1}}{u-1}$
of observations from group mean ("within" groups)	$\sum_{v=1}^{v=u} \sum_{i=1}^{i=r_v} w(k_{vi})(k_{vi} - \bar{k}_v)^2$ (s.s.2)	$r-u$	$\frac{\text{s.s.2}}{r-u}$

The computations are greatly simplified if the weights $w(k_{vi})$ are equal. For then the required sums of squares can be calculated from the quantities

$$T_v = \sum_{i=1}^{i=r_v} k_{vi}$$

$$T = \sum_{v=1}^{v=u} T_v$$

together with the values $\sum_{v=1}^{v=u} \sum_{i=1}^{i=r_v} k_{vi}^2$, r_v and r. Replacing each weighting factor $w(k_{vi})$ by unity, we have

$$\text{s.s.1} = \sum_{v=1}^{v=u} \sum_{i=1}^{i=r_v} (\bar{k}_v - \bar{k})^2 = \sum_{v=1}^{v=u} \frac{T_v^2}{r_v} - \frac{T^2}{r}$$

$$\text{s.s.2} = \sum_{v=1}^{v=u} \sum_{i=1}^{i=r_v} (k_{vi} - \bar{k}_v)^2 = \sum_{v=1}^{v=u} \sum_{i=1}^{i=r_v} k_{vi}^2 - \sum_{v=1}^{v=u} \frac{T_v^2}{r_v}$$

Finally, if the set of rate coefficients is homogeneous, the 'best' value is the weighted mean $\bar{k}$, *viz.*

$$\bar{k} = \frac{\sum_{v=1}^{v=u} \sum_{i=1}^{i=r_v} w(k_{vi}) k_{vi}}{\sum_{v=1}^{v=u} \sum_{i=1}^{i=r_v} w(k_{vi})}$$

The standard error of the weighted mean of the rate coefficients of the whole set can be estimated from the larger of the two estimates of the variance of the distribution. These two estimates are

$$ms_1 = \sum_{v=1}^{v=u} \sum_{i=1}^{i=r_v} w(k_{vi})(\bar{k}_v - \bar{k})^2/(u-1)$$

$$ms_2 = \sum_{v=1}^{v=u} \sum_{i=1}^{i=r_v} w(k_{vi})(k_{vi} - \bar{k}_v)^2/(r-u)$$

References p. 407

and so $s(\bar{k})$, the standard error of the weighted mean, is given by either

$$s(\bar{k}) = \left\{\frac{ms_1}{\sum_{v=1}^{v=u}\sum_{i=1}^{i=r_v} w(k_{vi})}\right\}^{\frac{1}{2}} \quad \text{or} \quad \left\{\frac{ms_2}{\sum_{v=1}^{v=u}\sum_{i=1}^{i=r_v} w(k_{vi})}\right\}^{\frac{1}{2}} \tag{74}$$

whichever is the larger. Confidence limits can be calculated, as explained previously, from the value of t appropriate to (a) the chosen confidence level and (b) the number of degrees of freedom used in estimating ms_1 or ms_2. The example in Appendix 2 should make this clearer.

(iv) Homogeneity of the rate coefficients of the replicate experiments

It is important to test the rate coefficients obtained from a group of replicate experiments for homogeneity. Such a test can provide information as to whether or not all factors influencing the rate of the reaction are completely controlled throughout the group. For this purpose, we compare the distribution of the k_i' values (not the rate coefficients themselves) about their mean with the distribution of the experimental values of $f(\alpha_{ij})$ about the fitted straight line

$$f(\alpha_{ij}) = l_i - k_i'(t_{ij} - t_{io})$$

In the case of the vth group of replicate experiments, we shall have r_v separate values of k_i'.

Firstly, we calculate a weighted average of the r_v values of k_i' using the expression

$$\bar{k}' = \frac{\sum_{i=1}^{i=r_v} W(k_i')k_i'}{\sum_{i=1}^{i=r_v} W(k_i')}$$

where the weighting factors $W(k_i')$ are given by

$$W(k_i') = \frac{D_i}{\sum_{j=1}^{j=n_i} w_{ij}}$$

The quantity D_i is given by eqn. (60) and $\sum_{j=1}^{j=n_i} w_{ij}$ can be calculated from the weights w_{ij} assigned to each of the n_i points in the ith run.

If the r_v values of the rate coefficients are derived from a single set of $f(\alpha_{ij})$ values characterized by the same value of σ (the standard deviation of a point of unit weight), and if the rate coefficients are distributed normally about some true value, it can be shown that

$$\frac{\sum_{i=1}^{i=r_v} W(k'_i)(k'_i - \bar{k}')^2}{r_v - 1}$$

and

$$\frac{\sum_{i=1}^{i=r_v} E_i}{\sum_{i=1}^{i=r_v} (n_i - 2)}$$

each provide an estimate of σ^2. The first quantity estimates σ^2 from the scatter of the individual values of k'_i about their weighted mean while the second quantity predicts σ^2 from the weighted sum of the squares of the residuals over all the lines [see eqns. (61) and (62)]. If we write the ratio F as

$$F = \frac{\sum_{i=1}^{i=r_v} W(k'_i)(k'_i - \bar{k}')^2/(r_v - 1)}{\sum_{i=1}^{i=r_v} E_i / \sum_{i=1}^{i=r_v} (n_i - 2)} \tag{75}$$

we can decide the maximum value of F which is consistent with the hypothesis that the replicate experiments constitute an homogeneous group for a chosen confidence level and given values of $(r_v - 1)$ and $\sum_{i=1}^{i=r_v} (n_i - 2)$. If $F < 1$, the group is homogeneous since the scatter of the gradients of the lines about their mean can be accommodated within the scatter of the individual points about the lines. Similarly, if $F > 1$ but is less than the critical value, the situation needs no discussion.

The real problem arises when the replicate experiments are found to constitute a non-homogeneous group for then the straight lines are so well-defined that the differences between the gradients are significant. In this case, it is probable that the reaction rate is influenced by some factor which, although remaining constant within any one run, varies from experiment to experiment within the group. For example, suppose the group of replicate experiments were carried out in a thermostat on separate days and suppose that, for some reason, the thermostat temperature shifted slightly from day to day though remaining constant over the time scale of any one experiment. Such a day-to-day temperature variation could give rise to a non-homogeneous group of k'_i values if the points on the graphs of $f(\alpha_{ij})$ against $(t_{ij} - t_{i0})$ happened to fall on particularly well-defined straight lines. Another example is afforded by the three experiments, 1, 2 and 3, of Table 3; the reason for their non-homogeneity is discussed in Appendix 2. If the data of a group of replicate experiments do not survive the test for homogeneity, the experimental techniques employed must be critically considered. It may be possible to redesign the procedures so that the uncontrolled factors are eliminated and satisfactory replica-

References p. 407

tion of the k_i' values is obtained. Alternatively, it may be decided that the uncontrolled factors operate in a purely random fashion from run to run and that the observed variations in the k_i' values are small for the required purpose; if this is so, the comparison of different groups of experiments carried out at different initial concentrations can be undertaken as described in the previous section. However, if the uncontrolled factors influence the k_i' values from replicate experiments in some systematic way, it is not useful to compare one group of experiments with another if the object is to decide whether or not the data fit a given rate expression. Such data would fail the first of the F-tests described previously [see eqn. (73)], but the reason would not necessarily be that the rate expression under investigation was an inappropriate representation of the data; the whole set would be subject to systematic errors just like the group of replicate experiments and a value of F greater than the critical could be obtained on this account alone. There is no choice but to eliminate factors which operate in this way. An example might be useful. Suppose a set of experiments were carried out in which known volumes of the reactant A were withdrawn from a stock solution. Furthermore, suppose that, unbeknown to the investigator, the reactant A was slowly decomposing in the stock solution giving inert products. Then, assuming that the quantity k_i' was dependent on the initial concentration of A, the values of k_i' obtained from a group of replicate experiments would vary systematically as the stock solution became weaker; if the experiments were performed in the order of decreasing formal initial concentrations, the values of k_i (not k_i') would show significant differences even though the rate expression under test were correct. If the decomposition of the reactant A were the sole factor affecting the results and if the experimental procedure could not be modified in any way to eliminate its influence, it is clear that merely randomizing the order in which the experiments were carried out would permit a valid comparison of the k_i values to be made. This is yet another example of the close link between the methods employed to process the data and the design of the experimental procedure.

It is obvious that these calculations are greatly facilitated by using a digital electronic computer. Indeed, it is because these machines have become so accessible in recent years that the procedures described have been given such prominence.

(v) Further comments on replication

It can be argued that the extensive replication of the kinetic experiments for all values of the initial concentrations is unnecessary. Suppose, for example, that no replicate experiments at all are carried out so that each run corresponds to different initial conditions. It is possible to modify the previous test for the homogeneity of the rate coefficients within a group of replicate experiments to examine the homogeneity of such a set. Since the experiments correspond to different initial conditions, we have to compare the distribution of the rate coefficients themselves about their weighted mean with the scatter of the points about the family of lines

$$f(\alpha_{ij}) = l_i - m_i k_i (t_{ij} - t_{i0})$$

The procedure is exactly analogous to that outlined previously except that we weight each rate coefficient thus

$$W(k_i) = \frac{m_i^2 D_i}{\sum_{j=1}^{j=n_i} w_{ij}}$$

F is calculated from the expression

$$F = \frac{\sum_{i=1}^{i=r} W(k_i)(k_i - \bar{k})^2/(r-1)}{\sum_{i=1}^{i=r} E_i / \sum_{i=1}^{i=r} (n_i - 2)} \tag{76}$$

If the data pass the F-test, there can be little doubt of the correctness of the assumed rate expression. The best value of the rate coefficient calculated in this way is $\bar{k}$, *viz.*

$$\bar{k} = \frac{\sum_{i=1}^{i=r} W(k_i) k_i}{\sum_{i=1}^{i=r} W(k_i)} \tag{77}$$

This estimate is negligibly different from that calculated using eqn. (72) where the weighting factors were $1/s^2(k_i)$, provided that the number of points defining each straight line is fairly large. The standard error of $\bar{k}$ follows as before from the larger of the two mean squares

$$ms_1 = \sum_{i=1}^{i=r} W(k_i)(k_i - \bar{k})^2/(r-1)$$

$$ms_2 = \sum_{i=1}^{i=r} E_i / \sum_{i=1}^{i=r} (n_i - 2)$$

$$s(\bar{k}) = \left\{\frac{ms_1}{\sum_{i=1}^{i=r} W(k_i)}\right\}^{\frac{1}{2}} \quad \text{or} \quad \left\{\frac{ms_2}{\sum_{i=1}^{i=r} W(k_i)}\right\}^{\frac{1}{2}}$$

The second of these two equations is analogous to eqn. (63). Confidence limits for $\bar{k}$ can be calculated from $s(\bar{k})$ and the value of the statistic t corresponding to the chosen confidence level and the number of degrees of freedom used in estimating ms_1 or ms_2.

References p. 407

This procedure is all very well, but a problem arises if the *F*-test suggests that the scatter of the rate coefficients is not consistent with the scatter of the points about the lines. If no replicate experiments have been performed, there are insufficient data to test the reproducibility of the procedure. In this case, failure of the data to pass the *F*-test is as likely to arise from inadequate control of the experimental conditions as from incorrect assumptions about the form of the rate equation. It follows, therefore, that a certain amount of replication is desirable.

The essential feature of the procedures described above is that the test employed to decide whether or not a given rate equation represents the kinetics of a particular reaction is based on the values of the rate coefficients obtained at a number of different values of $[A]_0$ and $[B]_0$. If the rate coefficients show no systematic variation with one or other of the initial concentrations, the assumed rate equation is applicable and the orders a and b postulated from the preliminary examination of the data are confirmed. Linearity of the graph of a particular form of $f(\alpha)$ against $(t-t_0)$ at one pair of values of $[A]_0$ and $[B]_0$ is not conclusive proof of the applicability of the chosen rate equation since it is a matter of fairly common experience that kinetic data may fit a particular equation quite well at one set of initial concentrations only to show significant deviations when the initial concentrations are changed.

(c) *A re-examination of the functions* f(α) *of Table 1*

We have already suggested that each equation of Table 1 can be written in the general form

$$f(\alpha_0)-f(\alpha) = k'(t-t_0)$$

With the exception of eqns. (34) to (38), this formulation is inexact since each integrated expression contains the parameter c; a better generalization of the equations of Table 1 is

$$f(\alpha_0, c)-f(\alpha, c) = k'(t-t_0) \tag{78}$$

For certain values of c, some of the equations given are inappropriate. More suitable forms are described in this section.

It will be recalled that

$$c = (I/r)-1$$

where I stands for the ratio of the initial concentrations, $[B]_0/[A]_0$, and r represents the ratio of the stoichiometric coefficients, ν_B/ν_A. Thus c can range from very large values ($I \gg r$) to values close to -1 ($I \ll r$). Difficulties arise in the case of eqns. (41), (43), (44), (47), (49), (50) and (53) when c is either zero or very close to zero. When c is zero, k and its standard error become indeterminate— in this situation, the special forms of the integrated expressions which are shown in Table 1 must be used. When c is small, it is advisable to transform the equations under discussion by suitable expansion in terms of c/α. A case in point is eqn. (50)

$$\ln\left\{\frac{\alpha+c}{\alpha(1+c)}\right\} = kc[\mathrm{A}]_0 r(t-t_0)$$

On expanding the logarithm as far as second order terms in c, we obtain

$$\left\{\frac{1}{\alpha}-1\right\} - \frac{c}{2}\left\{\frac{1}{\alpha^2}-1\right\} = k[\mathrm{A}]_0 r(t-t_0) \tag{79}$$

Eqn. (79) reduces to the special form (51) when $c = 0$. Thus, at sufficiently low values of c so that terms in c^3/α^3 can be neglected, we write eqn. (50) in the form

$$\mathrm{f}(\alpha) = \frac{1}{\alpha} - \frac{c}{2\alpha^2}$$

Similar transformations can be made for the other equations mentioned.

We should also note that the expressions for $\mathrm{f}(\alpha)$ simplify considerably when c is very large. In this case $I \gg r$ and so the concentration of B remains almost unchanged as A reacts; the integrated form of the rate expression, eqn. (32), reduces to

$$\int_\alpha^1 \frac{\mathrm{d}\alpha}{\alpha^a} = k[\mathrm{A}]_0^{a-1}[\mathrm{B}]_0^b(t-t_0)$$

The integral on the left-hand side of the above equation is particularly simple. For example, eqn. (50), in which $a = b = 1$, reduces to

$$\ln\frac{1}{\alpha} = k[\mathrm{B}]_0(t-t_0)$$

(d) Special procedures

The major part of our previous discussion has been cast in terms of the variable α defined by

$$\alpha = \frac{[\mathrm{A}]}{[\mathrm{A}]_0}$$

It happens in certain types of work that the absolute values of [A] and $[\mathrm{A}]_0$ cannot be determined with any accuracy; this occurs when the quantitative relation-

References p. 407

ship between [A] and the physical property used to follow the disappearance of A cannot be determined with any accuracy. For example, suppose we follow a physical property whose value ϕ_t at time t is linearly related to the concentration of A, [A], thus

$$\phi_t = \lambda[\mathrm{A}]+\mu \tag{80}$$

where λ and μ are constants. The initial and final values of the physical property ϕ, ϕ_0 and ϕ_∞, are given by

$$\phi_0 = \lambda[\mathrm{A}]_0+\mu$$

and

$$\phi_\infty = \mu$$

if all the A originally present disappears. Therefore,

$$\alpha = \frac{[\mathrm{A}]}{[\mathrm{A}]_0} = \frac{\phi_t-\phi_\infty}{\phi_0-\phi_\infty}$$

and clearly, in a simple case, α can be calculated from the observed value of ϕ at any time and the observed or calculated values of ϕ_0 and ϕ_∞. It sometimes happens that it is not possible to determine the values of ϕ_0 and/or ϕ_∞ with any accuracy. In such cases, special methods of treating the data have been devised and it is with these procedures that the present section is concerned.

(*i*) *Reactions of the type* $-\mathrm{d}[\mathrm{A}]/\mathrm{d}t = k[\mathrm{A}]$
For such reactions

$$\alpha = \mathrm{e}^{-k(t-t_0)}$$

and so, if the value of the physical property employed is related to [A] by eqn. (80), we obtain

$$\phi_t-\phi_\infty = (\phi_0-\phi_\infty)\mathrm{e}^{-k(t-t_0)} \tag{81}$$

Guggenheim's method.[1] For any particular run, consider a pair of values of ϕ, ϕ_{t_j} and $\phi_{t_j+\tau}$, observed at the two times t_j and $t_j+\tau$ as indicated by the subscripts. Then,

$$\phi_{t_j}-\phi_\infty = (\phi_0-\phi_\infty)\mathrm{e}^{-k(t_j-t_0)}$$
$$\phi_{t_j+\tau}-\phi_\infty = (\phi_0-\phi_\infty)\mathrm{e}^{-k(t_j+\tau-t_0)}$$

If we eliminate the unknown value ϕ_∞ from the left-hand side of the above equations, we obtain

$$\phi_{t_j} - \phi_{t_j+\tau} = (\phi_0 - \phi_\infty)\mathrm{e}^{-k(t_j - t_0)}\{1 - \mathrm{e}^{-k\tau}\} \tag{82}$$

Eqn. (82) is the exact analogue of eqn. (81) with $\phi_{t_j+\tau}$ replacing ϕ_∞ on the left-hand side and $(\phi_0 - \phi_\infty)(1 - \mathrm{e}^{-k\tau})$ replacing $(\phi_0 - \phi_\infty)$ on the right-hand side; the equivalence can be seen immediately by putting $\tau = \infty$ in eqn. (82). It follows from eqn. (82) that, if a series of pairs of values of ϕ, ϕ_{t_j} and $\phi_{t_j+\tau}$, are taken such that the time interval τ separating them is a constant, a plot of $\ln(\phi_{t_j} - \phi_{t_j+\tau})$ against $t_j - t_0$ has exactly the same slope as the more usual plot of $\ln(\phi_{t_j} - \phi_\infty)$ against $t_j - t_0$. We have

$$\ln(\phi_{t_j} - \phi_{t_j+\tau}) = \text{constant} - k(t_j - t_0)$$

where the constant is $\ln(\phi_0 - \phi_\infty) + \ln(1 - \mathrm{e}^{-k\tau})$. The best value of the slope of this line can be obtained by following the methods already described. The only problem concerns the most suitable value of τ. If a very large value of τ is used, only a limited number of pairs of values of ϕ can be used in the calculation of the rate coefficient. In this case, the value of k is determined primarily by the very early and very late observations of ϕ so that intermediate observations might as well not have been made. As a result of the small number of observations utilized, the standard error in k is unnecessarily inflated. On the other hand, a large number of pairs of observations can be obtained by choosing a very small value of τ. However, as the standard error in k is inversely proportional to the square root of the number of paired observations of ϕ used, there is clearly a point where increasing the number of paired observations by decreasing τ makes little impact on the precision of the measurement. Thus, once the plot of $\ln(\phi_{t_j} - \phi_{t_j+\tau})$ against $t_j - t_0$ has 10 to 20 points, there is little to be gained by reducing τ still further so as to obtain more. Indeed, if τ is made too small there is a chance that some observations will be used more than once in the calculation. To achieve a satisfactory compromise, it is essential to follow the reaction over three or four half-lives. If our data consist of a total of N values of ϕ made at equally spaced time intervals, we select the value of τ as follows. We divide the observations into two groups of $N/2$† *consecutive* observations and choose the value of τ as the time interval separating the first observations in each group. In this way, each observation in the first group is paired with an observation in the second group and no observation is used more than once. From the similarity of eqns. (81) and (82), it can be seen that this procedure is equivalent to making $N/2$ observations of ϕ and a further $N/2$ accurate determinations of ϕ_∞ spread over the same time as was required for the $N/2$ observations of ϕ.

† If N is odd, we omit the last observation of the whole sequence.

References p. 407

It is obvious from the preceding discussion that it is almost essential to determine the magnitude of the physical property, ϕ, at equally spaced time intervals. If the observations are irregularly spaced in time, only a limited number of pairs of experimental observations can be obtained which are separated by the same interval; an increased number of paired ϕ_{t_j}, $\phi_{t_j+\tau}$ values could be obtained by interpolation but this is less satisfactory than using experimental values.

It should also be clear that the same procedure can be applied to every run in a series of experiments at different initial concentrations. Consequently we obtain a series of values for the rate coefficients: $k_1, k_2 \ldots k_i \ldots$ etc., in just the same way as we would obtain had it been possible to treat the data by the more conventional method described earlier. The method of averaging these rate coefficients follows the lines already described.

Mangelsdorf's method.[2] Mangelsdorf has presented an alternative method of processing the data, again utilizing paired values of the physical property ϕ separated by a constant time interval, τ. As before, we write

$$\phi_{t_j}-\phi_\infty = (\phi_0-\phi_\infty)e^{-k(t_j-t_0)}$$
$$\phi_{t_j+\tau}-\phi_\infty = (\phi_0-\phi_\infty)e^{-k(t_j+\tau-t_0)}$$

Instead of subtracting these two equations to eliminate ϕ_∞ on the left-hand side as suggested by Guggenheim, we divide one by the other to eliminate the factor $(\phi_0-\phi_\infty)e^{-k(t_j-t_0)}$ and obtain

$$\phi_{t_j+\tau}-\phi_\infty = (\phi_{t_j}-\phi_\infty)e^{-k\tau}$$

or

$$\phi_{t_j+\tau} = \phi_{t_j}e^{-k\tau}+\phi_\infty(1-e^{-k\tau}). \tag{83}$$

Eqn. (83) shows that a plot of $\phi_{t_j+\tau}$ against ϕ_{t_j} is a straight line of slope $e^{-k\tau}$ and intercept at $\phi_{t_j} = 0$ of $\phi_\infty(1-e^{-k\tau})$. Instead of using the intercept, the value of ϕ_∞ can be found in a rather simpler way from the point of intersection of the above straight line with the line $\phi_{t_j+\tau} = \phi_{t_j}$; from eqn. (83), the co-ordinates of this point are clearly $(\phi_\infty, \phi_\infty)$. Strictly speaking, the least squares formulae previously given [eqns. (58) to (63)] cannot be used to obtain the best values of the slope, intercept, and their associated standard errors. The reason is that *both* ϕ_{t_j} and $\phi_{t_j+\tau}$ are subject to random errors whereas eqns. (58) to (63) pertain to the situation where *only the dependent variable* is subject to error. To take account of the presence of random errors in both variables in a proper fashion requires us to examine the statistics of the straight line in considerable detail. This would be out of place in an article of this type. However, since the two variables are characterized by the same standard error (assuming this to be independent of t_j), it is reasonable to proceed with the computation of the slope of the line as if only the observed $\phi_{t_j+\tau}$ values were subject to error.

A more important consideration is the choice of τ. If we choose a very large value of τ, we have the same problem as discussed under Guggenheim's method, namely that our slope is determined by the very early and very late values of ϕ, the intermediate values being unused; we also unnecessarily inflate the standard error of the slope by having too few points define the straight line. If we choose a very small value of τ, we run into a somewhat different problem to that which arises with the Guggenheim method. In the Mangelsdorf method, the magnitude of the observed slope depends on the value of τ selected—the smaller the value of τ, the smaller the slope. The implications can be seen by writing the slope of the Mangelsdorf plot as S and considering the relation of its standard error to the standard error of the rate coefficient. We write

$$S = \mathrm{e}^{-k\tau}$$

and therefore

$$\sigma(k) = \frac{1}{S\tau}\,\sigma(S)$$

Therefore

$$\frac{\sigma(k)}{k} = \frac{1}{k\tau}\,\frac{\sigma(S)}{S}$$

Thus if $k\tau \ll 1$, the *relative error* in the rate coefficient greatly exceeds the relative error in the computed slope. This is obviously undesirable. It is obvious that a value of $\tau \sim 1/k$ is the most suitable provided that we have a sufficient number of points to obtain a reliable estimate of $\sigma(S)$. This means that τ should be about one and a half times the half-life of the reaction and that the reaction should be studied over a time scale of at least three half-lives. In this way, we come to much the same conclusion as we reached in our discussion of the Guggenheim method. We divide our N observations of ϕ made at equally spaced time intervals over a time scale of about three half-lives or more into two groups of $N/2$ consecutive observations. We take our values of ϕ_{t_j} from the first group and the retarded values $\phi_{t_j+\tau}$ from the second group, the value of τ being the time interval separating the first observations in each group. In this way, we ensure that each observation is used once and once only and that the relative error in k is commensurate with the relative error in S.

(ii) Reactions of other types

With other types of reaction, only a limited amount of progress can be made if the *only* data available are the observations of the physical property ϕ at reac-

References p. 407

tion times intermediate between the initial and final values. The basic reason for this is that the integrated expressions for reactions other than the simple first order case discussed above require the values of the initial concentrations. Thus, if these data are not available, absolute values of the rate coefficients cannot be calculated.

However, if we consider reactions in which the order with respect to the second reactant B is zero or in which the concentration of the second reactant is in such large excess that its time dependence can be neglected, it is possible to manipulate the integrated rate equations so as to give a numerical value of the rate coefficient involving the dimensions of λ [see eqn. (80)] and/or $[B]_0$. Two simple examples will suffice to make this clearer. Firstly, suppose we have a reaction which is first order with respect to A and first order with respect to B and suppose we can arrange our conditions so that B is in considerable excess of A even though we may not actually know the value of $[B]_0$. Then we may apply either the Guggenheim or Mangelsdorf methods to obtain the product of the second order rate coefficient and $[B]_0$. For the second example, consider a reaction which is second order with respect to A and zero order with respect to B. The integrated form of the rate expression for this case is eqn. (38), namely

$$\frac{1}{\alpha} - 1 = k[A]_0(t-t_0)$$

which in terms of the physical property ϕ becomes

$$\frac{1}{\phi_t-\phi_\infty} - \frac{1}{\phi_0-\phi_\infty} = \frac{k}{\lambda}(t-t_0)$$

We may regard this equation as containing three unknowns: ϕ_∞, $(\phi_0-\phi_\infty)$, and k/λ. Therefore, by taking any three pairs of (t, ϕ_t) values, we can eliminate two of these unknowns to find the third. Thus if we eliminate ϕ_∞ and $(\phi_0-\phi_\infty)$, we obtain k/λ (but *not* k). The computations are simpler if we take three observations of ϕ separated by equal time intervals, thus: ϕ_t, $\phi_{t+\tau}$, $\phi_{t+2\tau}$. The value of k/λ is given[3,4] by

$$\frac{k}{\lambda} = \frac{(2\phi_{t+\tau}-\phi_t-\phi_{t+2\tau})^2}{2\tau(\phi_{t+2\tau}-\phi_t)(\phi_{t+\tau}-\phi_t)(\phi_{t+2\tau}-\phi_{t+\tau})}$$

Clearly from a succession of triads, we can calculate a series of values of k/λ if we wish. We shall not pursue this method any further except to point out that there are problems to be solved concerning the optimum value of τ and the method of averaging the set of k/λ values. Obviously there is no real need to use observations at equally spaced time intervals to eliminate two of our three unknowns and so obtain a value of k/λ; this problem has been examined by Sturtevant[5].

Other workers have considered the problems discussed in this section—for example, see references 6–9. A useful discussion of the statistical background is given by Guest[10].

2.2 REACTIONS WITH UNKNOWN STOICHIOMETRY

If we study the kinetics of a reaction between A and B and do not know the stoichiometry, it should be clear from our previous discussion that very little real progress can be made in analysing the data. Suppose for the sake of argument that we are unable to analyse our reaction mixture for the reactant B but are able to follow the disappearance of A. Even in the case of the simplest possible type of rate expression, namely the one under discussion

$$-\frac{\mathrm{d}[\mathrm{A}]}{\mathrm{d}t} = k[\mathrm{A}]^a[\mathrm{B}]^b$$

a knowledge of the concentration of B is required in general for the evaluation of the rate coefficient. If the reaction rate is characterized by one of the more complicated expressions, the instantaneous value of the concentration of B is very likely to be involved and so we have insufficient information to make any general statement regarding the form of the differential rate expression.

However, there are two situations in which the previous calculations can be fruitfully undertaken without requiring a knowledge of the concentration of B. These are when the order with respect to B is zero or when the disappearance of A can be studied in the presence of a considerable excess of B. If the preliminary examination of the data suggests that the order of the reaction with respect to B is zero, then the validity of the rate expression

$$-\frac{\mathrm{d}[\mathrm{A}]}{\mathrm{d}t} = k[\mathrm{A}]^a$$

can be demonstrated by showing that the rate coefficient k is independent of $[\mathrm{A}]_0$ and $[\mathrm{B}]_0$. Where the disappearance of A can be conveniently followed in the presence of a large excess of B, we first have to estimate the order with respect to B at the same time as we estimate the order with respect to A (see p. 360); we then compute k' in the equation

$$-\frac{\mathrm{d}[\mathrm{A}]}{\mathrm{d}t} = k'[\mathrm{A}]^a$$

and obtain a value for the rate coefficient, k, from the expression $k' = k[\mathrm{B}]_0^b$. If the rate coefficient is independent of $[A]_0$ and $[B]_0$, the rate of the reaction is given by

$$-\frac{d[A]}{dt} = k[A]^a[B]^b$$

It must be stressed that this conclusion is only valid under the concentration conditions employed, that is, when the reaction is followed in the presence of a large excess of B. It is quite possible that a more complicated expression for the rate is necessary when the reactant concentrations are close to stoichiometric equivalence.

Of course, both these attempts to formulate a simple expression for the rate of the reaction may fail. This should not cause any concern since, with access to such a limited amount of data, many of the features of the reaction cannot but remain obscure. In such a case, further experimental work must be carried out particularly with a view to determining the concentration of B.

3. Testing the data for consistency with complicated rate expressions

If the concentration–time data cannot be represented by a rate expression of the ype

$$-\frac{d[A]}{dt} = k[A]^a[B]^b$$

it is a very difficult problem to find the correct form utilizing solely these data. The difficulty arises from the fact that there are a wide variety of different types of rate expression that can be employed. These are summarized in eqns. (2), (3) and (4) and exemplified in Chapter 1, Vol. 2, where the kinetic consequences of complex reaction mechanisms are discussed in some detail. Because of the wide range of possible differential equations which might be investigated, it is not possible to formulate a general procedure as was done in the previous section. All that we can do here is to offer some guidance as to the best type of differential expression to be selected in any particular instance. We shall base our advice solely on the observable characteristics of the reaction. In general, however, it is usually possible to supplement these facts with an intelligent guess at the reaction mechanism so that the range of possible differential rate expressions is reduced to a small number. The danger of taking into account one's views of the reaction mechanism in selecting the type of rate expression to be investigated is that, if a particular form successfully represents the data, one is not inclined to investigate the fit given by alternative forms. As a result, one is liable merely to confirm one's own pre-conceptions unless additional experiments are devised to test the mechanism. The point is that concentration–time data which cannot be represented by a simple expression may be equally well described by quite different types of equation. For example, the two expressions

$$-\frac{d[A]}{dt} = k_1[A][B]-k_2[A]^2$$

and

$$-\frac{d[A]}{dt} = \frac{k_1[A][B]}{1+\dfrac{k_2}{k_1}\dfrac{[A]}{[B]}}$$

are equivalent when $k_1[B] \gg k_2[A]$ as can be seen by expanding

$$\left(1+\frac{k_2[A]}{k_1[B]}\right)^{-1}$$

and neglecting terms involving $[A]^2/[B]^2$ and above. Consequently data which fit the one may be equally well represented by the other over a limited range of concentrations and so it may not be possible to make a choice between the two on the basis of concentration–time data alone. Since these two expressions for the rate of the reaction correspond to quite different mechanisms, it follows that it is not possible in every case to make a choice between alternative mechanisms on the basis of concentration–time data alone; additional criteria (and possibly further experiments) are necessary in order to select one mechanism rather than another. These difficulties are, of course, well known to experienced kineticists; however, provided that the dangers of utilizing preconceived views of the reaction mechanism are realized, it is perfectly legitimate to use these as a guide to the type of rate expression to be investigated.

At this stage, it is useful to repeat the examples of complicated rate expressions given earlier in generalised form [eqns. (2), (3) and (4)],

$$-\frac{d[A]}{dt} = \sum k_i f_i([\text{reactants}], [\text{products}]) \tag{84}$$

or

$$-\frac{d[A]}{dt} = \sum k_i f_i([\text{reactants}], [\text{products}]) - \sum k_i' f_i'([\text{reactants}], [\text{products}]) \tag{85}$$

or

$$-\frac{d[A]}{dt} = \frac{k f([\text{reactants}], [\text{products}])}{1 \pm k' f'([\text{reactants}], [\text{products}])} \tag{86}$$

In principle we can have even more complicated expressions by taking linear

References p. 407

combinations of the form shown in eqn. (86) with the forms shown in the other two equations. The various functions of the reactant and product concentrations are in most cases analogous to the simple form discussed earlier, *e.g.*

$$f([\text{reactants}], [\text{products}]) = [A]^a[B]^b[P]^p[Q]^q$$

The exponents a, b, p, q are usually integers or half-integers in the range 0 to 2 though, unlike our simple formulation, *negative* values are allowed. Many examples of these forms are given in Chapter 1, Vol. 2, but, for our purposes, three examples will suffice:

(1) in aqueous alkaline solution, methyl iodide disappears at a rate given by

$$-\frac{d[CH_3I]}{dt} = k_1[CH_3I]+k_2[CH_3I][OH^-];$$

(2) in the gas phase, hydrogen iodide decomposes into hydrogen and iodine at a rate

$$-\frac{d[HI]}{dt} = k_1[HI]^2-k_2[H_2][I_2];$$

(3) in the gas phase, hydrogen and bromine react together to produce hydrogen bromide at a rate

$$-\frac{d[H_2]}{dt} = \frac{k_1[H_2][Br_2]^{\frac{1}{2}}}{1+k_2\dfrac{[HBr]}{[Br_2]}}$$

The problem with which we are concerned here is the procedure which is employed to recognize the various forms.

Before developing these criteria, we suggest that it is unnecessary to concentrate any attention on expressions of the type generalized by eqn. (86) at the beginning of the investigation. The reason is that eqn. (86) is formally equivalent to either eqn. (84) or eqn. (85). The procedure is to attempt to represent the data by linear combinations of terms such as

$$k[A]^a[B]^b[P]^p[Q]^q$$

as suggested in eqns. (84) and (85); if it is found that the data require a large number of such terms, it is then worthwhile attempting to fit the data to an expression of the type (86) which may be more economical. In fact, where no guidance as to the likely form of the rate expression can be obtained from the chemistry of the process under investigation, there is a good case for comparing the fit of the data to an expression of the type (84) or (85) with the fit obtained when it is represented by an equation of the type (86).

Let us return to the criteria that we used to recognize a complex reaction. We stated that a reaction is necessarily characterized by a complicated rate expression if, on a preliminary examination of the concentration–time data, we observe any one of the following three features:

(1) stoichiometric coefficients which vary with the extent of the reaction or with temperature;
(2) an initial rate which is affected by the presence of product;
(3) an initial rate which is less than a rate observed at a later stage in the reaction.

To these criteria, we might now add a fourth, namely, that a complicated rate expression is required if

(4) the concentration–time data cannot be adequately represented by the simple rate expression $k[A]^a[B]^b$.

The variation of the stoichiometric coefficients is not in itself of great utility in deciding upon one particular form of eqns. (84) or (85) for investigation. It simply means that the reactants are involved in more than one reaction in such a way that their rate of disappearance is the sum or difference of the rates of the separate reactions. For example, consider the simple mechanism

$$\begin{aligned} A &\rightarrow P \quad &\text{rate coefficient } k_1 \\ A+B &\rightarrow Q+R \quad &\text{rate coefficient } k_2 \end{aligned}$$

where the stoichiometric coefficients in the *elementary reactions are unity*. Anticipating the discussion of Chapter 1, Vol. 2, we have

$$-\frac{d[A]}{dt} = k_1[A]+k_2[A][B]$$

$$-\frac{d[B]}{dt} = k_2[A][B]$$

and so the instantaneous ratio of the stoichiometric coefficients, $(\nu_A/\nu_B)_t$, is given by

$$\left(\frac{\nu_A}{\nu_B}\right)_t = \frac{-d[A]}{-d[B]} = \frac{k_1+k_2[B]}{k_2[B]}$$

Clearly this ratio varies with the extent of the reaction (or time) as the concentration of B alters. This example illustrates one very important point: although the variation of the stoichiometric coefficients in itself is no guide to the form taken by the rate equation, a close examination of the time dependence of a particular stoichiometric ratio may reveal that the reaction is divided into two stages:

References p. 407

(*a*) a period in which the chosen stoichiometric ratio varies with time;
(*b*) a second stage in which the chosen stoichiometric ratio remains constant.
This second stage appears when there is only one reaction determining the rate, the other reactions which caused the complications having effectively died out owing to the disappearance of one of the reactants. Thus, in our example, the preliminary experiments would indicate that the reaction was of the form

$$\mathrm{A+B \rightarrow P+Q+R}$$

where the stoichiometric ratios $\nu_{\mathrm{A}}/\nu_{\mathrm{B}}$ and $\nu_{\mathrm{A}}/\nu_{\mathrm{P}}$ varied with time; however, by studying these ratios over a range of conditions, we should observe that $\nu_{\mathrm{A}}/\nu_{\mathrm{P}}$ eventually reached a constant value of unity at the same time as the concentration of B became negligible; with this information, we should deduce that the product P is formed by a reaction not involving B; similarly the time variation of $\nu_{\mathrm{A}}/\nu_{\mathrm{Q}}$ and $\nu_{\mathrm{A}}/\nu_{\mathrm{R}}$ would enable us to recognize that these products were formed from the reaction of A and B. A close study of the time dependence of the stoichiometric coefficients can thus be of value in separating a complicated reaction into its constituent parts. An equivalent separation of the constituent reactions making up certain types of complex reactions can sometimes be achieved by varying the temperature; that is to say, a reaction exhibiting variable stoichiometric ratios at one temperature may show constant stoichiometric ratios at another owing to the effective suppression of all but one of the reactions of the sequence. Clearly in such cases where the conditions can be adjusted to make the stoichiometric ratios time invariant, the kinetics of the reaction in this region can be examined using eqn. (1). These results can then serve as the basis for a more complete interpretation of the reaction kinetics. It should be appreciated, however, that the discovery of conditions in which the stoichiometric ratios remain time invariant for such a reaction is the exception rather than the rule.

Perhaps the most important single preliminary observation is the influence of the products on the initial rate. If the products have no influence, we have an immediate simplification of eqns. (84) or (85) and a limited number of possibilities for our rate expression. In this case, the reaction is most likely to be a set of *parallel* reactions with a rate given by eqn. (84), *e.g.*

$$-\frac{\mathrm{d[A]}}{\mathrm{d}t} = k_1[\mathrm{A}]^{a_1}[\mathrm{B}]^{b_1}+k_2[\mathrm{A}]^{a_2}[\mathrm{B}]^{b_2}\ldots \tag{87}$$

or a *chain reaction* (see Chapter 2, Vol. 2) in which the product does not participate. Both of these reaction types would normally exhibit variable stoichiometry, although, in the case of a chain reaction, this feature might be limited to an initial period of short duration. In fact, the only way of distinguishing parallel reactions from this group of chain reactions without using special tests (see Chapter 2, Vol. 2) is based upon the value of the initial rate. In the case of parallel reactions, the initial rate is the maximum rate observed as can be seen from eqn. (87); on the other hand, a chain reaction would normally exhibit an induction period in which the rate of the reaction increases from a small initial value to a maximum value. If the reaction is thought to be a chain reaction, no guidance can be given as to the likely form of the rate expression since, in these reactions, a wide range of kinetic behaviour is observed—probably the most practicable procedure is to postulate a mechanism, deduce its kinetic consequences and test the data against these. Of course, there are many chain reactions whose induction periods are so short that they are undetectable in the usual type of experimental procedure. However, those of this group which are unaffected by the addition of product would be unlikely to be confused with parallel reactions since their variable stoichiometry would be confined to the induction period; with apparently invariant stoichiometry and an initial rate the maximum observed and independent of the product concentration, this group of chain reactions would automatically be examined on the basis of eqn. (1) and would, therefore, not be considered as possible parallel reactions.

If the products affect the initial reaction rate, the reaction is either *reversible* or it involves a sequence of *consecutive* reactions in which the product enters as a reactant at one stage—this sequence of consecutive reactions could well be a chain reaction. If the reaction were a simple reversible reaction, *e.g.*

$$\mathrm{A+B \rightarrow P+Q}$$
$$\mathrm{P+Q \rightarrow A+B}$$

we should observe that the initial rate is the maximum rate observed, that the initial rate is reduced by the addition of the products and that the stoichiometry of the reaction is independent of time. In this case, the data should be examined on the basis of an expression of the type generalized by eqn. (85); a common form is

$$-\frac{\mathrm{d[A]}}{\mathrm{d}t} = k_1[\mathrm{A}]^a[\mathrm{B}]^b - k_2[\mathrm{P}]^p[\mathrm{Q}]^q \tag{88}$$

On the other hand, in the usual type of consecutive reaction involving product as a reactant, the initial rate is not the maximum observed since the rate increases as product is formed; for this reason, the initial rate is increased by the addition of the product at the beginning of the reaction and the stoichiometry alters as the reaction proceeds. For this type of reaction, we require an equation of the type set out in eqn. (84), namely

$$-\frac{\mathrm{d[A]}}{\mathrm{d}t} = k_1[\mathrm{A}]^{a_1}[\mathrm{B}]^{b_1} + k_2[\mathrm{A}]^{a_2}[\mathrm{P}]^{p_2} + \ldots$$

Such expressions are always extremely difficult to deal with.

References p. 407

Finally we come to the case where a reaction apparently exhibits invariant stoichiometry, an initial rate which is not exceeded at a later stage in the reaction and which is unaffected by the presence of products, and yet whose rate cannot be expressed according to the simple expression

$$-\frac{d[A]}{dt} = k[A]^a[B]^b$$

with a constant value of the rate coefficient, k. Assuming that we can discount purely physical effects arising from increases in such factors as the viscosity or ionic strength of the reaction system which may occur on the formation of the products, the reason almost certainly lies in the existence of minor parallel reactions which do not seriously influence the reaction stoichiometry. In other words, such problems are best dealt with by treating the data on the basis of eqn. (87), thus

$$-\frac{d[A]}{dt} = k_1[A]^{a_1}[B]^{b_1}+k_2[A]^{a_2}[B]^{b_2}$$

A summary of these general considerations is given in Appendix 3 in the form of a flow sheet. It is not suggested that the questions and answers cover all possible situations. Furthermore, the majority of the questions will not be capable of being answered with complete confidence in any experimental investigation. For example, to the question "Are all reactants and products identified at all stages of the reaction?", only a qualified "Yes" is possible; there may well be products formed, intermediate between the reactants and final products, which are present in very small concentrations and which are undetectable by the analytical methods employed. It follows that Appendix 3 is no more than a guide to the identification of reaction type. Finally, a word about the term, "simple reaction"; in the context of Appendix 3 and in this article, we use this term to describe a reaction, the rate of which is given by eqn. (1), *viz.*

$$-\frac{d[A]}{dt} = k[A]^a[B]^b$$

A simple reaction in the sense used here is, therefore, not necessarily an elementary reaction.

We shall not attempt a detailed account of the methods of processing the experimental data when the reaction is complicated since there are so many possible differential expressions; examples are given in Chapter 1, Vol. 2. The general procedure is as follows.

On the basis of the previous discussion and guided by Appendix 3, we select a possible form of equation for the rate of disappearance of the reactants. Where the stoichiometry of the reaction varies as the reaction proceeds, a separate expression for *each* of the reactants is necessary; in such cases, it is essential to follow the variation of the concentration of *each* reactant with time since it is not possible to relate one reactant concentration to another by eqn. (7).

These differential equations will usually be inexpressible in a simple integral form and will usually contain several unknown rate coefficients and exponents. To proceed further, we have to discern the experimental conditions under which the kinetic behaviour will be considerably simplified. Herein lies the skill. A fairly obvious procedure where equations such as (87) are likely to represent the data is to study the reaction in the presence of a large excess of B. For example, if

$$-\frac{\mathrm{d}[\mathrm{A}]}{\mathrm{d}t} = k_1[\mathrm{A}]^{a_1}[\mathrm{B}]^{b_1}+k_2[\mathrm{A}]^{a_2}[\mathrm{B}]^{b_2}$$

and if $[\mathrm{B}_0] \gg [\mathrm{A}]_0$ so that the time dependence of [B] can be neglected, we can write

$$-\frac{\mathrm{d}[\mathrm{A}]}{\mathrm{d}t} = k[\mathrm{A}]^{a_1}+k'[\mathrm{A}]^{a_2}$$

where k and k' are $k_1[\mathrm{B}]_0^{b_1}$ and $k_2[\mathrm{B}]_0^{b_2}$ respectively. Writing $[\mathrm{A}] = \alpha[\mathrm{A}]_0$ as before, we obtain

$$-\frac{\mathrm{d}\alpha}{\mathrm{d}\theta} = \alpha^{a_1}\{1+K\alpha^{a_2-a_1}\} \tag{89}$$

where

$$\theta = [\mathrm{A}]_0^{a_1-1}k(t-t_0)$$

and

$$K = \frac{k'[\mathrm{A}]_0^{a_2-a_1}}{k}$$

Clearly eqn. (89) can be integrated for given values of a_1, a_2 and K. It is possible, therefore, to construct families of curves of α against $\log_{10}\theta$ for, say, a series of values of K with a_1 and a_2 fixed. Two such sets of curves are shown in Figs. 3 and 4 which refer to a first order process perturbed by a zero order and second order process respectively; *i.e.*

$$-\frac{\mathrm{d}[\mathrm{A}]}{\mathrm{d}t} = k[\mathrm{A}]+k' \qquad \text{(Fig. 3)}$$

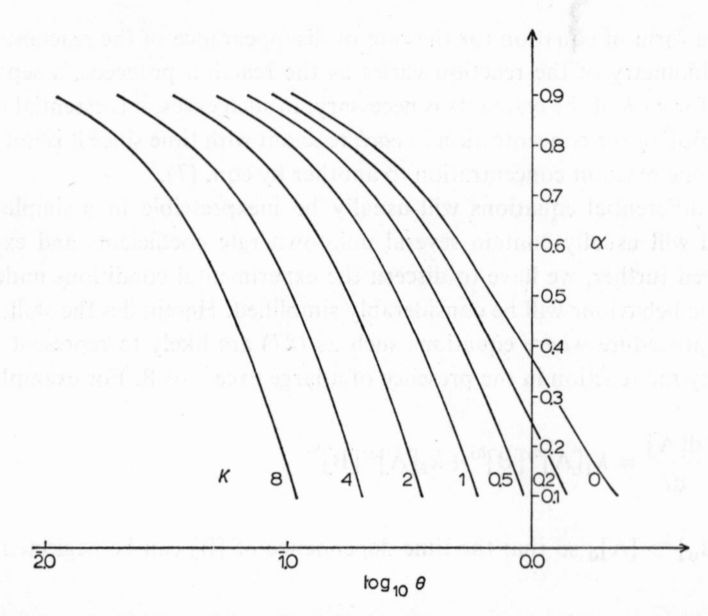

Fig. 3. Variation of α with $\log_{10}\theta$ for a first order reaction accompanied by a zero order reaction. As K increases, the proportion of the reactant consumed by the zero order process increases; since $K = k'/k[\mathrm{A}]_0$, the proportion of A disappearing by the zero order process in a particular instance increases as its initial concentration decreases.

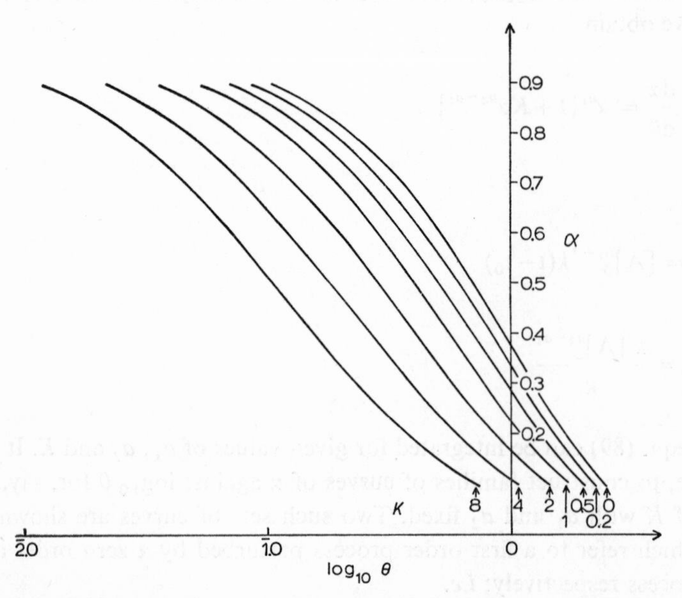

Fig. 4. Variation of α with $\log_{10}\theta$ for a first order reaction accompanied by a second order reaction. As K increases, the proportion of the reactant consumed by the second order process increases; since $K = k'[\mathrm{A}]_0/k$, the proportion of A disappearing by the second order process in a particular instance increases as its initial concentration increases.

and

$$-\frac{d[A]}{dt} = k[A]+k'[A]^2 \qquad \text{(Fig. 4)}$$

The numerical data for the construction of these curves are given in Appendix 4. The use of these and analogous curves to estimate the quantities a_1, a_2, and K should be obvious; furthermore, the dependence of K and the displacement of the $\log_{10}\theta$, $\log_{10}(t-t_0)$ scales on $[B]_0$ enables estimates of b_1 and b_2 to be made. It is not necessary to give full details, for this procedure merely exemplifies the point that, by studying the reaction under experimental conditions where the kinetic equations assume comparatively simple forms, it is possible to obtain estimates of the rate coefficients and exponents. As an alternative to the reaction being studied in the presence of a large excess of A or B, these special conditions may correspond to the position at which a particular product reaches its maximum concentration or to the position at which the reaction comes to equilibrium.

Once some idea has been obtained of the type of expression that is required, it is then possible to integrate the rate equations numerically. An iterative method is often convenient; the iteration would normally take the exponents in the various concentration terms as fixed and would proceed by substituting trial values of the rate coefficients (estimates of these having been obtained from prior experiments) in the differential equations and integrating; in this way, it is possible to find values which give a satisfactory match of calculated and observed data. The question of the criterion of fit is outside the scope of this chapter.

Some examples of the procedures employed when dealing with particular types of reaction mechanism are given in Chapter 1, Vol. 2.

4. Elucidating the dependence of the rate coefficient(s) on temperature, concentration of catalyst, and ionic strength

In general, the rate coefficient is a function of all factors which influence the reaction rate other than the concentrations of the reactants and products. We shall only consider three factors here, namely temperature, concentration of catalyst and ionic strength. The general experimental procedure is obvious. For example, suppose we wish to elucidate the effect of temperature on a particular rate coefficient. For this purpose, we perform a set of kinetic experiments at a particular temperature T_1 keeping all other factors which are liable to influence the rate coefficient constant. Following the procedures given earlier, we calculate the rate coefficient and its standard error (if possible). We then repeat the experiments at other temperatures T_2, T_3, ... etc., keeping the values of the other factors constant throughout the series. In this way, we obtain a set of values for k at the various temperatures T_1, T_2, etc. Clearly, we can study the influence of the other factors on the rate coefficient in an analogous fashion.

References p. 407

4.1 INFLUENCE OF TEMPERATURE

Theories of reaction rates suggest that the rate coefficient is related to absolute temperature by the expression

$$k = A(T) \exp(-E(T)/RT) \tag{90}$$

where $E(T)$ is termed the activation energy and $A(T)$ is the pre-exponential factor. In the majority of rate studies, the temperature variation of the activation energy and pre-exponential factor can be disregarded over the small temperature ranges normally employed and so eqn. (90) may be written

$$k = A \exp(-E/RT) \tag{91}$$

Our object in this section is to indicate how the best values of A and E together with their standard errors are obtained from the experimental data. We write

$$\ln k = \ln A - \frac{E}{RT}$$

and fit a weighted least squares line through the points $(1/T, \ln k)$ assuming that the temperatures are error-free; this assumption needs no defence for data obtained in the usual way since accurate thermostatting presents few problems. The equations are exactly similar to eqns. (58)–(64), namely

$$\frac{E}{R} = \frac{\left(\sum_{s=1}^{s=N} w_s x_s\right)\left(\sum_{s=1}^{s=N} w_s y_s\right) - \left(\sum_{s=1}^{s=N} w_s\right)\left(\sum_{s=1}^{s=N} w_s x_s y_s\right)}{D}$$

$$\ln A = \frac{\left(\sum_{s=1}^{s=N} w_s y_s\right)\left(\sum_{s=1}^{s=N} w_s x_s^2\right) - \left(\sum_{s=1}^{s=N} w_s x_s\right)\left(\sum_{s=1}^{s=N} w_s x_s y_s\right)}{D}$$

where N is the number of different temperatures used in defining the line

$$x_s = \frac{1}{T_s}$$

$$y_s = \ln k_s$$

and k_s represents the (weighted) mean value of the rate coefficients determined at the temperature T_s. The value of the weight, w_s, attached to each point is found in the usual way, thus

$$w_s = \frac{1}{\sigma^2(y_s)}$$
$$= \frac{k_s^2}{\sigma^2(k_s)}$$
$$\doteqdot \frac{k_s^2}{s^2(k_s)}$$

and D is calculated using the equation

$$D = (\sum_{s=1}^{s=N} w_s)(\sum_{s=1}^{s=N} w_s x_s^2) - (\sum_{s=1}^{s=N} w_s x_s)^2$$

The standard errors in E/R and $\ln A$ are estimated from the weighted sum of the squares of the residuals, δ_s, defined by

$$\delta_s = \text{observed value } \ln k_s - \text{calculated value } \ln k_s$$

$$= y_s - \left\{\ln A - \frac{E}{RT}\right\}$$

Thus

$$s(E/R) = \left\{\frac{e \sum_{s=1}^{s=N} w_s}{(N-2)D}\right\}^{\frac{1}{2}}$$

or

$$s(E) = R \left\{\frac{e \sum_{s=1}^{s=N} w_s}{(N-2)D}\right\}^{\frac{1}{2}}$$

and

$$s(\ln A) = \left\{\frac{e \sum_{s=1}^{s=N} w_s x_s^2}{(N-2)D}\right\}^{\frac{1}{2}}$$

or

$$s(A) = A \left\{\frac{e \sum_{s=1}^{s=N} w_s x_s^2}{(N-2)D}\right\}^{\frac{1}{2}}$$

where

$$e = \sum_{s=1}^{s=N} w_s \delta_s^2$$

The only comment necessary on these equations concerns the magnitude of the

References p. 407

estimated standard error of the activation energy. To a first approximation, the standard error of the slope of a straight line is proportional to the standard error of a point of unit weight and inversely proportional to the product of the range of the quantity x and the square root of the number of observations defining the line. In the determination of the activation energy of a reaction, it is usually only possible to cover a small range of values of $1/T$ since the normal experimental techniques employed are unable to follow much more than a 100-fold change in rate. Consequently the precision with which E can be measured is limited; it is for this reason that the temperature dependence of the activation energy cannot normally be measured. Incidently, the previous comment regarding the precision of the slope indicates that, as far as this quantity is concerned, it is more profitable to double the range of $1/T$ than to double the number of points within a particular temperature interval.

4.2 INFLUENCE OF THE CONCENTRATION OF CATALYST

In general, the rate coefficient is related to the concentration of catalyst by the simple expression

$$k = k'[\text{catalyst}]^c$$

where c is usually integral or half-integral and k' is a rate coefficient. To determine these quantities, we simply have to find c from the slope of the straight line of $\ln k$ against $\ln$ [catalyst] in the usual way and then calculate k'. Occasionally, the relationship between k and [catalyst] is of the form

$$k = k'[\text{catalyst}]^c + k''$$

Since there are but a limited number of values of c, namely 0.5, 1.0, 1.5, and 2.0, the four plots generalized in the above equation can be constructed; the values of c, k' and k'' can be found from that plot which gives the best straight line.

4.3 INFLUENCE OF IONIC STRENGTH

For reactions involving ions, the rate coefficient is a function of the ionic strength, μ, of the reaction system. If the two reactants A and B are ions and if the total ionic strength is small, the rate coefficient is exponentially related to the square root of the ionic strength, thus

$$k = k_0 \exp(2z_A z_B C\mu^{\frac{1}{2}})$$

In this expression, k is the observed rate coefficient at an ionic strength μ, k_0 is the limiting value of the rate coefficient at $\mu = 0$, z_A and z_B are the number of electronic charges carried by each ion, and C is a known constant for a given solvent at a particular temperature. This equation can be linearized by taking logarithms of each side, and the values of k_0 and $z_A z_B C$ obtained in an exactly analogous way to previous treatments.

For reactions between neutral molecules or between a neutral molecule and an ion at fairly high ionic strengths, the exponential dependence of the rate coefficient on the square root of the ionic strength may be replaced by an exponential dependence on μ alone.

The origin of these expressions and the evidence for them is outside the scope of this chapter. As far as we are concerned, they present no special problems.

Acknowledgement

The author would like to thank Dr. J. R. Green of the Department of Computational and Statistical Science, The University of Liverpool, for helpful discussion.

REFERENCES

1 E. A. Guggenheim, *Phil. Mag.*, 2 (1926) 538.
2 P. C. Mangelsdorf, *J. Appl. Phys.*, 30 (1959) 443.
3 A. A. Frost and R. G. Pearson, *Kinetics and Mechanism*, 2nd Edn., Wiley, New York, 1961.
4 W. E. Roseveare, *J. Am. Chem. Soc.*, 53 (1931) 1651.
5 J. M. Sturtevant, *J. Am. Chem. Soc.*, 59 (1937) 699.
6 S. W. Benson, *The Foundations of Chemical Kinetics*, McGraw-Hill, New York, 1960.
7 L. J. Reed and E. J. Theriault, *J. Phys. Chem.*, 35 (1931) 673.
8 L. J. Reed and E. J. Theriault, *J. Phys. Chem.*, 35 (1931) 950.
9 D. T. DeTar and V. M. Day, *J. Phys. Chem.*, 70 (1966) 495.
10 P. G. Guest, *Numerical Methods of Curve Fitting*, Cambridge University Press, Cambridge, 1961.
11 D. Margerison, D. M. Bishop, G. C. East and P. McBride, *Trans. Faraday Soc.*, 64 (1968) 1872.
12 R. C. P. Cubbon and D. Margerison, *Progress in Reaction Kinetics*, Vol. 3, Ed. G. Porter, Pergamon, Oxford, 1965.
13 S. Bywater, *Advances in Polymer Science*, 4 (1965) 66.

Appendix 1

Numerical data for the construction of the curves of α against $\log_{10} \theta$

1. VALUES OF θ FOR GIVEN VALUES OF α AND TOTAL ORDER

α \ order	0	0.5	1.0	1.5	2.0
0.9	0.1000	0.1026	0.1054	0.1082	0.1111
0.8	0.2000	0.2112	0.2232	0.2360	0.2500
0.7	0.3000	0.3266	0.3567	0.3904	0.4286
0.6	0.4000	0.4508	0.5108	0.5820	0.6667
0.5	0.5000	0.5858	0.6932	0.8284	1.000
0.4	0.6000	0.7350	0.9163	1.162	1.500
0.3	0.7000	0.9046	1.204	1.652	2.333
0.2	0.8000	1.106	1.610	2.472	4.000
0.1	0.9000	1.368	2.303	4.326	9.000

2. VALUES OF $\log_{10} \theta$ FOR GIVEN VALUES OF α AND TOTAL ORDER

α \ order	0	0.5	1.0	1.5	2.0
0.9	$\bar{1}.000$	$\bar{1}.011$	$\bar{1}.022$	$\bar{1}.034$	$\bar{1}.046$
0.8	$\bar{1}.301$	$\bar{1}.325$	$\bar{1}.348$	$\bar{1}.373$	$\bar{1}.398$
0.7	$\bar{1}.477$	$\bar{1}.514$	$\bar{1}.552$	$\bar{1}.592$	$\bar{1}.632$
0.6	$\bar{1}.602$	$\bar{1}.654$	$\bar{1}.708$	$\bar{1}.765$	$\bar{1}.824$
0.5	$\bar{1}.699$	$\bar{1}.768$	$\bar{1}.841$	$\bar{1}.918$	0.000
0.4	$\bar{1}.778$	$\bar{1}.866$	$\bar{1}.962$	0.065	0.176
0.3	$\bar{1}.845$	$\bar{1}.957$	0.081	0.218	0.368
0.2	$\bar{1}.903$	0.044	0.206	0.393	0.602
0.1	$\bar{1}.954$	0.136	0.362	0.636	0.954

Appendix 2

In this appendix, we discuss briefly some of the results of a typical kinetic investigation carried out in the author's laboratory by a number of collaborators[11]. This work illustrates several of the points raised in the text. For example, the work shows how complicated kinetic behaviour may be turned into something much easier of interpretation by identifying and eliminating the factors responsible for the complication. Furthermore, although the data yield an homogeneous set of first order rate coefficients covering a four-fold range of initial concentration, the data of a sub-set of replicate experiments are non-homogeneous indicating the presence of some partially uncontrolled factor. These points will be discussed and the calculations set out in some detail.

The data under discussion were obtained in the course of a study of the anionic polymerization of isoprene in benzene at 35.1° C. Earlier work on the polymerization using *n*-butyl lithium as initiator in a number of hydrocarbon solvents had shown that the kinetic behaviour was complicated. For example, the isoprene concentration–time curves were sigmoidal and the ratio of the number of moles of isoprene consumed to the number of moles of *n*-butyl lithium reacted increased as the reaction proceeded. These and other complications showed that the rate of the reaction could not possibly be represented by an equation of the type

$$-\frac{\mathrm{d[isoprene]}}{\mathrm{d}t} = k[\mathrm{isoprene}]^a[n\text{-butyl lithium}]^b$$

In principle, it would have been possible to fit the data to a more complex expression and to proceed, thereby, to propose a reaction mechanism. However, in this case—and this must be typical of many analogous situations—it was possible to suggest the reason for the appearance of many of the complex features already referred to. Without going into the detailed arguments [these are discussed in two review articles [12,13]], it was thought that the complications were due in part to the presence of two competitive reactions consuming isoprene, namely,

$$CH_2{=}CH{-}\underset{}{\overset{CH_3}{\overset{|}{C}}}{=}CH_2 + C_4H_9Li \xrightarrow{k_I} C_4H_9CH_2{-}CH{=}\overset{CH_3}{\overset{|}{C}}{-}CH_2^-Li^+$$

$$CH_2{=}CH{-}\overset{CH_3}{\overset{|}{C}}{=}CH_2 + C_4H_9\{CH_2{-}CH{=}\overset{CH_3}{\overset{|}{C}}{-}CH_2\}_i\, CH_2{-}CH{=}\overset{CH_3}{\overset{|}{C}}{-}CH_2^-Li^+$$

$$\xrightarrow{k_{II}} C_4H_9\{CH_2{-}CH{=}\overset{CH_3}{\overset{|}{C}}{-}CH_2\}_{i+1}\, CH_2{-}CH{=}\overset{CH_3}{\overset{|}{C}}{-}CH_2^-Li^+$$

It followed that, if this explanation were correct, the kinetic behaviour would simplify considerably once all the *n*-butyl lithium had reacted. This was found to be the case. The procedure adopted was as follows: *n*-butyl lithium was allowed to react with isoprene, the initial concentrations and the temperature being chosen so that no *n*-butyl lithium remained when all the isoprene had disappeared; the resulting mixture consisted of active centres

$$C_4H_9\{CH_2{-}CH{=}\underset{\displaystyle|}{\overset{CH_3}{C}}{-}CH_2\}_i\, CH_2{-}CH{=}\overset{CH_3}{\underset{\displaystyle|}{C}}{-}CH_2^-Li^+$$

in equilibrium with an associated complex, probably involving four such molecules, which is incapable of initiating the polymerization of isoprene. The important feature from our point of view is that the active centres were capable of initiating the polymerization of a further sample of isoprene. Following this argument one stage further, it was obviously possible in principle to study the rate of disappearance of this second charge of isoprene at a constant active centre concentration and so obtain particularly simple kinetic behaviour, thus

$$-\frac{d[\text{isoprene}]}{dt} = k_{II}[\text{AC}][\text{isoprene}]$$

$$= k[\text{isoprene}]$$

where k, the first order rate coefficient, includes the bimolecular rate coefficient, k_{II} and the constant active centre concentration, [AC].

To prove these assertions, the rate of disappearance of isoprene had to be shown to be first order with respect to isoprene at a given active centre concentration. For this purpose, the concentration of isoprene was measured dilatometrically at a series of times following its addition to a solution of active centres; a series of such experiments were carried out, some to test the reproducibility of the data and some at different initial concentrations of isoprene. These data were analysed on the basis of eqn. (36), the rate coefficients being obtained from the particular form of the general rate expression

$$f(\alpha) = l - k'(t - t_0)$$

namely,

$$\ln \alpha = l - k(t - t_0)$$

It should be noted that the rate coefficient k and the slope of the straight line of $f(\alpha)$ *versus* $t - t_0$, k', are identical in this case.

Table 4 gives the results of a typical run; the concentrations of isoprene are given to rather more figures than the basic experimental data warrant so that no errors from rounding up at too early a stage appear in the final result. The technique employed did not permit observations to be taken at the very early stages of the reaction; the time t_0 was the time at which the dilatometer containing the reaction mixture was placed in the thermostat and was not the time at which the reactants were mixed together. Since the first value of α which was measured was 0.7290, the fraction of the reactant which had disappeared prior to observations being taken was 0.2710; this is rather greater than would be the case in most kinetic studies and, of course, is undesirable in that valuable information can sometimes be obtained from a study of the initial period. However, the fact that only a portion of the reaction has been studied causes no difficulty since the arguments are based on the comparison of the slopes of the least squares lines which are not constrained to pass through the point $\{t_0, 0\}$ (the value of $f(\alpha_0)$ is ln 1, *i.e.* zero, for a first order reaction). Fig. 5 shows graphically the accuracy with which the data fit the equation

$$\ln \alpha = l - k(t - t_0)$$

TABLE 4

CONCENTRATION OF ISOPRENE AT A SERIES OF TIMES, AND COMPARISON OF OBSERVED AND CALCULATED VALUES OF $\ln \alpha$ FOR RUN 1 OF THE SET DISCUSSED IN THE TEXT

$\frac{t-t_0}{\text{min}}$	$\frac{[Isoprene]}{\text{mole . litre}^{-1}}$	α	$\ln \alpha$	$\ln \alpha_{calc}$ These values were calculated from least squares estimates of l and k, weighting each point according to $w_j = \alpha^2$	δ	$\ln \alpha_{calc}$ These values were calculated from least squares estimates of l and k weighting each point equally, *i.e.* $w_j = 1$	δ
2.25	1.2320	0.7290	−0.3161	−0.3218	0.0057	−0.3293	0.0132
3.50	1.1630	0.6882	−0.3737	−0.3751	0.0014	−0.3819	0.0082
6.25	1.0270	0.6077	−0.4981	−0.4922	−0.0059	−0.4976	−0.0005
7.75	0.9729	0.5757	−0.5522	−0.5560	0.0038	−0.5608	0.0086
9.50	0.8973	0.5309	−0.6331	−0.6305	−0.0026	−0.6344	0.0013
11.50	0.8258	0.4886	−0.7161	−0.7157	−0.0004	−0.7186	0.0025
13.00	0.7743	0.4582	−0.7805	−0.7796	−0.0009	−0.7818	0.0013
15.50	0.6864	0.4062	−0.9010	−0.8860	−0.0150	−0.8870	−0.0140
18.00	0.6253	0.3700	−0.9943	−0.9925	−0.0018	−0.9922	−0.0021
20.50	0.5631	0.3332	−1.0990	−1.0990	0.0000	−1.0974	−0.0016
22.00	0.5279	0.3124	−1.1636	−1.1628	−0.0008	−1.1606	−0.0030
26.00	0.4452	0.2634	−1.3340	−1.3332	−0.0008	−1.3289	−0.0051
28.00	0.4113	0.2434	−1.4132	−1.4183	0.0051	−1.4131	−0.0001
31.50	0.3506	0.2075	−1.5728	−1.5674	−0.0054	−1.5604	−0.0124
34.00	0.3166	0.1873	−1.6748	−1.6738	−0.0010	−1.6657	−0.0091
37.00	0.2811	0.1663	−1.7938	−1.8016	0.0078	−1.7919	−0.0019
40.00	0.2481	0.1468	−1.9187	−1.9293	0.0106	−1.9182	−0.0005
43.00	0.2150	0.1272	−2.0618	−2.0571	0.0047	−2.0445	−0.0173
46.00	0.1908	0.1129	−2.1813	−2.1848	0.0035	−2.1707	−0.0106
50.00	0.1654	0.0979	−2.3241	−2.3551	0.0310	−2.3391	0.0150
55.00	0.1347	0.0797	−2.5294	−2.5680	0.0386	−2.5496	0.0202
60.00	0.1078	0.0638	−2.7522	−2.7810	0.0288	−2.7600	0.0078

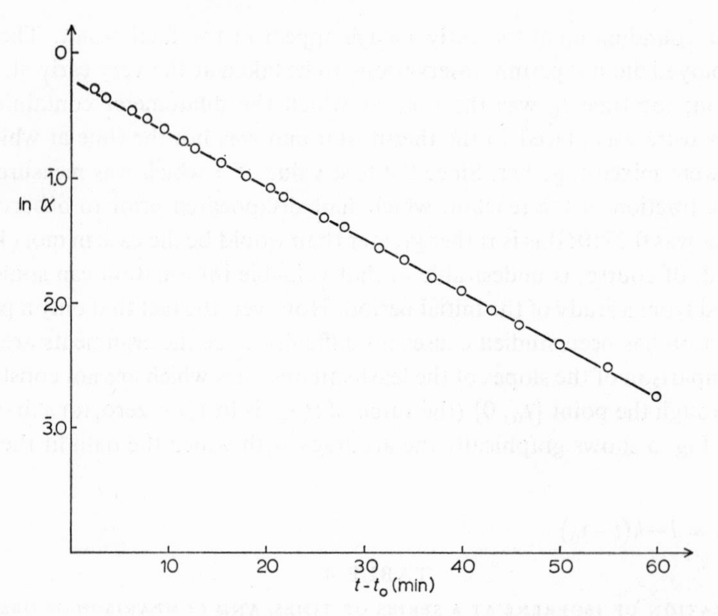

Fig. 5. A typical plot of ln α against $t - t_0$ for the reaction between isoprene and polyisoprenyl lithium when the concentration of the latter is held constant. The numerical data for this plot are given in Table 4.

Table 4 also shows the accuracy with which the data fit the above equation by listing the values of ln α and the residuals, δ, calculated from the least squares estimates of l and k; two cases are given corresponding to assigning weights α^2 and unity to each point. The residuals are defined by the expression

$$\delta = \ln \alpha - \ln \alpha_{\text{calc}}$$

TABLE 5

Run	*[Isoprene]$_0$* mole.litre^{-1}	*Calculated assigning* $w_{ij} = \alpha^2_{ij}$		*Calculated assigning* $w_{ij} = 1$	
		$10^4\,k$ sec^{-1}	$10^4\,s(k)$ sec^{-1}	$10^4\,k$ sec^{-1}	$10^4\,s(k)$ sec^{-1}
1		7.097	0.023	7.015	0.020
2	1.690	6.944	0.015	6.917	0.013
3		7.275	0.010	7.223	0.004
4	3.220	7.155	0.036	6.937	0.032
5		7.050	0.033	6.780	0.027
6	0.815	7.091	0.012	7.125	0.012

It will be observed that they are more randomly distributed about zero when each point is weighted by the factor α^2 than when each point is assigned the same weight. This fact suggests that the procedure in which each point is weighted by the factor α^2 is the superior of the two, a conclusion which agrees with the author's judgement that, in this work, $\sigma^2(\alpha)$ would be independent of α. For these two reasons, we shall not discuss in detail the parameters obtained when each point is given equal weight in the least squares equations although, for interest, we give in Table 5 a comparison of the rate coefficients and their standard errors for the two methods of weighting for each of the six experiments of our set. It can be seen immediately that the weighting procedure $w_{ij} = \alpha_{ij}^2$ improves the consistency of the set.

Although we do not need a statistical analysis to tell us that the six values of the rate coefficients (listed in column 3, Table 5) constitute an homogeneous set, we give in Table 6 the details of the calculations which would be done if the conclusion were less obvious. This table also shows how the best value of the rate coefficient is calculated and how an estimate of its standard error is obtained. The result should be written thus: "the mean value of the rate coefficient and its standard error are given by

$$\bar{k}(35.1° \text{ C}) = (7.139 \pm 0.069) \times 10^{-4} \text{ sec}^{-1} \text{ (D.F. = 3)"}$$

The final bracket containing D.F. = 3 shows the number of degrees of freedom used in estimating the quoted standard error in $\bar{k}$; if the number of degrees of freedom is not given, it is impossible to decide upon the significance to be attached to the value quoted. Alternatively, the rate coefficient and its 90% confidence limits could be given. In this case, the result should be written: "the mean value of the rate coefficient together with the 90% confidence limits are given by

$$\bar{k}(35.1° \text{ C}) = (7.14 \pm 0.16) \times 10^{-4} \text{ sec}^{-1}\text{"}$$

The former summary of the calculations is preferred since we may wish to use the estimates $s(\bar{k})$ in weighting the rate coefficients in a subsequent calculation (*e.g.* in evaluating the weighted least squares estimate of the activation energy).

Inspection of the calculations in Table 6 shows that the estimate $s(\bar{k})$ is determined almost entirely by the results of group 1 (runs 1, 2 and 3). Furthermore, $s(\bar{k})$ is considerably greater than the standard error of any individual rate coefficient estimated from the scatter of the experimental values of ln α about the fitted straight line. In other words, it is the inability to replicate the experimental conditions exactly rather than the definition of the straight lines which determines the precision of the final estimate of the rate coefficient. In fact, we show in the calculations set out in Table 7 that the replicate experiments in Group 1 differ significantly from one another. It is because these results are so discordant that we have

(Text continued on p. 417.)

TABLE 6

The calculations set out in this table show that the six rate coefficients listed in Table 3 determined at three different initial concentrations of isoprene constitute an homogeneous set

(a) CALCULATION OF GROUP AND SET MEANS

Original number of run	*Number of group* v	*Number of run within group* i	$\dfrac{10^4 k_{vi}}{\text{sec}^{-1}}$	$\dfrac{10^6 s(k_{vi})}{\text{sec}^{-1}}$	$\dfrac{10^{-12} w(k_{vi})}{\text{sec}^2}$ or $\dfrac{10^{-12}}{\text{sec}^2 s^2(k_{vi})}$	$\dfrac{10^{-8} w(k_{vi}) k_{vi}}{\text{sec}}$	$\dfrac{10^4\ \Sigma w(k_{vi}) k_{vi}}{\text{sec}^{-1}\ \Sigma w(k_{vi})}$
1		1	7.097	2.3	0.189	1.341	mean k for group 1
2	1	2	6.944	1.5	0.444	3.083	
3		3	7.275	1.0	1.000	7.275	$\dfrac{10^4 \bar{k}_1}{\text{sec}^{-1}} = \dfrac{11.699}{1.633}$
		sums over group			1.633	11.699	$= 7.164$
4	2	1	7.155	3.6	0.077	0.5509	mean k for group 2
5		2	7.050	3.3	0.093	0.6557	
		sums over group			0.170	1.2066	$\dfrac{10^4 \bar{k}_2}{\text{sec}^{-1}} = \dfrac{1.2066}{0.170}$
							$= 7.098$
6	3	1	7.091	1.2	0.694	4.9211	mean k for set
		sums over set			2.497	17.8271	$\dfrac{10^4 \bar{k}}{\text{sec}^{-1}} = \dfrac{17.8271}{2.497}$
							$= 7.139$

(b) DISTRIBUTION OF GROUP MEANS ABOUT SET MEAN

As shown above, the whole set comprises three groups; in group 1, there are 3 experiments, in group 2, 2 experiments and group 3, 1 experiment; therefore, $r_1 = 3$, $r_2 = 2$ and $r_3 = 1$. The object is to calculate

$$\sum_{v=1}^{v=3} \sum_{i=1}^{i=r_v} w(k_{vi})(\bar{k}_v - \bar{k})^2$$

and thence ms_1

Number of group v	$\dfrac{10^6(\bar{k}_v - \bar{k})}{\text{sec}^{-1}}$	$\dfrac{10^{12}(\bar{k}_v - \bar{k})^2}{\text{sec}^{-2}}$	$\dfrac{10^{-12}}{\text{sec}^2} \sum_{i=1}^{i=r_v} w(k_{vi})$	$\sum_{i=1}^{i=r_v} w(k_{vi})(\bar{k}_v - \bar{k})^2$
1	2.5	6.25	1.633	10.206
2	−4.1	16.81	0.170	2.858
3	−4.8	23.04	0.694	15.990
			sum over set	29.054

$$\therefore \text{s.s.1} = \sum_{v=1}^{v=3} \sum_{i=1}^{i=r_v} w(k_{vi})(\bar{k}_v - \bar{k})^2 = 29.054$$

$$u - 1 = 3 - 1 = 2$$

$$\therefore\ ms_1 = 14.53$$

TABLE 6 (continued)

(c) DISTRIBUTION OF INDIVIDUAL VALUES ABOUT GROUP MEANS

The object is to calculate

$$\sum_{v=1}^{v=3}\sum_{i=1}^{i=r_v} w(k_{vi})(k_{vi}-\bar{k}_v)^2$$

and thence ms_2.

Number within group i	$10^6(k_{vi}-\bar{k}_v)$ / sec^{-1}	$10^{12}(k_{vi}-\bar{k}_v)^2$ / sec^{-2}	$10^{12}w(k_{vi})$ / sec^2	$w(k_{vi})(k_{vi}-\bar{k}_v)^2$
1	− 6.7	44.89	0.189	8.48
2	−22.0	484.00	0.444	214.90
3	11.1	123.21	1.000	123.21
1	5.7	32.49	0.077	2.50
2	− 4.8	23.04	0.093	2.14
1	0	0	0.694	0
			sum over set	351.23

$$\therefore \text{s.s.2} = \sum_{v=1}^{v=3}\sum_{i=1}^{i=r_v} w(k_{vi})(k_{vi}-\bar{k}_v)^2 = 351.23$$

$$r-u = 6-3 = 3$$

$$\therefore ms_2 = 117.1$$

(d) CALCULATION OF F

$$F = \frac{ms_1}{ms_2} = \frac{14.53}{117.1}$$

Clearly $F < 1$ and hence the set is homogeneous. The maximum value that F could take in 95% of cases if the set were homogeneous is 9.55 (2 degrees of freedom in numerator, 3 in denominator).

(e) BEST VALUE OF THE RATE COEFFICIENT AND AN ESTIMATE OF ITS STANDARD ERROR

Since the set is homogeneous, the best value of the rate coefficient is the weighted mean, $\bar{k}$,

i.e. $7.139\times10^{-4}\ \text{sec}^{-1}$.

The best estimate of $s(\bar{k})$ is found from the larger of the two values ms_1 or ms_2, thus

$$s^2(\bar{k}) = \frac{ms_2}{\sum_{v=1}^{v=3}\sum_{i=1}^{i=r_v} w(k_{vi})}$$

$$= \frac{117.1}{2.497}\times10^{-12}\ \text{sec}^{-2}$$

$$\therefore s(\bar{k}) = 6.9\times10^{-6}\ \text{sec}^{-1}$$

(f) CALCULATION OF THE 90% CONFIDENCE LIMITS FOR THE RATE COEFFICIENT

Since the value ms_2 used above in calculating $s(\bar{k})$ was based on 3 degrees of freedom, we require the value of t appropriate to this number. From tables,

$$t = 2.35$$

The 90% confidence limits for the rate coefficient are, therefore,

$$(7.139\pm2.35\times0.069)\times10^{-4}\ \text{sec}^{-1}$$
$$= (7.139\pm0.16)\times10^{-4}\ \text{sec}^{-1}$$

In other words, we expect that the true value of the rate coefficient will lie between (6.98 and 7.30) $\times10^{-4}\ \text{sec}^{-1}$ with 90% confidence.

TABLE 7

The calculations set out in this table show that the three replicate experiments, runs 1, 2 and 3 constitute a non-homogeneous set

(a) THE BASIC DATA: CALCULATION OF WEIGHTED MEAN USING WEIGHTS $W(k_i)$ DEFINED IN TEXT

Original number of run	i	n_i	n_i-2	$\frac{10^4 k_i}{\text{sec}^{-1}}$	$10^4 E_i$	$\frac{10^{-6} D_i}{\text{sec}^2}$	$\sum_{j=1}^{j=n_i} w_{ij}$	$\frac{10^{-6} W(k_i)}{\text{sec}^2}$ *i.e.* $\frac{10^{-6} D_i}{\text{sec}^2 \sum_{j=1}^{j=n_i} w_{ij}}$	$\frac{10^{-2} W(k_i) k_i}{\text{sec}}$
1	1	22	20	7.097	1.0538	3.3665	3.2517	1.0353	7.3475
2	2	32	30	6.944	1.4485	12.0470	5.8362	2.0642	14.3338
3	3	51	49	7.275	1.2031	19.4609	7.4326	2.6183	19.0481
						sums over group		5.7178	40.7294

$$\text{Mean } k \text{ for group, } \bar{k} = \sum_{i=1}^{i=3} W(k_i)k_i \Big/ \sum_{i=1}^{i=3} W(k_i)$$

$$= \frac{40.7294 \times 10^2 \text{ sec}}{5.7178 \times 10^6 \text{ sec}^2}$$

$$= 7.123 \times 10^{-4} \text{ sec}^{-1}$$

(b) DISTRIBUTION OF RATE COEFFICIENTS ABOUT MEAN VALUE

i	$\frac{10^6(k_i-\bar{k})}{\text{sec}^{-1}}$	$\frac{10^{12}(k_i-\bar{k})^2}{\text{sec}^{-2}}$	$\frac{10^{-6} W(k_i)}{\text{sec}^2}$	$10^6 W(k_i)(k_i-\bar{k})^2$
1	− 2.6	6.8	1.035	7.0
2	−17.9	320.4	2.064	661.3
3	15.2	231.0	2.618	604.9
			sum over group	1273.2

$$\sum_{i=1}^{i=3} W(k_i)(k_i-\bar{k})^2 = 1273.2 \times 10^{-6}$$

$$r = 3$$

$$\therefore \sum_{i=1}^{i=3} W(k_i)(k_i-\bar{k})^2/r-1 = 636.6 \times 10^{-6}$$

(c) ESTIMATION OF BASIC ERROR

This is simply $\sum_{i=1}^{i=3} E_i \Big/ \sum_{i=1}^{i=3} (n_i-2)$

$$\sum_{i=1}^{i=3} E_i = 3.7054 \times 10^{-4}$$

$$\sum_{i=1}^{i=3} (n_i-2) = 99$$

$$\therefore \sum_{i=1}^{i=3} E_i \Big/ \sum_{i=1}^{i=3} (n_i-2) = 3.743 \times 10^{-6}$$

TABLE 7 (continued)

(d) CALCULATION OF F

$$F = \frac{\sum_{i=1}^{i=3} W(k_i)(k_i - \bar{k})^2 / r-1}{\sum_{i=1}^{i=3} E_i / \sum_{i=1}^{i=3} (n_i - 2)}$$

$$= \frac{636.6 \times 10^{-6}}{3.743 \times 10^{-6}}$$

$$= 170.1$$

The critical value of $F_{2,99}$ (*i.e.* F associated with 2 degrees of freedom in numerator and 99 degrees of freedom in denominator) is 3.09 (5 %) or 4.82 (1 %); these are the values which would be exceeded in 5 % or 1 % of cases if the group of rate coefficients were homogeneous. Clearly $F_{\text{calc}} \gg F_{\text{critical}}$ and so the replicate experiments constituting the group are significantly different from one another.

to estimate $s(\bar{k})$ from the residuals rather than from the equation sometimes used, *viz.*

$$s^2(\bar{k}) = \frac{1}{\sum_{v=1}^{v=u} \sum_{i=1}^{i=r_v} \frac{1}{s^2(k_{vi})}}$$

In our case, this would give a totally misleading impression of the precision of the derived mean value of the rate coefficient (the actual value for $s(\bar{k})$ calculated in this way would be 0.006×10^{-4} sec^{-1}). It should be noted that we have used the residuals $(k_{vi} - \bar{k}_v)$ as the basis of our calculation; it is quite permissible to use the residuals calculated from the set mean rather than the group means—this procedure gives a slightly lower estimate for $s(\bar{k})$ in our case but the general conclusion regarding our final value of $\bar{k}$ is not affected.

The analysis of the data of the three replicate experiments shown in Table 7 shows that significant differences exist between the rate coefficients. It is not difficult to deduce the cause. Every value of k is the product of two factors: (*a*) the rate coefficient k_{II} and (*b*) the active centre concentration [AC]. It is very likely that the concentration of active centres is not exactly the same for each experiment of the group† (and presumably the same comment applies to the whole set); within any one run, however, the value of [AC] remains constant giving a well-defined straight line of ln α against $t - t_0$. Consequently, the values of the various rate coefficients are distributed about their mean to a greater extent than the scatter of the points about the corresponding straight lines would suggest.

---

† The presence in the reaction vessel of proton donating impurities such as water which destroy the lithiated polyisoprenes is a possible source of the irreproducibilities observed; small irreproducible amounts of these impurities could give rise to a small variation in the active centre concentration from run to run.

The end result of this discussion can be summarized thus:

(1) the rate of disappearance of isoprene is first order with respect to isoprene at constant active centre concentration;

(2) the rate coefficient at the particular temperature of 35.1° C is

$$7.139 \times 10^{-4} \text{ sec}^{-1},$$

the true value lying with 90% probability between $(6.98 \text{ and } 7.30) \times 10^{-4} \text{sec}^{-1}$;

(3) the standard error of the rate coefficient is

$$0.069 \times 10^{-4} \text{ sec}^{-1} (\text{D.F.} = 3)$$

and is determined by the lack of precise control of the active centre concentration from run to run;

(4) the precision of the rate coefficient is best improved by refining the technique of obtaining a constant number of active centres in the reaction system.

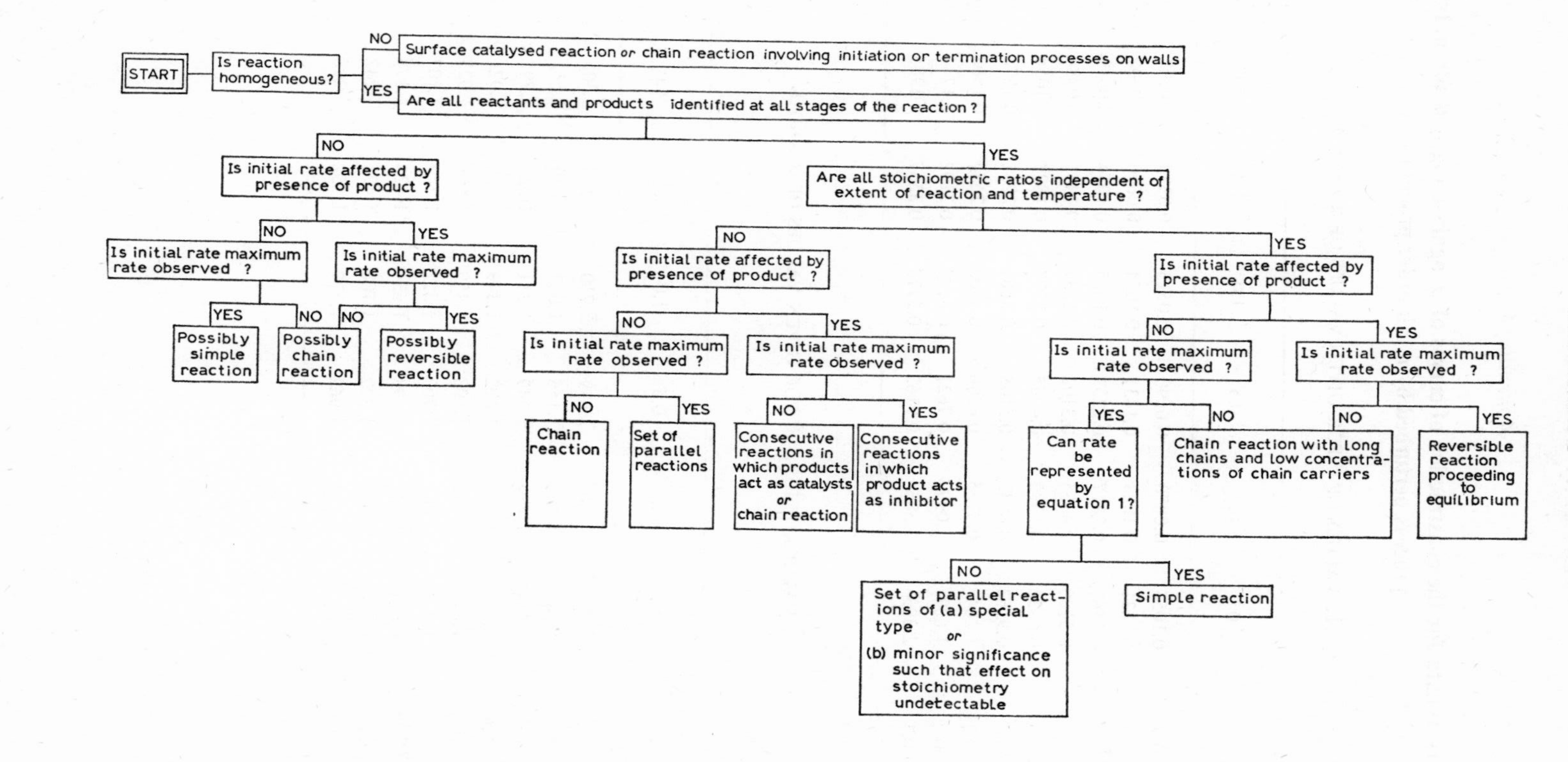
START
Is reaction homogeneous?
NO
Surface catalysed reaction or chain reaction involving initiation or termination processes on walls
YES
Are all reactants and products identified at all stages of the reaction?
NO
Is initial rate affected by presence of product ?
NO
Is initial rate maximum rate observed ?
YES
Possibly simple reaction
NO
Possibly chain reaction
YES
Is initial rate maximum rate observed ?
NO
Possibly reversible reaction
YES
Are all stoichiometric ratios independent of extent of reaction and temperature ?
NO
Is initial rate affected by presence of product ?
NO
Is initial rate maximum rate observed ?
NO
Chain reaction
YES
Set of parallel reactions
YES
Is initial rate maximum rate observed ?
NO
Consecutive reactions in which products act as catalysts or chain reaction
YES
Consecutive reactions in which product acts as inhibitor
YES
Is initial rate affected by presence of product ?
NO
Is initial rate maximum rate observed ?
YES
Can rate be represented by equation 1?
NO
Set of parallel reactions of (a) special type or (b) minor significance such that effect on stoichiometry undetectable
YES
Simple reaction
NO
Chain reaction with long chains and low concentrations of chain carriers
YES
Is initial rate maximum rate observed ?
NO
YES
Reversible reaction proceeding to equilibrium

Appendix 4

Numerical data for the construction of curves of α against $\log_{10} \theta$ for a 1st order process perturbed by a 0th order process

1. VALUES OF θ FOR GIVEN VALUES OF α AND K

α \ K	0	0.2	0.5	1.0	2.0	4.0	8.0
0.9	0.1054	0.0870	0.0690	0.0513	0.0339	0.0202	0.0112
0.8	0.2231	0.1823	0.1431	0.1054	0.0690	0.0408	0.0225
0.7	0.3567	0.2877	0.2231	0.1625	0.1054	0.0619	0.0339
0.6	0.5108	0.4055	0.3102	0.2231	0.1431	0.0834	0.0455
0.5	0.6932	0.5390	0.4055	0.2877	0.1823	0.1054	0.0572
0.4	0.9163	0.6932	0.5108	0.3567	0.2231	0.1278	0.0690
0.3	1.204	0.8755	0.6286	0.4308	0.2657	0.1508	0.0810
0.2	1.609	1.099	0.7621	0.5108	0.3102	0.1744	0.0931
0.1	2.303	1.386	0.9163	0.5978	0.3567	0.1985	0.1054

2. VALUES OF $\log_{10} \theta$ FOR GIVEN VALUES OF α AND K

α \ K	0	0.2	0.5	1.0	2.0	4.0	8.0
0.9	$\bar{1}.022$	$\bar{2}.940$	$\bar{2}.839$	$\bar{2}.710$	$\bar{2}.530$	$\bar{2}.305$	$\bar{2}.048$
0.8	$\bar{1}.348$	$\bar{1}.261$	$\bar{1}.156$	$\bar{1}.023$	$\bar{2}.839$	$\bar{2}.611$	$\bar{2}.352$
0.7	$\bar{1}.552$	$\bar{1}.459$	$\bar{1}.349$	$\bar{1}.211$	$\bar{1}.023$	$\bar{2}.792$	$\bar{2}.530$
0.6	$\bar{1}.708$	$\bar{1}.608$	$\bar{1}.492$	$\bar{1}.349$	$\bar{1}.156$	$\bar{2}.921$	$\bar{2}.658$
0.5	$\bar{1}.841$	$\bar{1}.732$	$\bar{1}.608$	$\bar{1}.459$	$\bar{1}.261$	$\bar{1}.023$	$\bar{2}.757$
0.4	$\bar{1}.962$	$\bar{1}.841$	$\bar{1}.708$	$\bar{1}.552$	$\bar{1}.349$	$\bar{1}.107$	$\bar{2}.839$
0.3	0.081	$\bar{1}.942$	$\bar{1}.798$	$\bar{1}.634$	$\bar{1}.424$	$\bar{1}.178$	$\bar{2}.908$
0.2	0.206	0.041	$\bar{1}.882$	$\bar{1}.708$	$\bar{1}.492$	$\bar{1}.242$	$\bar{2}.969$
0.1	0.362	0.142	$\bar{1}.962$	$\bar{1}.777$	$\bar{1}.552$	$\bar{\ }.298$	$\bar{1}.023$

Appendix 4 (continued)

Numerical data for the construction of curves of α against $\log_{10}\theta$ for a 1st order process perturbed by a 2nd order process

1. VALUES OF θ FOR GIVEN VALUES OF α AND K

α \ K	0	0.2	0.5	1.0	2.0	4.0	8.0
0.9	0.1054	0.0886	0.0715	0.0541	0.0364	0.0220	0.0123
0.8	0.2231	0.1892	0.1542	0.1178	0.0800	0.0488	0.0274
0.7	0.3567	0.3054	0.2513	0.1942	0.1335	0.0822	0.0465
0.6	0.5108	0.4418	0.3677	0.2877	0.2007	0.1252	0.0715
0.5	0.6932	0.6061	0.5108	0.4055	0.2877	0.1823	0.1054
0.4	0.9163	0.8109	0.6932	0.5596	0.4055	0.2624	0.1542
0.3	1.204	1.080	0.9383	0.7732	0.5754	0.3830	0.2305
0.2	1.609	1.466	1.299	1.099	0.8473	0.5878	0.3677
0.1	2.303	2.140	1.946	1.705	1.386	1.030	0.6932

2. VALUES OF $\log_{10}\theta$ FOR GIVEN VALUES OF α AND K

α \ K	0	0.2	0.5	1.0	2.0	4.0	8.0
0.9	$\bar{1}.022$	$\bar{2}.947$	$\bar{2}.854$	$\bar{2}.733$	$\bar{2}.561$	$\bar{2}.342$	$\bar{2}.089$
0.8	$\bar{1}.348$	$\bar{1}.277$	$\bar{1}.188$	$\bar{1}.071$	$\bar{2}.903$	$\bar{2}.688$	$\bar{2}.438$
0.7	$\bar{1}.552$	$\bar{1}.485$	$\bar{1}.400$	$\bar{1}.288$	$\bar{1}.126$	$\bar{2}.915$	$\bar{2}.668$
0.6	$\bar{1}.708$	$\bar{1}.645$	$\bar{1}.566$	$\bar{1}.459$	$\bar{1}.303$	$\bar{1}.098$	$\bar{2}.854$
0.5	$\bar{1}.841$	$\bar{1}.783$	$\bar{1}.708$	$\bar{1}.608$	$\bar{1}.459$	$\bar{1}.261$	$\bar{1}.023$
0.4	$\bar{1}.962$	$\bar{1}.909$	$\bar{1}.841$	$\bar{1}.748$	$\bar{1}.608$	$\bar{1}.419$	$\bar{1}.188$
0.3	0.081	0.033	$\bar{1}.972$	$\bar{1}.888$	$\bar{1}.760$	$\bar{1}.583$	$\bar{1}.363$
0.2	0.206	0.166	0.114	0.041	$\bar{1}.928$	$\bar{1}.769$	$\bar{1}.566$
0.1	0.362	0.330	0.289	0.232	0.142	0.013	$\bar{1}.841$

Index

B

C

D

E

F

I

N

O

P

Q

R

S

T